Interstellar Boundary Explorer (IBEX)

David McComas • Gary Zank • Nathan Schwadron
Editors

Interstellar Boundary Explorer (IBEX)

Previously published in *Space Science Reviews* Volume 146, Issues 1–4, 2009

Springer

David McComas
Space Research Division
Southwest Research Institute (SWRI)
6220 Clebra Road
P.O. Box 28510
San Antonio, TX 78228
USA

Gary Zank
Center for Space Plasma
Aeronomy & Astrophysics Research
University of Alabama, Huntsville
S101 Technology Hall
Huntsville, AL 35899
USA

Nathan Schwadron
College of Arts & Science
Department Astronomy
Boston University
725 Commonwealth Ave.
Boston, MA 02215
USA

ISBN 978-1-4899-9924-5
Springer Dordrecht Heidelberg London New York

Printed on acid-free paper

Springer is part of Springer Science+Business Media (www.springer.com)

Contents

Space Sci Rev (2009) 146: 1–3
DOI 10.1007/s11214-009-9569-7

Foreword

Dave McComas

Received: 9 July 2009 / Accepted: 13 July 2009 / Published online: 5 August 2009

The Interstellar Boundary Explorer (IBEX) is a remarkable mission of exploration and discovery. IBEX is currently producing the first global images of the heliosphere's interaction with the local interstellar medium—these images provide an entirely new framework for understanding the relationship between our heliosphere and the galactic environment. IBEX achieves this by making observations of the energetic neutral atoms (ENAs) produced at the edge of the heliosphere via charge exchange between the solar wind and embedded pickup ions with the cold neutral atoms drifting in from the local interstellar medium. IBEX measures ENAs using two independent single pixel cameras: IBEX-Lo, which measures ENAs from ~10 eV and ~2 keV, and IBEX-Hi, which measures them from ~300 eV to ~6 keV. The combination of extremely high sensitivity, excellent spatial resolution (~1800, 6° pixels covering the sky), energy spectral information across a very broad energy range, and simultaneous and independent measurements over the most crucial energy range, makes IBEX ideally suited to provide these unique, new observations. Finally, IBEX also directly measures the interstellar neutral atoms drifting in from the interstellar medium—these include not just the previously observed interstellar He, but also the first direct observations of interstellar H and O as well as the first observations of both the primary and secondary populations of these neutrals.

The IBEX mission grew out of a community consensus about the critical importance of outer heliospheric physics and a call for a mission to somehow provide the global picture of the heliosphere's interaction with the local interstellar medium. For example, the National Academy of Science's 2002 "Decadal Survey" entitled *The Sun to the Earth—and Beyond: A Decadal Research Strategy in Solar and Space Physics*, pointed out that "The boundary between the solar wind and the local interstellar medium (LISM) is one of the last unexplored regions of the heliosphere. Very little is currently known about this boundary or the nature of the LISM that lies beyond it. ... certain aspects of these regions can be studied by a combination of remote sensing and in situ sampling techniques. This investigation could

D. McComas (✉)
NASA/Goddord Space Flight Center, Southwest Research Institute, San Antonio, TX, USA
e-mail: dmccomas@swri.edu

be accomplished by a mission ... to obtain energetic neutral atom images ... of the heliospheric boundary. Such a mission is gauged to be feasible within the resource limits of the Explorer program ..." Similarly, NASA's *Sun–Earth Connection Roadmap 2003–2028: Understand how the Sun, Heliosphere, and the Planetary Environments are Connected in a Single System* (*January 2003*) argued that "An entirely new level of information on the interface between the heliosphere and the LISM is required. ... Imaging of the outer boundaries ~100 AU away can also be done from orbits near 1 AU. ... Remote-sensing techniques such as energetic neutral atom (ENA) imaging of energetic protons... near the termination shock and in the heliosheath should provide additional diagnostics of the interfaces with the galaxy." Finally, even the top-level *National Aeronautics and Space Administration 2003 Strategic Plan* included the comment that "Even the interstellar gas that enters the solar system can be analyzed using remote sensing."

Unfortunately, it was clear from two previous proposals for larger Mid-sized Explorer (MIDEX) missions, which received excellent reviews and ratings but no funding, that NASA was not able to support such an expensive mission as a MIDEX for this work. Thus it was early in 2002 when we contacted Orbital Sciences Corporation (Orbital) to see if there was some way to design and launch a very small (~100 kg) spacecraft into an orbit that would reach far outside the Earth's magnetosphere from the low Earth orbit launcher typically provided by NASA for SMEX missions—a Pegasus launch vehicle. Several of the engineers from Orbital jumped at the opportunity to help develop something so revolutionary and, along with others at Southwest Research Institute (SwRI), we collectively came up with a detailed plan to add our own Solid Rocket Motor (SRM) to the top of the Pegasus, effectively making it an air-launched, four-stage rocket for the first time.

In parallel with the rocketry work, I assembled a team of the world's experts in the science of the outer heliosphere and the interstellar interaction, as well as in the development of highly specialized ENA imaging instrumentation. This outstanding group pulled together into a single, seamless team, sharing ideas and optimizing the mission concept as a single collective effort in a way that I have never seen in any other space mission. This effort led to the IBEX proposal in response to NASA's 2003 Small Explorer (SMEX) Announcement of Opportunity.

After initial selection and a detailed Phase A study, we were finally told in January 2005 that IBEX was selected for development and flight. IBEX has been on the fast track ever since selection with Mission Preliminary Design Review (PDR) in January 2006, Mission Critical Design Review (CDR) in September 2006, and Payload delivery to the spacecraft in September 2007. Delivery to the Vandenberg Air Force Base was scheduled for March 2008 and the entire spacecraft and payload were ready when the IBEX team learned about a new problem with excessive vibration loads from the Pegasus launch vehicle. We ended up having to put our spacecraft in storage for four months and run a crash program to build a shock-absorbing ring, which acted as the interface between our flight hardware and the Pegasus, effectively reducing its vibration loads. In July 2008, after just four quick months, our interface ring was designed, built, tested, and installed in the flight stack and we were out at Vandenberg preparing for launch.

With integration complete, IBEX and the Pegasus rocket were loaded onto the bottom of the L1011 airplane and ferried out to Kwajalein Atoll in the south Pacific where, near the equator, the rotation of the Earth provided additional energy for our launch. On 19 October 2008, IBEX was successfully launched. Our SRM and spacecraft's internal hydrazine propulsion system worked perfectly and ultimately delivered the IBEX spacecraft from the ~200 km, low Earth orbit that Pegasus provided into a highly elliptical orbit that extends out to ~300,000 km, roughly 80% of the way to the Moon. The orbit raising and spacecraft commissioning took up the remainder of 2008 and in January 2009, we transitioned over to nominal operations and began making our groundbreaking science observations. Remarkably, the IBEX project was able to do all this work including developing an entirely new launch capability, building and flying a unique and highly specialized spacecraft and instrument suite, and maintaining full funding for our Education and Public Outreach and Phase E science activities, while still under-running our original cost cap (as modified by NASA-directed changes), by roughly three-quarters of a million dollars.

This book comprises a set of papers that describe the IBEX science, instruments, and mission and put these in the context of the existing knowledge of the interstellar interaction at the time of the launch. The book sets the stage for research that will be based on data from the IBEX mission. We sincerely hope that future researchers, authors and students will use this information to help in their studies. Chapter 1 [McComas et al.] provides an overview of the entire IBEX program including the IBEX science, hardware, and mission. Chapter 2 describes the IBEX spacecraft and flight system [Scherrer et al.]. Chapters 3–4 provide the details of the IBEX-Hi instrument [Funsten et al.] and background monitor that is built into it [Allegrini et al.], while Chapters 5–7 describe the IBEX-Lo instrument [Fuselier et al.], how IBEX-Lo can measure the interstellar neutrals directly entering the heliosphere [Möbius et al.] and the backgrounds and signal-to-noise ratios expected for IBEX prior to launch [Wurz et al.]. The IBEX Science Operations Center (ISOC) is described in Chapter 8 [Schwadron et al.], which includes how we get data from the IBEX spacecraft and how it will be made available to the science community. Chapters 9–12 summarize the current knowledge of the local galactic environment [Frisch et al.] and various perspectives on the pre-IBEX theory of the interstellar interaction [Lee et al.; Zank et al.; and Izmodenov et al.]. Finally, Chapter 13 describes IBEX's extensive Education and Public Outreach program [Bartolone et al.].

The team of truly outstanding men and women who made the IBEX mission such a great success is given in the following (Acknowledgements) section; this volume is dedicated to every one of them with my sincerest thanks!

Seattle, Washington
4 July 2009

Space Sci Rev (2009) 146: 5–9
DOI 10.1007/s11214-009-9570-1

Acknowledgements

Published online: 25 July 2009

This mission would not have been possible without the truly outstanding efforts, contributions, and dedication of the literally hundreds of team members, reviewers, and participants from institutions such as NASA, universities, national laboratories, non-profits, and corporations across the country and around the world. I am delighted to acknowledge and thank all of these important contributors:

A. Abdallah, M. Acevedo, S. Adamick, T. Adamietz, M. Adrian, J. Aguilera, L. Ahr, M. Ahr, M. Al-Dayeh, J. Alexander, J.E. Alexander, N. Alexander, D. Alexashov, D. Alicia, F. Allegrini, J. Alquiza, J. Ambrose, E. Anderson, J. Andrews, S. Apgar, R. Araujo, P. Archuleta, I. Arevalos, A. Arguello, R. Arine, E. Arlington, J. Arlington, B. Armand, M. Arnold, J. Arrell, A. Arriaga, S. Asselin, C. Ayalacomayajula

A. Baca, B. Bachir-Bouiadjra, D. Baker, R. Baldwin, B. Baldwin, M. Balint, J. Ball, J. Banh, K. Barclay, A. Barnes, Z. Barricklow, J. Barrowman, S. Bartell, W. Barth, J. Bartlett, L. Bartolone, J. Basile, D. Bates, R. Bates, J. Battcher, P. Baumgartner, J. Bayless, H. Bayne, L. Beattie, R. Beaudoin, B. Beaver, J. Beavers, L. Becker, C. Beebe, L. Belaidi, K. Bell, W. Benson, L. Berger, J. Bernardin, J. Bevilacqua, S. Beyer, D. Bintner, D. Bird, A. Biskner, J. Bittle, M. Bitzer, R. Black, P. Bland, C. Blum, J. Bobbett, P. Bochsler, G. Bodmer, R. Bolanos, J. Bolek, A. Bolton, S. Bolton, J. Bonn, S. Borovikov, M. Borrelli, S. Boupha, R. Bowman, D. Bradford, R. Bramlett, P. Brandt, B. Breech, M. Breslof, J. Bretton, C. Briggs, L. Briggs, J. Brigoli, R. Brockdorf, T. Broiles, D. Brown, D.C. Brown, M. Brysch, C. Bumbaugh, J. Burch, J. Burgess, R. Burley, J. Burnett, M. Busch, J. Byrnes, M. Bzowski

J. Calvert, R. Calvo, M. Camillieri, J. Campbell, E. Canada, C. Cannon, B. Carder, D. Carmen, K. Carney, M. Carney, K. Carr, J. Carrico, E. Carrig, T. Casale, S. Casazza, T. Case, P. Casey, S. Casteel, C. Castillo, J. Cavallo, G. Cavazos, A. Ceton, S. Chalov, L. Chance, F. Chandler, P. Chen, P. Chenault, D. Chenette, D. Chevers, F. Chmara, D. Chornay, K. Chow, N. Chrissotimos, E. Christian, J. Christmas, M. Chu, G. Chudej, J. Chung, M. Cibula, J. Cirone, F. Cisneros, V. Cisneros, P. Clair, C. Clark, G. Clark, C. Clingan,

K. Clinton, M. Coers, S. Cohen, M. Collier, A. Comberiate, T. Comunale, M. Concha, C. Conemac, K. Cook, S. Cooley, J. Cooper, V. Coppola, W. Corley, A. Cormier, D. Cornwell, M. Cortazzo, A. Cortez, P. Cothren, E. Cotton, E. Cover-Randall, D. Covert, R. Cox, J. Craft, S. Craig, R. Cramer, T. Craychee, G. Crew, J. Crock, H. Croft, D. Cronk, W. Cross, P. Crouse, C. Cullum, D. Cumings, D. Curiel, J. Curley, A. Czechowski

K. Davidson, J.L. Davis, J. Davis, W. Davis, J. Dean, K. Deane, S. DeBoda, M. Del Puerto, M. Delatorre, L. Delgado, A. DeLosSantos, D. DeLuna, R. DeMagistre, P. Demaine, D. Demkee, J. Demko, E. Denbrook, W. Denier, C. Dent, T. Denver, J. Deprez, M. Desai, J. Dezio, L. DiBiasi, J. Dickinson, G. Diller, S. Dion, G. Dirks, S. Doctolero, E. Dodge, B. Dodson, H. Dornbusch, C. Dovale, W. Downing, J. Doyle, M. Draves, M. Dugan, S. Dumas, C. Dumlao, E. Dunlap, R. Dunn, A. Dunn, G. Dunn, W. Dziegiel

P. Easley, R. Eason, D. Eberle, R. Ebert, M. Echaniz, P. Eckard, E. Edlund, H. Edrisi, K. Edrisi, R. Edwards, T. Edwards, J. Egger, J. Ehrsam, J. Eitnier, D. Eldeen, V. Elizondo, G. Ellers, H. Elliott, S. Ellis, G. Ellison, V. Elsbernd, M. Epperly, S. Epstein, J. Erickson, T. Errigo, J. Esquivel, J. Eterno, J. Etter, S. Eury, N. Evans, D. Everett

J. Fabunmi, M. Fagan, H. Fahr, M. Faller, R. Farlkey, J. Farnan, W. Faulconer, M. Feibig, J. Feiger, J. Feil, W. Feimer, H. Feldmesser, N. Fell-Murray, M. Fendall, D. Ferguson, J. Ferren, M. Ferris, L. Fetterman, H. Fichtner, B. Fields, T. Finley, J. Fischer, R. Fisher, S. Fleming, Y. Flom, A. Flores, V. Florinski, E. Floyd, F. Foerst, P. Ford, J. Ford, D. Foss, L. Foster, E. Fowler, F. Fox, R. Foxwell, J. Francis, S. Francois, J. Franks, G. Frazier, R. Frederic, C. Freitas, T. Friedmann, P. Frisch, C. Frost, R. Funk, H. Funsten, S. Fuselier

E. Gaddy, B. Gahagan, T. Gaio, A. Galan, D. Gallup, R. Ganley, V. Ganley, A. Garcia, C. Garcia, L. Garcia, R. Garcia, R. Garnica, R. Garza, D. Gates, J. Gearhart, D. Gell, T. Gerber, J. Gerhardus, M. Gerry, B. Gewirz, A. Ghielmetti, R. Giacobone, K. Giberson, W. Gibson, D. Gibson, D. Gittle, R. Gitzendanner, J. Glass, G. Gloeckler, M. Goans, G. Gochar, M. Golightly, E. Gonzales, P. Gonzales, S. Gonzales, M. Gonzalez, R. Gonzalez, R. Gooch, S. Gooch, K. Goodrich, S. Goodson II, J. Googins, S. Gordon, T. Gordon, H. Grabowski, R. Graham, C. Grandjean, M. Granoff, C. Gray, J. Greagrey, D. Green, T. Greggory, R. Griffin, M. Griffin, A. Griffith, M. Griffiths, K. Griswold, F. Gross , W. Grossman, E. Grotheer, M. Grove, E. Grün, M. Gruntman, G. Guerra, D. Guerrero, R. Guerrero, M. Guhathakurta, M. Guillen, P. Guldager, Y. Guo, P. Gupta, V. Gupta, R. Gurnee, S. Gurst, J. Gutbezahl, A. Guthrie, L. Guzman

W. Haag, R. Haenke, C. Haiyan, J. Hamilton, J. Hanley, T. Hannon, R. Harbaugh, W. Harding, J. Harper, R. Harper, R. Harris, D. Harris Jr, S. Harrison, A. Hashemi, A. Hassan, S. Häusler, B. Hautamaki, D. Hawes, T. Hayden, J. Hayes, J. Heerikuisen, D. Heihn, D. Heirtzler, C. Helferstay, A. Henry, R. Henry, E. Henry, C. Hernandez, M. Hernandez, E. Herrera, E. Hertzberg, R. Hevner, K. Hills, L. Hixson, D. Hobel, C. Hobson, J. Hochstedtler, S. Hodgson, S. Hoge, W. Holemans, D. Hollenbach, S. Hollo, B. Hong, R. Hornstein, H. Hougon, C. Howard, J. Hoyos, H. Hsin, C. Hubert, V. Hubert, A. Hudgins, D. Hughes, J. Hughes, L. Hughes

A. Ingegneri, S. Ingemi, D. Isaac, V. Izmodenov

S. Jackson, J. Jacobs, A. Jacques, T. Jaeckle, J.-M. Jahn, P. Janzen, S. Jaskulek, W. Jenkins, K. Jennings, M. Jensen, P. Jensen, L. Jimenez, B. Johnson, D. Johnson, J. Johnson,

K. Johnson, M.D. Johnson, M.A. Johnson, M. Johnson, T. Johnson, D. Jones, J. Jones, N. Jones, E. Jones, W. Jones, G. Jones, C. Jordan, J. Jørgensen, P. Jorgensen, J. Jost, J. Joyce, K. Judge, F. Jung

P. Kacenski, R. Kallenbach, R. Kalmbach, O. Katushkina, D. Kauthen, R. Keedy, B. Keegan, J. Keeney, M. Keeter, J. Kelley, D. Kelly, C. Kempe, J. Kenawell, M. Kenny, D. Ketchum, K. Kihara, R. Killough, S. Kinaman, B. King, D. King, R. Kinna-Cuddahee, L. Kipple, P. Kirchman, S. Kirn, R. Klar, G. Kleve, E. Kline, P. Knappenberger, U. Knauss, C. Kofoed, J. Kota, J. Krage, C. Kraus, R. Kraus, H. Krüger, I. Krykov, S. Ksor, H. Kucharek, F. Kudirka, P. Kurilchik, V. Kurt

A. Lacey, J. Lackovich, C. Laferriere, T. LaForge, R. Lallement, K. Lambrecht, J. Lander, M. Lang, E. Lange, M. Lara, M. Larkin, R. Lasky, A. Lawhead, M. Leahy, R. Lebois, K. Ledbetter, B. Lee, B.R. Lee, F. Lee, M. Lee, R. Lee, I. Leja, M. Lessard, C. Leto, S. Levalley, J. Levasseur, M. Lew, A. Lewis, W. Lewis, D. Lewis, T. Lewis, Y. Li, C. Lilly, J. Linsky, E. Lira, D. Little, G. Livadiotis, C. Livesay, R. Livi, S. Livi, T. Livingston, M. Llond, J. Lobell, C. Lockhart, C. Loeffler, S. Longworth, F. Lopez, J. Lord, M. Loucks, J. Loughlin, P. Lozano, M. Lucero, S. Lucero, J. Luna, M. Luther, K. Lysiak, J. Lyver

T. Macaulay, R. MacDowall, D. Mackler, M. Magoffin, M. Maher, Y. Malama, H. Mallory, A. Manka, I. Mann, M. Manzo, M. Maple, M.E. Martinez, M. Martinez, O. Martinez Jr, K. Mashburn, S. Massetti, D. Massey, J. Matejcik, P. Matull, L. Mauch, D. May, T. May, T. Mayces, K. Maynard, T. Mayves, M. Mazur, A. McCain, C. McCarty, J. McCoig, R. McComas, V. McCormick, L. McCullough, V. McDermott, W. McGrady, R. McGrath, W. McLaren, M. McLelland, J. McLeod, M. McManus, D. McNerney, R. McNutt, P. McPike, G. Melino, R. Menchaca, C. Mendelsohn, G. Mendenilla, L. Mendez, D. Menendez, R. Merry, A. Merwarth, L. Mesa, R. Metz, T. Michalek, R. Mikulas, A. Miller, C. Miller, R. Miller, D. Miller, G. Miller, K. Miller, M. Miller, W. Mills, C. Minerva, B. Mischel, H. Mischler, D. Mitchell, J. Mitchell, A. Mitskevich, H. Mittelman, J. Mobilia, E. Möbius, J. Monk, M. Monteleone, S. Montgomery, A. Moore, C. Moore, D. Moore, K. Moore, M. Moore, S. Moore, T. Moore, J. Moreno, J. Morin, P. Morinelli, K. Morris, G. Morrow, J. Moryl, W. Motley III, X. Moussas, J. Mueller, J. Mukherjee, H.-R. Müller, M. Mullin, N. Murphy, A. Murray, B. Mutchler, L. Myers

H. Nakagawa, L. Natalino, N. Neal, M. Negrete, K. Nelson, R. Nemanich, M. Newman, G. Newton, A. Nguyen, C. Nguyen, K. Nguyen, T. Nguyen, M. Nichols, R. Nielsen, L. Niemeyer, T. Niemeyer, W. Nipa, S. Niswander, J. Nolin, P. Noonan Jr, C. Nunez II, R. Nussbaumer

R. Obenschain, I. O'dea, S. Odendahl, C. Ogas, K. Ogasawara, K. Ogilvie, M. Olenick, M. O'Malley, M. O'Neill, M. Opher, C. Orduna, T. Orr, S. Orsini, S. Ortega, M. Ortiz, J. Osche, A. Osovets, A. Oullet, J. Oxley

L. Pacini, J. Packer, G. Palacios Jr, C. Pandipati, R. Pare, T. Parfet, M. Parker, K. Parsons, N. Paschalidis, J. Passuite, M. Paternostro, J. Patton, J.S. Payne, J.D. Payne, T. Payne, J. Pearson, N. Pelletier, Q. Peng, A. Perez, R. Perison, W. Perry, T. Perry, K. Persson, S. Persyn, L. Petersen, L. Peterson, M. Peterson, R. Pfisterer, M.L. Phillips, M. Phillips, C. Philyaw, D. Piazza, W. Pickett-Otto, R. Picone, P. Piconi, B. Piepgrass Jr, A. Pimentel, T. Pirone, F. Pitman, R. Pitman, N. Pogorelov, J. Polesel, L. Policastri, S. Pollard, C. Pollock,

E. Poole, S. Pope, M. Popecki, A. Posner, F. Poujardieu, K. Powers, M. Powers, C. Prested, S. Preston, D. Pride, E. Provornikova, A. Puente

S. Quarderer, E. Quemerais, D. Quinn

B. Rafferty, G. Rahal, G. Raine, A.G. Ramirez, A.K. Ramirez, B. Rammelkamp, B. Randol, J. Ranger, A. Rao, R. Rasmison, S. Rasmussen, A. Rausch, N. Rawlings, C. Rawlins, J. Raymond, J. Redfern, S. Redfield, J. Regalado, T. Regiec, T. Reilly, D. Reisenfeld, J. Remez, R. Rendon, M. Reno, R. Rerko, M. Resnick, R. Reyes, W. Reyle, R. Reynolds, C. Rhoad, S. Richard, J. Richards, J. Richardson, S. Richardson-Moore, E. Riedl, J. Rinehart, D. Roberts, I. Robertson, J. Roddy, A. Rodriquez, B. Rodriquez, C. Rodriquez, E. Roelof, J. Roese, S. Rogillio, S. Rohrbacher, D. Roman, J. Roman, P. Rosmarynowski, A. Rouche, S. Roy, M. Ruderman, J. Rudzki III, H. Runge, C.T. Russell, C.A. Russell, R. Russell, B. Ryder, J. Ryder

D. Salazer, M. Saldana, J. Salgado, M. Sampson, J. Sanborn, D. Sanders, J. Sanders, R. Sardar, M. Sattler, L. Saul, A. Sawka, M. Sawka, H. Schattenberg, E. Scheer, J. Scheer, D. Scheidt, K. Scherer, K. Scherm, J. Scherrer, D. Scheve, T. Schilling, C. Schlemm, D. Schmid, C. Schneck, P. Schneider, R. Schnurr, E. Schram, A. Schroder, S. Schumacher, N. Schwadron, A. Schwartz, M. Schwinger, S. Scott, B. Seale, M. Secunda, S. Sekira, D. Shaffer, A. Shamkovich, R. Shelton, A. Sherer, S. Sherman, D. Shiderly, C. Shinohara, C. Shoemaker, J. Short, R. Sibley, A. Siegel, J. Siegel, A. Sierra, M. Sigrest, M. Silver, H. Silvus Jr, D. Simpson, J. Simpson, R. Skoug, M. Slankard, D. Slater, J. Slavin, S. Small, E. Smart, A. Smith, A.H. Smith, C. Smith, K. Smith, V. Smith, D. Smith, W. Smithson, E. Smoke, K. Sneddon, A. Snively, M.-A. Soehl, D. Sollberger, V. Soloviev, M. Somers, J. Sommerer, B. Song, V. Sosa, A. Souksamlane, M. Specht, J. Speck, H. Spence, J. Spencer, R. Spies, J. Sprencel, J. Stack, R. Staley III, E. Stasiunas, V. Stavast, J. Stepovich, M. Sterba, A. Stern, F. Stocklin, B. Stone, F. Stone, J. Stone, E. Stoneking, J. Stowers, S. Stoyanova, R. Strain, J. Stratton, D. Strickland, J. Strycharske, B. Stuart, E. Suarez, S. Suess, H. Sykes III

M. Tapley, R. Taylor, C. Taylor, J. Taylor, J. Teller, K. Thal, B. Thaxton, L. Theis, J. Thomas, B. Thompson, D. Thompson, J. Thompson, R.L. Thompson, R.A. Thompson, R. Thorpe, C. Tillier, M. Titerance, W. Tittley, A. Tolbert III, W. Tolson, W. Tomlinson, K. Torkaman, V. Torres Jr, B. Tossman, B. Trantham, T. Trbovich, S. Turco, R. Turner, D. Tutera, J. Tyler, R. Tyler, Y. Tyler

B. Ungar, S. Urman

P. Vachon, P. Valek, E. Valles, M. Van Hecke, R. Vanderspek, R. Varela, T. Vaughn, S. Vaughn, A. Vedder, N. Veliz, M. Vincent, R. Visscher, S. Voglewede, F. Volpe, N. Von-Bank, J. Voos, V. Voos, J. Vorreiter, M. Vosbury

D. Wagar, S. Wagner, N. Walbridge, H. Walker Jr, P. Walpole, D. Walter, E. Walter, D. Wampler, T. Ward, H. Washimi, M. Watkins, W. Weaver, M. Webb, R. Weber, S. Weidner, G. Weigle II, E. Weiler, S. Wells, W. Wells, W.L. Wells, J. Wesdock, S. Wesley, J. Westfall, L. Westine, J. Westlake, M. Whalen, D. White, D. Whitney, S. Whittaker, K. Whitton, M. Widholm, T. Wiegand, M. Wieser, V. Wikander, D. Wiles, R. Williams, B. Williams,

C. Williams, D. Williams, R.L. Williams, R. Williams, G. Willingham, R. Wingard, C. Witherow, M. Witte, S. Wojtulewicz, R. Wolak, G. Wolf-Chase, M. Woltman, H. Wong, B. Wood, J. Wood, J. Woodburn, D. Wright, P. Wu, P. Wurz, R. Wyckoff, R. Wynn

M. Yeager, M. Young, R. Young, T. Young, R. Youngblood

S. Zaffke, A. Zajac, G. Zank, L. Zelenyi, Y. Zheng, and Z. Zhou

Finally, it is impossible to generate such a list, even with the help of lead team members from a variety of institutions as we have done here, without missing a least a few important names—for those of you we have missed, I am truly sorry and thank you also.

With respect to this book, I want to acknowledge my co-editors, Nathan Schwadron and Gary Zank, as well as Jim Burch, who acted as editor for the one paper that all three of the editors were authors (McComas et al., Chap. 1). Finally, I especially thank Wendy Mills, who did most of the detailed editing work for this volume.

Space Sci Rev (2009) 146: 11–33
DOI 10.1007/s11214-009-9499-4

IBEX—Interstellar Boundary Explorer

D.J. McComas · F. Allegrini · P. Bochsler · M. Bzowski · M. Collier · H. Fahr · H. Fichtner · P. Frisch · H.O. Funsten · S.A. Fuselier · G. Gloeckler · M. Gruntman · V. Izmodenov · P. Knappenberger · M. Lee · S. Livi · D. Mitchell · E. Möbius · T. Moore · S. Pope · D. Reisenfeld · E. Roelof · J. Scherrer · N. Schwadron · R. Tyler · M. Wieser · M. Witte · P. Wurz · G. Zank

Received: 17 November 2008 / Accepted: 23 March 2009 / Published online: 18 April 2009

Abstract The Interstellar Boundary Explorer (IBEX) is a small explorer mission that launched on 19 October 2008 with the sole, focused science objective to *discover the global interaction between the solar wind and the interstellar medium*. IBEX is designed to achieve this objective by answering four fundamental science questions: (1) What is the global strength and structure of the termination shock, (2) How are energetic protons accelerated

D.J. McComas (✉) · F. Allegrini · S. Livi · S. Pope · J. Scherrer
Southwest Research Institute, San Antonio, TX 78228, USA
e-mail: dmccomas@swri.edu

P. Bochsler · M. Wieser · P. Wurz
University of Bern, Physikalisches Institut, Bern, Switzerland

M. Bzowski
Space Research Centre of the Polish Academy of Sciences, Warsaw, Poland

M. Collier · T. Moore
NASA Goddard Space Flight Center, Greenbelt, MD 20771, USA

H. Fahr
University of Bonn, Bonn, Germany

H. Fichtner
Ruhr-Universität Bochum, Bochum, Germany

P. Frisch
University of Chicago, Chicago, IL 60637, USA

H.O. Funsten · D. Reisenfeld
Los Alamos National Laboratory, Los Alamos, NM 87545, USA

S.A. Fuselier
Lockheed Martin Advanced Technology Center, Palo Alto, CA 94304, USA

G. Gloeckler
University of Michigan, Ann Arbor, MI 48109, USA

at the termination shock, (3) What are the global properties of the solar wind flow beyond the termination shock and in the heliotail, and (4) How does the interstellar flow interact with the heliosphere beyond the heliopause? The answers to these questions rely on energy-resolved images of energetic neutral atoms (ENAs), which originate beyond the termination shock, in the inner heliosheath. To make these exploratory ENA observations IBEX carries two ultra-high sensitivity ENA cameras on a simple spinning spacecraft. IBEX's very high apogee Earth orbit was achieved using a new and significantly enhanced method for launching small satellites; this orbit allows viewing of the outer heliosphere from beyond the Earth's relatively bright magnetospheric ENA emissions. The combination of full-sky imaging and energy spectral measurements of ENAs over the range from ~10 eV to 6 keV provides the critical information to allow us to achieve our science objective and understand this global interaction for the first time. The IBEX mission was developed to provide the first global views of the Sun's interstellar boundaries, unveiling the physics of the heliosphere's interstellar interaction, providing a deeper understanding of the heliosphere and thereby astrospheres throughout the galaxy, and creating the opportunity to make even greater unanticipated discoveries.

Keywords Interstellar boundary · Termination shock · Heliopause · Energetic Neutral Atom · ENA · LISM

1 Introduction

The galaxy is filled with the ancient debris of exploded stars (novae and supernovae) and fossil stellar winds. This interstellar medium is composed of neutral gas, ionized and magnetized plasma, and dust. The stellar winds that expand out from other stars carve out and

M. Gruntman
University of Southern California, Los Angeles, CA 90089, USA

V. Izmodenov
Moscow State University, Moscow, Russia

P. Knappenberger
Adler Planetarium, Chicago, IL 60605, USA

M. Lee · E. Möbius
University of New Hampshire, Space Science Center, Morse Hall, Durham, NH 03824, USA

D. Mitchell · E. Roelof
Applied Physics Laboratory, Johns Hopkins University, Laurel, MD 20723, USA

N. Schwadron
Boston University, 725 Commonwealth Ave, Boston, MA 02215, USA

R. Tyler
Orbital Sciences Corporation, Dulles, VA 20166, USA

M. Witte
Max Planck Institute Aeronomie, Katlenburg Lindau, Germany

G. Zank
University of Alabama, Huntsville, AL 35899, USA

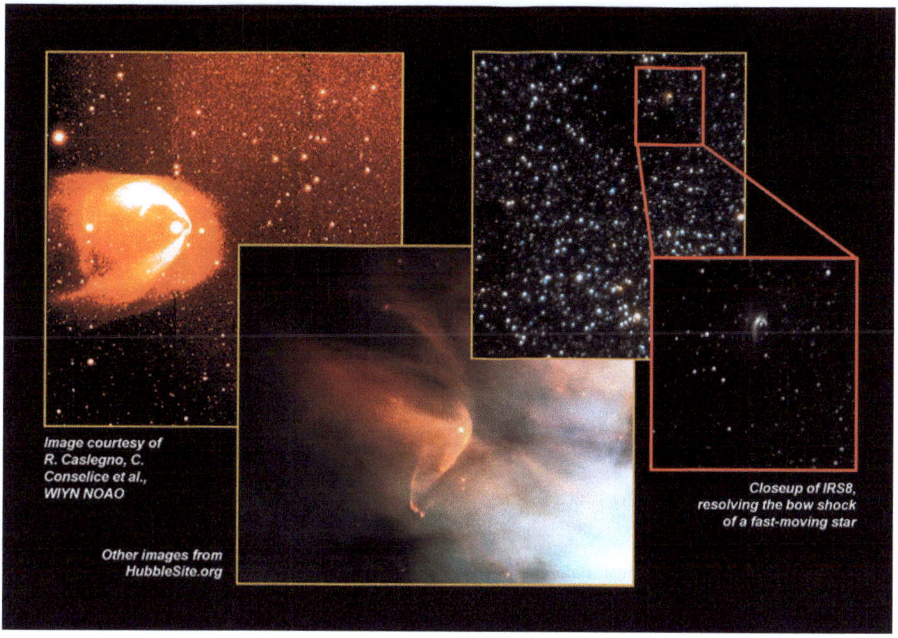

Fig. 1 Stars produce powerful stellar winds that clear out and interact with their nearby interstellar medium. This complex interaction creates the emissions seen in the images here from the Hubble telescope and WIYN NOAO observatory

interact with part of their nearby interstellar medium to form "astrospheres" (Fig. 1) around these stars. Our own astrosphere, the heliosphere, is inflated by the Sun's supersonic solar wind and its embedded Interplanetary Magnetic Field (IMF). Because the Sun moves with respect to the local interstellar medium (LISM) at $\sim$26 km s^{-1}, the interaction is highly asymmetric, being both compressed on the nose and stretched out into a long "heliotail" on the opposite side. (e.g., see recent review by Fahr et al. 2007 and references therein). Figure 2 schematically represents our heliosphere and identifies its primary boundaries.

The neutral portion of the LISM is unaffected by plasma interactions and continually drifts into the heliosphere. However, some fraction of these interstellar neutral atoms become ionized via charge-exchange with ions that pile-up ahead of the heliopause in the so-called hydrogen wall. Some of the interstellar neutrals that drift into the heliosphere become ionized, creating "pickup" ions (PUIs), which gyrate about the IMF and are swept outward from the Sun. The pickup process also endows PUIs with up to twice the solar wind speed (four times the energy per nucleon); these higher-energy ions can be preferentially accelerated at shocks, and diffusive shock acceleration at the termination shock (TS) (Pesses et al. 1981) has long been the standard model to energize these PUIs into anomalous cosmic rays (ACRs) with energies of $\sim$10–100 MeV/nuc.

In 2003, when IBEX was proposed, our limited knowledge of the interstellar interaction had been gleaned largely from a variety of indirect methods, including observations well inside the TS of interstellar neutral H and He atoms, pickup ions, ACRs, and radio waves. These observations were combined with theory and relatively simple modeling in order to try and understand the distant heliospheric interactions. (See for example the review by Zank 1999 and references therein.) Since then, the study of the outer heliosphere has burst into the forefront of Heliophysics research with the crossing of the TS by Voyager 1 on

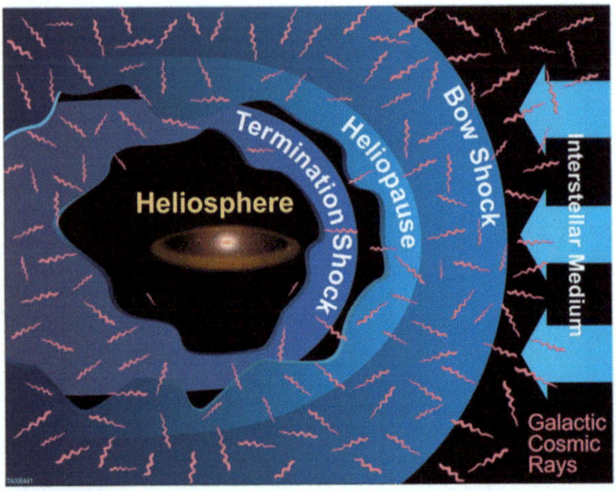

Fig. 2 Our current understanding is that the interstellar interaction creates three distinct interstellar boundaries schematically shown here. From inside out, these are: (1) the TS where the solar wind is slowed and heated and begins to divert away from the inflowing LISM, (2) the heliopause, which separates the solar wind plasma from the ionized LISM material, and, furthest out, (3) a bow shock (BS) or bow wave, which begins to divert the upstream LISM around the heliosphere. The wiggly red lines represent galactic cosmic rays, the vast majority of which are shielded out by the heliospheric interaction, particularly in the inner heliosheath, between the BS and heliopause

16 December 2004 (Stone et al. 2005; Burlaga et al. 2005; Cummings and Stone 2001; Decker et al. 2005) and then several TS crossings by Voyager 2 between 30 August and 1 September 2007 (Stone et al. 2008; Burlaga et al. 2008; Decker et al. 2008; Richardson et al. 2008). Many of the fundamental ideas that existed in 2003 about this interesting boundary have not survived the reality of the Voyagers' direct observations.

Three major results from the Voyager observations were especially surprising. First, the ACRs were not accelerated at the TS, at least at the times and locations where the two Voyager spacecraft crossed it. Since the Voyager 1 crossing, several possible explanations have been proposed to explain the paucity of observed ACRs. McComas and Schwadron (2006) pointed out that the asymmetric shape of the TS has critical implications for the locations where particles can be accelerated up to ACR energies. ACR acceleration should not be expected near the nose where both Voyagers crossed, but instead should occur back along the flanks and tail of the TS where the IMF has been connected to the TS for very long times (~1 year) and diffusive acceleration would have persisted long enough to accelerate PUIs to ACR energies (McComas and Schwadron 2006; Schwadron et al. 2008). Another idea is that the TS may not accelerate ACRs at all and instead they might be energized by distributed stochastic acceleration beyond the TS, out in the heliosheath (Fisk et al. 2006).

A second major surprise from the Voyagers was the huge difference in radial distance between the Voyager 1 crossing at ~94 AU and Voyager 2 crossings ~10 AU closer at ~84 AU. While some of this difference may well be caused by a prolonged decrease in the solar wind dynamic pressure (McComas et al. 2008), some of it likely reflects a north-south asymmetry in the TS. Such an asymmetry could be caused by the interstellar magnetic field distorting the TS and heliopause (Linde et al. 1998; Lallement et al. 2005). Subsequent modeling efforts had grappled with the large asymmetry observed by the Voyagers (e.g., Opher et al. 2007; Pogorelov et al. 2007; Heerikhuisen et al. 2008; Sternal et al. 2008).

A third major surprise was not observed until Voyager 2, with its functioning plasma instrument, reached the TS (Richardson et al. 2008). These observations showed that the bulk solar wind plasma received only about a fifth of the heating available from crossing the TS. Instead, the bulk of the available energy appears to go into higher energy PUIs. Thus, these PUIs play a major role in moderating the TS physics, and detailed physical models of this region need to account for both the lower-energy solar wind and higher-energy PUI populations.

The many mysteries surrounding the Voyager observations show just how exciting and dynamic the rapidly evolving field of outer heliospheric physics is today. However, they also point to the limitations of single-point (or even dual-point) *in situ* observations in the heliosheath. Because of the size, scale, complexity, and evidently even variability of the interaction, it is extremely difficult to divine the outer heliosphere's global configuration and properties without global information from all portions of this interaction. The Interstellar Boundary Explorer (IBEX) is designed and optimized precisely to provide this global, all-sky information.

IBEX was designed to make the first all-sky observations of the heliosphere's interaction with the LISM by imaging energetic neutral atoms (ENAs) produced largely beyond the TS, in the inner heliosheath, where the solar wind and imbedded PUIs have been slowed and heated. In the heliosheath, these populations of predominantly hydrogen ions produce a significant flux of detectable inward moving ENAs via charge-exchange with local interstellar neutrals. Only one potential observation of ENAs produced from PUI populations in the heliosphere has been reported to date. Wurz et al. (2008, and references therein), reported ENAs with limited spatial and temporal coverage in the energy range from 200 eV to 80 keV. In contrast, IBEX was optimized to provide all-sky imaging of the ENA fluxes, which are spatially resolved in both latitude and longitude, and does so over the critical energy range that covers both the solar wind and much of the PUI population in the inner heliosheath.

IBEX carries two very large geometric factor ENA cameras: IBEX-Lo, which measures ENAs from ~10 eV to 2 keV and IBEX-Hi, which measures them from ~300 eV to 6 keV. Both sensors have angular resolutions of ~6.5° × 6.5°, enabling well-resolved images with ~1800 pixels covering the whole sky. Furthermore, IBEX-Lo has eight energy-resolved channels covering its energy range and IBEX-Hi six, thus producing images at many different energies, and even more importantly, energy spectral information for each direction in the sky. The two IBEX sensors' energy ranges also overlap from ~300 eV to 2 keV to provide independent observations across the critical energy range. Thus, IBEX is optimized to globally image ENAs from the outer heliosphere for the first time. Figure 3 provides a summary of the charge-exchange process and how IBEX rotation and motion naturally generate all-sky maps of the ENAs propagating in from the inner heliosheath each six months.

IBEX is central to NASA's Heliophysics program, since it aims to discover the ultimate fate of the flow of energy and matter from the Sun, and more fundamentally, how the Sun and solar wind interact with the galactic medium. IBEX was developed to provide global, fundamental, and direct insights into the interactions at the boundary of our Sun's astrosphere, the heliosphere. These insights should also improve our understanding of the broader problem of how galaxies and stars interact and evolve. IBEX was designed to discover the global properties of energetic protons near the TS, which also allows us to explore the very important problem of how charged particles are accelerated at shocks in space plasmas more generally.

Prior to the selection of IBEX, such observations had been called for by a broad community consensus that was documented in prior Decadal Survey—*The Sun to the Earth—and Beyond: A Decadal Research Strategy in Solar and Space Physics* (*2003*); the prior

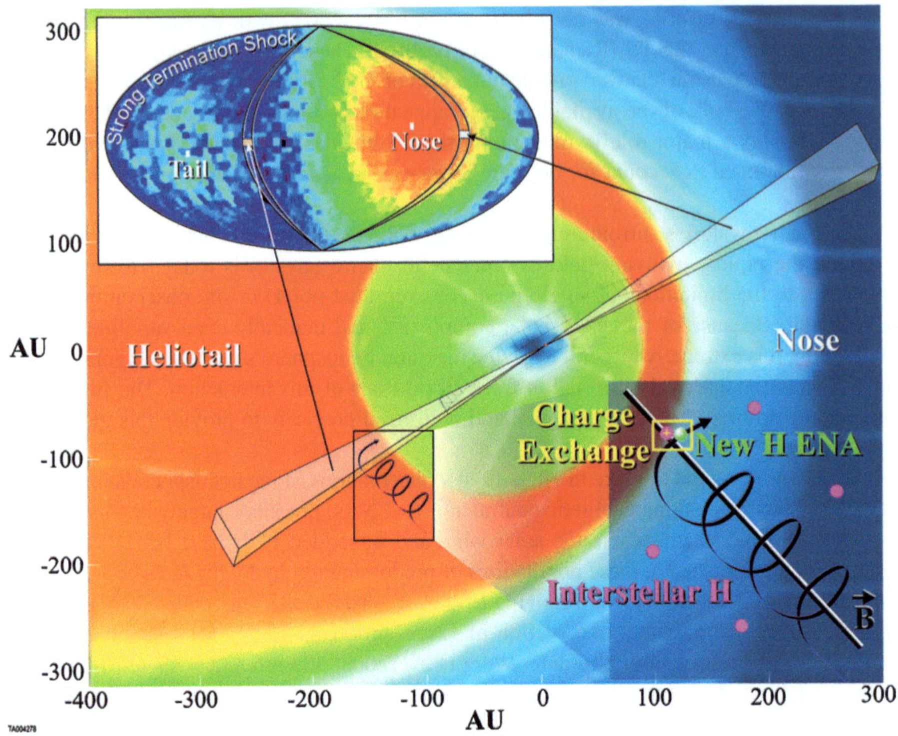

Fig. 3 Charge exchange between hot ions and cold interstellar neutrals (*lower right inset*) produce ENAs, some of which happen to be directed inward and propagate all the way into the inner heliosphere where they can be detected by the IBEX spacecraft in a very high-altitude Earth orbit. The two instantaneous $\sim 6.5^\circ \times 6.5^\circ$ fields-of-view of the IBEX-Hi and Lo sensors are depicted in the main image; these map to the two indicated pixels in the all-sky map (*top left*). As the sun-pointed IBEX spacecraft spins, the pixels viewed move repeatedly around the two indicated crescents in the sky map. Then, as IBEX repoints at the end of each orbit, the viewed crescents move across the sky producing all-sky maps each six months

Roadmap—*Sun-Earth Connection Roadmap 2003–2028: Understand how the Sun, Heliosphere, and the Planetary Environments are Connected in a Single Connected System (January 2003*); and even NASA's 2003 agency-wide plan—*National Aeronautics and Space Administration 2003 Strategic Plan*. In the case of the National Research Council's Decadal Survey, the Explorer Program was explicitly recommended: "The boundary between the solar wind and the local interstellar medium (LISM) is one of the last unexplored regions of the heliosphere. Very little is currently known about this boundary or the nature of the LISM that lies beyond it. . . .certain aspects of these regions can be studied by a combination of remote sensing and in situ sampling techniques. This investigation could be accomplished by a mission. . . to obtain energetic neutral atom images. . . of the heliospheric boundary. Such a mission is gauged to be feasible within the resource limits of the Explorer program. . . ." The IBEX Small Explorer mission proposal responded to these urgent calls from the community for global imaging of the heliospheric interaction.

IBEX was developed on a very fast timeline, moving from selection on 26 January 2005 to the fully integrated and tested payload and spacecraft being ready to deliver to the launch vehicle at Vandenberg Air Force Base (VAFB) in mid-April 2008. Such a delivery would have supported the planned June 2008 launch. Unfortunately, launch vehicle loads problems

Table 1 Schedule of major milestones for the IBEX mission

Proposal submitted	2 May 2003
Concept Study Report	18 June 2004
Mission selected	26 January 2005
Mission PDR	17–18 January 2006
Confirmation Review	13 March 2006
Mission CDR	13–18 September 2006
Payload Delivery to Spacecraft	8 October 2007
Final Delivery to VAFB	28 July 2008
Launch	19 October 2008

forced the IBEX team to design, build, and test a "Shock Ring" in order to reduce the loads the launch vehicle would have driven into our flight system down to acceptable and previously tested-to levels. That process, led by the IBEX project, was fast-tracked and completed in about three months, far shorter than the original estimate of 9–12 months for the launch vehicle to provide a "Soft Ride" to bring the loads down. Table 1 summarizes the critical milestones for the IBEX mission.

This paper briefly summarizes the IBEX mission and serves as an overview and introduction to the other chapters in the IBEX book. The bulk of those papers provide detailed discussions of the IBEX flight hardware and other aspects of the IBEX mission; references to those chapters are provided in the appropriate sections throughout this paper. In addition, we included four chapters that summarize the state of knowledge of the LISM and outer heliospheric interaction at the time of the IBEX launch (Frisch et al. 2009, this issue; Izmodenov et al. 2009, this issue; Lee et al. 2009, this issue; Zank et al. 2009, this issue).

2 Scientific Objective, Questions, and Closure

IBEX's sole, focused science objective is to *discover the global interaction between the solar wind and interstellar medium.* IBEX is designed to achieve this objective by answering four fundamental science questions:

Question I: What is the global strength and structure of the termination shock?
Question II: How are energetic protons accelerated at the termination shock?
Question III: What are the global properties of the solar wind flow beyond the termination shock and in the heliotail?
Question IV: How does the interstellar flow interact with the heliosphere beyond the heliopause?

Because so little was known about the heliosphere's global interaction at the time of the IBEX proposal, we developed a broadly scoped science strategy to address these questions. Now, with the Voyager observations opening new mysteries about the interstellar interaction, our broad approach is even more critical. We consider the science return in terms of three levels of study: *Discovery*, *Exploration*, and *Deep Understanding*. At the *Discovery* level, fundamental properties of the interstellar interaction can be directly gleaned from the IBEX images, energy spectra, and interstellar neutral fluxes. At the *Exploration* level, we will combine the observations with simple physics-based calculations, theory and limited 2D and 3D modeling to explore the more detailed properties of the outer heliosphere. Finally, at the

Fig. 4 The IBEX H ENA energy range is designed to provide the critical distributions needed to reveal the global properties of the proton populations of the inner heliosheath. Shown here are predicted ENA energy distributions near the nose of the heliosphere for a strong (*black curve*) and weak (*green curve*) TS (Gruntman et al. 2001). These curves are for a nominal, slow (1 keV) solar wind. The blue curve shows the predicted ENA flux due to energetic protons inside the TS. Energetic ENA distributions >1 keV (*black and green curves*) are predicted from observed energetic proton tails (Gloeckler et al. 1994, 2000; Schwadron et al. 1996) assuming that the intensity of the tails scale with the intensity of interstellar pickup protons (Vasyliunas and Siscoe 1976)

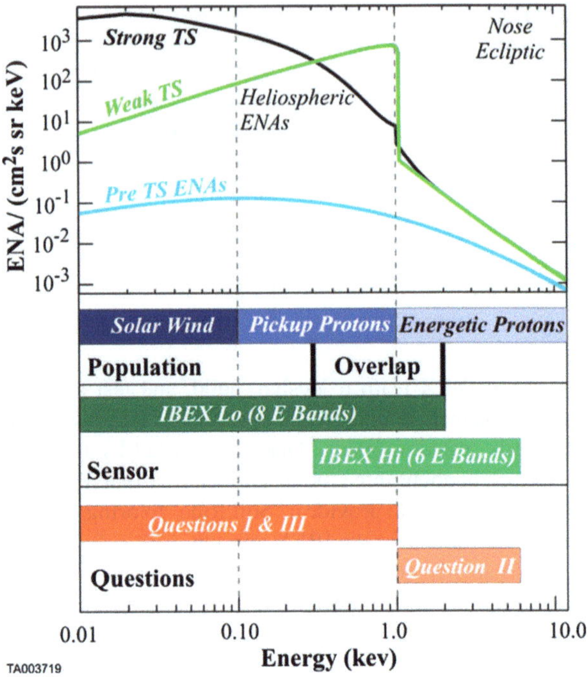

Deep Understanding level, we will extract the detailed global properties of the interstellar interaction through iterative analyses using IBEX data observations in concert with detailed 3D models of the heliosphere.

At the *Exploration* and *Deep Understanding* levels, we will make increasingly extensive use of theory and modeling to gain further insight into the global properties of the interstellar interaction. Many IBEX team members have a strong background in heliospheric theory and modeling. However, in order to expand the team and pull in an even broader segment of the community, we set aside $2M from our proposal cost-cap to fund a NASA-selected Guest Investigator (GI) program that specifically targets the coordinated use of IBEX data products to iteratively refine 3D models of the heliosphere. NASA has already made a first solicitation for such proposals and we look forward to welcoming the selected scientists into the IBEX Team.

The next four subsections very briefly summarize our four science questions and how IBEX observations were designed to uniquely address them. For illustration, Fig. 4 summarizes two possible extreme shapes of the H ENA energy distribution and compares them to the energy coverage of the IBEX-Hi and Lo sensors. More detailed discussion of the four science questions and the approach that our team proposed to answer them was given in McComas et al. (2004b). Much more detail on the final, complete IBEX capabilities and our ability to make the needed measurements to answer these questions is provided in the other detailed papers throughout this volume.

2.1 Question 1: What Is the Global Strength and Structure of the TS?

When we proposed IBEX, before the Voyagers observed the TS directly, we asked: "*Is there a termination shock? If the solar wind carries most of the pressure, the TS should be a strong gas-dynamic shock producing a large, abrupt speed decrease. The presence of pickup and energetic protons, however, provides additional pressure that weakens the shock, or possibly, in an extreme case, causes the shock to dissolve into a wave.*" Now with the *in situ* observations from Voyager 1 and especially Voyager 2, we know that there is a TS, at least at times, near the nose of the heliosphere where the Voyagers crossed it. However, it is not a strong shock, and it is highly moderated by the addition of PUIs. Analysis of the combination of global images and energy-spectral information from IBEX will make it possible to derive the proton-energy distributions in the inner heliosheath and determine the shock strength as a function of position. Such global observations are the only way to answer the fundamental questions of the existence and strength of the TS in all directions of the sky.

As an example of the types of observations IBEX was designed to make, Fig. 5 shows simulated all-sky maps of ENA fluxes over the energy range examined by IBEX. The model (Heerikhuisen et al. 2008) includes neutrals, the suspected external magnetic field orientation in the LISM, and a realistic κ-function ion distribution (they used $\kappa = 1.63$, based on the Voyager 1 data at higher energies Decker et al. 2005). Several features stand out in these simulations, including the highest emissions from down the tail and the bright, diffuse, and apparently banded emissions from across the nose (see Heerikhuisen et al. 2008 for details). While the real heliospheric interaction is undoubtedly much more complex than this simulation, IBEX's energy-resolved maps will clearly provide the detailed all-sky observations needed to understand the global configuration.

2.2 Question 2: How Are Energetic Protons Accelerated at the Termination Shock?

The TS is a nearly perpendicular shock (shock normal perpendicular to the local magnetic field) owing to the winding up of the IMF in the outer heliosphere. This geometry leads to an "injection problem" (Lee 2000; Rice et al. 2000; Giacalone 2001) where the particles can only be efficiently accelerated at the shock if they already have very high speeds along the magnetic field. Furthermore, the TS is highly moderated by the large numbers of PUIs, suggesting that the shock acceleration is a highly non-linear process where the TS accelerates protons, and the energetic protons in turn modify the shock, thereby changing the very nature of the acceleration.

IBEX will infer the properties of accelerated protons near the TS by measuring their energy distributions via the H ENAs produced from these accelerated protons up to 6 keV. While these protons are much lower energy than ACRs or even the injected particles, they feed these higher energy protons. By measuring the intensity and energy dependence at lower energies, IBEX will infer the injection and acceleration of protons at higher energies. In addition, IBEX's energy spectral measurements from all directions in the sky are designed to enable the exploration of the variability of the TS, further informing the complicated, iterative mechanisms of shock physics. IBEX will directly observe the intensity of ENAs from accelerated protons relative to the solar wind and pickup protons below 1 keV. These observations, in concert with modeling to deconvolve the line-of-sight (LOS) integration and to extrapolate the measurements over an energy range needed to estimate the energetic proton pressure, should make it possible to determine how the TS is moderated at various locations. Finally, more advanced models of shock acceleration in the future will ultimately have to tally with all of the detailed IBEX spectral observations.

2.3 Question 3: What Are the Global Properties of the Solar Wind Flow beyond the Termination Shock and in the Heliotail?

After crossing the TS, the solar wind and PUIs become swept back in the inner heliosheath by the interaction with the LISM. Ultimately, nearly all this material must flow back and

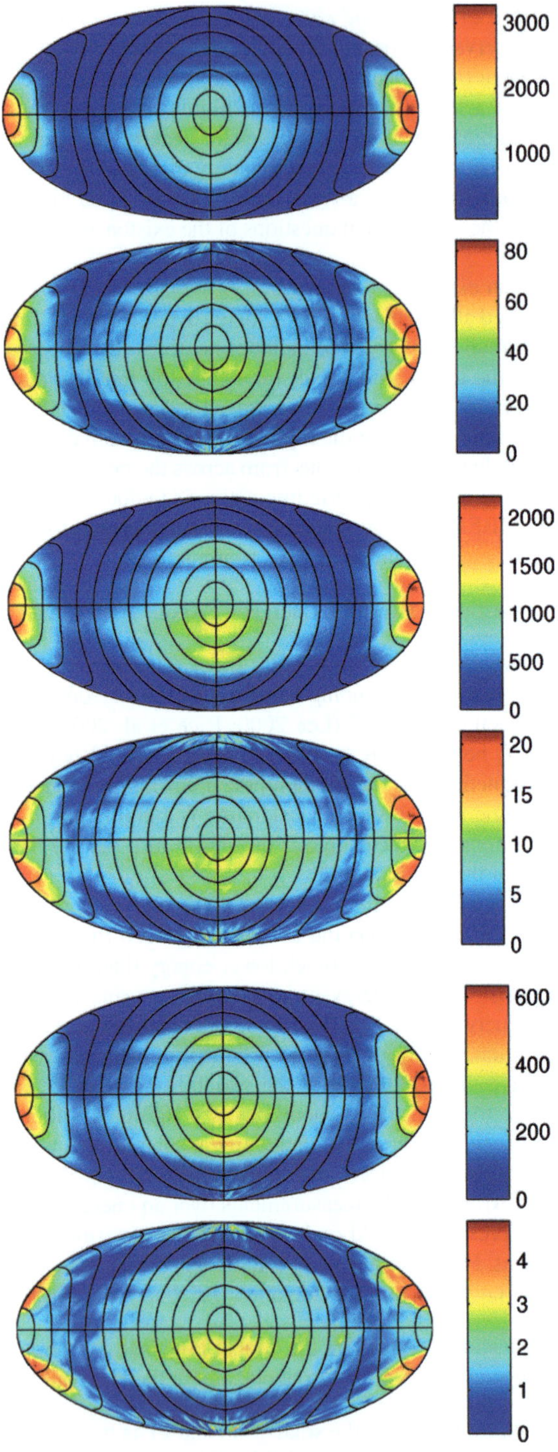

Fig. 5 Simulated all-sky maps of ENAs at various energies taken from Heerikhuisen et al. (2008). The maps plot color-coded ENA fluxes in units of $\mathrm{cm^2\,s\,sr\,keV^{-1}}$ for energy bands of 8–12 eV, 40–60 eV, 180–220 eV, 900–1100 eV, 2200–2600 eV and 5600–6400 eV, respectively from *top* to *bottom*. In each panel, the nose of the heliosphere is in the center with the poles at the *top* and *bottom* and tail on the *far left* and *right sides*

down the heliotail. Various models predict the flow patterns in the heliosheath and heliotail (e.g., Baranov and Malama 1993, 1996; Zank et al. 1996; Linde 1998; Linde et al. 1998; Müller et al. 2000) with differences that depend critically on the model approach and assumptions about both the solar wind and LISM.

Global ENA observations from IBEX should provide extremely sensitive measures of the asymmetries in the properties of the ions in the inner heliosheath as well as the thickness of this region. By making all-sky observations over the full energy range of the bulk populations, IBEX will measure the thermalization and energy partition of the solar wind and PUIs and the global flow patterns of the solar wind beyond the TS.

As a simple example, Fig. 4 compares model energy spectra for a strong gas-dynamic shock (black) with no contribution from PUIs and a shock weakened by PUIs (green). Differences in the source ion energy distributions generate significant differences in the ENA emissions even in this one energy band. The combination multiple images across the broad range covered by IBEX will very strongly constrain the properties and flow patterns in the inner heliosheath.

2.4 Question 4: How Does the Interstellar Flow Interact with the Heliosphere beyond the Heliopause?

Because hydrogen and oxygen ions and atoms readily charge-exchange with each other, the filtration process that modifies the inflowing interstellar neutral H similarly modifies the interstellar neutral O. Thus, interstellar O measured by IBEX comprises two populations: an unmodified or "primary" population that reflects the undisturbed properties of the LISM and a "secondary" population, which generally reflects the ion population in the outer heliosheath. IBEX is designed to provide the first direct measurement of filtered interstellar neutrals (see Fig. 6). This is possible because the Sun's gravity focuses the inflowing interstellar O atoms in the direction opposite to the interstellar flow. Careful measurements of the detailed distribution of the incoming neutrals' directions by IBEX will allow differentiation between and separate quantification of the primary and secondary populations. Measurements of the primary population should provide additional direct information about the LISM, while measurements of the secondary population should allow us to measure the heating, deceleration and depletion associated with the interstellar interaction near the heliopause at the hydrogen wall. Detailed discussion of this topic is provided by Möbius et al. (2009, this issue).

2.5 Scientific Closure

The process we followed to define the IBEX capabilities and requirements flowed naturally from deciding what was needed to answer the above four science questions. As summarized in Table 2, for each question, we identified answers that could be provided at each of the three levels of examination: *Discovery*, *Exploration*, and *Deep Understanding*. Then we specifically identified what measurement requirements would allow us to fully answer these questions with sufficient margin to span unanticipated discovery science because of the extremely limited knowledge at the start of the IBEX mission of the structure and dynamics of the interaction region. As examples, the required angular and energy resolutions and ranges are shown at the bottom of Table 2. These and many other requirements went into our baseline mission requirements, which drove the entire design and development of the IBEX mission. In addition to baseline requirements, we also developed minimum mission requirements that would still provide acceptable science return for the mission if we

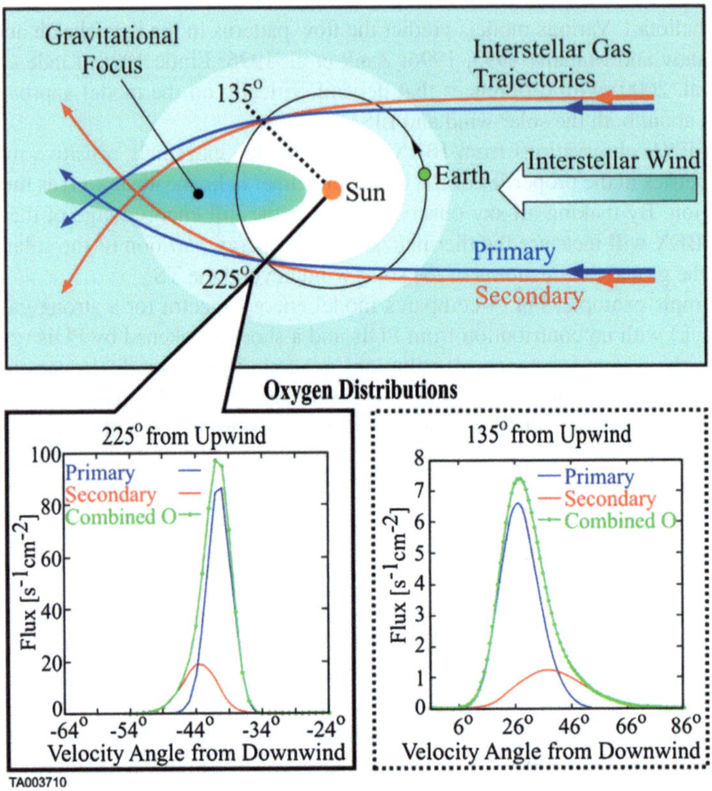

Fig. 6 Schematic diagram of neutral O trajectories along hyperbolic orbits and locations in the IBEX orbit when it will be able to make direct measurements of the inflowing interstellar oxygen (*top*). The *bottom panels* show model fluxes as a function of velocity angle measured during the two optimum times of the year. The filtered secondary population is slower, hotter and more strongly deflected than the primary population

were unable to fully meet the baseline requirements in some area. We are delighted to report that the IBEX mission as built and flown meets, and in most areas exceeds, our full baseline requirements.

3 IBEX Flight System

The IBEX flight system comprises all the mission components that launched into space on board the Pegasus launch vehicle. These include the IBEX payload, which consists of two science sensors and a combined electronics box; the IBEX spacecraft bus, which carries the payload; and a solid rocket motor (SRM) and adapter cone, which are used to help boost the spacecraft into a highly elliptical, high-altitude Earth orbit.

3.1 The IBEX Payload

The IBEX payload was designed to meet or exceed all baseline requirements and answer all four science questions described in the previous section. Because technical resources in general and mass in particular, were very limited, we needed to develop the most efficient

Table 2 IBEX science questions, levels of study, and derived top-level measurement requirements

	The IBEX approach and measurement requirements				
	The sole IBEX objective to discover the global interaction between the solar wind and the interstellar medium is achieved by answering four fundamental science questions				
	Fundamental Science Questions / **Levels**	**I:** *TS Strength/ Structure*	**II:** *Energetic Protons Near TS*	**III:** *Solar Wind Flow Patterns of Inner Heliosheath*	**IV:** *Interstellar Flow and Interaction*
Raw Images	**Discover** (Straightforward interpretation of IBEX data products)	All-sky survey: strong vs. weak TS	Intensity and spectra of energetic protons near TS	Nose/tail asymmetries	First direct measurement of filtered interstellar neutral, O
	Explore (Interpretation based on simple physics-based calculations and limited modeling)	All-sky map of shock strength	Modification of TS dynamics due to energetic particle pressure	Solar wind flow direction beyond TS vs. angle from the nose	Bulk speed, direction, and temperature of interstellar O inside TS
Detailed Models	**Understand** (use data products to iteratively define, revise, and refine 3D models of the heliosphere)	-3D TS configuration incl. distance scale and strength -Energy partition of solar wind, pickup protons, and energetic protons	Bound injection processes at TS	-3D solar wind flow patterns beyond TS -LISM B	-Filtration -Ionized fraction of O in the LISM -Interstellar flow patterns beyond heliopause -Bow shock existence
	Measurements Requirements				
	Global heliospheric ENA imaging	$7^{o}x7^{o}$ pixels	$21^{o}x21^{o}$ pixels	$7^{o}x7^{o}$ pixels	Determine O Direction to 2.0^{o}FWHM
	H energy spectra: Solar wind through energetic particle ENAs	0.01-1 keV in 7 energy bands	1-6 keV in 4 energy bands	0.01-1 keV in 7 energy bands	
	O fluxes and velocity direction vs. time of year				0.01-0.5 keV in 6 energy bands

TA003718_b

science payload possible, making full use of all of the heritage experience and expertise from across the IBEX team. This process led to an optimized payload consisting of two very large aperture single pixel ENA cameras: IBEX-Lo (Fuselier et al. 2009, this issue), which measures ENAs with energies from ~10 eV up to 2 keV, and IBEX-Hi (Funsten et al. 2009, this issue), which measures ENAs with energies from ~300 eV up to 6 keV. Both sensors are served by a single combined electronics unit (CEU). Critical parameters for IBEX-Hi, Lo and the CEU are summarized in Table 3.

Both the IBEX-Hi and Lo sensor designs are based on the same physical principles, but each is tailored to optimize ENA measurements for their respective energy ranges. Figure 7 shows schematic diagrams of the two sensors. Each is comprised of the same four subsystems: an entrance system, a charge-conversion system, an electrostatic analyzer (ESA), and a detection system.

ENAs enter the IBEX sensors through entrance systems, which are comprised of a sun shield, electron rejection ring, pre-collimator, and collimator. A slanted sunshade shields the rest of the entrance system components from any direct solar illumination during normal operations. The electron rejection ring imposes a negative potential across the aperture without any grids or other structures that could generate neutral particles. This potential excludes all except the highest-energy electrons (>600 eV) from reaching the aperture.

The pre-collimators and collimator set the intrinsic angular resolution of our measurements to ~6.5° FWHM in both sensors. In addition, IBEX-Lo has a higher angular resolution quadrant (~3.2° FWHM) that is used for direct detection of the low-energy interstellar oxygen (science question 4 above). The entire collimator is biased to +10 kV, which rejects positive ions from the surrounding space environment with energies <10 keV/q. UV light, which can be a significant source of background for these measurements, passes unaffected through the collimators as do the ENAs.

Table 3 Key IBEX payload parameters and resources

	IBEX-Hi	IBEX-Lo	CEU
Energy Range	300 eV–6.0 keV	10 eV–2 keV	N/A
Energy resolution ($\Delta E_{\text{FWHM}}/E$)	0.47–0.66	0.8	N/A
Number of Energy Steps	6	8	N/A
FOV (FWHM)	$6.5^\circ \times 6.5^\circ$	$6.5^\circ \times 6.5^\circ$ $3.2^\circ \times 3.2^\circ$	N/A
Geometric Factor for H near 1 keV (double coincidence)	3×10^{-3} cm^2 sr eV/eV at 1.1 keV	8.1×10^{-4} cm^2 sr eV/eV at 0.78 keV	N/A
Geometric Factor for H near 1 keV (triple coincidence)	7.3×10^{-4} cm^2 sr eV/eV at 1.1 keV	2.9×10^{-4} cm^2 sr eV/eV at 0.78 keV	N/A
Mass	7.70 kg	12.09 kg	5.42 kg
Volume	43.94 cm Dia × 23.75 cm	40 cm Dia × 31.45 cm	43.82 cm × 27.9 cm × 9.5 cm
Power	0.7 W	3.5 W	11.2 W
Telemetry	102.6 bps	122.8 bps	11.8 bps

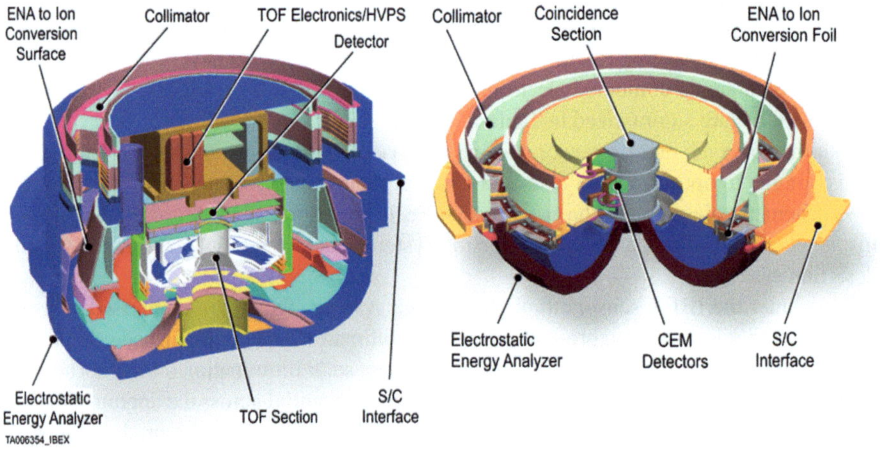

Fig. 7 Schematic diagrams of the IBEX-Lo (*left*) and IBEX-Hi (*right*) sensors. Both are based on large "bundt pan" type ESAs, which focus huge aperture areas onto their central detector sections. While optimized differently for the higher and lower energy ranges, both are comprised of the same four basic subsystems

The next step in detecting the ENAs is to convert them back into charged particles. ENAs are detected in the IBEX sensors using the same process that produced them in the outer heliosphere—charge exchange. In the case of IBEX-Hi some fraction of the neutrals are converted to positive ions as they pass through a set of ultra-thin carbon foils, which are only ~0.5 μg cm^{-2}, or ~50–100 atoms thick (see McComas et al. 2004a and references therein). At even lower energies, the ENAs can't make it through even these very thin foils. Instead, in IBEX-Lo, conversion produces negative ions, as they reflect off of special CVD diamond surfaces (Gruntman 1997 and references therein). The efficiency for both of these processes is low: ~one percent to a few tens of percent, so one of the biggest drivers for the sensor design was to find a way to maximize the aperture area.

The aperture areas of both IBEX-Lo and Hi were maximized by using "Bundt pan" shaped ESAs (Moestue 1973). With this design, truly immense open areas can be arrayed around the perimeters of the sensors, and ions produced by the incident ENAs are bent around into reasonable sized, central detector sections. Thus, in addition to removing the UV that enters the sensors and setting a voltage-dependant energy pass band as all ESAs do, the IBEX ESAs also concentrate the particles from a very large aperture onto a small, central detector in a way that greatly increases both sensors' signals and signal-to-noise ratios.

The final stage of measuring the heliospheric ENAs is the detection process. In both detector subsystems, the converted incident ions pass through a pair of ultra-thin carbon foils, which then emit secondary electrons. In IBEX-Lo, both the primary particle and the secondary electrons are detected on a microchannel plate (MCP) detector, and the time-of-flight (TOF) of the primary is measured, allowing determination of its species (e.g., H, He, or O). In the case of IBEX-Hi, the ionized ENA transits three stacked detection chambers separated by the two foils, and secondary electrons generated in each chamber are detected by the channel electron multiplier (CEM) in that chamber. Both sensors have both double and triple coincidence data products, with the triple coincidence ones having exceptionally low background (false positive) rates. Both of these designs are optimized to suppress the intrinsic detector section backgrounds. The backgrounds of both sensors and the overall mission noise and background have been carefully studied and painstakingly minimized (Wurz et al. 2009, this issue).

Finally, both IBEX sensors have small additional components that support their science observations. For IBEX-Lo, a small star sensor has been added that allows extremely precise knowledge of this sensor's pointing. These measurements greatly increase the precision with which the interstellar neutrals can be measured and the primary and secondary populations can be separated (Möebius et al. 2009, this issue). In IBEX-Hi, we added a small background monitor to provide independent measurements of the local energetic ion environment (Allegrini et al. 2009, this issue). These observations enhance our ability to identify and remove times when the local plasma environment could be contaminating the ENA observations.

Both sensors are served by the CEU. This box provides the single spacecraft interface, data processing, low-voltage power supplies, and high-voltage power supplies for both sensors. As part of the spacecraft interface, the CEU accepts, parses and routes commands and stores and plays back spacecraft and science telemetry. The command and data handling is based on a 8051 core processor. As the CEU steps the various sensor high voltages, it gathers, bins, and processes the raw science data. A two-gigabit solid state recorder makes it possible to store multiple orbits' worth of science data and spacecraft telemetry.

The CEU's low-voltage power supplies create and distribute separate custom low voltages for the two sensors and provide monitoring as well as solid-state relay switching for all of these outputs. The high-voltage system comprises four independent high-voltage supplies, which have a resonant (at ~100 kHz) fly-back-type topology and contain collimator supplies and combined bulk supplies separately for each sensor. Other than one additional high-voltage supply located inside IBEX-Lo, the CEU supplies all 16 sensor high-voltage outputs on coaxial high-voltage cables.

Fig. 8 The IBEX spacecraft is just under 1 m across and approximately 0.5 m tall. The sun-pointed surface is covered with solar cells; IBEX spins around an axis perpendicular to this plane. The white protrusion on the bottom is a blanket that covers one of the two symmetric hydrazine tanks; the gold-plated instrument on the right is IBEX-Lo

3.2 The IBEX Spacecraft

The IBEX spacecraft (Fig. 8) was designed and built by Orbital Sciences Corporation, based on their earlier "MicroStar" line flown on the 35 ORBCOMM spacecraft launched between 1995 and 1999. Although single string, the MicroStar bus has proven to be reliable in its previous missions; it was also part of NASA's Rapid Spacecraft Development Office catalog that we reviewed at the time of the IBEX proposal.

For IBEX, the largely unchanged MicroStar avionics were put into a new thrust-tube-based mechanical design structure. These electronics feature a distributed Motorola 68302 microprocessor-based architecture, with the Command and Data Handling (C&DH) system, Attitude Control System (ACS), and Electrical Power Subsystem (EPS) each controlled by a dedicated processor. The Flight Computer is the spacecraft controller; the Attitude Control Electronics interface to the ACS components and hosts the ACS software; and the Battery Charge Regulator provides power for all spacecraft loads and performs ancillary functions including heater control, SRM fire, and separation-system actuation. The Mission Interface Unit (MIU) provides the interface between the spacecraft and the payload CEU. (The CEU, briefly described above, stores all mission data, including spacecraft housekeeping telemetry.) The MIU also provides the interface to the S-band transceiver to receive uplinked commands and downlink telemetry and science data. On-board propulsion for perigee raising, spin down, and orbital maneuvers is provided by the hydrazine propulsion system (HPS). The spacecraft block diagram is shown in Fig. 9. More details about the CEU, spacecraft, and mission design are provided by Scherrer et al. (2009, this issue).

The IBEX flight system and spacecraft key parameters are summarized in Table 4. The IBEX flight system was the heaviest object ever launched by Pegasus with a total mass of ~458 kg. Of that, the fueled (wet) spacecraft was only ~105 kg, less than a quarter of the total launch mass. The remainder of the mass was used for our SRM, adapter cone, and associated separation systems and harnessing. By developing and flying these additional components, the IBEX team was able to lift our spacecraft from its low Earth orbit provided by Pegasus to a very high-altitude orbit required for our science mission; this process is described in Sect. 4.

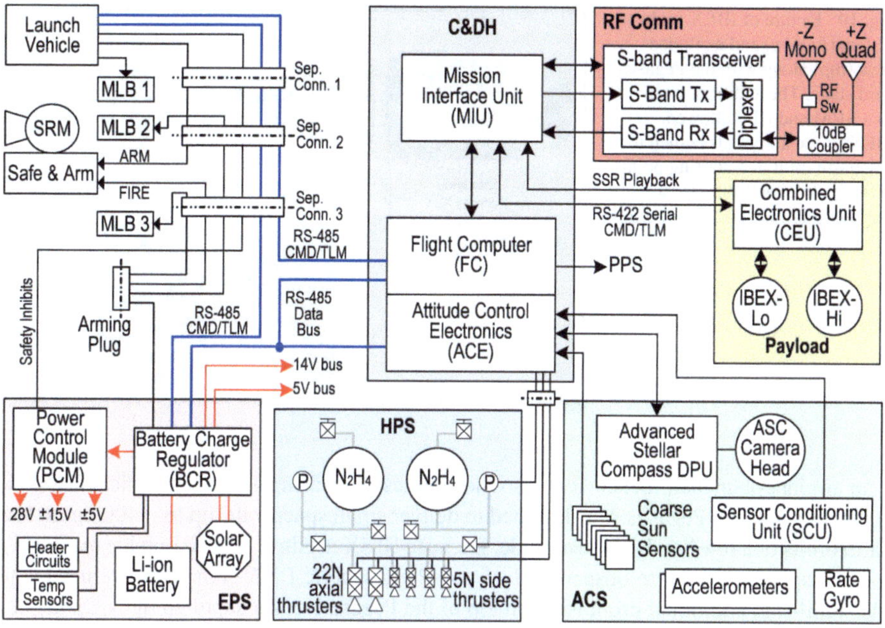

Fig. 9 Block diagram of the IBEX flight system

Table 4 Key parameters for IBEX Spacecraft and Flight System

Resource	Value
Dry Mass (Spacecraft)	78.6 kg
Wet Mass (Spacecraft)	104.9 kg
Wet Mass (Flight System)	457.7 kg
Volume (Spacecraft)	95 × 89 × 49 cm (124 cm with antennas)
Volume (Flight System)	95 × 89 × 166 cm (221 cm with antennas)
Energy Consumption (Nominal Ops)	14.0 kW-Hr
Solar Array Generation (Nominal Ops)	15.9 kW-Hr
Data Generation (Per Orbit)	257.9 Mbits
Data Storage Capacity	1074 Mbits
Science Spin Rate	4 RPM (23 for maneuvers, 60 at launch)
Pointing Knowledge	0.12°
Pointing Control	1.7°
Command Protocol	S-Band
Telemetry Downlink Rate	2, 40, 64, 160 or 320 kbps
Command Rate	2 kbps

4 A New Launch Capability and Mission Design

One of the most challenging aspects of the IBEX mission was to develop a totally new way to achieve a very high-altitude Earth orbit starting from a Pegasus rocket launch. The Pegasus

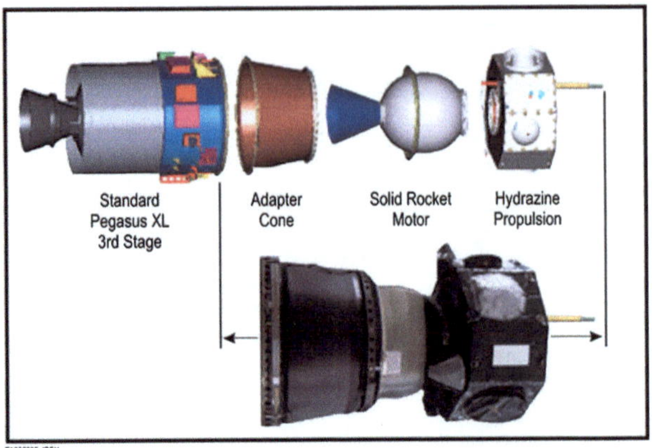

Fig. 10 Picture of IBEX flight system (*bottom*) and exploded view (*top*) along with the Pegasus third stage. The addition of our own SRM and HPS allowed IBEX to use a Pegasus launch to low Earth orbit as a starting point to boost our spacecraft into a very high-altitude ($\sim 50\ R_E$) orbit

is an airplane-launched rocket. It is also the smallest and cheapest launch vehicle available in NASA's arsenal. Pegasus was designed to deliver small spacecraft (up to ~400 kg) to low Earth orbits of a few hundred km altitude. Because IBEX needs to be well outside the Earth's magnetosphere in order to observe the tenuous heliospheric ENA emissions, we needed to add significant additional propulsion on top of the Pegasus rocket supplied by NASA.

After considerable study and optimization, we developed a new approach that combines the addition of another disposable SRM and a highly capable HPS on the IBEX spacecraft. Our SRM was an Alliant Techsystems (ATK) STAR 27 motor optimized with a STAR 30C nozzle (McComas et al. 2006). We also added a well-balanced adapter cone to connect our SRM to the Pegasus and replaced the usual Pegasus separation band with three lighter-weight Motorized Light Band (MLB) separation rings, one between the Pegasus and our adapter cone, one between the adapter cone and our SRM, and finally one to separate the SRM from the spacecraft after burnout. Figure 10 shows this configuration, which we call the full IBEX flight system.

After completion of its third stage burn out, Pegasus pointed the IBEX flight system along the track of its trajectory, spun the flight system up to ~60 RPM, and sent the release pulse that initiated its separation from the bottom of the adapter cone. This separation, which occurred at ~200 km altitude, marked the end of the Pegasus launch and initiated the timer that ran the rest of IBEX's internal launch sequence. Within the next 5 minutes, we ejected the adapter cone, ignited the SRM, performed active nutation control during the SRM burn, and released the burned out SRM. This sequence put IBEX into a high-altitude orbit with apogee at $\sim 35 R_E$ and perigee still at ~200 km.

Over the following several weeks we spun the spacecraft down to ~23 RPM, raised both apogee and perigee with a series of hydrazine burns, and finally spun down to ~4 RPM and repointed the spacecraft toward the Sun. This orbit-raising activity was followed by commissioning of the spacecraft and then the science payload. Finally, in early December 2008, IBEX entered its science phase and began routinely collecting IBEX-Hi and Lo observations of heliospheric ENAs.

Once we moved past the admittedly new and challenging way of getting IBEX into its high-altitude orbit, the remainder of the mission design is extremely simple. Science observations are taken throughout the orbit whenever the spacecraft is above $\sim 10\ R_E$. Below that altitude, on each orbit, we reduce and safe the instrument high voltages, downlink the data from the just-completed orbit, and repoint the spacecraft spin axis from ~4° east of the Sun to ~4° west of the Sun; this must be done because the Earth moves ~1° per day in its orbit about the Sun (360° in 365 days), so over IBEX's ~8-day orbit, the Sun has an apparent motion of ~8° with respect to the inertially fixed spin axis. As IBEX passes perigee and rises past ~10 R_E, the high voltages are ramped back up and science observations begin again. IBEX spends over 7 out of each ~7.6 day orbit at high altitude in science data collection mode, and each orbit provides excellent viewing of a single swath of the sky. As IBEX repoints repeatedly over each six months, those swaths move across the sky allowing the generation of two all-sky maps per year (see Fig. 3); of course, the swaths will also be combined to produce a single all-sky map with even better angular resolution and statistics.

5 Science Operations

Science operations are carried out through the IBEX Science Operations Center (ISOC) (Schwadron et al. 2009, this issue). The ISOC is responsible for supporting analysis of IBEX data, generating special payload command procedures, delivering the IBEX data products and building the global heliospheric ENA maps in collaboration with the IBEX team. The software provides considerable flexibility in the building of the global flux maps. This flexibility is essential since the global properties of the interstellar interaction are virtually unknown. As the data analysis proceeds and we discover the physical properties associated with heliospheric ENA production, the ISOC will iteratively improve its software and data products.

IBEX analysis can be viewed as a processing pipeline connecting the principle science data products. This pipeline is a part of a larger framework for receiving telemetry, trending the payload, delivering Science Tasking Files when needed, archiving data, and distributing data to the scientific community and public (Fig. 11). Because the raw telemetry (Level 0 data) may contain corrupted, garbled communications, data dropouts or repeats, the first stage of the processing produces a "clean copy" of the science data (Level 0.5) as it was stored in the solid state recorder on orbit. If necessary, commands can be sent to the spacecraft to resend portions of the recorded data until it has all been received on the ground. The CEU flight algorithm encodes the data rather compactly, and for subsequent processing expanded forms of this data (Level 1.0 data products) are needed. The most significant change is the correlation of event times (or spin phase) with the corresponding IBEX attitude to determine the arrival direction of each event.

At this point in the processing pipeline, with a magnetospheric model and knowledge of background sources, we remove obvious non-heliospheric signal events and perform data quality checking (magnetospheric ENA data are culled and assembled into a separate data set for use by that community). With the resulting data, we construct ENA count maps or simple spectral estimates for various regions of the sky (Level 2.0 data products). A "toolbox" of methods for the selection and display of data along with some degree of human interaction is used at this processing level. At the next level, we include sensor calibration data and produce estimations of the true ENA flux into the sensor from the direct-event counts. Because these calculations require the inclusion of detailed instrument models, the maps and spectra of ENA flux are higher-level (3.0) products.

Level 3 maps include both the incident maps of ENA flux onto the sensors near 1 AU and heliospheric maps of ENA flux from the outer heliosphere. (The ENA flux from an arbitrary outer boundary at 100 AU is used in existing software, but can be easily modified to any closed surface in the outer heliosphere.) An important component in the generation of the heliospheric ENA maps is understanding the transmission of ENAs from their points of origin to observation at the IBEX spacecraft (Bzowski 2008;

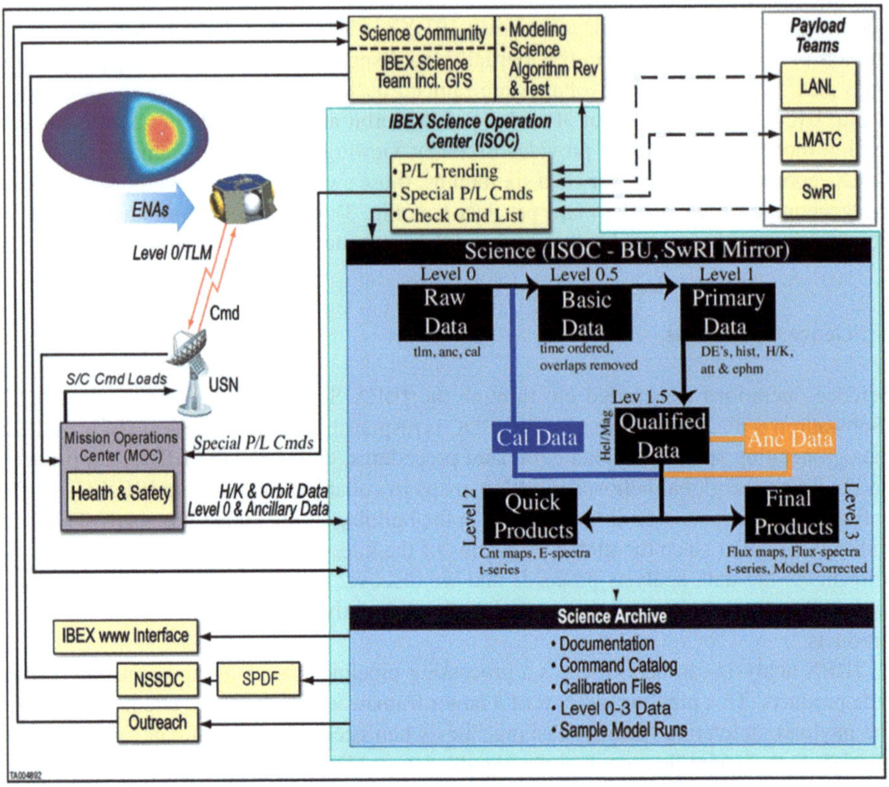

Fig. 11 IBEX data flow from the spacecraft and through the ISOC. The IBEX raw telemetry data and ancillary data are received and processed into high-level data products, archived and distributed to the scientific community

Roelof and Bzowski 2009, this issue). To solve for the ENA transmission, we take into account loss by ionization (predominantly photo-ionization and charge-exchange) and the effects of gravitation and radiation pressure. The final destination of IBEX data products is the SPDF (Space Physics Data Facility) at the National Space Science Data Center in a form suitable for use by the community.

6 Conclusions

The IBEX mission's sole focused science objective is to discover the global interaction between the solar wind and the interstellar medium. This interaction encompasses the structures, dynamics, energetic particle acceleration and charged particle propagation in the complex region where the solar wind meets the interstellar medium. IBEX was designed to provide the first global observations of the interstellar interaction—disclosing its fundamental nature and providing the observations needed for detailed modeling and in-depth understanding.

In order to make heliospheric ENA observations from high above the contamination of Earth's magnetosphere, we developed a new method for launching small satellites to extremely high-altitude orbits, starting from the low Earth orbit achieved by the NASA-supplied Pegasus rocket. In order to achieve this we included our own SRM and highly capable HPS on the IBEX spacecraft. Using this newly developed scheme, future missions could also use a Pegasus rocket as the starting point for propelling other small satellites, up to ~100 kg, to high Earth orbit, the Moon, the L1 point between the Earth and Sun, or even further out in the solar system or to other planets (McComas et al. 2006).

The IBEX-Hi and Lo sensors collectively measure ENAs from ~10 eV up to 6 keV, with redundant, overlapping measurements from 300 eV to 2 keV. These two sensors were optimized to maximize the open aperture area while simultaneously providing exceptionally low background rates from triple coincidence detector sections. The result is extremely high sensitivity and low background observations of the heliospheric ENAs, coming predominantly from the inner heliosheath. In addition, because of IBEX's viewing, throughout much of the year, IBEX will also make the highest sensitivity observations of ENAs coming from various regions of the magnetosphere ever made. While not a formal part of IBEX science, these data are culled from the heliospheric observations and used for additional magnetospheric studies by both members of the IBEX team and the broader magnetospheric community.

Another important part of the IBEX mission is our Education and Public Outreach (EPO) program (Bartolone et al. 2009, this issue). This ambitious and comprehensive program develops exciting EPO themes from Astronomy and Heliophysics and includes substantial formal and informal education components as well as a diverse array of public outreach approaches. The program is implemented by the Adler Planetarium & Astronomy Museum and includes a nationally distributed planetarium show about IBEX with accompanying informal education materials that are also accessible to individuals with special needs. The IBEX EPO program also helped develop a national Space Science Core Curriculum program for grades 6-8 in collaboration with other NASA missions; a professional development program for teachers; and workshops that engage Hispanic and Native American students. Materials are made available for download or for order via our website (www.ibex.swri.edu).

IBEX is a remarkable mission of exploration and discovery. With a launch on 19 October 2008 and the highly successful subsequent orbit-raising and commissioning activities, IBEX is now on station, collecting unique and first-ever observations of the global interaction at the edge of our heliosphere—the region that separates our solar system from the galactic environment.

Acknowledgements We are deeply indebted to all the outstanding men and women who made the IBEX mission a reality. This work was funded by the National Aeronautics and Space Administration as a part of its Explorer program.

References

F. Allegrini, G.B. Crew, D. Demkee, H.O. Funsten, D.J. McComas, B. Randol, B. Rodriguez, P. Valek, S. Weidner, The IBEX background monitor. Space Sci. Rev. (2009, this issue)

V.B. Baranov, Y.G. Malama, Model of the solar wind interaction with the local interstellar medium: Numerical solution of self-consistent problem. J. Geophys. Res. **98**, 15157 (1993)

V.B. Baranov, Y.G. Malama, Axisymmetric self-consistent model of the solar wind interaction with the LISM: Basic results and possible ways of development. Space Sci. Rev. **78**, 305 (1996)

L. Bartolone, S. Cohen, J. Erickson, J. Gutbezahl, P. Knappenberger Jr., IBEX Education and public outreach. Space Sci. Rev. (2009, this issue)

L.F. Burlaga, N.F. Ness, M.H. Acuña, R.P. Lepping, J.E.P. Connerney, E.C. Stone, F.B. McDonald, Crossing the termination shock into the heliosheath: magnetic fields. Science **309**, 2027 (2005)

L.F. Burlaga, N.F. Ness, M.H. Acuña, R.P. Lepping, J.E.P. Connerney, J.D. Richardson, Magnetic fields at the solar wind termination shock. Nature **454**, 78 (2008)

M. Bzowski, Astron. Astrophys. **488**, 1057 (2008)
A.C. Cummings, E.C. Stone, Inferring energetic particle mean free paths from observations of anomalous cosmic rays in the outer heliosphere at solar maximum, in *Proc. ICRC 2001*, p. 4243 (2001)
R.B. Decker, S.M. Krimigis, E.C. Roelof, M.E. Hill, T.P. Armstrong, G. Gloeckler, D.C. Hamilton, L.J. Lanzerotti, Voyager 1 in the foreshock, termination shock, and heliosheath. Science **309**, 2020 (2005)
R.B. Decker, S.M. Krimigis, E.C. Roelof, M.E. Hill, T.P. Armstrong, G. Gloeckler, D.C. Hamilton, L.J. Lanzerotti, Mediation of the solar wind termination shock by on-thermal ions. Nature **454**, 67 (2008)
H.J. Fahr, H. Fichtner, K. Scherer, Theoretical aspects of energetic neutral atoms as messengers from distant plasma sites with emphasis on the heliosphere. Rev. Geophys. **45**(4), RG4003 (2007)
L.A. Fisk, G. Gloeckler, T.H. Zurbuchen, Astrophys. J. **644**, 631 (2006)
P. Frisch, J. Linsky, S. Redfield, R. Shelton, J. Slavin, The galactic environment of the sun: interstellar matter inside and outside of the heliosphere. Space Sci. Rev. (2009, this issue)
H. Funsten, F. Allegrini, D. Everett, S. Fuselier, R.W. Harper, P. Janzen, A. Guthrie, D. McComas, E. Moebius, S. Pope, D. Reisenfeld, S. Weidner, Interstellar boundary explorer high energy (IBEX-Hi) neutral atom imager. Space Sci. Rev. (2009, this issue)
S.A. Fuselier, A.G. Ghielmetti, E. Hertzberg, A. Moore, D. Isaac, J. Hamilton, C. Tillier, E. Moebius, M. Granoff, D. Heirtzler, B. King, H. Kucharek, S. Longworth, J. Nolin, S. Turco, P. Wurz, M. Wieser, J. Scheer, L. Saul, C. Schlemm, D.J. McComas, D. Chornay, J. Lobell, T. Moore, P. Rosmarynowski, R.J. Nemanich, T. Friedmann, H. Funsten, The IBEX-Lo sensor for the IBEX mission. Space Sci. Rev. (2009, this issue)
J. Giacalone, in *The Outer Heliosphere: The Next Frontiers*, ed. by K. Scherer, H. Fichtner, H.J. Fahr, E. Marsch. COSPAR Colloquia Series, vol. 11 (Pergamon, Elmsford, 2001), p. 377
G. Gloeckler et al., Acceleration of interstellar pickup ions in the disturbed solar wind observed on Ulysses. J. Geophys. Res. **99**, 17637 (1994)
G. Gloeckler, L.A. Fisk, T.H. Zurbuchen, N.A. Schwadron, Sources, injection and acceleration of heliospheric ion populations, in *Acceleration and Transport of Energetic Particles Observed in the Heliosphere: ACE 2000 Symposium*, ed. by R.A. Mewaldt et al. AIP Conference Proceedings, vol. 528 (AIP, New York, 2000), p. 221
M. Gruntman, Energetic neutral atom imaging of space plasmas. Rev. Sci. Instrum. **68**(10), 3617 (1997)
M. Gruntman, E.C. Roelof, D.G. Mitchell, H.J. Fahr, H.O. Funsten, D.J. McComas, Energetic neutral atom imaging of the heliospheric boundary region. J. Geophys. Res. **106**, 15,767–15,781 (2001)
J. Heerikhuisen, N.V. Pogorelov, V. Florinski, G.P. Zank, J.A. le Roux, The effects of a κ-distribution in the heliosheath on the global heliosphere and ENA flux at 1 AU. Astrophys. J. **682**(1), 679–689 (2008)
V. Izmodenov, D.B. Alexashov, S.V. Chalov, O.A. Katushkina, Y.G. Malama, E.A. Provornikova, Kinetic-gasdynamic modeling of the heliospheric interface: global structure, interstellar atoms and heliospheric ENAs. Space Sci. Rev. (2009, this issue)
R. Lallement, E. Quemerais, J.L. Bertaux et al., Deflection of the interstellar neutral hydrogen flow across the heliospheric interface. Science **307**(5714), 1447–1449 (2005)
M. Lee, Acceleration of energetic particles on the Sun, the heliosphere, and in the galaxy, in *Acceleration and Transport of Energetic Particles Observed in the Heliosphere*, ed. by R.A. Mewaldt, J.R. Jokipii, M.A. Lee, E. Möbius, T.H. Zurbuchen. AIP Conf. Proceedings, vol. 528 (AIP, New York, 2000), p. 3
M.A. Lee, H.J. Fahr, H. Kucharek, E. Moebius, C. Prested, N. Schwadron, P. Wu, Physical processes in the outer heliosphere. Space Sci. Rev. (2009, this issue)
T.J. Linde, A three-dimensional adaptive multifluid MHD model of the heliosphere. Thesis (Ph.D.) The University of Michigan, Ann Arbor, Source DAI-B 59/02, p. 709 (1998)
T.J. Linde, T.I. Gombosi, P.L. Roe, K.G. Powell, D.L. Dezeeuw, Heliosphere in the magnetized local interstellar medium—Results of a three-dimensional MHD simulation. J. Geophys. Res. **103**, 1889 (1998)
D.J. McComas, N.A. Schwadron, An explanation of the Voyager paradox: Particle acceleration at a blunt termination shock. Geophys. Res. Lett. **33**, L04102 (2006). doi:10.1029/2005GL025437
D.J. McComas, F. Allegrini, C.J. Pollock, H.O. Funsten, S. Ritzau, G. Gloeckler, Ultra-thin ($\sim$10 nm) carbon foils in space instrumentation. Rev. Sci. Instrum. **75**(11), 4863–4870 (2004a)
D.J. McComas, F. Allegrini, P. Bochsler, M. Bzowski, M. Collier, H. Fahr, H. Fichtner, P. Frisch, H. Funsten, S. Fuselier, G. Gloeckler, M. Gruntman, V. Izmodenov, P. Knappenberger, M. Lee, S. Livi, D. Mitchell, E. Möbius, T. Moore, D. Reisenfeld, E. Roelof, N. Schwadron, M. Wieser, M. Witte, P. Wurz, G. Zank, The interstellar boundary explorer (IBEX), in *Physics of the Outer Heliosphere, Third Annual IGPP Conference*, ed. by V. Florinski, N.V. Pogorelov, G.P. Zank. AIP CP719 (AIP, New York, 2004b), pp. 162–181
D.J. McComas, F. Allegrini, L. Bartolone, P. Bochsler, M. Bzowski, M. Collier, H. Fahr, H. Fichtner, P. Frisch, H. Funsten, S. Fuselier, G. Gloeckler, M. Gruntman, V. Izmodenov, P. Knappenberger, M. Lee, S. Livi, D. Mitchell, E. Moebius, T. Moore, S. Pope, D. Reisenfeld, E. Roelof, H. Runge, J. Scherrer, N. Schwadron, R. Tyler, M. Wieser, M. Witte, P. Wurz, G. Zank, The interstellar boundary explorer

(IBEX): Update at the end of Phase B, in *Physics of the Inner Heliosheath, CP 858, Proceedings of 5th Annual International Astrophysics Conference*, pp. 241–250 (2006)

D.J. McComas, R.W. Ebert, H.A. Elliott, B.E. Goldstein, J.T. Gosling, N.A. Schwadron, R.M. Skoug, Weaker solar wind from the polar coronal holes and the whole Sun. Geophys. Res. Lett. (2008). doi:10.1029/2008GL034896

H. Moestue, The electric field and geometrical factor of an annular curved plate electrostatic analyzer. Rev. Sci. Instrum. **44**, 1709–1713 (1973)

E. Moebius, H. Kucharek, L. Petersen, M. Bzowski, L. Saul, P. Wurz, S. Fuselier, V. Izmodenov, D. McComas, H. Mueller, D. Alexashov, Diagnosing the neutral interstellar gas flow at 1 AU with IBEX-Lo. Space Sci. Rev. (2009, this issue)

H.-R. Müller, G.P. Zank, A.S. Lipatov, Self-consistent hybrid simulations of the interaction of the heliosphere with the local interstellar medium. J. Geophys. Res. **105**(A12), 27419–27438 (2000)

M. Opher, E.C. Stone, T.I. Gombosi, The orientation of the local interstellar magnetic field. Science **316**(5826), 875 (2007)

M.E. Pesses, J.R. Jokipii, D. Eichler, Cosmic-ray drift, shock wave acceleration, and the anomalous component of cosmic rays. Astrophys. J. **256**, L85 (1981)

N.V. Pogorelov, E.C. Stone, V. Florinski, G.P. Zank, Termination shock asymmetries as seen by the voyager spacecraft: The role of the interstellar magnetic field and neutral hydrogen. Astrophys. J. **668**(1), 611–624 (2007)

W.K.M. Rice, G.P. Zank, J.A. Le Roux, An injection mechanism for shock waves of arbitrary obliquity. Geophys. Res. Lett. **27**, 3793–3796 (2000)

J.D. Richardson, J.C. Kasper, C. Wang, J.W. Belcher, A.J. Lazarus, Cool heliosheath plasma and deceleration of the upstream solar wind at the termination shock. Nature **454**, 63 (2008)

E. Roelof, M. Bzowski, Motion of neutrals and ENA inversion. Space Sci. Rev. (2009, this issue)

J. Scherrer et al., The IBEX spacecraft. Space Sci. Rev. (2009, this issue)

N.A. Schwadron, L.A. Fisk, G. Gloeckler, Statistical acceleration of interstellar pickup ions in corotating interaction regions. Geophys. Res. Lett. **23**, 2871 (1996)

N.A. Schwadron, M.A. Lee, D.J. McComas, Diffusive acceleration at the blunt termination shock. Astrophys. J. **675**, 1584–1600 (2008)

N.A. Schwadron, G. Crew, R. Vanderspek, K. Goodrich, K. Maynard, C. Prested, P. Wu, M. Reno, K. Mashburn, G. Dunn, J. Hanley, C. Loeffler, D. McComas, B. Randol, M. Bzowski, R. DeMagistre, E. Roelof, M. Gruntman, J. Heerikuisen, H. Funsten, P. Janzen, D. Reisenfeld, S. Fuselier, E. Moebius, D. Heirtlzer, H. Kucharek, The interstellar boundary explorer science operations center. Space Sci. Rev. (2009, this issue)

O. Sternal, H. Fichtner, K. Scherer, Calculation of the energetic neutral atom flux from a 3D time-dependent model heliosphere. Astron. Astrophys. **477**, 365–371 (2008)

E.C. Stone, A.C. Cummings, F.B. McDonald, B.C. Heikkila, N. Lal, W.R. Webber, Voyager 1 explores the termination shock region and the heliosheath beyond. Science **309**, 2017 (2005)

E.C. Stone, A.C. Cummings, F.B. McDonald, B.C. Heikkila, N. Lal, W.R. Webber, An asymmetric solar wind termination shock. Nature **454**, 71 (2008)

V.M. Vasyliunas, G.L. Siscoe, On the flux and the energy spectrum of interstellar ions in the solar system. J. Geophys. Res. **81**, 1247 (1976)

P. Wurz, A. Galli, S. Barabash, A. Grigoriev, Energetic neutral atoms from the heliosheath. Astrophys. J. **683**, 248–254 (2008)

P. Wurz, J. Scheer, L. Saul, M. Wieser, S.A. Fuselier, A.G. Ghielmetti, E. Hertzberg, E. Moebius, H. Kucharek, P. Brandt, D. McComas, F. Allegrini, H. Funsten, IBEX Backgrounds and signal to noise ratio. Space Sci. Rev. (2009, this issue)

G.P. Zank, Interaction of the solar wind with the local interstellar medium: A theoretical perspective. Space Sci. Rev. **89**, 413 (1999)

G.P. Zank, H.L. Pauls, L.L. Williams, D.T. Hall, Interaction of the solar wind with the local interstellar medium: A multifluid approach. J. Geophys. Res. **101**, 21639 (1996)

G.P. Zank, N. Pogorelov, J. Heerikhuisen, H. Washimi, V. Florinski, S. Borovikov, I. Krykov, Physics of the solar wind-Local interstellar medium interaction: Role of magnetic fields. Space Sci. Rev. (2009, this issue)

Space Sci Rev (2009) 146: 35–73
DOI 10.1007/s11214-009-9514-9

The IBEX Flight Segment

J. Scherrer · J. Carrico · J. Crock · W. Cross · A. DeLosSantos · A. Dunn · G. Dunn · M. Epperly · B. Fields · E. Fowler · T. Gaio · J. Gerhardus · W. Grossman · J. Hanley · B. Hautamaki · D. Hawes · W. Holemans · S. Kinaman · S. Kirn · C. Loeffler · D.J. McComas · A. Osovets · T. Perry · M. Peterson · M. Phillips · S. Pope · G. Rahal · M. Tapley · R. Tyler · B. Ungar · E. Walter · S. Wesley · T. Wiegand

Received: 9 February 2009 / Accepted: 21 April 2009 / Published online: 11 June 2009

Abstract IBEX provides the observations needed for detailed modeling and in-depth understanding of the interstellar interaction (McComas et al. in Physics of the Outer Heliosphere, Third Annual IGPP Conference, pp. 162–181, 2004; Space Sci. Rev., 2009a, this issue). From mission design to launch and acquisition, this goal drove all flight system development. This paper describes the management, design, testing and integration of IBEX's flight system, which successfully launched from Kwajalein Atoll on October 19, 2008. The payload is supported by a simple, Sun-pointing, spin-stabilized spacecraft with no deployables. The spacecraft bus consists of the following subsystems: attitude control, command and data handling, electrical power, hydrazine propulsion, RF, thermal, and structures. A novel 3-step orbit approach was employed to put IBEX in its highly elliptical, 8-day final orbit using a Solid Rocket Motor, which provided large delta-V after IBEX separated from the Pegasus

J. Scherrer (✉) · A. DeLosSantos · G. Dunn · M. Epperly · J. Gerhardus · J. Hanley · C. Loeffler · D.J. McComas · M. Phillips · S. Pope · M. Tapley
Southwest Research Institute, 6220 Culebra Road, San Antonio, TX 78238-5166, USA
e-mail: JScherrer@swri.edu

J. Carrico
Applied Defense Solutions, Inc., 8171 Maple Lawn Boulevard, Suite 210, Fulton, MD 20759, USA

J. Crock
Alliant Techsystems, Inc. (ATK), 55 Thiokol Road, Elkton, MD 21921, USA

W. Cross · B. Fields · T. Gaio · W. Grossman · B. Hautamaki · D. Hawes · S. Kinaman · S. Kirn · A. Osovets · T. Perry · M. Peterson · G. Rahal · R. Tyler · B. Ungar · E. Walter · S. Wesley · T. Wiegand
Orbital Sciences, 21839 Atlantic Boulevard, Dulles, VA 20166, USA

A. Dunn
Comtech AeroAstro Inc., 20145 Ashbrook Place, Ashburn, VA 20147, USA

E. Fowler
CSA Engineering, 2565 Leghorn Street, Mountain View, CA 94043, USA

W. Holemans
Planetary Systems Corporation, 2303 Kansas Avenue, Silver Spring, MD 20190, USA

launch vehicle; an adapter cone, which interfaced between the SRM and Pegasus; Motorized Lightbands, which performed separation from the Pegasus, ejection of the adapter cone, and separation of the spent SRM from the spacecraft; a ShockRing isolation system to lower expected launch loads; and the onboard Hydrazine Propulsion System. After orbit raising, IBEX transitioned from commissioning to nominal operations and science acquisition. At every phase of development, the Systems Engineering and Mission Assurance teams supervised the design, testing and integration of all IBEX flight elements.

Keywords IBEX · Spacecraft · Launch

Acronyms

ACE:	Attitude Control Electronics
ACS:	Attitude Control System
ASC:	Advanced Stellar Compass
ATK:	Alliant Techsystems, Inc.
BCR:	Battery Charge Regulator
BOOT:	Build Own Operate and Transfer
BPSK:	Binary Phase Shift Keying
CCSDS:	Consultative Committee on Space Data Systems
C&DH:	Command and Data Handling
CEU:	Combined Electronics Unit
CLA:	Coupled Loads Analysis
DOORS®:	Dynamic Object-Oriented Requirements System
DTU Space:	Technical University of Denmark
EDAC:	Error Detection and Correction
EEPROM:	Electronically Erasable Programmable Read Only Memory
EMC:	Electromagnetic Compatibility
EMI:	Electromagnetic Interference
ENA:	Energetic Neutral Atom
EPS:	Electrical Power Subsystem
ESA:	Electrostatic Analyzer
FC:	Flight Computer
FCDCs:	Flexible Confined Detonation Cords
FEM:	Finite Element Model
FORMOSAT:	Taiwan's Formosa Satellite Mission
FPGA:	Field-Programmable Gate Array
GSE:	Ground Support Equipment
HASO:	High Altitude Science Operations
HPS:	Hydrazine Propulsion Subsystem
HTPB:	Hydroxy-Terminated Polybutadiene
HVENG:	High-voltage engineering
HVPS:	High-Voltage Power Supplies
HVSCI:	High-voltage science
HVSTANDBY:	High-voltage standby
IBEX:	Interstellar Boundary Explorer
IBEX-Hi:	IBEX sensor measuring ENAs from ~300 eV to 6 keV)
IBEX-Lo:	IBEX sensor measuring ENAs from ~10 eV to 2 keV)
I/O:	Input/Output
LAHO:	Low Altitude Housekeeping Operations

LVENG: Low-voltage engineering
LVSCI: Low-voltage science
LUT: Look-Up Table
MA: Mission Assurance
MIU: Mission Interface Unit
MLB: Motorized Lightband
MLI: Multilayer Insulation
MPPT: Maximum Power Point Tracking
MSE: Mission Systems Engineer
NASA: National Aeronautics and Space Administration
NRZ-M: Non-Return to Zero Mark
PCM: Power Converter Module
PDU: Pyro Driver Unit
PLFW: Payload Firmware
PIMS: Project Information Management System
PROM: Programmable Read-Only Memory
PSC: Planetary Systems Corporation
RF: Radio Frequency
RTS: Reagan Test Site
SCB: Spacecraft Bus
SCU: Sensor Conditioning Unit
SDRAM: Synchronous Dynamic Random Access Memory
SMEX: NASA Small Explorers
SRAM: Static Random Access Memory
SRM: Solid Rocket Motor
SSR: Solid State Recorder
SE: Systems Engineering
SwRI: Southwest Research Institute
TBI: Through Bulkhead Initiators
TDRSS: Tracking and Data Relay Satellite System
USN: Universal Space Network
UTC: Coordinated Universal Time
VAFB: Vandenberg Air Force Base
VCID: Virtual Circuit Identification
VEM: Viscoelastic Material
WFF: Wallops Flight Facility

1 Mission Overview

1.1 Mission Summary

The Interstellar Boundary Explorer (IBEX) explores the region where the solar wind and interstellar medium interact by providing the first global observations of the interstellar interaction—disclosing its fundamental nature and providing the observations needed for detailed modeling and in-depth understanding (McComas et al. 2004, 2009a, this issue). IBEX makes these measurements using two, high-sensitivity, single-pixel energetic neutral atom (ENA) sensors mounted on a Sun-pointing spinning spacecraft. The spacecraft was launched on a Pegasus XL launch vehicle and is now in a highly eccentric, $\sim$8-day orbit.

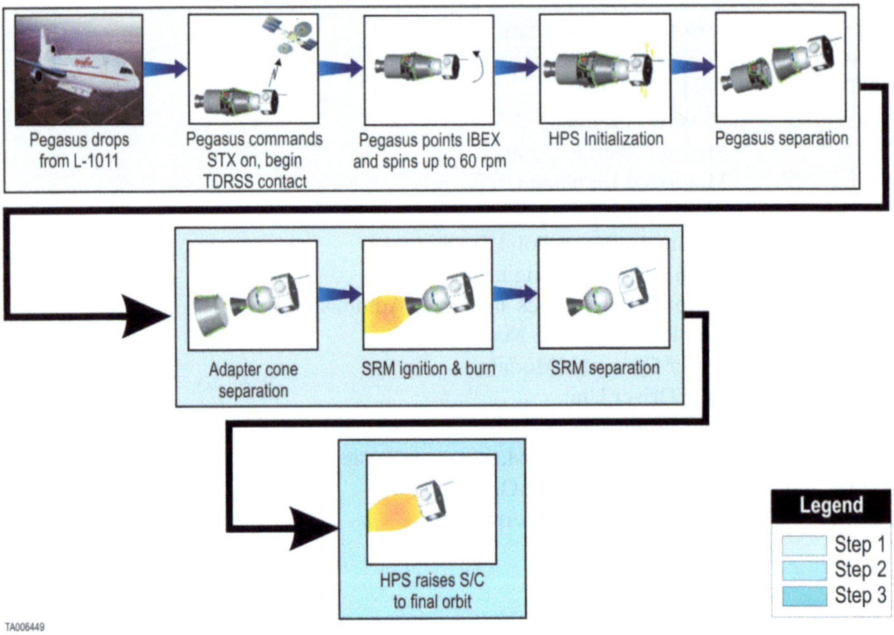

Fig. 1 Three-step orbit-raising procedure

Once the spacecraft achieved nominal orbit, the two radial-pointing sensors started to sweep out a swath of the sky with each spin of the spacecraft. Each orbit near perigee, the spin axis of the spacecraft is adjusted to 4 degrees "left" of the Sun; in eight days, through the orbital motion of the Earth around the Sun, the spin axis ends up 4 degrees to the "right" of the Sun, and IBEX is ready for the next spin axis pointing maneuver.

1.2 Mission Design

The signal (i.e. ENA flux) that IBEX measures from the interaction of the solar wind and the interstellar medium is very low. Conversely, the Earth's magnetosphere is a strong source of ENAs and swamps the prime signal when it is in one of the sensor's field of view. Because of this, the driving requirement for the mission design was to attain an orbit as high as possible to minimize IBEX's time in the magnetosphere. The nominal orbit for IBEX is 18,000 km by 40–50 R_E altitude. To achieve this orbit, which is 5/6 of the way to the Moon, IBEX used a novel 3-step, orbit-raising procedure never before performed on a Pegasus-launched vehicle (Fig. 1).

1. The Pegasus launch vehicle inserted the IBEX flight system into a nominal 200-km circular orbit. The third stage of Pegasus then pointed the flight system in the proper direction and spun the system up to 60 rpm.
2. The IBEX flight system included a Star 27 Solid Rocket Motor (SRM) from Alliant Techsystems, Inc. (ATK). After separation from the Pegasus third stage, the SRM was ignited and boosted the flight system to a parking orbit with an apogee altitude of 32 R_E, where the SRM was released from the IBEX spacecraft.
3. A series of apogee and perigee burns by the spacecraft using the onboard Hydrazine Propulsion Subsystem (HPS) raised IBEX to its final mission orbit.

IBEX is the heaviest payload ever launched on a Pegasus, but eventually reached the highest orbit ever achieved using a Pegasus because it carried its own SRM. The 3-step orbit-raising approach outlined above provided a key advantage by allowing the HPS to trim the final orbit to the exact desired parameters, overcoming the large performance dispersions of the Pegasus and the SRM. IBEX's use of an SRM and the relatively low-cost Pegasus launch vehicle has opened the door for low-cost missions beyond the Earth's orbit; Pegasus missions to the Moon, L1 or other planets are within reach using the 3-step strategy described above.

Why 18,000 km by 50 R_E Orbit? The 18,000-km perigee altitude was chosen to be above the inner radiation belt, minimizing the total radiation dose exposed to the IBEX electronics during the nominal 2-year mission. Several of the avionics subsystems on the spacecraft bus (SCB) have heritage to the Orbital MicroStar spacecraft line, which were designed for limited radiation exposure. Fifty R_E was a somewhat arbitrary compromise between maximizing the time outside of the Earth's magnetosphere and establishing a reasonable altitude that allowed the communications system to conservatively close the link with margin. The 18,000 km by 50 R_E orbit is $\sim$7.5 days, which also works well with the $\sim$7-degree field of view (full width half max) of the sensors and the $\sim$7.5-degree spin-axis pointing maneuver performed at each perigee, allowing contiguous swaths of the sky to be imaged.

1.3 IBEX Launch and Acquisition

IBEX was launched on October 19, 2008 from the Reagan Test Site (RTS) at Kwajalein Atoll. This launch site was chosen because its near-equatorial launch trajectory put more of the Earth's rotational energy into reaching orbit, spending less energy and allowing a higher launch mass. The Pegasus launch vehicle carrying IBEX was dropped from its L-1011 aircraft at 17:47:40 (UTC), beginning a sequence of events that culminated in the SRM burn and initiation of transmitter beaconing (see Table 1 for timeline).

Following the launch and SRM burn events, there was (as planned) significant uncertainty in the spacecraft orbit due to a combination of uncertainties in the insertion orbit provided by Pegasus, the flight system dynamics during the SRM burn, and the performance of the SRM. While each of these effects was within its expected range, the resulting tens of thousands of kilometers of orbit uncertainty were the focus of the next couple days, during which the ground team performed acquisition and tracking to pinpoint an accurate orbit estimate.

In order to perform acquisition, the flight team performed ground contacts using Universal Space Network (USN) ground stations in Hawaii, Alaska and Australia, as well as NASA ground stations in Virginia and Santiago, Chile. Through contacts with these stations, the team accumulated ground antenna angle and Doppler data used to characterize the orbit. As the spacecraft drifted away from its pre-launch orbit estimate, these data were used to develop a new orbit estimate; this process was repeated until it converged to a new orbital solution.

Once the solution was reached, and basic spacecraft functionality verified, the orbit-raising burns commenced. These burns not only raised IBEX to its final mission orbit but also phased the IBEX orbit relative to the Moon and Sun in order to avoid detrimental lunar orbit perturbations or long eclipse periods over at least the first two years of operations. Table 2 lists the orbit-raising burns performed and their effect on the spacecraft orbit.

Once the final orbit-raising burn was completed, the spacecraft was spun-down to its $\sim$4 rpm mission spin rate, and full spacecraft commissioning activities commenced.

Table 1 Timeline of IBEX Launch & SRM Burn Events

Time (UTC, 10/19/08)	Event description
17:47:40	Pegasus drops from L-1011 aircraft.
17:47:45	Pegasus stage 1 ignition: Pegasus arms the safe & arm device for the IBEX SRM.
17:53:55	Pegasus stage 3 burnout: IBEX transmitter turns on to begin TDRS contact.
17:55:00 to 17:55:50	Pegasus points IBEX in proper direction for SRM burn and spins the flight system up to ~ 59 rpm. IBEX initializes its HPS.
17:55:55	IBEX separates from Pegasus.
17:56:14	Adapter cone separates from IBEX flight system.
17:56:17	SRM ignition.
17:57:00	SRM burnout.
18:00:08	SRM separates from IBEX spacecraft.
18:00:14	Spacecraft aft antenna turns on (previously off to prevent radiating into Pegasus 3rd stage).
18:24:19	Spacecraft begins transmitter beacon cycle with 10-minute beacon on time.
18:34:19	Spacecraft transmitter turns off, ending the TDRS contact.

Table 2 Orbit-raising burns performed to reach final IBEX mission orbit

Burn time (UTC)	Thruster used	Effect on orbit
10/22/08, 03:00:00	Forward	Raised perigee altitude to 2070 km
11/01/08, 17:09:32	Aft	Raised apogee altitude to 321,400 km
11/06/08, 01:40:00	Forward	Raised perigee altitude to 18,480 km

1.4 Mission Phases and Nominal Operations

IBEX has five mission phases: 1) Launch, 2) SRM burn, 3) Orbit raising, 4) Commissioning, and 5) Nominal operations. Phases 1–3 were described in Sects. 1.2 and 1.3. The commissioning phase occurred for several orbits following the orbit raising. It consisted of the HPS spinning the spacecraft down from 23 rpm to the operational spin rate of 4 rpm, turning on the autonomous star camera and re-pointing the IBEX spin axis to ~ 4 degrees left of the Sun in the ecliptic plane. Payload commissioning consisted of turning on and commissioning the Combined Electronics Unit (CEU) and two sensors, IBEX-Lo and IBEX-Hi. Because both sensors utilize high voltage (up to ~ 16 kV) in their optics and detectors, the ramping of the voltages was slow and methodical to avoid corona or arcing.

While commissioning utilized a large number of ground passes, IBEX nominal operations minimizes the number of ground contacts to reduce mission cost (Fig. 2). Nominal operations is divided into two operational segments: High Altitude Science Operations (HASO) and Low Altitude Housekeeping Operations (LAHO). HASO occurs when IBEX

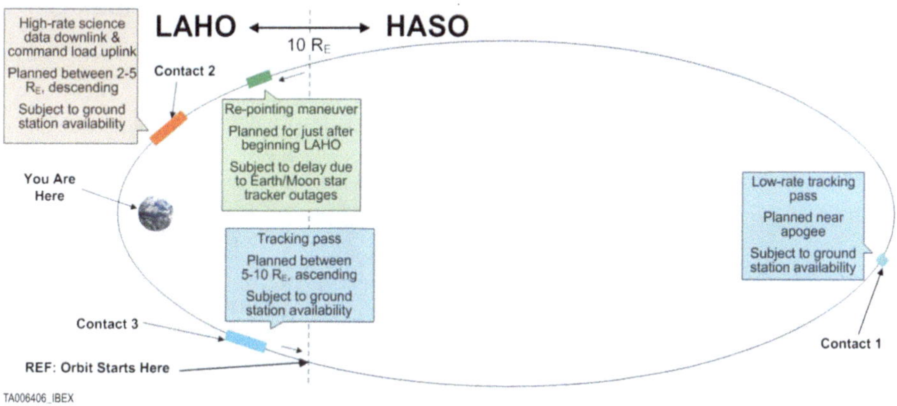

Fig. 2 IBEX's nominal operations segments: High Altitude Science Operations (HASO) and Low Altitude Housekeeping Operations (LAHO)

is above 10 R_E altitude (∼7.5 days of the orbit) and consists of nominal science operations. There is one contact planned near each apogee to check on the health and status of the spacecraft. When IBEX crosses 10 R_E on the descending leg of the orbit, the spacecraft, through time-tagged stored commands, enters LAHO. Sensor high voltages are ramped down to protect against radiation and the higher count rates from the magnetosphere, the spin axis re-pointing maneuver is performed, science and back-orbit housekeeping telemetry is downloaded, and time-tagged commands for the orbit after next are uploaded (the commands for the subsequent orbit are already in the on-board queue). While only one contact is needed to complete the above tasks, two are planned to provide the needed tracking precision for the propagation of the IBEX orbit, which is highly influenced by lunar perturbations. When IBEX crosses ∼10 R_E on the ascending leg of the orbit, it again enters HASO, voltages are ramped up to operational levels, and nominal science operations resume. IBEX operations are deliberately designed to be repetitive and simple, minimizing the number of operations personnel and ground contacts necessary thus reducing the cost of operations.

The Tracking and Data Relay Satellite System (TDRSS) was used during the first three mission phases to monitor critical events, including the SRM burn, IBEX separation from Pegasus, and the perigee burns. The USN ground stations in Hawaii, Perth and Alaska are used for the nominal mission. The NASA stations in Santiago and at Wallops Flight Facility (WFF) were used to close the 4-hour gap between TDRSS and first acquisition of IBEX by USN in Hawaii and are now available for contingency operations.

2 Flight System Overview

The IBEX flight system shown in Fig. 3 consists of the following elements:

- Spacecraft bus: the structure and avionics that support the IBEX science payload. See Sect. 3 for a full description of the SCB elements.
- Payload: a single instrument that consists of a CEU that handles payload data processing and two sensor heads, IBEX-Hi and IBEX-Lo. See Sect. 4 and subsequent papers in this book (Funsten et al. 2009, this issue; Fuselier et al. 2009, this issue) for more details concerning the payload.
- Spacecraft: the combination of the SCB and payload. Section 5 discusses the integration of the payload and the SCB.

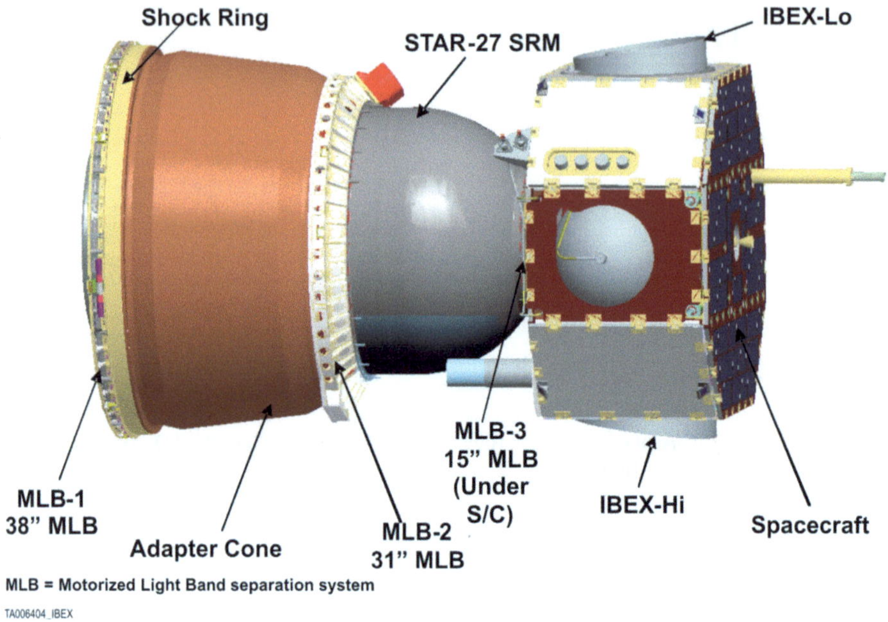

Fig. 3 IBEX flight system

- Flight System: the full launch stack including the spacecraft, adapter cone, separation systems, and SRM. Section 6 describes the final integration and testing of the final full-up stack.

The other parts of the IBEX flight system (i.e. the SRM, adapter cone, ShockRing, and Motorized Lightband (MLB) separation systems) are briefly described here.

2.1 Solid Rocket Motor (SRM)

The ATK STAR 27H SRM (Fig. 4) provided the bulk of the orbit adjustment impulse after IBEX separation from the Pegasus launch vehicle. The motor featured a machined head-end web grain design with 274.1 kgs of HTPB-based TP-H-3340A propellant. The SRM was designed to fire for approximately 40 seconds, with a nominal effective specific impulse of 291.1 seconds and an average thrust level of 19,350 N. The maximum acceleration level, given the mass of the IBEX flight system, was less than 10.5 g, and the SRM delivered a total nominal impulse of 784,399 N·sec. The case of the motor, machined from a 6AI–4V solution-treated and aged titanium forging, included an integrated flange on the forward dome for attaching the MLB-15 and a ~ meridinal flange for attaching the MLB-31.

Two identical SRMs were manufactured. One was used for qualification testing and ultimately for a test firing to validate the SRM performance (Fig. 5). Post test firing, the qualification motor was cleaned, filled with inert propellant, static load tested, then used for IBEX flight system random and sine vibration and MLB separation tests. The other SRM was used for the IBEX launch.

The SRM was fired by the Battery Charge Regulator (BCR) located on the IBEX spacecraft. The BCR sent an electrical signal to the safe & arm device, which converted the signal to an ignition source that was routed by redundant flexible confined detonation cords

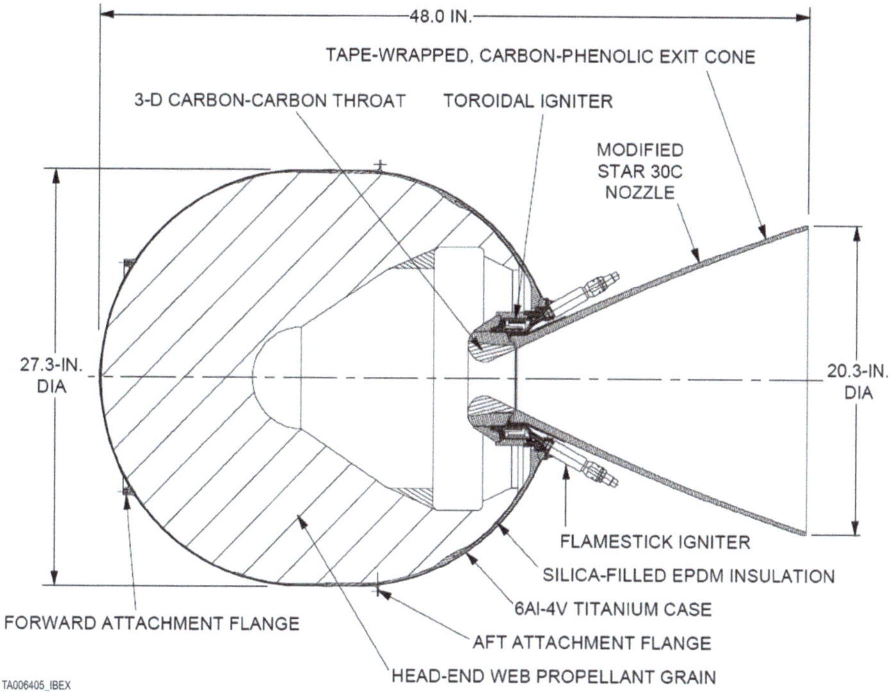

Fig. 4 Schematic of the STAR 27H Solid Rocket Motor

Fig. 5 Test firing to validate Solid Rocket Motor performance

(FCDCs) to the through bulkhead initiators (TBIs) located on the STAR 27 SRM flame sticks that ignited the motor. The safe & arm device is mounted on the MLB-31, which is part of the IBEX flight system, but the actual ARM command came via the Pegasus interface. Figure 6 shows a schematic of the ordnance train.

2.2 Adapter Cone

The IBEX adapter cone manufactured by Canyon Composites in Anaheim, CA provided the interface between the SRM and the Pegasus launch vehicle. The height of the cone was set to

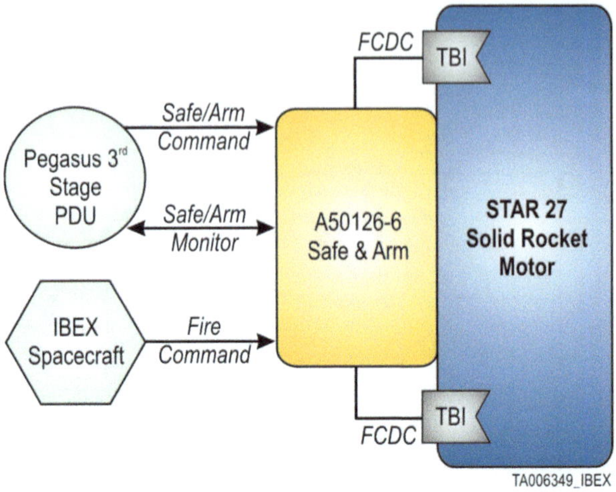

Fig. 6 Solid Rocket Motor ordnance train

minimize the distance the SRM nozzle extended below the interface plane with the Pegasus and still allow the tall flight system to fit within the fairing envelope. The cone design was based upon the Pegasus avionics structure using an IM7 fiber/M76 resin composite system. The honeycomb core is 2.0 pounds per cubic foot, 1/2-inch thick with EY-3010 syntactic foam used to stabilize the core at the end tapers. The skin is 4-ply unidirectional tape with a build up at the 0.18-inch thick flanges of 22 plies of fabric. The film adhesive is FM73U .03 pound per square foot. After the core assembly was completed, the entire cone was laid up and co-cured in a single piece, then trimmed and drilled at each mating interface.

The cone itself was inherently radially balanced. However, with the installation of cable ties and flight harnesses, the integrated cone required spin balancing at WFF.

2.3 Motorized Lightbands (MLBs) and Separation

The separation from the Pegasus launch vehicle, the ejection of the adapter cone, and the separation of the spent SRM from the spacecraft were performed by Planetary Systems Corporation's (PSC) MLBs. The first separation from the Pegasus third stage was accomplished with a 38-inch MLB (MLB38.810-60) driven by the Pegasus Pyro Driver Unit (PDU). The PDU included some mission unique electronics that allowed it to drive the MLB motors instead of a pyro. A 31-inch MLB (MLB31.600-48) driven by the IBEX BCR performed the second separation, ejecting the adapter cone off the SRM prior to SRM ignition. The final separation between the IBEX spacecraft and the SRM was accomplished with a 15-inch (Fig. 7) MLB (MLB15.000-24) driven by the BCR and timed to occur ∼3 minutes after SRM burn out. The timing was designed to minimize the duration of thermal soak back and still ensure that all SRM chuffing was complete in order to eliminate the possibility of re-contact between the spent SRM and spacecraft. The entire orbit injection process, from separation with the Pegasus third stage to separation of the SRM from the IBEX spacecraft, occurred within 250 seconds.

Planetary Systems Corporation's MLBs were chosen over traditional Pegasus V-band separation systems early in the IBEX project for several reasons. First and foremost, MLBs are substantially less massive than their V-band counterparts. Because of the need for the high apogee, mass was always the most critical technical resource. Secondly, MLBs have the distinct advantage over V-Bands because they are re-settable without having to replace

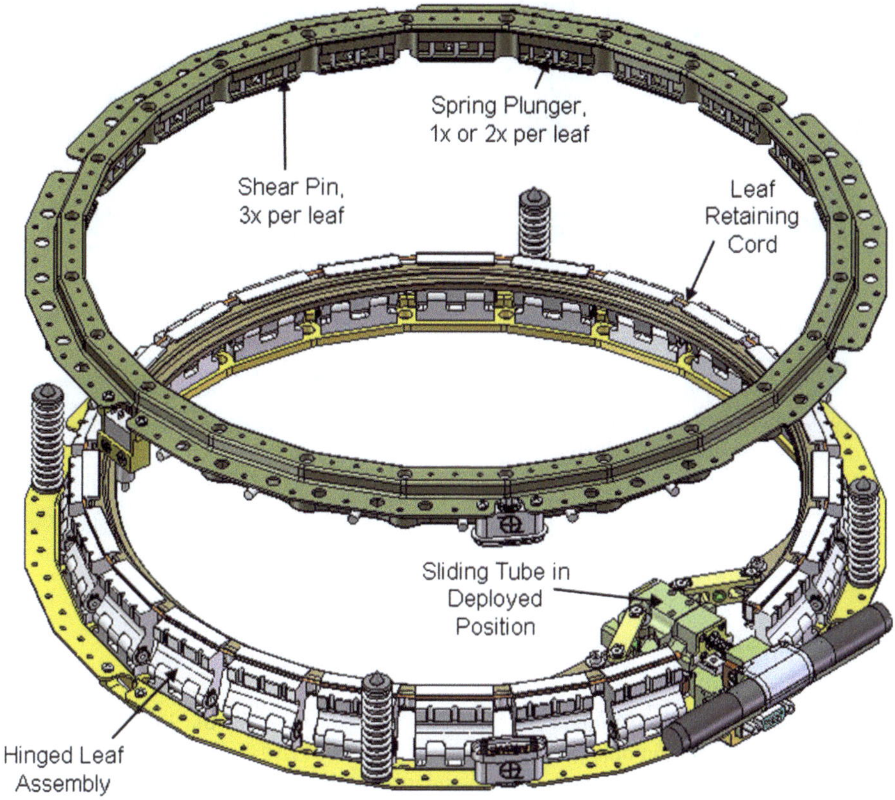

Fig. 7 Fifteen-inch Motorized Lightband

pyros. The redundant motors are simply driven in reverse to reset the retaining ring, thus stowing the MLB. This allows the MLB to be flown in the same configuration in which it is tested. Motorized Lightbands also have tighter tip-off angle specifications than traditional V-bands, again helping to ensure clean separation. During deployment, the MLBs use redundant motors to drive the gear train to push a sliding tube inward, allowing the retaining ring to contract. This allows the leaves of the lower ring to disengage from the groove in the upper ring (Fig. 8). While deployment requires only one motor, stowing requires the use of both motors working in tandem (requiring significantly more energy than deployment); therefore, if the unit properly stows, it has a very high likelihood of successful deployment.

2.4 ShockRing

As with all spacecraft integration and test flows, IBEX produced a correlated finite element model (FEM) after flight system vibe testing. The correlated IBEX flight system FEM was about 12 Hz stiffer than the one used in preliminary analyses. While normally "stiffer is better" for flight hardware (which is why there is no requirement on the upper bound on stiffness), the increased IBEX stiffness produced a coupling with the Pegasus stage 1 ignition transient. The predicted loads from the final Coupled Loads Analysis (CLA) performed by the launch vehicle team were significantly higher than the IBEX design limit (i.e. 16.7 G vs. 12.8 G).

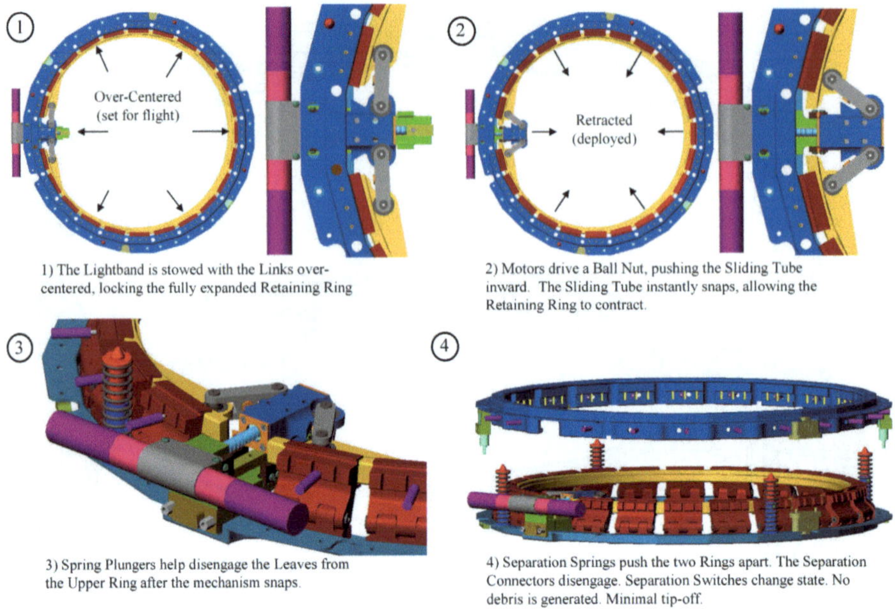

1) The Lightband is stowed with the Links over-centered, locking the fully expanded Retaining Ring

2) Motors drive a Ball Nut, pushing the Sliding Tube inward. The Sliding Tube instantly snaps, allowing the Retaining Ring to contract.

3) Spring Plungers help disengage the Leaves from the Upper Ring after the mechanism snaps.

4) Separation Springs push the two Rings apart. The Separation Connectors disengage. Separation Switches change state. No debris is generated. Minimal tip-off.

Fig. 8 Motorized Lightband separation system operation

The IBEX project transitioned into emergency mode to solve the loads problem as quickly as possible to minimize cost impact to the project. CSA Engineering, Inc. of Mountain View, CA was engaged to develop an isolation system that could be installed in the flight stack to lower the expected launch loads. The most efficient method for this mission to reduce the loads was to "detune" the axial mode of the system by moving its frequency away from the excitation frequency of the stage 1 ignition event and to add damping. CSA, in coordination with the IBEX project, chose to implement their proprietary ShockRing Isolation System (Patent # 6,202,961) to achieve this detuning and thereby reduce the large axial responses induced by the Pegasus stage 1 ignition transient. The ShockRing achieves response reduction through highly engineered stiffness characteristics in combination with a constrained layer damping treatment. Conceptually, the ShockRing is a spring that operates mechanically in parallel with a damper. Physically, the device is composed of a stack of stainless steel rings, viscoelastic material (VEM) pads, and aluminum constraining layers. When the stack of rings displaces in the axial direction, the rings bend and cause the VEM to shear. This action converts the mechanical energy of motion across the assembly into heat, which dissipates in the VEM and produces vibration damping. The IBEX ShockRing has a 38.81-inch diameter bolt circle and is 1.5 inches high. It consists of five stainless-steel rings: three flexible rings in the center and two outer interface rings. Each internal ring is separated with a series of spacers (offset circumferentially) and fastened together with #10-32 pan head machine screws into Heliocoil threaded inserts. The ShockRing is illustrated in Fig. 9. CSA designed, fabricated and delivered the ShockRing to IBEX in record time. It was then static load proof tested by PSC and subjected to random and sine vibration by Orbital. Data from these tests correlated so well with the FEM that no modifications to the ShockRing FEM were required for the final CLA. With the ShockRing located in the flight stack between the MLB-38 and adapter cone, the resultant loads from the final CLA were well within IBEX specification.

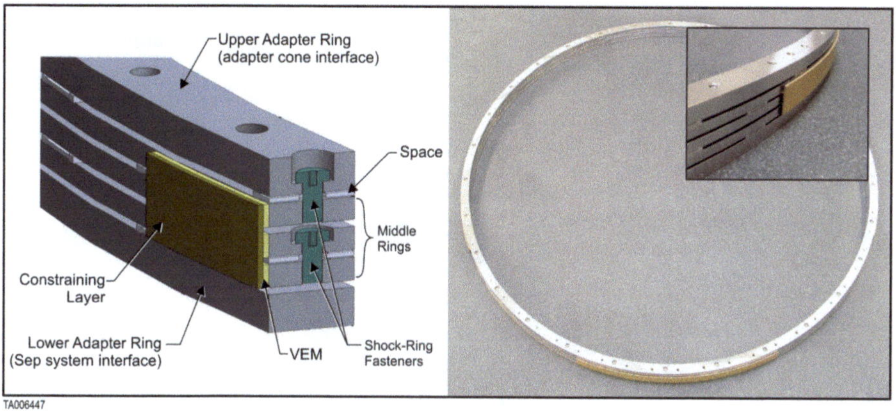

Fig. 9 IBEX ShockRing

Fig. 10 The IBEX spacecraft

3 Spacecraft Bus (SCB) Overview

The IBEX spacecraft is a simple, Sun-pointing, spin-stabilized spacecraft with no deployable appendages (Fig. 10). The IBEX SCB is based on Orbital's MicroStar architecture, initially developed for the ORBCOMM constellation and most recently used on the FORMOSAT program.

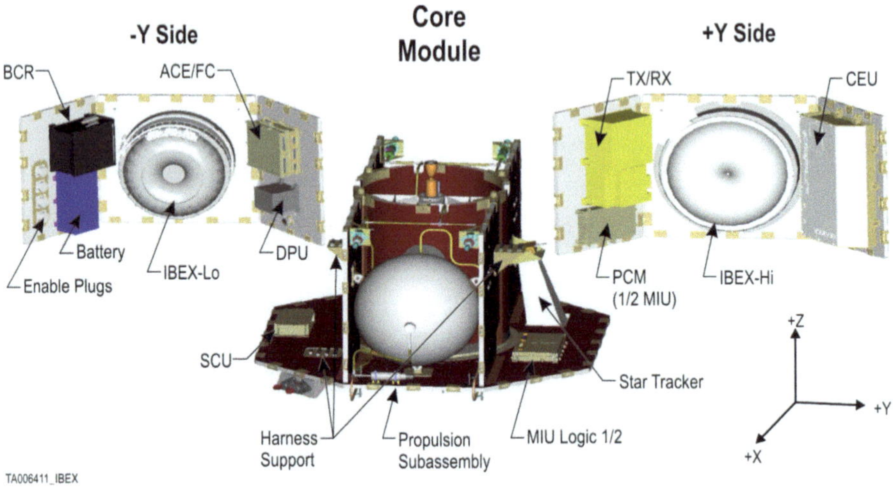

Fig. 11 Component locations on the IBEX spacecraft

For IBEX, the MicroStar avionics were put into a new thrust-tube structure, but the avionics architecture remained largely unchanged. The SCB features a distributed Motorola 68302 microprocessor-based architecture, with the Command and Data Handling (C&DH) Subsystem, Attitude Control Subsystem (ACS), and Electrical Power Subsystem (EPS) each featuring a dedicated processor. An expanded view of the spacecraft showing component location is given in Fig. 11.

The Flight Computer (FC) is the spacecraft controller; the Attitude Control Electronics (ACE) interfaces to the ACS components and hosts the ACS software. The BCR provides power for all spacecraft loads and performs ancillary functions including heater control, SRM fire, and separation system actuation. The Mission Interface Unit (MIU) is the interface between the SCB and the CEU in the payload. The CEU provides power and data connections to the IBEX-Hi and IBEX-Lo sensors and stores all mission data, including SCB housekeeping telemetry. The MIU also provides the interface to the S-band transceiver to receive uplinked commands and downlink telemetry and science data. On-board propulsion for perigee and apogee raising, spin down, and orbital maneuvers is provided by the HPS. The following sections describe the subsystems of the SCB.

A block diagram of the IBEX spacecraft is shown in Fig. 12.

3.1 Spacecraft Bus Structure

The IBEX spacecraft structure is an irregular octagon approximately 3 feet across and 19.5-inch tall composed of a core module, two payload equipment panels, four bus equipment panels, two close out panels, and the solar array panel. The six equipment panels were constructed from .012-inch thick aluminum 6061-T6 skins, 5056 aluminum honeycomb core and FM73U film adhesive. The remainder of the structure was constructed from 3-ply graphite-reinforced cyanate ester prepreg M55J/RS-3C skins, 5056 aluminum honeycomb core, and FM73U film adhesive, except the solar array deck, which used RS-4A film adhesive for the higher service temperature. The inner facesheets and honeycomb core of all panels were perforated for venting. The core module consists of the thrust tube, base panel and four core panels bonded together with pre-cured graphite clips. The other structural panels and solar array deck were assembled to each other and to the core module

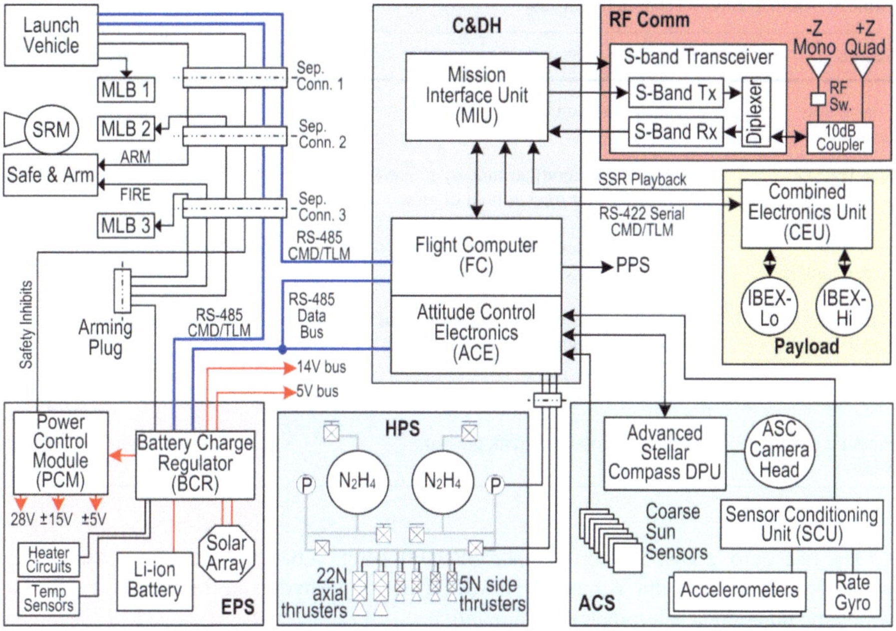

Fig. 12 IBEX spacecraft block diagram

with bonded aluminum clips and threaded fasteners. The clips were match bonded with EA9309NA epoxy for a shear interface. Inserts to attach components were potted into the panels using EA9396.6MD syntactic foam. Thermal isolators between the solar array and core structure (0.2-inch thick G-10 washers) isolate the core structure from the Sun-pointing solar array panel. Likewise, thermal isolators between the SRM and MLB-15 (0.25-inch thick titanium washers) isolate the MLB from SRM thermal soak back. The primary structure was fabricated by Alliance Spacesystems in Signal Hill, CA.

3.2 Attitude Control Subsystem (ACS)

The ACS maintains stability and control throughout all mission phases, beginning with separation from Pegasus. The ACS uses one processor contained in the ACE, which drives core ACS algorithms and interfaces to the ACS sensors and actuators.

The IBEX spacecraft uses a simple spin-stabilized ACS. Therefore, once the SRM was separated from the spacecraft, the spacecraft was passively stable in its approximately Sun-pointed attitude. The ACS is responsible for determining the current spacecraft attitude and only rarely uses control authority to maintain the Sun-pointed attitude.

3.2.1 Attitude Determination Sensors

Attitude determination is provided by the sensors listed in Table 3. The two Honeywell QA2000-030 accelerometers are mounted on different axes such that one measures spin rate and the other measures nutation. The spin rate accelerometer saturates at around 27.5 rpm. Coarse Sun sensor data can be used to approximate a value, but at high spin rates only the accelerometer reading is available for onboard spin rate determination.

Table 3 Attitude control subsystem sensors

Sensor	Function	Operational range
Coarse Sun Sensor	Measure the Sun vector	±70° Half Cone
Star Tracker	Measure spacecraft attitude by referencing local stars in the star tracker field of view	Any attitude, spin rate up to 4.5 rpm
Spin Rate Accelerometer	Measure acceleration in the spin axis to determine spacecraft spin rate	0 to 27.5 rpm (2.88 rad/sec)
Nutation Accelerometer	Measure acceleration in the transverse axes to determine vehicle nutation	±0.08 rad/sec
Rate Gyro	Measure angular rate to determine nutation	±50 deg/sec in X–Y
Pressure Sensors	Measure propulsion tank pressure	0 to 422 psi

The rate gyro, a BEI Systron Donner QRS100, offers a backup nutation measurement, which is needed for delta-V maneuvers (SRM burn and hydrazine orbit raising), during which the nutation accelerometer is saturated.

The star tracker, the Advanced Stellar Compass from the Technical University of Denmark (DTU Space) is only used at a low (<4.5 rpm) spin rate and is the high accuracy sensor used during the nominal science mission. When on, the star tracker also measures spin rate and nutation (more accurately than the accelerometer or rate gyro).

The coarse Sun sensors are used as a backup to the star tracker, providing a reference to perform maneuvers to point the solar array at the Sun.

3.2.2 Attitude Control Actuators

Following SRM separation, the spacecraft is a passively stable spinner (spinning at 60 rpm for SRM operations, 22 rpm for orbit-raising operations, and 4 rpm for all other on-orbit operations). The only active attitude control is actuated using the 5 N thrusters in the HPS. Attitude control using these thrusters includes spin rate control, nutation control, and spin axis precession.

Spin rate corrections are performed by firing opposing thrusters to change spin rate without imparting a delta-V. Nutation control and precession maneuvers are performed by pulsing pairs of thrusters on the same side of the spacecraft (pointed in the same direction) simultaneously at certain points in the spacecraft rotation in order to get the desired effect.

The ACS uses a high rate and a low rate controller at different points in the mission. ACS High Rate Mode uses a 10 Hz controller for attitude determination and control at high spin rates, including during the SRM burn and all hydrazine delta-V burns. The low rate controller operates at 2 Hz and is used for the nominal mission.

3.3 Command and Data Handling (C&DH) Subsystem

The C&DH Subsystem provides command decoding and distribution, performs data collection and storage, performs health and maintenance functions, and provides the electrical interface to the payload through the CEU. The C&DH Subsystem consists of the FC, the ACE, and the MIU. A block diagram of the C&DH Subsystem is shown in Fig. 13.

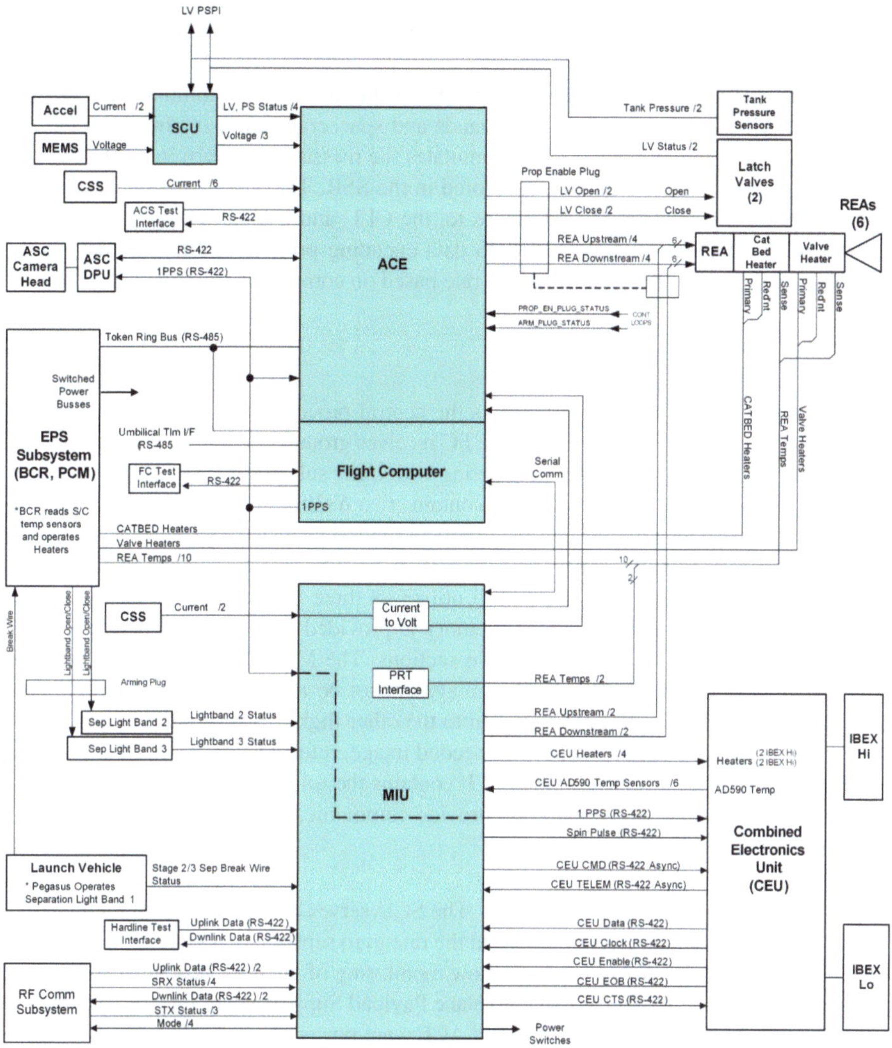

Fig. 13 Command and Data Handling subsystem block diagram

The FC and ACE serve as the core of the subsystems, with internal architectures based on the radiation-tolerant version of the Motorola 68302 multi-protocol communications processor. The FC communicates with the BCR and ACE via a multi-drop RS-485 bus and with the MIU via a dedicated RS-422 interface; the MIU communicates with the payload CEU via another dedicated RS-422 interface. The FC also includes eight discrete I/O ports for resetting systems, power switching, and other functions. A 256 kB boot ROM stores code for processor startup and boot load functions, and a 256 kB EEPROM holds the application code at startup.

The FC executes real-time commands, as well as time-based stored commands. Its 3 MB EDAC-protected RAM stores spacecraft state-of-health telemetry received from the BCR, ACE, MIU, and CEU. All stored telemetry is also sent to the CEU to be stored in its much larger Solid-State Recorder (SSR) until the downlink ground pass. Several orbits of telemetry can thus be stored in the CEU, while the FC maintains telemetry from only the most recent day or so.

The MIU is the interface between the FC, payload CEU, and RF Communications Subsystem. The MIU receives payload commands and spacecraft telemetry from the FC and forwards those to the CEU, which de-commutates the message to determine if it is a command to be executed or telemetry to be stored in the SSR. The MIU also collects discrete telemetry from, and controls the interfaces to, the CEU and S-band transceiver. During a data downlink, the MIU performs CCSDS data encoding prior to sending the data to the transmitter and controls the downlink data rate based on commands from the FC.

3.3.1 C&DH Hardware

3.3.1.1 Flight Computer (FC) The FC is the central processor of the spacecraft and receives +5.0 Vdc power from the BCR. The FC receives ground commands via the MIU and performs the appropriate actions by interacting with other subsystems.

The FC is a single card assembly that contains two 68302 multi-protocol microprocessors, one configured as a master and the other configured as a slave. Each 68302 has three high-speed serial data ports. All flight software in the FC executes on the master 68302 processor. The slave CPU is configured to utilize its three I/O ports only; no software is executed on the slave CPU. EEPROM memory is provided for storage of boot code, and the application memory is divided into two sections. The MON2 flight software image is provided 256 Kbytes of EEPROM. This image cannot be modified on-orbit but contains functionality to allow on-orbit modification to the other flight software image contained in the second 256 Kbytes of EEPROM. This second image, referred to as the MON3 image, is modifiable on-orbit by ground commands. It contains the full flight software application or mission software image. Six serial ports provide communication to and from other spacecraft subsystems and processors (Fig. 14).

3.3.1.2 Sensor Conditioning Unit (SCU) The SCU serves as a conditioning interface between the ACE unit and accelerometers and the rate gyro sensors. In addition, the SCU provided cold sparing tolerant buffering to allow monitoring of latch valve status and pressure sensor readings by the L-1011 Carrier Airplane Payload Support Panel Interface equipment during L-1011 ferrying, while the SCU and ACE were powered on or off.

3.3.1.3 Mission Interface Unit (MIU) The MIU provides the spacecraft-to-ground interface for commanding, telemetry, and science data. The MIU interfaces with the RF Communications Subsystem, the FC, the CEU, and the ACE. For the RF communications interface, the MIU receives and decodes data from the S-band receiver and forwards that data to the FC. It also performs CCSDS encoding on telemetry data from both the FC and CEU and sends that data to the transmitter for downlink.

The MIU is the sole electrical interface between the SCB and the payload. All payload commands are forwarded by the MIU from the FC to the CEU. The MIU also sends pulse-per-second and spin pulse signals to the CEU as commanded by the FC and ACE. In addition, the MIU provides switched +28 V power and six switched heater interfaces to the CEU.

The MIU also provides the interface to several other components, including the SRM safe and arm, separation breakwires for safety inhibits, two coarse Sun sensors, the accelerometers, eight temperature sensors, and both delta-V thrusters.

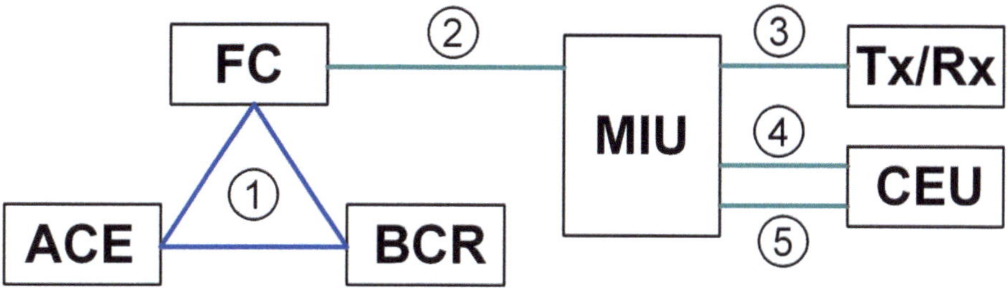

Ref #	Bus	Data Rate	Function(s)
1	Inter-Processor Token Ring Bus	38.4 Kbps RS-485, Half Duplex	Messages between FC, ACE, and BCR
2A	FC to MIU (Async)	57.6 Kbps RS-422, Full Duplex	MIU Commanding, CEU Commanding, Data transfers from FC to CEU Mass Memory, FC to MIU D/L Transfer Frames
2B	MIU to FC (Async)	57.6 Kbps RS-422, Full Duplex	MIU Telemetry Collection, CEU Telemetry Collection
3	MIU to Transmitter Receiver to MIU/FC	2 to 320 Kbps – Dnlink 2 Kbps – Uplink	Downlink / Uplink
4	MIU to CEU (Async)	57.6 Kbps RS-422, Full Duplex	CEU Commanding, Data transfers from FC to CEU Mass Memory, CEU Telemetry
5	CEU to MIU (Sync)	1 Mbps RS-422	High Speed SSR Data. Gated by MIU 'Clear to Send' (CTS) signal.

Fig. 14 Command and Data Handling interfaces

3.4 Electrical Power Subsystem (EPS)

The EPS provides and controls the electrical power that operates the IBEX spacecraft. Its functions include power generation and storage, load management, voltage conversion and regulation, heater control, safety inhibits, and launch/deployment sequence control. A block diagram of the EPS is shown in Fig. 15.

3.4.1 EPS Components

The EPS is comprised of the BCR, Power Converter Module (PCM), lithium-ion battery, and solar array panel. These components operate collectively to supply optimal power to the spacecraft.

3.4.1.1 Battery Charge Regulator (BCR) The BCR is the hardware core of the EPS, converting solar array power into unregulated (+14 V) and regulated (+5 V) power and distributing that power to the SCB; the power outputs are switchable. The EPS software architecture integrates the functionality of the major BCR components: microcontroller, telemetry card, accessory card, central regulator, and charge regulator. Through these components, the BCR controls the maximum power point tracking (MPPT), battery charging, battery protection, load switching, load management, the +5 V regulated bus, and the 14 V unregulated bus.

The BCR also collects voltage, current, and temperature data from several spacecraft sensors and controls battery charging/discharging based on battery cell voltage monitoring.

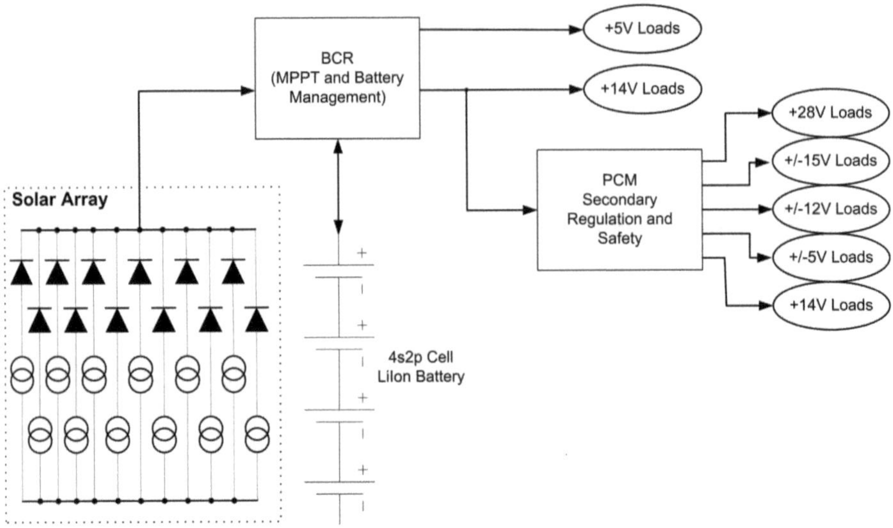

Fig. 15 Electrical Power Subsystem block diagram

Additional BCR functions include controlling load shedding and heaters and managing separation and deployment sequence. The EPS communicates with the FC and the ACE via the avionics bus, and the BCR is the token master.

The central regulator provides regulated +5 Vdc power to the spacecraft components, excluding the BCR internal electronics. Power provided by the central regulator is mainly used by the digital electronics components in other subsystems.

The solar array feeds into the charge regulator, which converts the 20–40 V solar array voltage to the unregulated 14 V bus. The charge regulator is responsible for managing power between the solar array and battery.

The Phoenix Regulator powers the BCR microprocessor and its supporting logic at +5 Vdc. The BCR electronics are the only spacecraft load that is not switchable. The Phoenix Regulator provides a means by which the spacecraft can recover from a complete loss of power by allowing the BCR electronics to power up in a stable manner before powering up the rest of the bus. The Phoenix Regulator receives power from the batteries or the solar arrays, whichever power source has a higher voltage.

3.4.1.2 Power Converter Module (PCM) The PCM receives power from the BCR unregulated bus and uses it to generate secondary voltages, including +28 V, ±5 V, ±15 V, and ±12 V. In addition to secondary power conversion, the PCM is also responsible for the following: sequencing the startup of the ACE, accelerometers, gyro, and SCU; providing triple breakwire based inhibits for the S-band transmitter and all hydrazine thrusters; providing spare switching capability; and controlling the aft antenna RF switch.

PCM power is separated into three power buses that are controlled by switches in the BCR:

- Power Bus A includes conversions to +28 V, ±5 V, ±15 V, and ±12 V, which are used to power the S-band receiver, S-band transmitter, accelerometers, gyro, 12 V ACE power supply, RF switch, and temperature sensors. While the transmitter, gyro, and RF switch can all be inhibited, basically serving as power switching to those components, the BCR switch for PCM Bus A is always powered on, and the remaining components are on whenever PCM Bus A is on. Because PCM Bus A powers the receiver, it cannot be turned off or communication with the spacecraft will be impossible (until it would be automatically turned back on following the 18-day master reset).

- Power Bus B provides switched +28 V power to the CEU, star tracker, and the two 22 N axial delta-V thrusters. Unlike Power Bus A, the BCR switch for Power Bus B is off by default. It is turned on for orbit raising to power the delta-V thrusters and again from commissioning through the end of the mission to power the CEU and star tracker.
- Power Bus C provides safety inhibited +14 V power to the 5 N hydrazine thrusters, latch valves, and pressure sensors controlled by the ACE. Like Power Bus A, the BCR switch for Power Bus C is always powered on.

3.4.1.3 Lithium-Ion Battery IBEX uses a 300 Watt-hour lithium-ion battery manufactured by Yardney (NCP8-2 cells) for energy storage. The battery features eight cells (4s2p) and balancing electronics. Battery management electronics are incorporated within the battery pack, with the monitor equalizer board (MEQ) ensuring that charge is equalized between cells and counteracting the effects of cell internal leakage currents on internal cell balance.

The battery operates over a range of 12 V to 16.4 V, with each cell going from 3 to 4.1 V. For ideal battery maintenance, the battery cells are generally kept at 3.9 V in full sunlight, except in cases of a long eclipse when the full 4.1 V charge is used. As a lithium-ion battery, battery state of charge is determined by battery voltage; temperature is not used.

3.4.1.4 Solar Array Power generation for the spacecraft comes from a single body-mounted solar array on the $+Z$ deck of the spacecraft. The spacecraft is nominally a Sun-pointed (within 8°) spinner, keeping the solar array illuminated. Between launch and the end of orbit raising, however, the spacecraft can remain power positive with the Sun as much as 55° off of the $+Z$-axis.

The solar array generates 114 W EOL at 110°C and 8° angle of incidence with 168 standard Emcore ATJ cells laid out in 12 strings of 14 cells each. The solar array is thermally isolated from the rest of the spacecraft and operates within a range of −160°C to 145°C, depending on the solar angle of incidence or if the vehicle is in sunlight or eclipse. Indium Tin Oxide (ITO) coating on the front surface of the solar cell coverglasses was used to minimize the solar array static charge build-up. The cells are laid out to reduce the number of strings affected at one time by shadowing from the $+Z$ thruster or antenna mast when the Sun is off at an angle from the $+Z$ direction.

3.5 Hydrazine Propulsion Subsystem (HPS)

The IBEX HPS is a simple monopropellant (hydrazine) blowdown system. Propellant is contained within two outboard spherical tanks; the location and shape of the tanks were selected to minimize nutation effects due to propellant sloshing. Four radial thrusters (5 N nominal) are oriented to provide impulse in the $\pm X$ direction for active nutation control during the SRM burn. During normal spacecraft operation these thrusters perform spin-rate adjustment, spin-axis repointing, and general attitude control. Two axial thrusters (22 N nominal) are oriented to provide impulse in the $\pm Z$ direction for orbit-raising maneuvers. The HPS design includes service and service valves, filters, and pressure transducers.

A functional schematic of the HPS is shown in Fig. 16. The thruster orientation on the spacecraft, with thruster plumes identifying the firing direction is shown in Fig. 17. The thruster locations were deliberately chosen to prevent the sensor apertures from observing the plumes.

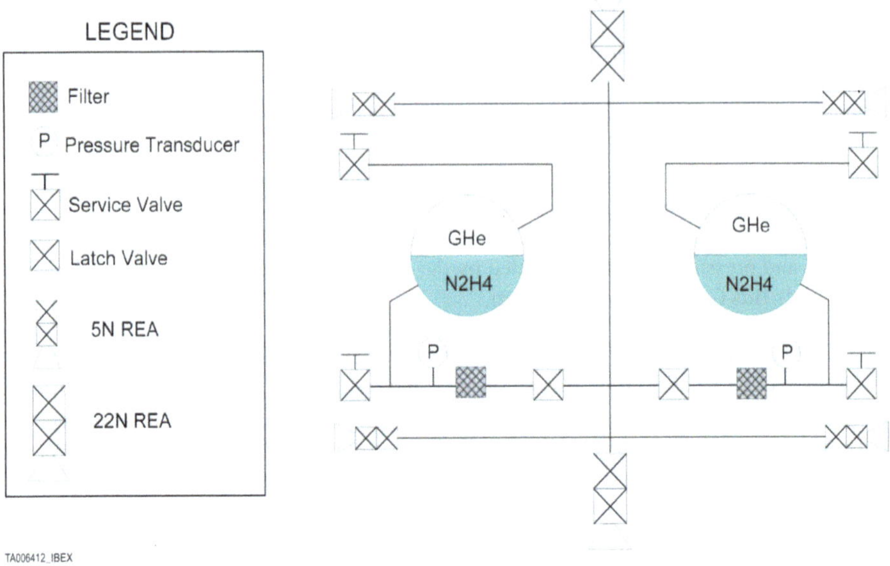

Fig. 16 Hydrazine Propulsion Subsystem schematic

Fig. 17 Hydrazine Propulsion Subsystem thruster plumes

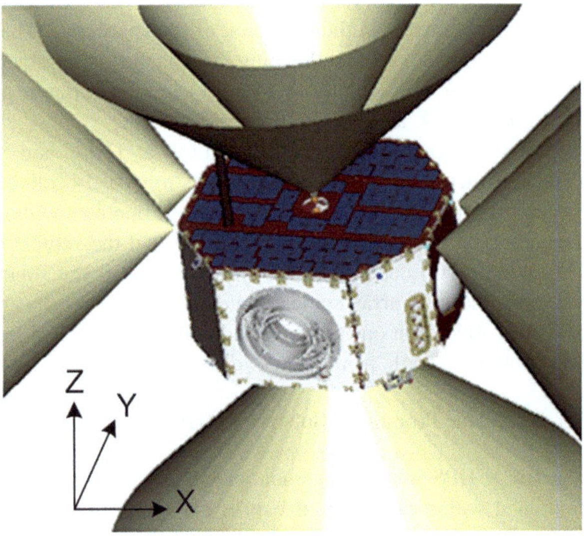

Propellant is held within two ARDE, 32-cm (12.6 inches) diameter, spherical Inconel tanks. Rather than a traditional diaphragm or propellant management device, the tank design features a gas arrestor and vortex suppressor. Each tank carries 17.4 liters, or 13.13 kg of hydrazine, with a MEOP of 2.76 MPa (400 PSIA). The outlet port for each tank is located on the outboard side, such that centripetal acceleration of the spinning spacecraft keeps propellant over the outlet port. Four 5 N Aerojet MR 111C thrusters each feature a dual-series, redundant-propellant control valve. Dual catalyst bed heaters require a maximum of 60 minutes to heat up prior to firing the thrusters, although pulse warming of the thrusters

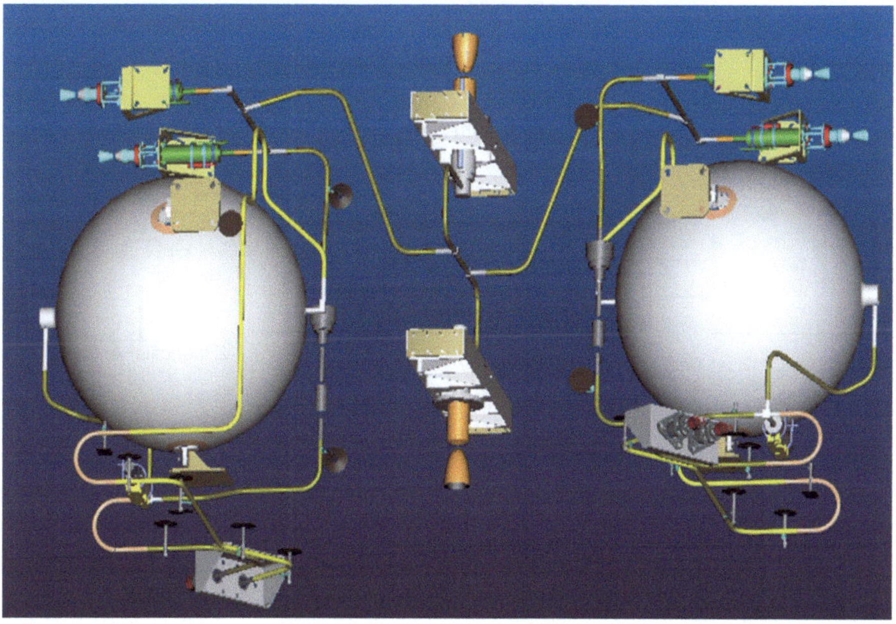

Fig. 18 Hydrazine Propulsion Subsystem tubing layout

is possible on a limited basis. Thrust levels range from 5.25 to 0.93 N, depending on pressure. Steady-state-specific impulse ranges from 229 to 207 seconds, although that drastically decreases when used at the minimum impulse bit, which is how the ACE operates these thrusters. Two 22 N Aerojet MR 106 L thrusters are used for delta-V maneuvers, and each features a dual series redundant propellant control valve. The catalyst bed heaters operate at 14 V, requiring a maximum of 75 minutes of warming prior to thruster use, and thrust levels range from 34.3 to 10.2 N, depending on pressure. Steady-state-specific impulse ranges from 235 to 228 seconds.

Other HPS components include: two Vacco V1E10728 latch valves with dual latching capability; two Taber P2911C401AEEX4L7 pressure transducers with a range of 0 to 2.76 MPa; two Vacco F1D10755 propellant filters that provide 15 microns absolute filtration; and four Vacco V1E10701 service valves used for propellant and pressurant gas loading. Figure 18 shows a layout of the HPS. The HPS was manufactured by bench welding individual sections, then assembling those sections on the spacecraft structure. The HPS is a fully welded system and was installed on the structural core module prior to spacecraft integration and test.

3.6 RF Communications Subsystem

The RF Communications Subsystem provides communications to and from the spacecraft during the IBEX mission. A block diagram of the RF Communications Subsystem is shown in Fig. 19. The system features a Thales Alenia Space S-band transponder that provides not only ground communications but also orbit-determination capability in the form of coherent Doppler measurements. Table 4 gives parameters for the uplink and downlink.

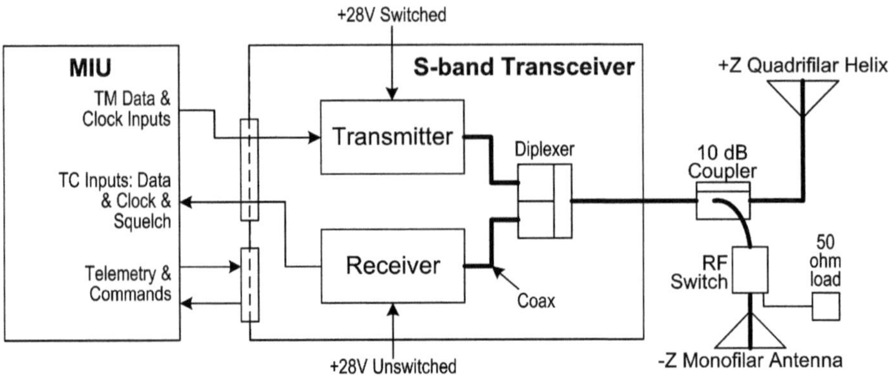

Fig. 19 RF Communications Subsystem block diagram

Table 4 Uplink and downlink parameters

Parameter	Downlink (from spacecraft)	Uplink (to spacecraft)
Data Rate (total symbols per second post-encoding)	320, 160, 64, 40, or 2 ksps	2 ksps
Data Rate Accuracy	$\leq$1%	$\leq$1%
Data Transition Density	$\geq$1 per 50 bits	$\geq$1 per 50 bits
Error Control Coding	CCSDS Reed-Solomon (255,223) code, 1/2 rate Viterbi (K = 7)	N/A
Modulation Format	NRZ-M	NRZ-M
Modulation Type	BPSK	PM/BPSK

3.6.1 RF Communications Subsystem Components

3.6.1.1 S-Band Transponder The S-band transponder includes a transmitter, receiver, and diplexer. The receiver accepts command signals from the ground while rejecting signals from other spacecraft transmitters, ground radars, and other ground emitters. A low noise figure (<5 dB) is required for the receiver to prevent a reduction in the spacecraft uplink sensitivity. The uplink center frequency is ~2 GHz. The receiver will acquire and track the carrier and data transitions within the commanded signal when the frequency is ±160 kHz. The receiver is always powered on and receives data at a 2 kbps rate.

The transmitter transmits spacecraft telemetry and scientific data to the space and ground segments. The transmitter will accept asynchronous data with symbol rates of 2, 40, 64, 160, or 320 ksps, and BPSK modulates the S-Band downlink carrier with limited bandwidth. Operation of the transmitter does not affect the receiver during the transmission. The transmitter uses an internal clock with ±35 kHz frequency stability, but upon receipt of a transmission through the receiver, the transmitter will lock onto that frequency if coherency is turned on (or will continue to use the internal clock if coherency is off). Coherency must be on in order to perform Doppler-based orbit determination. The default transmit rate for

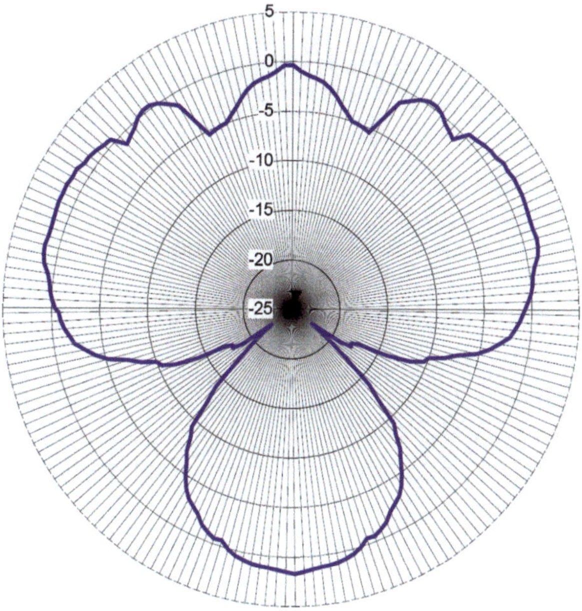

Fig. 20 Combined IBEX antenna pattern

the transmitter is 2 ksps. After turning on the transmitter, the desired transmit rate can be selected by sending the appropriate commands.

3.6.1.2 Antennas, Coupler, and RF Switch IBEX features an omni-directional antenna system that consists of two antennas, a quadrifilar in the $+Z$ direction and a monofilar in the $-Z$ direction, which are joined by a 10 dB coupler. The quadrifilar has a cardiod pattern to get full $+Z$ and X–Y plane coverage, while the monofilar helix has a narrow spectral pattern to augment coverage in the $-Z$ direction with minimal interference on the $+Z$ quadrifilar helix (see Fig. 20). The quadrifilar antenna is mounted on a boom to maximize the field of view, allowing coverage of the X–Y plane of the spacecraft, which is required for communications during orbit-raising burns performed at apogee. The combined antenna pattern is designed to provide >85% spherical coverage for 2 ksps uplink and downlink at 50 R_E altitude (USN Hawaii ground station) as well as opportunities every orbit to downlink to any of the USN stations at 320 ksps for at least 30 minutes.

An RF switch in the path between the 10 dB coupler and monofilar antenna allowed the spacecraft to turn on its transmitter (with RF switch open, preventing transmission through the $-Z$ monofilar antenna) while still attached to Pegasus. This made it possible to initiate the TDRSS link prior to spin-up and SRM burn without radiating into the Pegasus fairing, which is not allowed. The RF switch was closed (essentially turning on the monofilar antenna) by the BCR as a final step in the SRM burn and separation sequence.

3.7 Thermal Control Subsystem

Thermal control on IBEX is maintained by passive thermal components as well as heaters. Passive features include isolation of the top (solar array) deck with multilayer insulation (MLI) and thermal washers, MLI blankets with patch radiators on the side panels, black paint on internal surfaces (except battery), and titanium isolators between the SRM and

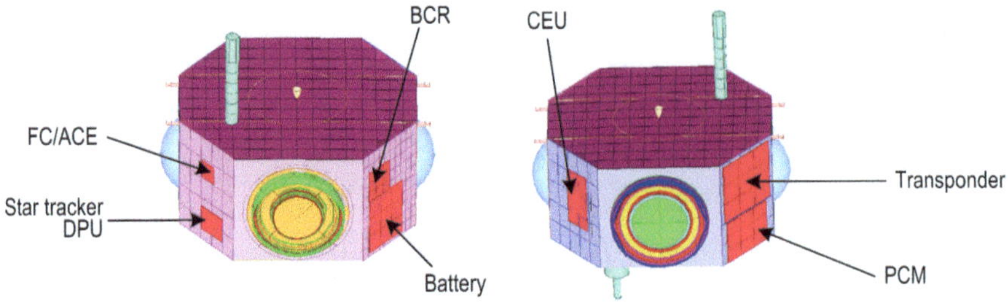

Panel	Component	Radiator Area (in^2)	Hot Case Dissipation (W)	Percent of Panel Covered
+X, +Y	CEU	48	14.7	20%
-X, +Y	PCM	70	17.2	77%
	Transponder	115	9.1*	
+X, -Y	BCR	25	12.8	40%
	Battery	72	2.4	
-X, -Y	FC/ACE	9	2.2	12%
	Star tracker DPU	20	8.4	
Base	SCU	0	1.0	1%
	MIU	9	3.0	

*S-band transponder dissipates 34.6W when transmitting

Fig. 21 Parameters of the spacecraft radiators and their placement on the IBEX spacecraft bus

MLB-15. Several heaters are also used, most on the payload sensors and HPS, and others on the transponder and star tracker electronics.

The spacecraft is hot-biased to minimize make-up heater power in eclipse. Patch radiators are sized such that the components are near maximum allowable temperatures in the hot case. The payload sensors are conductively coupled to the spacecraft, while the battery is weakly coupled to a radiator panel. The battery and star tracker baffle are blanketed inside the spacecraft.

The nominal temperature limits for the battery are 0 to 30°C. It is mounted on the same panel as the BCR, which has temperature limits from −25 to 50°C. Given the tighter restrictions on temperature, the battery is mounted on G-10 thermal isolators to weakly couple (0.15 to 0.31 W/°C) it to the panel and blanketed to radiatively isolate it from the spacecraft. An oversized radiator allows the battery to reject heat in a hot case. Weak coupling and radiative isolation serve to minimize heater power when in eclipse.

Figure 21 describes where radiators are employed on the IBEX spacecraft and gives the basic parameters for those radiators.

4 Payload Description

The IBEX payload consists of two ENA sensors and a CEU. Details of the two sensors are provided in other papers in this book (Funsten et al. 2009, this issue; Fuselier et al. 2009, this issue). This section will briefly describe the CEU and overall payload operations.

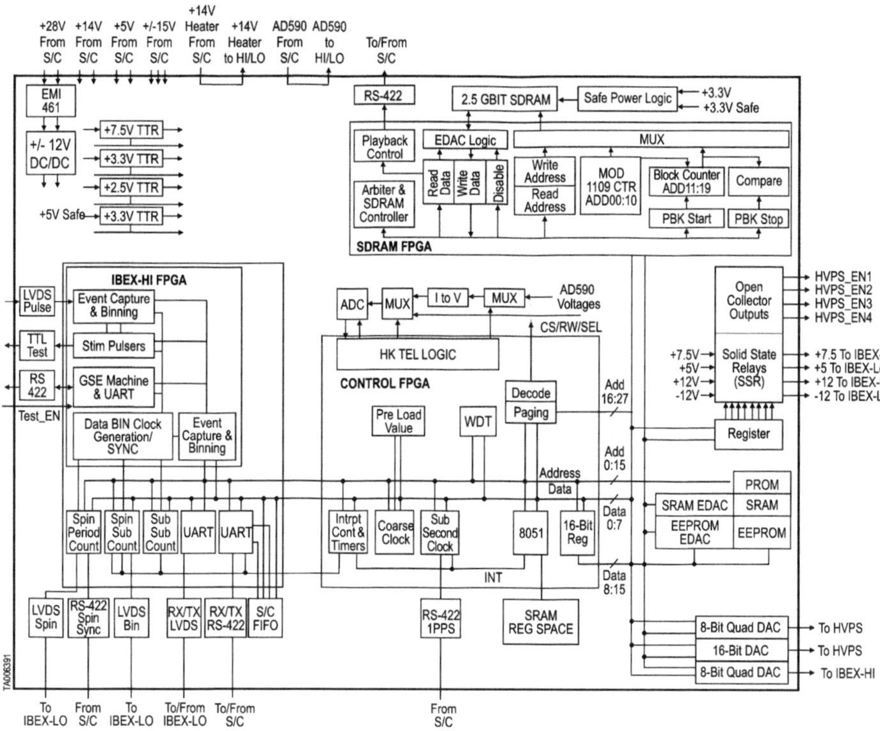

Fig. 22 Combined Electronics Unit digital board block diagram

4.1 Combined Electronics Unit (CEU)

The CEU serves as the single electrical interface between the payload and the SCB. This system architecture was chosen early in the project primarily to minimize overall system mass as well as to simplify the interface between the payload and spacecraft. The CEU is comprised of 2 circuit boards (the digital board and the high-voltage power supply (HVPS) board) and the mechanical chassis.

4.1.1 CEU Digital Board

The digital board accepts, parses and routes commands from the spacecraft to the two sensors and serves as the Solid State Recorder (SSR) where spacecraft and science data is stored and played back for downlink. The digital board creates and distributes custom low voltage for the sensors and gathers housekeeping telemetry from the sensors and the high-voltage board in the CEU. The digital board is responsible for sweeping high-voltage supplies and gathering, binning, processing and storing all raw science data. The multilayer digital board is based on an 8051 core in an Actel FPGA. It includes 2 Gbit of SSR memory for science and SCB back-orbit telemetry. Solid state relays are included for low-voltage distribution. Figure 22 shows a block diagram of the CEU digital board. Figure 23 shows a photograph of the top side of the CEU digital board.

Fig. 23 Top side of the CEU digital board

4.1.2 CEU High-Voltage Power Supply (HVPS) Board

The CEU HVPS board provides 16 high-voltage outputs ranging in voltage from 300 V to 11 kV. The board includes standalone independent power supplies for each of the two sensor positive collimators and two independent bulk supplies for the remainder of the high voltages. The supplies use a resonant flyback topology running at approximately 100 kHz. High-voltage pigtail cables are connected directly to and strain-relieved to the HVPS. No HV connectors are used anywhere on the payload. Custom lug connections are used at the sensor end of the high-voltage cables. The CEU includes a "V/16" mode that limits the output of the high-voltage supplies, when the payload safing plug is installed, to a level safe for turn on at ambient conditions. Figure 24 shows a block diagram of the HVPS board, and Fig. 25 shows a photograph of the HVPS board.

4.2 Payload Modes of Operation

The payload has several operating modes, each serving a different primary function as outlined below. Figure 26 shows the mode transition diagram. The "SAFING" arrows show the case when an anomalous situation is detected by the payload or if communications with the spacecraft is lost.

- BOOT—Initial mode when the payload is turned on; synonymous with payload firmware (PLFW) running PROM code. After one minute, if the PLFW PROM has not been commanded to stay in BOOT mode and does not detect any faults (memory checksums, watchdog timeouts, etc.), the PLFW PROM reads the permanent EEPROM software code and

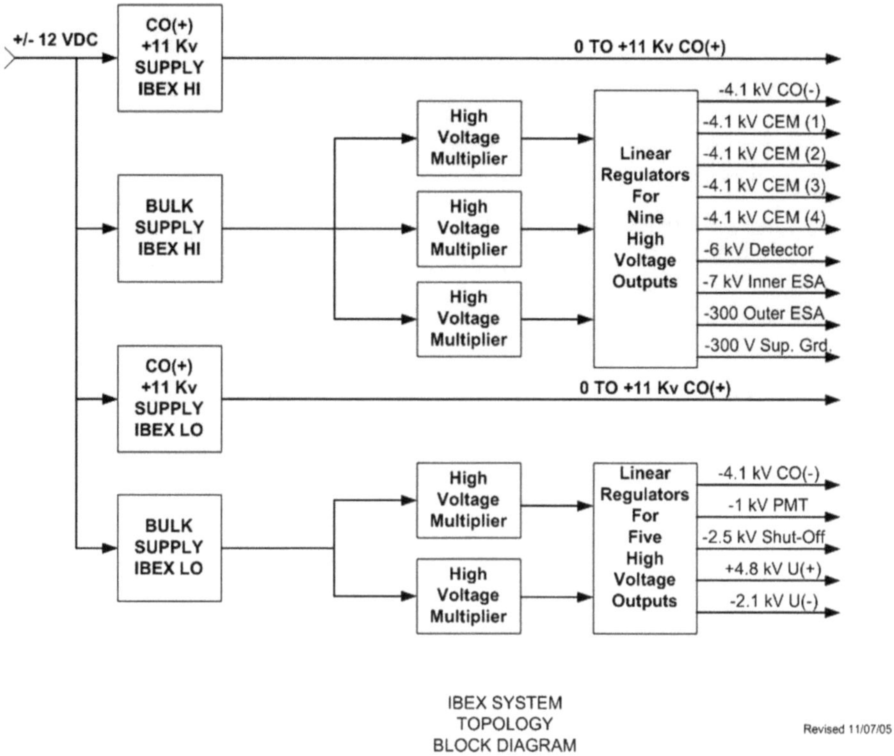

Fig. 24 High-Voltage Power Supply block diagram

copies it to the temporary SRAM then transfers control to SRAM. When this occurs, the mode transitions to low-voltage engineering (LVENG).

- LVENG—Transition mode to power up sensors; also for diagnostics and memory uploads. In LVENG high voltage cannot be enabled or commanded. This is also the mode that the CEU will revert to in response to a payload safing event. This was the primary mode used during LV Commissioning.
- Low-voltage science (LVSCI)—Low-voltage mode that behaves like high-voltage science (HVSCI, nominal science acquisition mode) by outputting science data sets using stim pulsers rather than collected ENA events; the stim patterns are defined in the stim table section of the look-up table. High voltage cannot be enabled or commanded, and this mode varies from HVSCI because sensor electrostatic analyzer (ESA) voltages are not swept.
- High-voltage engineering (HVENG)—Used for enabling and manual control of the HVPS. High voltage can be enabled and commands can be executed to ramp the HV supplies. This mode is primarily used for ramping up HV supplies when preparing for science mode, and ramping them down when preparing for standby mode or turn off. This was the primary mode used during HV Commissioning.
- HVSCI—Used for HASO in flight ($>10\ R_E$). High-voltage ESA sweeping supplies (IBEX-Lo U+/−, Lo USO, and IBEX-Hi INNER/OUTER_ESA) are automatically swept according to values stored in the look-up table. Science data products are generated and stored in the SSR.

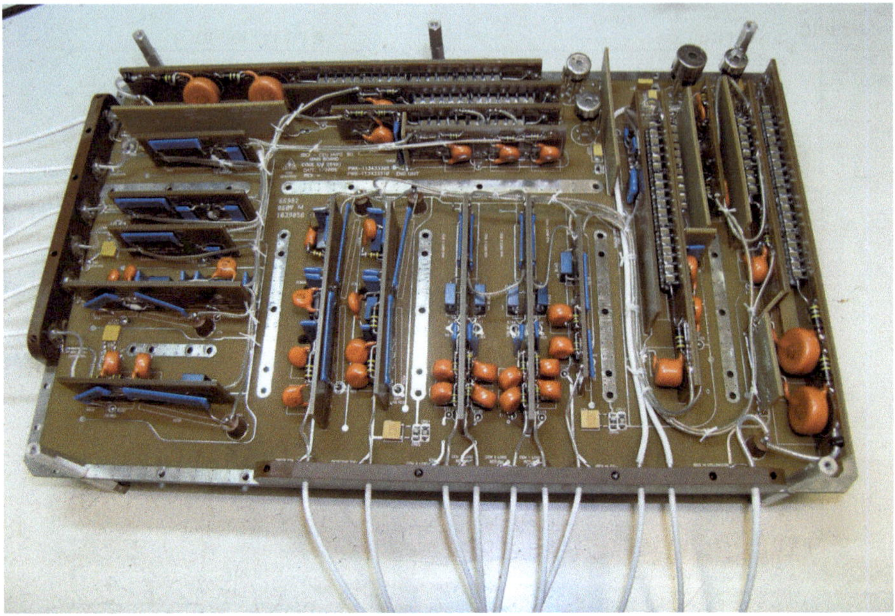

Fig. 25 High-Voltage Power Supply board

- High-voltage standby (HVSTANDBY)—Used for LAHO in flight during perigee pass ($<10\ R_E$). High voltages are on at a reduced level. In order to prevent accidental HV commanding while below $10\ R_E$, no HV commanding is allowed.

4.3 Solid State Recorder (SSR) Management

The IBEX SSR resides on the CEU digital board within SDRAM. Collectively, the SDRAM parts are organized as 128 K (i.e., 131072) blocks of 2 kB (2048 bytes) each for a total of 256 MB (268,435,456 bytes) SDRAM capacity. Each transfer frame is 1109 bytes. To make digital hardware design more straightforward, there is one transfer frame per 2k block, so the actual usable SSR capacity is 128 K $\times$ 1109 bytes = 145,358,848 bytes rather than the full 256 MB. The remainder of each 2 kB block is available for PLFW general purpose use but cannot be played back using the SSR Playback Interface.

When the payload receives an SSR dump command to downlink and the SCB issues the "Clear to Send" message to the payload, SSR playback is performed by transfer frame (i.e., units of 1109 bytes). These transfer frames are sent directly to the FC and MIU, which add headers (transfer frame header, sync pattern and Reed-Solomon) and pass these on for eventual downlink. All SSR packets are virtual circuit identification (VCID) VC2.

The SSR is a circular buffer with enough storage space to hold over 4 orbits' worth of data at the storage rates allotted in the telemetry budget. Position in the SSR is referenced by the SSR block number.

5 Integration and Test

Virtually all components of the IBEX flight system were subjected to protoflight qualification testing at the box level prior to delivery to the IBEX integration and test team. This in-

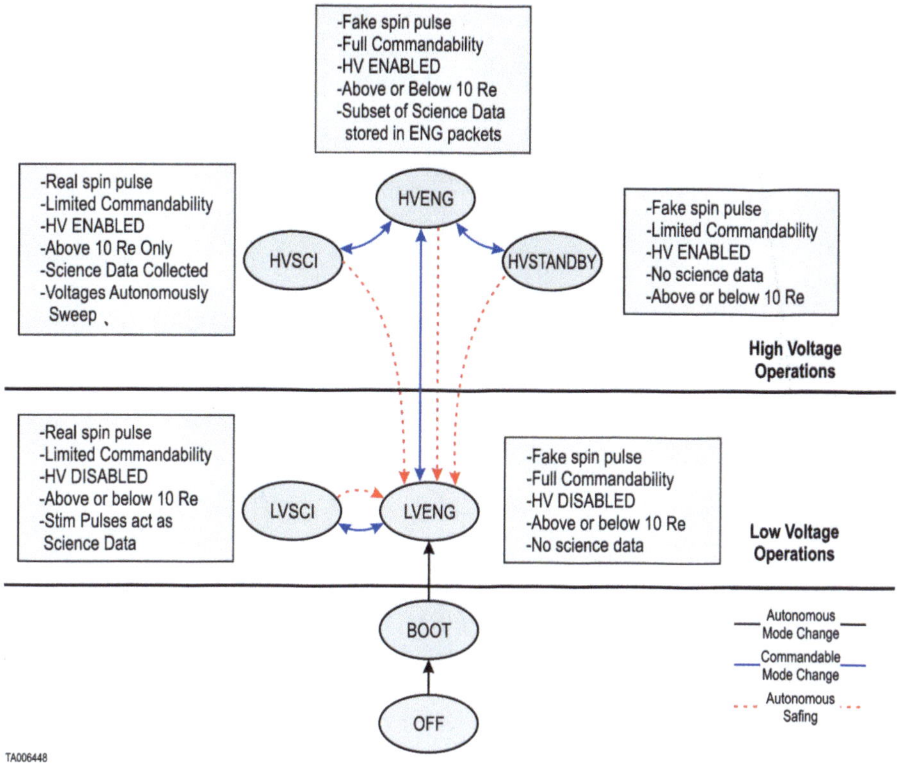

Fig. 26 Payload modes of operation

cluded functional, vibration and thermal vacuum testing for virtually all components. A few heritage components were subjected to either acceptance level testing only (when prior testing was demonstrated to envelop IBEX levels) or thermal testing at ambient pressure (due to similarity to existing flight components). The IBEX sensors were subjected to acoustic testing at the component level to verify that their internal foils would survive the acoustic launch environment.

One aspect of assembly, integration and test that was especially important to IBEX was contamination control because of its potential direct impact on the performance of the IBEX payload. The IBEX sensors contain contamination-sensitive components, such as microchannel plates and channel electron multipliers, and use high-voltage power supplies to operate the sensors. Like other spaceflight missions with similar contamination-sensitive instruments, IBEX performed assembly and testing only in class 10 K clean rooms, used low outgassing materials, purged the sensors with clean nitrogen, double and triple bagged the spacecraft when not in a clean environment, and utilized special covers for sensors as needed for the environment. The unique aspect of IBEX is that the ambient pressure in front of the sensor collimators would directly affect the sensor background rate. (See Wurz et al. 2009, this issue.) Because of this, special attention was made to directing all venting of the spacecraft away from the sensors (i.e. towards the solar array and aft deck), ensuring that there were no trapped volumes that would represent a virtual leak, providing super clean materials in the sensor collimators, and utilizing sensor designs with extremely open pump-out paths.

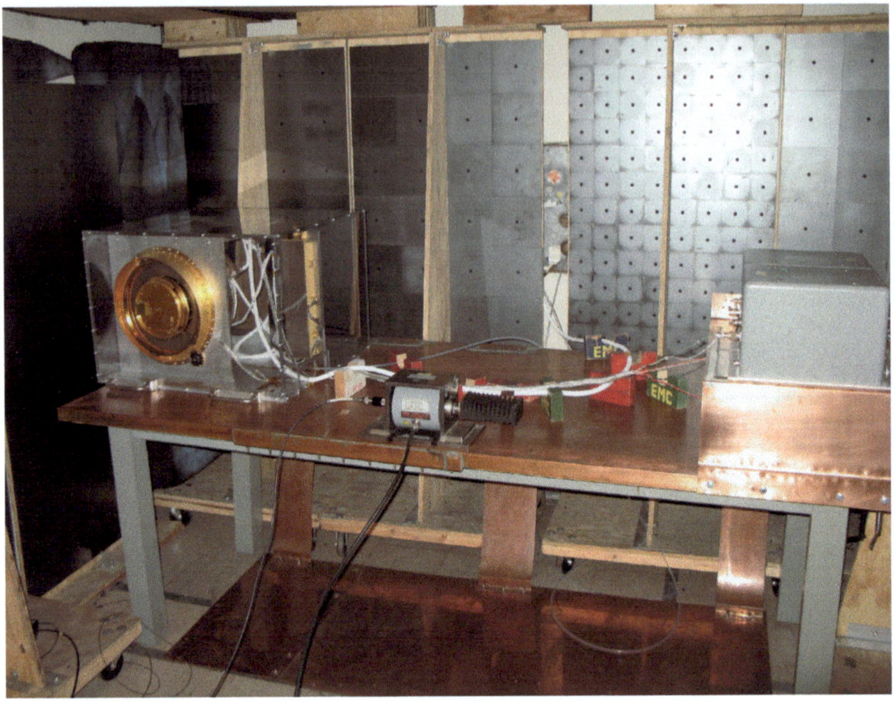

Fig. 27 Payload EMI/EMC testing

5.1 Payload

Each of the IBEX sensors was fully qualified and calibrated individually prior to delivery to Southwest Research Institute (SwRI). Likewise, the CEU was fully qualified at the box level. Because the IBEX payload is so tightly coupled it was decided early in the planning process to test the integrated payload prior to delivery and integration to the SCB. Integrated payload testing included the integration and functional testing of IBEX-Hi, IBEX-Lo and the CEU. This included mounting the three components in a mockup of the spacecraft, routing and terminating all high- and low-voltage cables, and performing the first payload Comprehensive Performance Test (CPT). The testing of the integrated payload included full MIL-STD 461E electromagnetic interference and compatibility (EMI/EMC) testing, thermal vacuum testing, and cross calibration. EMI/EMC testing was performed at SwRI; for contamination control this involved mounting the payload in an RF-transparent, static-dissipative, Polycarbonate PC 300 enclosure that was purged with clean dry Nitrogen throughout the testing (Fig. 27). Thermal vacuum testing was performed with the two sensors, the CEU and all flight cables. The test configuration included Ground Support Equipment (GSE) heaters and blankets to simulate the thermal boundary conditions that the payload encounters in flight on the IBEX spacecraft (Fig. 28). Cross calibration of the two sensors included mounting the payload on the positioning system in SwRI's calibration facility (McComas et al. 2009b), which includes a stage that allowed rotation of the sensors into and out of the ENA calibration beam. The cross calibration largely focused on the energy range of overlap between the two sensors (Fig. 29).

TAO06408_IBEX

Fig. 28 Payload thermal vacuum testing

5.2 Spacecraft

Like the payload components, all SCB components were delivered to Orbital fully qualified at the box level. Spacecraft bus integration and test started with developmental testing of the IBEX flight stack. The test configuration included the flight SCB structure, flight MLBs, flight adapter cone, mass simulators of the payload and SCB components, and a mockup of the SRM. Testing included random vibration and shock transmissibility, with these results used to correlate the preliminary FEM. The SCB structure was then disassembled to mount the HPS to the core module. Components and harness were then installed on the SCB and functionally tested. Once the SCB was baked-out, the payload was installed and functionally tested with the SCB prior to delivery of the transceiver, the critical path component for the mission, in order to minimize schedule impact.

Figure 30 details the IBEX spacecraft integration and test flow. The flow was fairly standard, with the exception of the two different vibration tests and the delay to solve the loads issue (Sect. 2.4). Two different vibration configurations were used: 1) Configuration A, which consisted of the dry spacecraft and MLB-15 and included 3-axis random vibe and shock testing. Spin balance of the spacecraft was also performed in this configuration; and 2) Configuration B, which included the dry spacecraft, all flight MLBs, plus the flight adapter cone and inert SRM. Three-axis random and axial sine vibe was performed, and MLB separation was also demonstrated in this configuration. See Fig. 31 for photos of the two different configurations.

Fig. 29 Payload cross calibration

6 Launch Site Operations

While IBEX was launched from the RTS on the Kwajalein Atoll, the bulk of the launch site integration and test was performed at Vandenberg Air Force Base (VAFB). Figure 32 details the integration and test flow that IBEX performed at VAFB, the ferry flight to RTS, and RTS operations. The IBEX spacecraft and SRM were shipped separately to VAFB and initial processing of the spacecraft and SRM was performed in parallel. The IBEX launch and separation sequence, as well as the prime mission, required the flight system to be spin balanced. The initial activities at VAFB performed at the Astrotech facility included final processing of the SRM, spin balancing of the various components of the flight stack and fueling the spacecraft. Figure 33 shows the final full-up flight stack spin balance. The components of the flight stack were then disassembled and transported to NASA building 1555 for integration to the Pegasus launch vehicle.

7 Systems Engineering (SE)

The highest priority in structuring the IBEX program was to produce a mission that fully meets the IBEX science objective at the lowest risk within NASA's Small Explorer (SMEX) cost cap. To obtain this goal, one of SE's and Mission Assurance's (MA) prime responsibilities was to minimize risk throughout the project. Systems Engineering typically has three main functions: 1) mission and system design, which includes the identification, allocation, documentation, management and verification of technical requirements; 2) management of

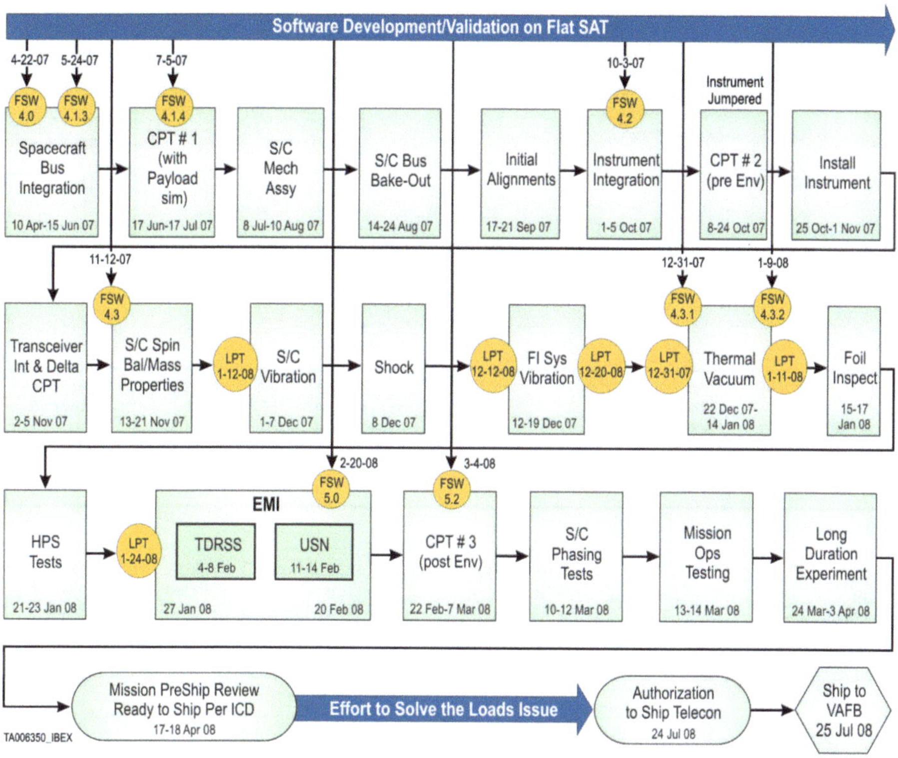

Fig. 30 IBEX spacecraft integration and test flow

Fig. 31 Vibration configurations: A (*left*) consisted of the dry spacecraft and MLB-15 and included 3-axis random vibe and shock testing; B (*right*) included the dry spacecraft, all flight MLBs, plus the flight adapter cone and inert SRM

technical resources; and 3) oversight and control of system interfaces and design trades. For IBEX, the SE team was also a key component in the risk reduction program.

The Mission Systems Engineer (MSE) led the IBEX SE team (i.e. SCB SEs, sensor and payload SEs, and ground segment SEs) in the management and verification of all IBEX technical requirements. The DOORS® database was used across the project by all IBEX in-

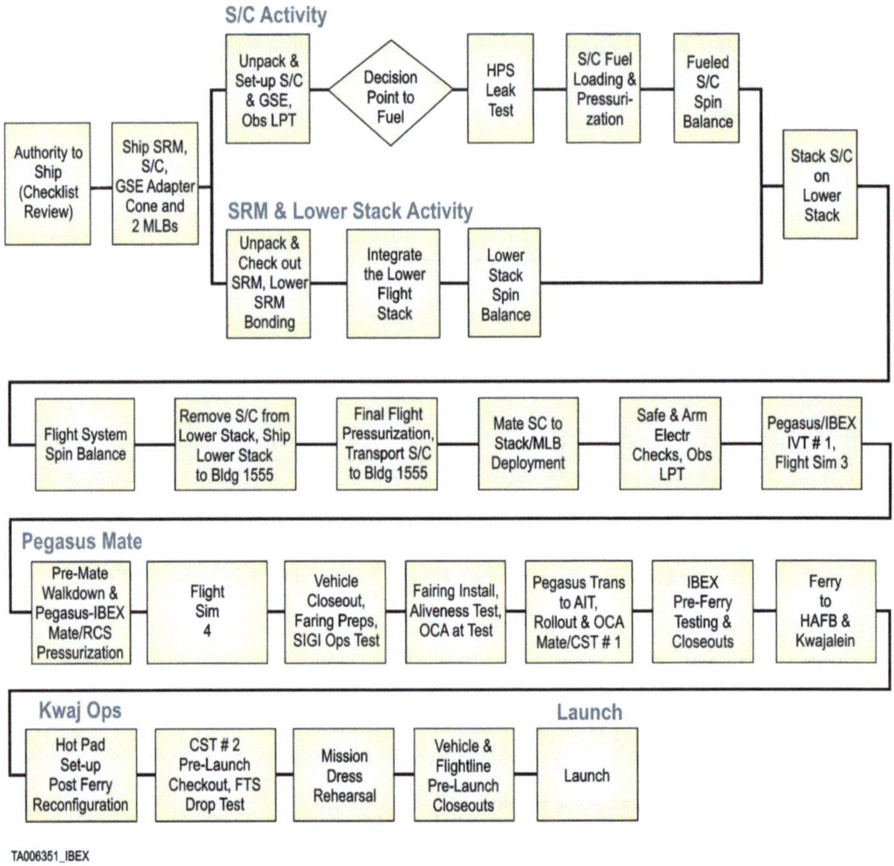

Fig. 32 Flow of integration and testing performed at VAFB, ferry flight to RTS, and RTS operations

stitutions to capture requirements from level 1 (IBEX Project Level Requirements) to level 2 (Mission Definition Requirements Agreement) to levels 3, 4 and beyond (Flight System and SCB Specification, IBEX-Lo Specification, IBEX-Hi Specification, etc.) On IBEX, the activity of identifying, documenting and verifying the closure of requirements started early and was given high priority throughout the development phase. The requirements were defined before NASA's final selection of IBEX and fully developed throughout the project. The SE team used the DOORS® database to document where all of the requirements were verified in addition to showing the flow-down of the requirements at all levels of verification. By launch all requirements that could be verified before flight (over 1800) were fully verified and documented.

The MSE also tracked and allocated technical resources and margins including mass, power, data rate, and memory. From IBEX's inception, mass was identified as the driving technical resource. The project appointed a Mass Czar early in the development phase to review, question and ultimately minimize excess mass across the system. The Mass Czar worked across the project to ensure that all subsystems were doing their part to minimize mass. This effort was so successful that the IBEX project gave back to Pegasus a significant amount of mass (over ~20 kg) that they could use to reduce launch vehicle risks. The summary of IBEX technical resources at launch are given in Table 5.

Fig. 33 Final full-up stack spin balance

The SE team also managed, documented, and controlled all system interfaces and trade studies throughout the project. To minimize cost, these trades were typically documented using the unique configuration management and documentation systems of the IBEX institutions (i.e. design notes, engineering memos, etc.) rather than forcing the team to use a single standardized system.

IBEX used the SwRI-developed, web-based Risk Management System in PIMS (Project Information Management System) to track all project risks. Risks were identified by IBEX team members, reviewers, NASA, etc. The risk status and mitigation plans were developed and statused by IBEX personnel as the risks were mitigated and modified during the project. The IBEX Project Manager and MSE worked together to assign consequence and likelihood for each risk since their positions in the project allowed them to "see the big picture". Risk status was updated and reported at least monthly.

IBEX MA will only be briefly discussed here due to page constraints—it could be the subject of its own paper. Like everything on IBEX, MA had to fit within the constraints of the SMEX cost cap. This required maximizing the efficiency of MA across the project by utilizing existing configuration control and non-conformance systems at IBEX institutions. The IBEX MA Lead ensured that these different systems worked together and correctly captured and documented the IBEX as-built configuration. More importantly, MA was the responsibility of all personnel who worked on IBEX not just MA personnel.

Table 5 Technical resources summary

Resource	Requirement	Value	Margin /Value	Notes
Flight System Dry Mass	163.61 kg	160.07 kg	(1)	Based on 461.60 kg LV capability, 271.38 kg SRM Propellant, and 26.3 kg hydrazine
Power	60 W	140 W	233%	Requirement is with transmitter off and no battery charging needed
Energy Margin, Nominal Mission	13945.0 W-hrs	15873.6 W-hrs	13.8%	Based on S/A Energy Generation
Pointing Knowledge	0.30 degrees	0.12 degrees	150%	By analysis
Pointing Control	2.50 degrees	1.79 degrees	40%	By analysis
Downlink Margin	See note	See note	>3 dB	Dependent on symbol rate & downlink altitude
Uplink Margin	See note	See note	>3 dB	Dependent on U/L altitude
Propellant, delta-V	See note	See note	10%	On top of delta-V margin
Delta-V (mean)	175.5 m/s	514.0 m/s	192.9%	Amount of delta-V available
Delta-V (3 sigma low)	200.1 m/s	514.0 m/s	156.8%	Amount of delta-V available
Data Storage (in CEU)	515.68 Mbits	1073.74 Mbits	108%	Based on 2 orbits of data
Commands, Nominal Mission	82	240	66%	Based on 2 orbits of commands

(1) 3.54 kg of ballast was added to IBEX at the launch site to bring the flight stack up to the launch vehicle requirement

8 Conclusion

In order to achieve and even exceed IBEX's scientific mission goal, the IBEX flight system was designed to be simple and robust. The Systems Engineering team compiled, tracked, and verified all subsystem requirements while drawing on personnel experience, hardware heritage and new, innovative solutions in order to efficiently manage all resources at every level of development. The mission design, driven by the orbit required for maximum science within the constraints of the less-expensive Pegasus launch vehicle, directed the overall design of the flight system. The innovative use of a Solid Rocket Motor to boost a Pegasus-launched spacecraft into its highest orbit yet was a pivotal part of putting IBEX in an orbit where it successfully images the interstellar boundary. The payload is supported by a spacecraft bus that uses attitude control, command and data handling, electrical power, hydrazine propulsion, RF communications and thermal control to stabilize, power, and direct the IBEX spacecraft in its nominal 2-year, data-gathering mission. Other flight system elements for separation, ejection and load isolation ensured IBEX's successful launch on October 19, 2008 and initiation of science operations at the completion of payload commissioning.

Acknowledgements The authors would like to acknowledge and thank Ms. W. Mills of SwRI for her tremendous help in compiling and editing this paper. The IBEX project was funded under NASA Contract No. NNG05EC85C.

References

D.J. McComas, F. Allegrini, P. Bochsler, M. Bzowski, M. Collier, H. Fahr, H. Fichtner, P. Frisch, H. Funsten, S. Fuselier, G. Gloeckler, M. Gruntman, V. Izmodenov, P. Knappenberger, M. Lee, S. Livi, D. Mitchell, E. Möbius, T. Moore, D. Reisenfeld, E. Roelof, N. Schwadron, M. Wieser, M. Witte, P. Wurz, G. Zank, The interstellar boundary explorer (IBEX), in *Physics of the Outer Heliosphere, Third Annual IGPP Conference*, ed. by V. Florinski, N.V. Pogorelov, G.P. Zank. AIP CP719 (AIP, New York, 2004), pp. 162–181

D.J. McComas, F. Allegrini, P. Bochsler, M. Bzowski, M. Collier, H. Fahr, H. Fichtner, P. Frisch, H. Funsten, S. Fuselier, G. Gloeckler, M. Gruntman, V. Izmodenov, P. Knappenberger, M. Lee, S. Livi, D. Mitchell, E. Möbius, T. Moore, S. Pope, D. Reisenfeld, E. Roelof, J. Scherrer, N. Schwadron, R. Tyler, M. Wieser, M. Witte, P. Wurz, G. Zank, IBEX—Interstellar boundary explorer. Space Sci. Rev. (2009a, this issue)

D.J. McComas, F. Allegrini, J. Baldonado, B. Blake, P.C. Brandt, J. Burch, J. Clemmons, W. Crain, D. Delapp, R. DeMajistre, D. Everett, H. Fahr, L. Friesen, H. Funsten, J. Goldstein, M. Gruntman, R. Harbaugh, R. Harper, H. Henkel, C. Holmlund, G. Lay, D. Mabry, D. Mitchell, U. Nass, C. Pollock, S. Pope, M. Reno, S. Ritzau, E. Roelof, E. Scime, M. Sivjee, R. Skoug, T.S. Sotirelis, M. Thomsen, C. Urdiales, P. Valek, K. Viherkanto, S. Weidner, T. Ylikorpi, M. Young, J. Zoennchen, The TwoWide-angle Imaging Neutral-atom Spectrometers (TWINS) NASA mission-of-opportunity. Space Sci. Rev. **142**(1), 157–231 (2009b)

H. Funsten, F. Allegrini, P. Bochsler, G. Dunn, S. Ellis, D. Everett, M. Fagan, S. Fuselier, M. Granoff, M. Gruntman, A. Guthrie, J. Hanley, R. Harper, D. Heirtzler, P. Janzen, K. Kihara, B. King, H. Kucharek, M. Manzo, M. Maple, K. Mashburn, D.J. McComas, E. Moebius, J. Nolin, D. Piazza, S. Pope, D.B. Reisenfeld, B. Rodriguez, E.C. Roelof, L. Saul, S. Turco, P. Valek, S. Weidner, P. Wurz, S. Zaffke, The Interstellar Boundary Explorer High Energy (IBEX-Hi) neutral atom imager. Space Sci. Rev. (2009, this issue)

S.A. Fuselier, P. Bochsler, D. Chornay, G. Clark, G.B. Crew, G. Dunn, S. Ellis, T. Friedmann, H.O. Funsten, A.G. Ghielmetti, J. Googins, M.S. Granoff, J.W. Hamilton, J. Hanley, D. Heirtzler, E. Hertzberg, D. Isaac, B. King, U. Knauss, H. Kucharek, F. Kurdirka, S. Livi, J. Lobell, S. Longworth, K. Mashburn, D.J. McComas, E. Moebius, A.S. Moore, T.E. Moore, R.J. Nemanich, J. Nolin, M. O'Neal, D. Piazza, L. Peterson, S.E. Pope, P. Rosmarynowski, L.A. Saul, J.A. Scheer, J.R. Scherrer, C. Schlemm, N.A. Schwadron, C. Tillier, S. Turco, J. Tyler, M. Vosbury, M. Wieser, P. Wurz, S. Zaffke, The IBEX-Lo sensor. Space Sci. Rev. (2009, this issue)

P. Wurz, J. Scheer, L. Saul, M. Wieser, S.A. Fuselier, A.G. Ghielmetti, E. Hertzberg, E. Moebius, H. Kucharek, P. Brandt, D. McComas, F. Allegrini, H. Funsten, IBEX Backgrounds and signal to noise ratio. Space Sci. Rev. (2009, this issue)

Space Sci Rev (2009) 146: 75–103
DOI 10.1007/s11214-009-9504-y

The Interstellar Boundary Explorer High Energy (IBEX-Hi) Neutral Atom Imager

H.O. Funsten · F. Allegrini · P. Bochsler · G. Dunn · S. Ellis · D. Everett · M.J. Fagan · S.A. Fuselier · M. Granoff · M. Gruntman · A.A. Guthrie · J. Hanley · R.W. Harper · D. Heirtzler · P. Janzen · K.H. Kihara · B. King · H. Kucharek · M.P. Manzo · M. Maple · K. Mashburn · D.J. McComas · E. Moebius · J. Nolin · D. Piazza · S. Pope · D.B. Reisenfeld · B. Rodriguez · E.C. Roelof · L. Saul · S. Turco · P. Valek · S. Weidner · P. Wurz · S. Zaffke

Received: 25 November 2008 / Accepted: 26 March 2009 / Published online: 23 April 2009

Abstract The IBEX-Hi Neutral Atom Imager of the Interstellar Boundary Explorer (IBEX) mission is designed to measure energetic neutral atoms (ENAs) originating from the interaction region between the heliosphere and the local interstellar medium (LISM). These ENAs are plasma ions that have been heated in the interaction region and neutralized by charge exchange with the cold neutral atoms of the LISM that freely flow through the interaction region. IBEX-Hi is a single pixel ENA imager that covers the ENA spectral range from 0.38

H.O. Funsten (✉) · M.J. Fagan · A.A. Guthrie · R.W. Harper · K.H. Kihara · M.P. Manzo
Los Alamos National Laboratory, PO Box 1663, Los Alamos, NM 87545, USA
e-mail: hfunsten@lanl.gov

F. Allegrini · G. Dunn · D. Everett · J. Hanley · M. Maple · K. Mashburn · D.J. McComas · S. Pope · B. Rodriguez · P. Valek · S. Weidner
Southwest Research Institute, 6220 Culebra Road, San Antonio, TX 78238, USA

P. Bochsler · D. Piazza · L. Saul · P. Wurz
Physikalisches Institut, University of Bern, Sidlerstrasse 5, 3012, Bern, Switzerland

S. Ellis · M. Granoff · D. Heirtzler · B. King · H. Kucharek · E. Moebius · J. Nolin · S. Turco · S. Zaffke
University of New Hampshire, 39 College Road, Morse Hall, Durham, NH 03824, USA

S.A. Fuselier
Lockheed Martin Advanced Technology Center, 3251 Hanover St., Palo Alto, CA 94304, USA

M. Gruntman
Astronautics and Space Technology Division, Viterbi School of Engineering, University of Southern California, Los Angeles, CA 90089-1192, USA

P. Janzen · D.B. Reisenfeld
Department of Physics & Astronomy, University of Montana, 32 Campus Drive, Missoula, MT 59812, USA

E.C. Roelof
Applied Physics Laboratory, 11100 Johns Hopkins Road, Laurel, MD 20723-6099, USA

to 6 keV and shares significant energy overlap and overall design philosophy with the IBEX-Lo sensor. Because of the anticipated low flux of these ENAs at 1 AU, the sensor has a large geometric factor and incorporates numerous techniques to minimize noise and backgrounds. The IBEX-Hi sensor has a field-of-view (FOV) of 6.5° × 6.5° FWHM, and a 6.5° × 360° swath of the sky is imaged over each spacecraft spin. IBEX-Hi utilizes an ultrathin carbon foil to ionize ENAs in order to measure their energy by subsequent electrostatic analysis. A multiple coincidence detection scheme using channel electron multiplier (CEM) detectors enables reliable detection of ENAs in the presence of substantial noise. During normal operation, the sensor steps through six energy steps every 12 spacecraft spins. Over a single IBEX orbit of about 8 days, a single 6.5° × 360° swath of the sky is viewed, and re-pointing of the spin axis toward the Sun near perigee of each IBEX orbit moves the ecliptic longitude by about 8° every orbit such that a full sky map is acquired every six months. These global maps, covering the spectral range of IBEX-Hi and coupled to the IBEX-Lo maps at lower and overlapping energies, will answer fundamental questions about the structure and dynamics of the interaction region between the heliosphere and the LISM.

Keywords Interstellar boundary · Termination shock · Heliopause · Energetic neutral atom · ENA · LISM

PACS 96.50.-e · 96.50.Ek · 96.50.Xy · 96.50.Zc

1 Introduction

The scientific objective of the Interstellar Boundary Explorer (IBEX) mission, launched on Oct. 19, 2008, is to discover the global interaction between the heliosphere and the local interstellar medium (LISM) (McComas et al. 2006, 2009b, this issue). By acquiring global maps of the spectral distribution of energetic neutral atom (ENA) fluxes emitted from this interaction region, IBEX will answer four fundamental questions:

I: What is the global strength and structure of the termination shock?
II: How are energetic protons accelerated at the termination shock?
III: What are the global properties of the solar wind flow beyond the termination shock and in the heliotail?
IV: How does the interstellar flow interact with the heliosphere beyond the heliopause?

Hydrogen ENAs are generated when solar wind or pickup protons, which are heated at or near the shocked interface between the solar wind and the local LISM, are neutralized by charge exchange with the cold neutral atoms that constitute a majority of the LISM (Gloeckler and Geiss 2004) and freely flow through the interaction region. A fraction of these ENAs successfully travel from the point of their neutralization to 1 AU, avoiding photoionization from solar UV and ionization by either impact with solar wind electrons or charge exchange with solar wind ions. Because the ENAs detected by IBEX-Hi have followed ballistic trajectories from their emission region beyond the edge of the termination shock, their flux and spectral distribution provide key signatures of the global nature of the structure and dynamics of the interaction between the heliosphere and the LISM.

The IBEX science payload consists of two ENA imaging sensors that act as highly sensitive single pixel cameras. They utilize the Common Electronics Unit (CEU) for sensor control, conditioned power, data processing, and telemetry management. While similar in functional design, these sensors are optimized to measure the tenuous ENA flux across two overlapping energy ranges: the IBEX-Lo sensor (Fuselier et al. 2009, this issue) measures ENAs from ~10 eV to 2 keV in eight energy steps, and the IBEX-Hi sensor measures ENAs from ~380 eV to 6 keV in six energy steps. The difference in energy ranges covered by the sensors is enabled by the different mechanisms by which they ionize ENAs before the ionized ENAs are electrostatically energy-analyzed and detected. In the higher energy range covered by the IBEX-Hi sensor, ENA ionization (specifically $H^0 \rightarrow H^+$) by transmission through an ultrathin charge-stripping foil provides higher ionization probability, whereas in the low energy range covered by IBEX-Lo higher ionization probability (specifically $H^0 \rightarrow H^-$ and $O^0 \rightarrow O^-$) is enabled by atomic reflection from a diamond-like carbon surface. A second difference is IBEX-Lo's ability to measure ENA mass and utilize a high angular resolution collimator section to measure oxygen and uniquely address Question IV.

The ENA flux originating from beyond the termination shock and measured at 1 AU by IBEX is anticipated to lie within the range of 1 to 500 $cm^{-2}\,s^{-1}\,sr^{-1}\,keV^{-1}$ at ~1 keV (Gruntman et al. 2001; Wurz et al. 2008). Because of this low ENA flux, the designs of the IBEX sensors have been driven toward maximizing the sensitivity and minimizing the noise and the backgrounds that would otherwise masquerade as ENAs. The numerous sources of background and noise (see Wurz et al. 2009, this issue) include ENAs of magnetospheric origin, ionization and acceleration of ambient gas molecules within the sensor, detection of ambient ions that are beyond the maximum rejection energy of the entrance subsystem, and coincidence events generated by penetrating radiation.

A second unique aspect of measuring these ENAs is their energy-dependent transit time from their formation beyond the termination shock to their detection by IBEX at 1 AU. This is illustrated in Fig. 1, which shows the time for a hydrogen ENA to travel 100 AU as a function of its energy. The transit times for ENAs within the energy range of IBEX-Hi to travel 100 AU range from 196 days at the central energy (4.09 keV) of the highest energy passband to 591 days at the central energy (0.45 keV) of the lowest energy passband. Furthermore, the range of travel times of ENAs emitted from a point source at 100 AU from the Earth that are detected within the full width at half maximum (FWHM) of a single energy passband ranges from 62 days at the highest energy passband to 143 days at the second lowest energy passband. This time uncertainty for an individual energy passband corresponds to ~1/3 and ~2/3, respectively, of the 6-month period over which IBEX generates a complete sky map. These time resolutions represent a lower limit because the ENA source region may be quite thick and extend significantly beyond the termination shock and because ENAs from source regions at the flanks and tail of the interaction region travel distances that can be substantially farther than 100 AU.

ENA imaging of the Earth's space environment was demonstrated serendipitously using energetic particle instruments on the ISEE-1 mission (Roelof et al. 1985) and then the Polar mission (Henderson et al. 1997). Neither these instruments nor subsequent dedicated ENA imagers (e.g., Gruntman 1997; Pollock et al. 2000; Mitchell et al. 2000; Moore et al. 2000) were optimized for the large aperture area, low noise and background, energy range, or energy resolution required to measure the dim ENA emission from the inner heliosheath.

IBEX-Hi is a high sensitivity, single pixel sensor. A critical priority of the IBEX-Hi sensor design and development was maximizing the sensitivity to ENAs while minimizing noise and backgrounds. This drove the detailed designs of each subsystem: the collimator is biased at +10 kV to reject ions up to 10 keV/q and is fabricated using nonlinearly stacked thin plates having collinear aperture holes to minimize ion scattering; the charge conversion subsystem ionizes ENAs so they can be electrostatically removed from the UV and electron background; the energy analysis subsystem projects the enormous entrance aperture area of the sensor onto a small detector area and is serrated to prevent UV and ions >10 keV/q from

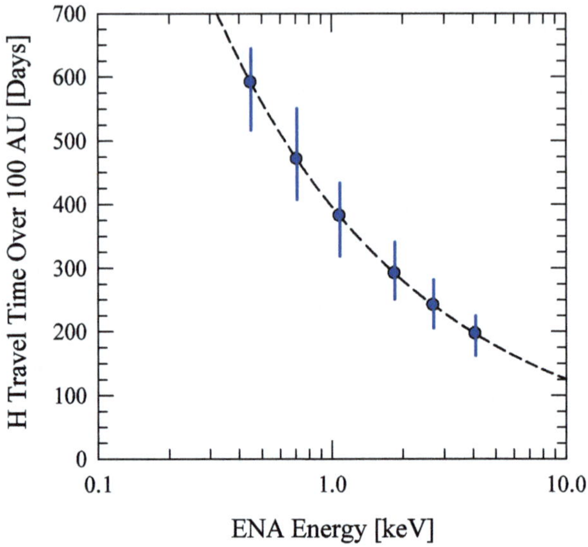

Fig. 1 The *points* show the travel time for a hydrogen ENA to transit 100 AU at the central energy of each of the six IBEX-Hi energy passbands. The *error bars* at these *points* represent the travel time uncertainty due to the energy FWHM of each of the energy passbands for events detected through triple coincidence

Table 1 IBEX-Hi sensor parameters and resources

Energy range	0.38–6.0 keV
Energy resolution ($\Delta E_{FWHM}/E$)	0.47–0.70
Number of energy steps	6
Field-of-view	6.5° FWHM (0.0147 sr)
Mass	7.37 kg
Power (W)	0.65 W
Telemetry	99 bps

reflecting into the detector subsystem; and the detector subsystem uses channel electron multiplier (CEM) detectors to minimize noise from penetrating radiation and to measure coincidence between these detectors to discriminate between noise and a true ENA that is detected in multiple CEMs as it transits the subsystem.

The summary of IBEX-Hi performance and resources is listed in Table 1. The next section describes in more detail the IBEX-Hi sensor subsystems and their unique designs that maximize sensitivity and minimize noise and background sufficient to view the faint ENA signal.

2 The IBEX-Hi Sensor

Figure 2 shows a cross-sectional illustration of the IBEX-Hi sensor. The sensor is divided into four subsystems that are sequentially encountered by an ENA. The entrance subsystem serves multiple purposes, including limiting the angular field-of-view (FOV) to 6.5° FWHM and electrostatic rejection of ambient electrons with energies up to 0.6 keV and ions with energies up to 10 keV/q. The ENA then encounters the charge conversion subsystem that utilizes ultrathin carbon foils to positively ionize a fraction of ENAs that transit a foil. The ionized ENAs then enter the electrostatic energy analysis subsystem, which consists of

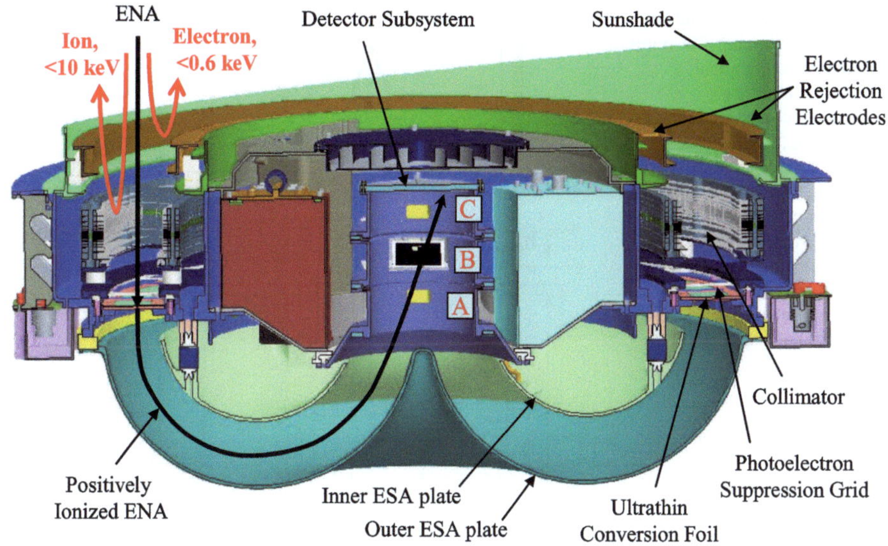

Fig. 2 This cross-sectional view of the IBEX-Hi sensor illustrates the subsystems and the trajectory of an ENA through the sensor

nested toroidal analyzer plates that project the large entrance aperture onto a small detector subsystem. The bias of the inner ESA plate sets the energy passband for ionized ENAs to enter the detector subsystem. ENAs entering the detector subsystem are accelerated by a bias of −6 kV for increased detection efficiency. The detector subsystem consists of three stacked cylindrical chambers (designated A, B, and C as shown in Fig. 2) with each chamber separated by an ultrathin carbon foil. Each chamber has a CEM detector that detects secondary electrons generated by the interaction of the ENA with a foil or an interior wall of a chamber. An ionized ENA can transit all three chambers and register a pulse in multiple detectors, generating a double (AB, BC, AC) or triple (ABC) coincidence event. The following sections describe in detail these subsystems.

2.1 Entrance Subsystem

The IBEX-Hi collimator subsystem, which is nearly identical to that of IBEX-Lo, serves multiple purposes. First, it defines the instantaneous FOV of 6.5° FWHM. Second, appropriately biased electrodes successively prevent electrons and ions from entering the sensor aperture. Third, it limits the access of light into the sensor. The collimator has been optimized to provide maximum ENA transmission for the size and geometry of the sensor. The collimators for IBEX-Hi and -Lo are identical in their design except for small differences in inner and outer radii and a high angular resolution quadrant in IBEX-Lo that is not present in IBEX-Hi.

The collimator consists of stacked plates, each with arrays of collinear hexagonal, photo-etched apertures. The use of plates minimizes the exposed surface area from which particles can scatter into the charge conversion subsystem. IBEX-Hi has the same FOV over its entire aperture.

Figure 3 shows a 2-dimensional cut through the collimator. An intrinsic complication using multiple plates with collinear apertures that form channels is leakage, in which particles with trajectories beyond the desired FOV enter one channel and can pass through to

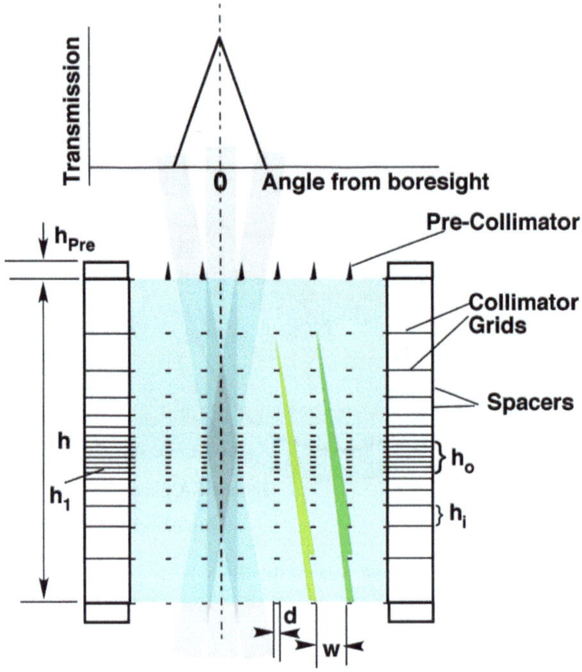

Fig. 3 The cross-sectional cut through the collimator illustrates the sequence of identical collimator plates that successively clip trajectories of particles (e.g., ENAs, ions >10 keV, electrons >0.6 keV, and UV light) incident at higher angles. For example, the *yellow* and *green trajectories* represent rejection of particles that are outside of the collimator FOV. The pre-collimator prevents particles from entering at very high angles

a neighboring channel. To prevent such leakage, a series of six plates with the same small separation ($h_1 = d \tan\theta_{\mathrm{Max}}$, where θ_{Max} is the largest possible angle of incidence of particles that must be rejected) is placed at the center of the collimator. From this set of plates in the center, additional plates are alternately stacked toward the entrance and exit ends of the collimator with the plate spacing increasing in a geometrical sequence according to $h_{i+1} = h_i(w + d)/w$ and the largest spacing at the exit plate. In the final collimator design, plate separations are slightly less than the theoretical geometric progression to account for potential leakage from manufacturing tolerances and deviations from plate planarity. The angle θ_{Max} is limited to $\le 50°$ at the collimator entrance by a precision-milled pre-collimator with trapezoid-shaped hexagon ribs, whose width is not larger than $d - 50\ \mu\mathrm{m}$.

Assuming no leakage, the angular response of the collimator is solely determined by the width w of the hexagonal openings and the total distance h between the entry and exit plates, and the FWHM is approximately $\theta_{\mathrm{FWHM}} = \tan^{-1}(w/h)$. For a geometry with close-packed hexagonal openings in each plate, the transparency T of the collimator at normal incidence is $T = 1/(1 + d/w)^2$, where d is the width of the plate between adjacent apertures.

2.1.1 Collimator Field-of-View (FOV)

Figure 4 shows the hexagonal point spread function $P(\theta, \phi)$ of IBEX-Hi as derived using a Monte-Carlo simulation of a single hexagonal channel. The maximum transmission of 0.67 lies at $P(0, 0)$, and the integrated FOV is 0.0147 sr. Modeling of the collimator based on worst-case manufacturing and plate alignment tolerances of 100 μm indicates a cumulative transmission of $<10^{-6}$ for particle trajectories over all angles beyond the nominal FOV. Testing of the IBEX-Lo and IBEX-Hi collimators with an Ar^+ beam showed that the combined effects of leakage through neighboring channels and scattering from edges of the collimator plates resulted in a transmission of $<10^{-5}$.

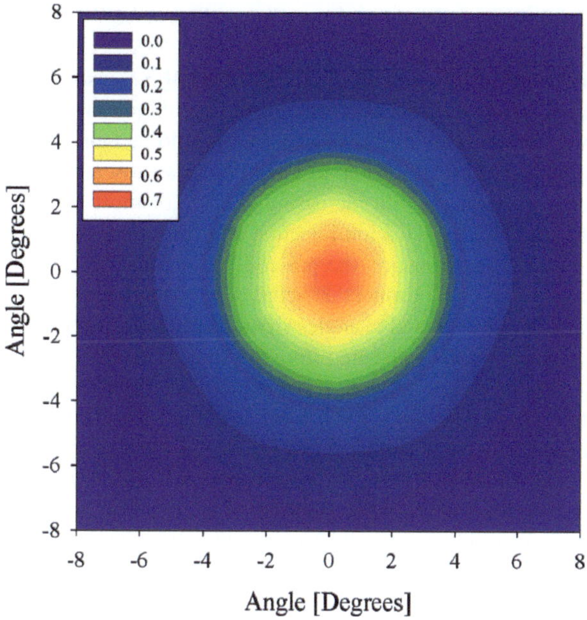

Fig. 4 The IBEX-Hi collimator transmission as a function of incident ENA angle is derived using Monte-Carlo simulations of the geometric response of a single hexagonal channel with an aspect ratio of length h to major hexagonal cell width w of $26.83:1$. This collimator point spread function $P(\theta, \phi)$ has a maximum transmission of 0.67 and an integrated FOV of 0.0147 sr

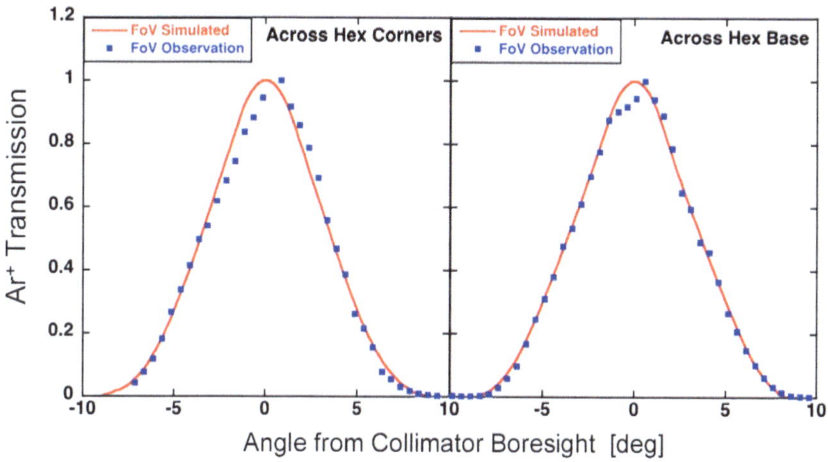

Fig. 5 Measured 20 keV Ar^+ transmission through the collimator as a function of incident beam angle θ, where $\theta = 0°$ corresponds to the boresight direction, for angular scans across the hexagon corners (*left*) and across the hexagon base (*right*). The collimator was grounded during the tests. The simulated angular response is shown in *red* and agrees well with the data except near 0°, which is an artifact of non-uniform illumination of the ion beam over a small region of the collimator

The FOV and the overall transparency of the collimator were determined using an optical comparator, which measured the total line width of the collimator stack. The FOV was measured to be $6.44 \pm 0.15°$ FWHM across the hexagon base and $6.5 \pm 0.15°$ FWHM across the hexagon corners. The transparency of the IBEX-Hi collimator was found to be $66.85 \pm 0.75\%$. Angular and linear scans with a narrow 20 keV Ar^+ beam across the base and the corners of the hexagon have verified these optical measurements as shown in Fig. 5.

2.1.2 Suppression of Ambient Charged Particles

In order to minimize background, the IBEX sensors must reject ambient (positive) ions up to 10 keV/q and electrons, including ubiquitous photoelectrons, up to 0.6 keV. If ions were allowed to enter IBEX-Hi, they would be indistinguishable from ENAs of similar energy. Rejection of ions up to 10 keV/q was chosen because ion fluxes at energies >10 keV decrease rapidly with increasing energy for most plasma environments encountered over the IBEX orbit. This ion rejection is achieved by biasing the entire collimator structure to +10 kV. Furthermore, the pre-collimator imposes electrostatic troughs at the entrance of each hexagonal channel that act as defocusing lenses, so most incident ions between 10 and 16 keV are deflected to angles significantly outside the collimator acceptance range and thus are substantially prevented from reaching the conversion foil.

Figure 6 shows the observed ion transmission through the IBEX-Hi collimator as a function of incident angle for 7.5 keV Ar^+ in a test configuration in which the collimator was biased to +8 kV and used to validate the collimator electro-optic model. Maximum transmission values of $\sim 2 \times 10^{-4}$ were obtained. This represents an upper bound to the actual ion transmission due to neutralization of a small fraction of the ion beam by charge exchange with the residual gas in the vacuum chamber; these neutrals can pass through the collimator and be detected.

If left exposed to the space environment, the collimator biased to +10 kV would attract and collect ambient electrons, including copious photoelectrons from the sunlit side of the spacecraft, resulting in a total electron current approaching 100 μA. Furthermore, many of these electrons would transit the collimator and ionize the residual gas in the space between the collimator and charge conversion subsystem; these ionized atoms and molecules would then be accelerated to the conversion foil and could masquerade as legitimate ENAs. To avoid both effects, negatively biased electron repeller ring electrodes are placed in front of the inner and outer radii of the collimator but completely outside of the collimator FOV. Figure 7 shows a cut through the entrance subsystem with simulated equipotential lines. To achieve a closed potential surface of −0.6 kV in front of the +10 kV collimator, the electron repeller electrodes are biased to −3.125 kV. Tests with an electron beam and the collimator biased to +10 kV demonstrated that rejection of electrons up to 0.6 keV is achieved with this configuration. We note also that photoelectrons and secondary electrons emitted from the electron repeller electrodes are accelerated to the collimator but only at high incident

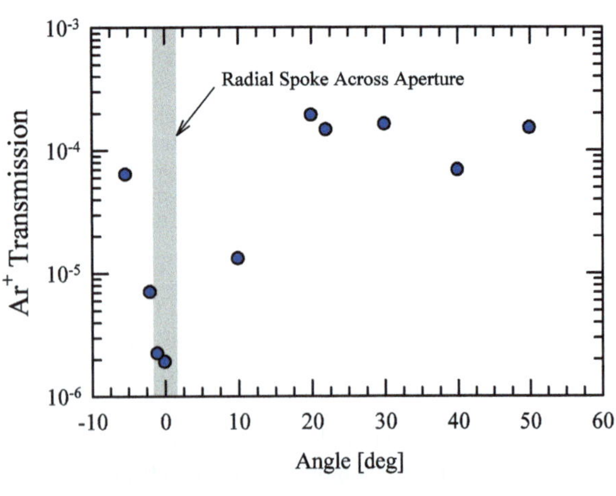

Fig. 6 Transmission of 7.5 keV Ar^+ as a function of the azimuthal angle around the annular collimator entrance. The collimator is biased to +8 kV and the electron repeller electrode is biased to −2.5 kV. A radial spoke spanning the aperture at 0° completely blocks the incident beam. Except for measurements at the spoke, these measured transmission values represent an upper limit because of the small fraction of Ar^0 in the incident beam that passes through the collimator and is detected

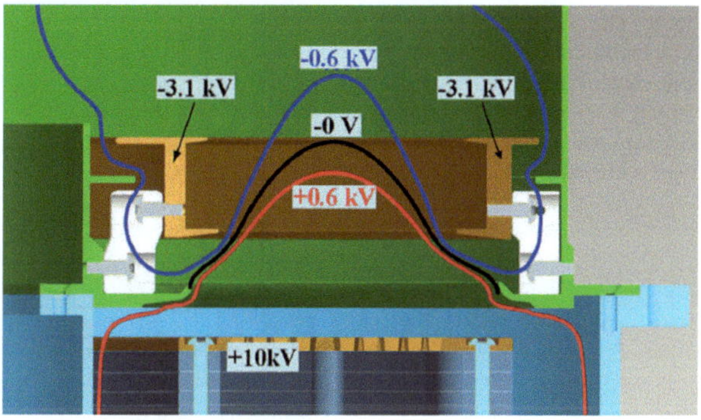

Fig. 7 The radial cut through the IBEX entrance subsystem illustrates equipotential field lines based on the subsystem's SIMION electro-optic model. When the collimator and electron repeller electrodes are biased to +10 kV and −3.1 kV, respectively, a closed −0.6 kV equipotential surface spans the entrance region, thereby rejecting all electrons with energies ≤0.6 keV

angles significantly outside of the collimator FOV, thus preventing these electrons from reaching the critical volume between the back of the collimator and the charge conversion subsystem.

2.2 Charge Conversion Subsystem: ENA Ionization

ENAs that transit the collimator enter the charge conversion subsystem whose purpose is to convert the ENAs to positive ions by transmission through an ultrathin foil. A total of 15 foils are distributed in a circular array as shown in Fig. 8 with inner and outer radii of 14.41 cm and 16.56 cm, respectively. These foils are nominal 0.6 $\mu g\,cm^{-2}$ carbon from ACF Metals and are mounted on 333 line-per-inch (lpi) Ni grids that have a typical transmission of 73.4% as measured using a 50 keV H^+ beam. Each foil is mounted on a two-window foil frame that was determined by analysis and testing to be the optimal geometry for acoustic survival during launch. The foil frames each have an open area of 10.44 cm^2 for a total conversion foil area of 156.6 cm^2. The conversion foil grid transmission was measured to be 73.4%, and the transmission of the photoelectron suppression grid located immediately in front of the conversion foils was 95%.

The probability of H^0 exiting the foil as H^+ is shown in Fig. 9 (Funsten et al. 2001). The ionization probability ranges from 0.013 at 0.33 keV to 0.13 at 6 keV, and the laboratory data infer two different ionization regimes by their different energy dependence. ENAs also lose energy in the foil as measured by Allegrini et al. (2006).

The foil constant $k_F = E_0\psi_{1/2}$, in which E_0 is the energy of an incident H^+ ion beam and $\psi_{1/2}$ is the angular halfwidth of the scattered beam after transit through the foil, was measured for each foil as a proxy for foil thickness (Funsten et al. 1993) and to quantify its energy-dependent angular scattering, which is an important part of the end-to-end sensor model. The angular halfwidth $\psi_{1/2}$ was determined by integrating the 2-D scatter distribution in one (x, y) dimension and dividing by 1.19 to correct for integration of a Lorentzian distribution. The measured foil constant, frame serial number, and location in IBEX-Hi of each foil are listed in Table 2. The average foil constant for all foils is 18.5 keV deg. The foils were additionally analyzed for pinholes using Ar^+ beam analysis (Funsten et al. 1992), and the pinhole fraction was less than 1% for all foils.

One possible source of significant background is electron impact ionization of outgassing atoms and molecules in the region between the conversion foils, which lie at ground potential, and the collimator, which lies at +10 kV. Ionization is caused by photoelectrons or secondary electrons that are emitted from the entrance surface of a foil or a foil frame surface and then accelerated toward the collimator. Ionized atoms and molecules are accelerated in

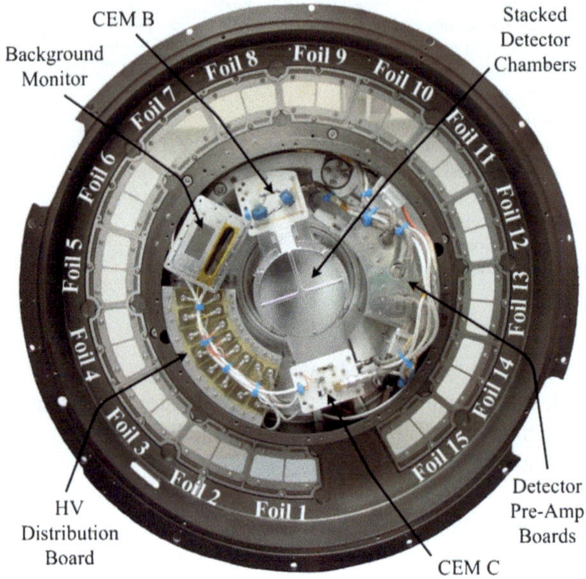

Fig. 8 The charge conversion foil locations are noted in this front view of the IBEX-Hi sensor that has its entrance subsystem, ESA subsystem, and detector cover removed

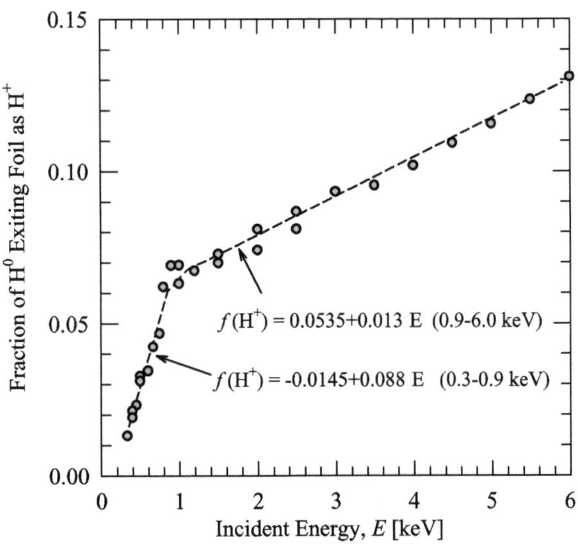

Fig. 9 The measured ionization probability of H^0 transmitted through an ultrathin carbon is shown as a function of incident energy

the opposite direction of electron acceleration, directly toward the conversion foils. These ions could exit the foil within the energy range of IBEX-Hi, therefore masquerading as heliospheric ENAs. We have minimized this "ion gun effect" background in two ways. First, the charge conversion foils act as a seal to block gas in the energy analysis and detector subsystems from entering the collimator region, which minimizes the pressure of the ambient gas in the "ion gun" region. Second, the charge conversion subsystem incorporates a 25 lpi nickel grid with 95% transmission located 0.48 cm in front of the charge conversion foils and biased to −300 V to suppress the photoelectrons and secondary electrons. Simulations show complete suppression of photoelectrons and secondary electrons emitted from

Table 2 IBEX-Hi foil placement and measured foil constant k_F. The requirement for IBEX-Hi is $k_F \leq 25$ keV deg

Foil location	Foil number	Foil constant, k_F [keV deg]
1	27	16.8
2	30	19.3
3	19	19.3
4	31	16.8
5	42	20.2
6	43	17.6
7	17	18.5
8	38	18.5
9	41	18.5
10	48	18.5
11	33	19.3
12	49	16.8
13	40	20.2
14	36	19.3
15	50	16.8
Average		18.4

the conversion foils and foil frames with energies up to at least 30 eV; this value is significantly lower than might be expected from a −300 V grid bias because of the large field leakage due to the high geometrical grid transmission (Read et al. 1998). The suppression grid reduces the "ion gun" background by a factor of ∼20.

2.3 Energy Analysis Subsystem

Positively ionized ENAs exiting the charge conversion foil enter the electrostatic energy analyzer (ESA), whose inner and outer analyzer plates are shaped like a "Bundt" baking pan (Moestue 1973). This geometry was successfully used on the Russian Interball mission (Vaisberg et al. 1995) and on the ISTP/POLAR mission (Shelley et al. 1995). The radius of the electro-optic surface of the inner (outer) ESA plate is 5.32 cm (8.34 cm), with the center of this radius located 8.76 cm (8.83 cm) from the central symmetry axis of the ESA plates and 0.82 cm (1.15 cm) below the plane of the conversion foils.

The inner ESA plate can be biased up to −7 kV and is the primary control for the central energy of each of the six energy passbands of IBEX-Hi, where we define the central energy as the energy of the maximum of the sensor response function for the passband. The ESA plates had serrations of hemispherical troughs of radius 0.76 mm machined such that the troughs are concentric relative to the symmetry axis of the sensor and, therefore, are aligned perpendicular to the direction of electrostatic deflection of ionized ENAs in the ESA. The serrations minimize scattering of many background particles into the detector subsystem, including UV light, ENAs exiting the foil as neutrals, ionized ENAs outside of the set energy passband, and ions >10 keV that can successfully transit the entrance and charge conversion subsystems.

The outer ESA plate is biased to −90 V, −80 V, and −20 V for the lowest three energy passbands (passbands 1, 2, and 3, respectively) for two purposes, both of which increase the energy analysis subsystem throughput. First, ENAs scatter in the foil, especially at lower energies, to an angle at which they cannot transit the ESA even though their energy resides within the energy passband. For the lowest energy passband, the −90 V on the outer ESA plate coupled with the −780 V on the inner ESA plate accelerate and proximity focus ionized ENAs so they have a higher probability of transiting the ESA. Second, ENAs lose proportionately more energy in the foil at lower energies, so fewer ENAs that would otherwise transit the ESA actually do. Because all ionized ENAs are accelerated by the same energy into the ESA, they are subsequently analyzed at a higher central energy and, importantly, at a broader energy passband. This shift of ionized ENAs to higher energies in the ESA and the wider energy passband at these higher energies therefore enable a higher throughput to compensate for the larger energy loss in the foil at lower incident ENA energies.

The intrinsic geometry of the toroidal ESA plates provides azimuthal focusing throughout the first 90° of deflection and azimuthal defocusing beyond 90°. The detector subsystem is biased to −6 kV to accelerate ionized ENAs into the detector subsystem to counteract this defocusing and also to increase the detection efficiency of the ionized ENAs. Electro-optic simulations of the coupled ESA and detector subsystems optimized the design so that the focal points of ENA trajectories at the central energy of each passband are located near the center of the detector subsystem and along its central axis.

2.4 Detector Subsystem

The detector subsystem consists of three nearly identical, stacked cylindrical chambers, each 5.6 cm in diameter and 2.6 cm tall and each having a CEM detector as shown in Fig. 10. The chambers, designated A, B, and C as sequentially encountered by an ionized ENA, are separated by two thin (nominal 2 μg/cm^2) carbon foils (McComas et al. 2004). The ionized ENAs transit the foils and generate secondary electrons (e.g., Ritzau and Baragiola 1998; Allegrini et al. 2003) at their entrance and exit surfaces. Similarly, the ionized ENAs generate secondary electrons when they impact the aluminum interior walls of the chambers or the back wall of Chamber C. As shown in Fig. 10 secondary electrons generated within one chamber are electrostatically attracted towards the CEM detector of the same chamber by the potential difference between the CEM funnel (−1.7 kV) and the chamber (−6.0 kV). The secondary electrons subsequently generate a pulse in the CEM, and detection of pulses in more than one CEM detector within a prescribed time window is registered as a coincident event.

Penetrating radiation (e.g., >5 MeV electrons or >100 keV photons) can generate a coincidence signal that can be mistaken for a detected heliospheric ENA. To minimize this effect, the CEMs are positioned so that no straight penetrating particle trajectory can go

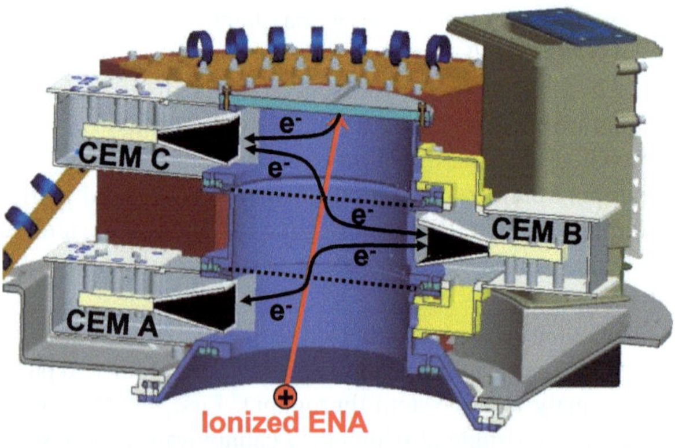

Fig. 10 Ionized ENAs enter the IBEX-Hi detector subsystem and generate secondary electrons (e$^-$) at interior chamber surfaces including the foils between Chambers A and B and between Chambers B and C. These secondary electrons are detected in each chamber, enabling double (two chamber) or triple (three chamber) coincidence detection of a single ENA

through all three CEMs. Thus, a triple coincidence is limited to penetrating particles actually crossing the two carbon foils. Careful characterization of the background in the detectors and detector subsystem is described in Sect. 4 (Calibration and Performance) and in more detail in Wurz et al. (2009).

The Sjuts CEMs have a rectangular funnel with a sensitive area of 1.20 × 2.47 cm. The carbon foils, procured from ACF Metals, are mounted on electro-formed 200 lpi nickel grids with a transmission of ∼78%. The entrance grid of Chamber A is covered by a 70 lpi electroformed nickel grid, with a transmission of ∼90% but no foil.

When a secondary electron is detected by any of the three CEM detectors, short (3 ns) and long (96 ns) electronic coincidence windows are opened. During these intervals the electronics are triggered to accept and record events detected in the other CEMs. At the end of the long coincidence window, 18 unique combinations of events are possible. We note that an ionized ENA may not be detected for at least three reasons: the secondary electron yield from carbon foils is statistical (Poisson) with the non-zero probability that no secondary electron is produced (Gruntman et al. 1990); the electro-optic model shows that not all secondary electrons impact the sensitive area of the CEM detector; and the CEM detection efficiency for electrons at ∼4 keV is ∼70–80% (Paschmann et al. 1970).

The shortest travel time for a 6 keV ENA to traverse the two foils of the detector subsystem is about 17 ns, which is much longer than the short coincidence window duration of 3 ns. Because an ionized ENA enters Chamber A first and reaches Chamber C last, a coincident event with CEM C registering a pulse in the short time window when an event in CEM A is also detected is unlikely to be a real ENA. Therefore, we developed two qualification schemes, discussed in the next section, to prioritize coincidence combinations that are most likely associated with ENAs rather than penetrating radiation or accidental coincidences.

2.5 Electronics

The IBEX-Hi electronics are designed to capture coincident events between any of the three CEM detectors and to record counts from the CEM detector in the IBEX Ion Background Monitor (IBaM) (Allegrini et al. 2009, this issue). The electronics are distributed between the Combined Electronics Unit (CEU) and the IBEX-Hi sensor. Within the IBEX-Hi sensor, high voltage filters eliminate noise and charge amplifier electronics convert CEM charge outputs into digital signals for processing on the Digital Board in the CEU.

The CEU also contains the high voltage power supplies (HVPS) for both sensors that generate and control voltages for the CEM detectors, collimator electrodes, and electro-optic elements. While the HVPS for both IBEX-Lo and IBEX-Hi reside on the same CEU board, their physical layouts are completely separated so that a fault on the HVPS of one sensor will not affect the HVPS performance of the other. The IBEX-Hi HVPS, their maximum and nominal operating voltages, and maximum output currents are listed in Table 3. Except for the Collimator (+) voltage, the high voltages for IBEX-Hi are generated by linearly regulating the output voltage from bulk high voltage supplies, which increases the overall efficiency. The step settling time of the inner ESA plate supply is 0.2 s for a step from −7.0 kV to −0.5 kV and 0.05 s for a step from −0.5 kV to −7.0 kV, both of which are much shorter than the measurement time at a single energy step of two consecutive spacecraft spins.

The high voltages applied to the CEM detectors and the chamber stack are filtered to prevent the introduction of noise in the signal electronics. Each CEM detector has a one-pole low-pass filter with a cut-off frequency of 72 Hz. The coaxial return line for each CEM HVPS has a "zap-trap" of back-to-back diodes and a capacitor to provide a chassis

Table 3 IBEX-Hi high voltage power supplies (HVPS)

High voltage power supply	Max. voltage (kV)	Nominal operating voltage (kV)	Max. current (μA)
Collimator (+)	11.0	10.0	4
Collimator (−)	−4.6	−3.1	4
CEM_A	−4.6	−1.7	23
CEM_B	−4.6	−1.7	23
CEM_C	−4.6	−1.7	23
CEM_D	−4.6	−1.7	23
Detector Chambers	−6.0	−6.0	1
Inner ESA	−7.0	Energy dependent	1
Outer ESA	−0.30	Energy dependent	1
Suppression Grid	−0.30	−0.30	1

return path to dissipate the large filter capacitor in the event of a high voltage discharge, thus protecting the electronics within the sensor. The high voltage to the chamber stack uses a two-pole filter with a cut-off frequency of 127 Hz.

Amptek A121 fast hybrid charge amplifiers convert the electronic pulse outputs from the CEMs into digital pulses. The Amptek A121 was selected because of its relatively low power, voltage-adjustable threshold, and small package. The amplifier electronics were housed in separate, grounded enclosures for each CEM to maximize isolation and therefore minimize crosstalk between the CEMs' electronics chains. The anode output of each CEM is connected to the input of the charge amplifier by an RG-178 coaxial cable that is ~15 cm long. The charge amplifier is protected from high voltage discharge by a 100 Ω carbon-composition resistor that is mounted away from other board components and followed by back-to-back input protection diodes.

The threshold for the A121 is voltage controllable within the range 5×10^4–5×10^6 electrons and commanded through software at a nominal operating value of 3×10^5 electrons. This is well above the analog noise floor, ensuring that no electronic noise can trigger a false coincidence in the measurement, and well below the centroid of the CEM pulse height distribution, ensuring no loss of valid counts. A filtered, differential-receive circuit processes the analog threshold that is generated in the CEU, reducing noise and re-referencing the threshold to the charge amplifier's ground. The output pulse width is set to 75 ns. The dead-time is 525 ns, implying a maximum theoretical periodic throughput for each channel of 1.9 MHz.

The TTL-level output of the charge amplifier is converted to Low Voltage Differential Signal (LVDS) levels before being transmitted to the CEU. This current-steering output produces small voltage swings that are equal and opposite, which minimizes noise on the power and ground lines. This is an important feature which greatly reduces the possibility of cross-talk between channels as well as feedback into the front-end of the amplifiers.

Besides the three CEM detectors (A, B, and C) in the IBEX-Hi detector subsystem, a fourth CEM detector (D) is used in the IBaM. CEM D plays no role in the coincidence measurement and simply accumulates counts to measure the ion background. The detector type

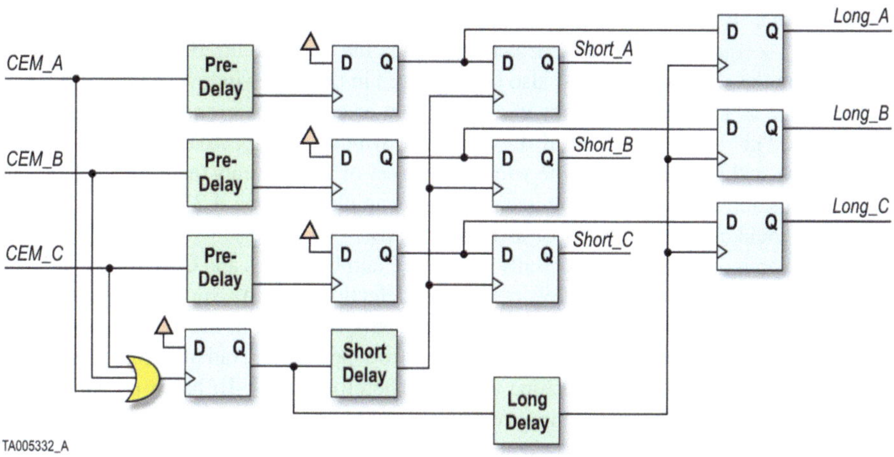

Fig. 11 The coincidence circuit within the CEU FPGA provides a novel polling scheme for capturing whether an event was detected in any of the three CEM detectors at the end of short and long time windows

and analog electronics (including input filtering, zap-trap protection, and charge amplifier board) for CEM D are identical to the other CEM detectors.

Test pulsers enable testing and exercising of the charge amplifier channels and the downstream coincidence circuitry. Each test pulse is equivalent to a CEM pulse magnitude of 6.8×10^6 electrons. The test pulsers are controlled by registers within the CEU field programmable gate array (FPGA). Various pulse patterns and coarse time delays can be generated to simulate all possible coincidence event types, providing tremendous flexibility to test the electronics chain.

The CEU receives the digital signals from CEM detectors A, B, and C and uses a novel polling scheme at the ends of two time windows to classify coincident events rather than more traditional and much more complex time-of-flight (TOF) measurement. As shown in Fig. 11, the digital signals from each CEM are split. The first lines are each input into a 4 ns fixed pre-delay and subsequently into a gated D-latch. When a pulse is detected in any CEM, its individual state is latched "hi" for the duration of the coincidence measurement.

The second lines split from the CEMs are used to poll the state of each CEM at two prescribed times (designated as "short" and "long") using the following method. The CEM outputs are input into an OR gate to register if any CEM detected an event, and the output of the OR gate is latched for the duration of the coincidence measurement. The latch output is then split and input into a "short" time delay (controllable within a range of 1–8 ns in steps of 1 ns) and a "long" time delay (controllable within a range of 30–100 ns in 10 ns steps). The outputs of the time delays are used as the clock inputs, or triggers, for the final gated D-latches to record snapshots of which detectors register an event at the ends of the short and long time windows. The contribution of the pre-delay of 4 ns combined with the nominal (but adjustable) settings for the short and long time delays of 7 ns and 100 ns, respectively, correspond to nominal short and long coincidence windows of 3 ns and 96 ns, respectively.

When an event is detected, the latched outputs of Short_A, Short_B, and Short_C are registered at the end of the short time window, and the latched outputs of Long_A, Long_B, and Long_C are registered at the end of the long time window. If a latched state is "hi", which corresponds to a detected event, then this state is recorded for the appropriate time window and detector. A "hi" event state captured at the end of the short window is designated as lowercase “a”, “b”, or “c” and at the end of the long window is designated as uppercase “A”, “B”, or “C” according to the CEM detector in which the event occurred. Note that an event registered in the short window will also be registered in the long window.

We define a “Long” event as any non-coincident or coincident event that has registered at least one “hi” state at the end of the long time window, irrespective of whether it was registered as a “hi” in the short time window. The set of Long events therefore includes all single events and coincidence combinations. For example, a “Long AB” corresponds to all possible coincident AB events: aAB, bAB, and abAB.

Because an ionized ENA sequentially traverses Chambers A, B, and C, we find that some coincidence combinations more likely result from detection of ENAs, such as an A event first observed in the short time window and a C event only observed in the long time window. Conversely, other combinations more likely result from penetrating radiation, such as both A and C events detected in the short time window. Therefore, IBEX-Hi has two qualification schemes, either of which can be invoked by the CEU for any coincidence combination, to prioritize individual coincident events for the histogram data and the direct event data stream.

The first qualification scheme, denoted as Qual(Not_C), retains all coincident events of a coincidence combination in which the short time window never includes an event in Detector C. For example, “Qual(Not_C) ABC” retains only the triple coincidence events aABC, abABC, and bABC.

The second qualification scheme, denoted as Qual(Not_Equal), retains all events in which the events in the short and long windows are different. For triple coincidences, “Qual(Not_Equal) ABC” includes all triple coincidence events except for abcABC; in another example, “Qual(Not_Equal) AB” includes aAB and bAB but excludes abAB.

We note two additional features of this implementation. First, the pre-delay of 4 ns mentioned above combined with the adjustable short time delay provides a trim range to set a short coincidence window that includes 0 ns and record a snap-shot of the combination of CEM signals that caused the event to start. Second, the coincidence measurement can statistically identify and quantify the abundance of heavy ions in the measurement (Allegrini et al. 2008).

3 Sensor Model

An end-to-end performance model of IBEX-Hi was developed and refined throughout development of the sensor to optimize the sensor design and to simulate and evaluate performance of the subsystems individually and the sensor as a whole. The model is constructed with a combination of analytic modeling, electro-optic simulations, and physics modeling of specific sensor elements. Electro-optic design and simulations were performed using SIMION, a commercially available charged-particle optics simulation package (Dahl 2000).

Testing of IBEX-Hi subsystems and calibration of the fully assembled sensor has been used to validate most components of this model. The calibration results alone do not constitute a complete characterization of IBEX-Hi; in fact, only a limited subset of all possible ion or neutral atom energies, species (H and O), incident angles, and foil locations on the sensor could be tested in the time available. A high-fidelity sensor model allows us to interpolate between these data points to derive an integrated response function and to predict sensor response across the full possible range of operational conditions expected throughout the mission, allowing the flexibility to modify the measurement strategy if needed for discovery science.

The end-to-end model includes a geometric ray-tracing of the collimator, an empirical determination of foil transmission and ionization fraction, and 3-D electro-optic SIMION models of the energy analysis and detector subsystems. The end-to-end simulation is performed by propagating a large number ($N \sim 10^6$) of ENAs distributed in energy (E) and angle (θ, ϕ) through the model for each of the 6 energy settings (j) of the ESA. As the ENAs "fly" through each sensor element (k), they are appropriately propagated in a manner that reflects the physical action of that particular element. At each stage, the transmission efficiency ($T_{j,k} = N^{\mathrm{out}}_{j,k}/N^{\mathrm{in}}_{j,k}$) is determined. Note that T, N^{in}, and N^{out} are all functions of energy, angle, and position. We consider each stage in turn below.

Collimator As previously described, the IBEX-Hi collimator is composed of a stacked array of hexagonal cells that restrict the FOV to 6.5° FWHM. Because the FOV is based solely on the geometry of a hexagonal channel, the angular dependence of the collimator transmission is accurately modeled using the collimator response function $P(\theta, \phi)$, which is illustrated in Fig. 4 and is indistinguishable from the measured performance. The sensor simulation is initiated here, with virtual ENAs uniformly spread across the cell entrance and uniformly distributed in angle up to $\pm 10°$ relative to the collimator boresight.

Conversion Foil Ultrathin carbon foils are used to convert a fraction of the incident ENAs into positive ions that are then energy-analyzed and accelerated into the detector section. Although the desired function of the foil is to simply strip an electron from an ENA, the ENA also undergoes statistical processes of angular scattering, energy loss, and ionization. These are simulated via Monte-Carlo sampling of energy-dependent empirical functions for each of the above factors.

- The scattering angle for a given ENA is determined by sampling a 2-D Lorentzian-squared, having an angular width determined by the ENA's initial energy and laboratory measurement of the foil constant k_F. The use of a Lorentzian-squared is based on empirical best fit to the measured laboratory distribution.
- Energy loss of ENAs at energies >1 keV is based on measurements by Allegrini et al. (2006) and uses an asymmetric Gaussian energy distribution having a mean and width that are functions of foil thickness, ion species, and ion energy. These results were extrapolated for ENA energies <1 keV.
- The probability for ENA transmission through the foil τ_{Foil} was derived using SRIM (Ziegler et al. 1985) Monte-Carlo simulations of protons incident on a 1.7 μg cm^{-2} carbon foil. The results were fit to $\tau_{\mathrm{Foil}} = \exp(-9.594 \times 10^4/E^2)$, where E is in eV.
- The probability of an ENA exiting the foil as H^+ is determined from the formulas in Fig. 9.

Electrostatic Energy Analyzer (ESA) A SIMION electro-optic model of the energy analysis subsystem incorporates the geometry of all relevant electrostatic elements, and the trajectories of H^+ exiting the foil through the ESA are subsequently computed. Figure 12 compares the simulation and calibration (Cal 1) results of the absolute transmission of the combined conversion foil and ESA for each of the six energy steps. The model clearly reproduces the measured behavior, including the energy passband centers and widths, especially for the higher energy passbands. The good correspondence, which is slightly worse at lower energies where the conversion foil behavior is less well known, gives us confidence in our general understanding of the foil physics and ESA electro-optics. We have employed a scaling factor to the modeled transmission values to account for the differences between the simulations and the calibration data; the scaling factors range from 1.45 at energy step 1 to 0.88 at energy step 6.

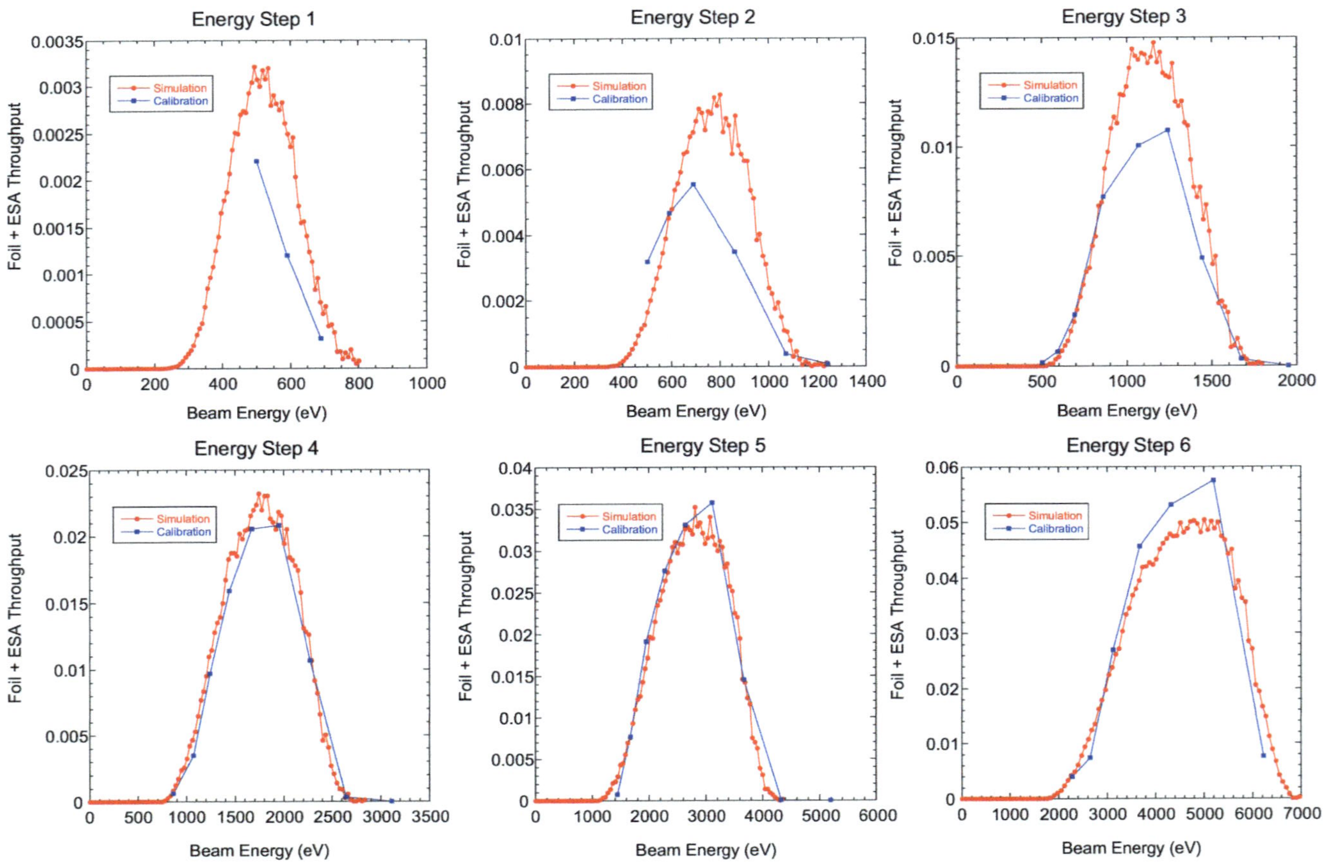

Fig. 12 Comparison of simulations and Cal 1 measurements of the absolute transmission of the combined charge conversion and energy analysis subsystems. Agreement is generally good, although slightly worse at lower energies where the conversion foil behavior is less well known

Detector Subsystem The detector model uses a separate electro-optic model for each chamber and a single carbon foil model for the foils between Chambers A and B and Chambers B and C. Simulated H^+ ions are initially incident at the entrance grid of Chamber A and sequentially propagate through the electro-optic model of Chamber A, the foil model, the electro-optic model of Chamber B, the foil model, and finally the electro-optic model for Chamber C. At the location of ion impact on each foil, the probability for secondary electron emission from each surface is calculated using a Poisson distribution and the measurements of Kozochkina et al. (1993) and Ritzau and Baragiola (1998). These secondary electrons, in turn, are propagated through the appropriate chamber electro-optic model by following their trajectory to the CEM detector, and their detection probability is 0.7. The detector represents the most complex and uncertain component of the end-to-end model, and at the time of the IBEX launch the sensor response function is primarily based on the empirical, scaled detector response that is discussed in Sect. 4 (Calibration and Performance).

4 Calibration and Performance

IBEX-Hi had four phases of calibration: Cal 1, which included calibration without the collimator and photoelectron suppression grids for quantification and electro-optic validation of the integrated performance of the charge conversion and energy analysis subsystems with ion and neutral atom beams; Cal 2, which verified performance for a small subset of energies and foils and was performed between vibration testing and thermal-vacuum testing (Jones and Bernardin 2007); Cal 3, which used the spare CEU that was functionally identical to the flight CEU for detailed, end-to-end performance characterization; and Cal 4, which was cross calibration with IBEX-Lo.

Cals 1, 2, and 3 were performed at Los Alamos National Laboratory (LANL) using a magnetically mass-resolved H^+ beam generated by a microwave ion source having an intrinsic energy spread <2 eV and accelerated to an energy from 0.45 to >10 keV. A neutral hydrogen (H^0) beam was created by charge exchange of the H^+ beam with the residual gas in the beam line between the mass-analyzing magnet and the calibration chamber, and the remaining H^+ in the H^0 beam was magnetically deflected before it could enter the chamber. The beam transited a 2.0-mm-diameter aperture and had a measured diameter at the IBEX-Hi entrance of $\sim$2.2 mm for energies $\geq$1 keV and 2.8 mm at 450 eV. The wider beam at the sensor entrance was due to angular divergence of the beam. The typical chamber pressure was $\leq 2 \times 10^{-7}$ torr, and Residual Gas Analysis (RGA) spectra were obtained at least once each day and every 20 minutes when IBEX was under vacuum but not powered. Cal 4 was performed at SwRI; details of the SwRI calibration facility are provided in McComas et al. (2009a).

In all calibrations, the ion or neutral atom beam flux was measured using an absolute beam flux monitor that is schematically illustrated in Fig. 13. The monitor utilizes a coincidence scheme (Funsten et al. 2005) in which an ion or neutral atom beam generates secondary electrons at the entrance surface of a foil, and these electrons are accelerated to and detected by detector D1. A fraction of the beam transits the foil and generates secondary electrons on the back surface of the foil and at the back plate of the monitor; these secondary electrons are accelerated toward and registered by detector D2. The absolute detection efficiencies for each detector are derived from measurement of the single (non-coincident) count rates C_1 and C_2 of CEM detectors D1 and D2, respectively, and the coincidence count rate C_{COIN} between them. The probability (and therefore detection efficiency) that an ion that enters the monitor is detected by D1 is $\varepsilon_1 = C_{\mathrm{COIN}}/(C_{\mathrm{COIN}} + C_2)$. Similarly, the D2 detection efficiency is $\varepsilon_2 = C_{\mathrm{COIN}}/(C_{\mathrm{COIN}} + C_1)$. Because these detection efficiencies include

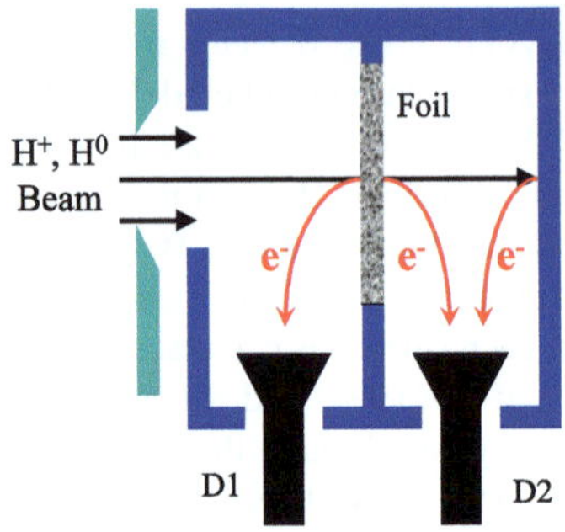

Fig. 13 The absolute beam flux monitor was used to measure the H^+ or H^0 beam flux throughout the IBEX-Hi calibrations

Table 4 Average background count rates for exclusive (non-coincident) single events and a subset of coincidence types measured over 19.6 hours of quiescent operation during cross calibration

Measurement		Background rate
Singles	CEM A	89 mHz
	CEM B	115 mHz
	CEM C	69 mHz
	CEM D	12 mHz
Double coincidences	Long AB	2.8 mHz
	Long BC	4.2 mHz
	Qual(Not_C) AC	0.13 mHz
Triple coincidences	Qual(Not_C) ABC	0.58 mHz
	Long ABC	1.04 mHz

the secondary electron yields, the fraction of the beam that transits the grid-mounted foil, and the probabilities that secondary electrons are accelerated into the detectors and generate pulses, we do not need to know these quantities individually. The resulting coincidence efficiency is therefore $\varepsilon_{\mathrm{COIN}} = \varepsilon_1 \varepsilon_2$, and the incident beam flux is therefore $C_{\mathrm{COIN}}/(A_{\mathrm{B}} \varepsilon_{\mathrm{COIN}})$, where A_{B} is the cross-sectional area of the beam.

The beam flux and beam stability were measured throughout the calibrations. The stability was typically 1–3% over a single IBEX-Hi measurement. The beam was also monitored for drift over time, and typical drift for $>80\%$ of the calibration data was $\leq 5\%$. Importantly, the IBEX-Hi sensor telemetry stream reports single (non-coincident) count rates for each of its three detectors as well as all coincidence count rates, enabling *in situ* monitoring of detection efficiencies of each detection chamber throughout the IBEX mission using this method.

The average background count rates, listed in Table 4, were measured over 19.6 hours during Cal 4 (cross calibration) when the sensor was fully operational but had no incident ion or ENA beam. The singles count rates in each CEM detector were <0.2 Hz. While the coincident count rates are higher than expected based on random coincidence of the background singles rates in the detectors, we have found that ambient gamma rays are associated with a majority of the background coincidence events observed in the detector subsystem. This is based on the relative frequency of different coincidence combinations in the detector subsystem from a ^{137}Cs source placed next to the sensor, from a 3.5 keV H^+ beam, and under a quiescent (unstimulated) condition. We used laboratory TOF electronics to measure the TOF between detected events in all coincidence combinations. While most coincident events from the 3.5 keV H^+ beam involved a first pulse from CEM A and no events in which

CEM C registered the first pulse, both the ^{137}Cs source and quiescent background generated approximately an equal number of coincidences in which CEM A and CEM C registered the first pulse. We note that ^{137}Cs also emits 1.175 MeV electrons that have a range of ~2.3 mm in Al, but these were completely blocked by the Al walls of minimum thickness >2.5 mm surrounding the interior of the detector chambers and CEM detectors.

We subsequently measured the background γ-ray environment in the LANL calibration facilities, showing a background flux of 6.3 γ cm^{-2} s^{-1} between 0.2–3 MeV and specific γ-ray lines corresponding to ^{40}K (K is used in concrete) and daughters associated with ^{222}Rn. Therefore, because a majority of coincident counts listed in Table 4 result from the ambient γ-ray environment of the calibration laboratory, the measured background count rates are significant overestimates of the background rates expected in space from this mechanism.

IBEX-Hi was also tested for response to UV light, in particular to study the "ion gun" effect in which photoelectrons generated at the conversion foil are accelerated toward the +10 kV collimator and ionize ambient atoms or molecules, which in turn are accelerated into the conversion foil and can masquerade as ENAs. During Cal 1, an Ar-purged deuterium lamp, followed by two notch filters used to maximize the fraction of H Ly α (1216 Å) and a MgF_2 window, directly illuminated the conversion foils. The photon rate at the foils was $\sim 4 \times 10^{10}$ s^{-1} as measured using a calibrated UV photodiode (Korde et al. 2003). The ion gun effect was observed when the collimator was biased to +10 kV and no voltage was applied to the photoelectron suppression grid, yielding individual count rates in CEMs A, B, and C of 17, 12, and 1.6 Hz, respectively; a total double coincidence rate of ~0.1 Hz; and a total triple coincidence rate of <0.05 Hz. The double and triple coincidence rates dropped to background levels when the collimator voltage was switched from +10 kV to 0 V.

The IBEX-Hi FOV alignment was measured during Cals 1 and 3 using a theodolite located at a viewport at the rear of the calibration chamber, collinear with the beam axis, and boresighted with two beam-defining apertures upstream of the chamber. A double-sided mirror was mounted flush to the IBEX-Hi baseplate to an estimated accuracy of <0.1°, and the alignment was measured using autocorrelation. The reproducibility of the autocorrelation and movement of the motion stages was ≤0.02°. Alignment accuracy of the theodolite with the upstream apertures is estimated to be ~0.1°. Finally, angular divergence of the H^0 beam was measured to be 0.2° FWHM. Using this alignment scheme, the angular response of the sensor to a 2.7 keV H^0 beam incident on Foil 6a was measured at Energy Passband 5 at several points in the azimuthal and polar directions, and the results are shown in Fig. 14. An angular offset of $0.17 \pm 0.07°$ in polar and $-0.20 \pm 0.07°$ in azimuth is likely due to a combination of the tolerance errors mentioned above and possibly a slight angular dependence of the overall sensor response. After correcting for beam divergence, the collimator acceptance angle was measured to be $6.5 \pm 0.1°$ FWHM, in agreement with the collimator tests described previously.

Using the Cal 1 results of the combined throughput of the conversion foils and energy steps as shown in Fig. 12, the central energies and energy resolutions were derived for double and triple coincidence events and are shown in Fig. 15. The FWHM energy resolution ranges from approximately 0.45 at the lowest energy passband to about 0.65 at the highest energy passband. Although limited to only the charge conversion and energy analysis subsystems (and therefore not representative of the flight energy passbands or the end-to-end IBEX-Hi energy response), this data was used to validate the physics model of the conversion foil and the electro-optic design of the ESA.

The energy-dependent trajectories through the ESA govern the location and angle at which an ionized ENA enters the detector subsystem. Furthermore, there is a known spatial dependence of the detection efficiency for each detector chamber because the CEM detectors

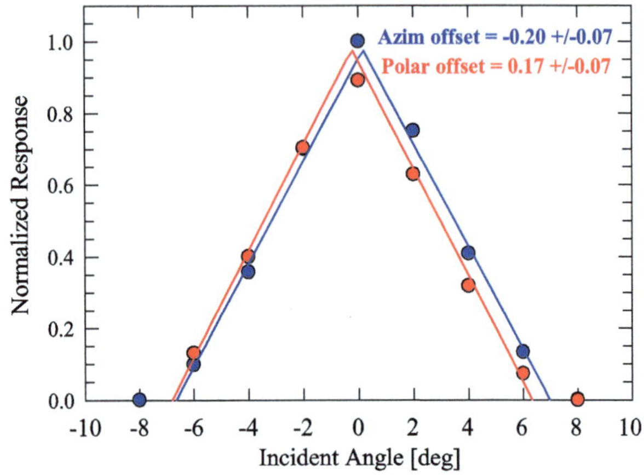

Fig. 14 The IBEX-Hi angular response as measured using a 2.7 keV H^0 beam

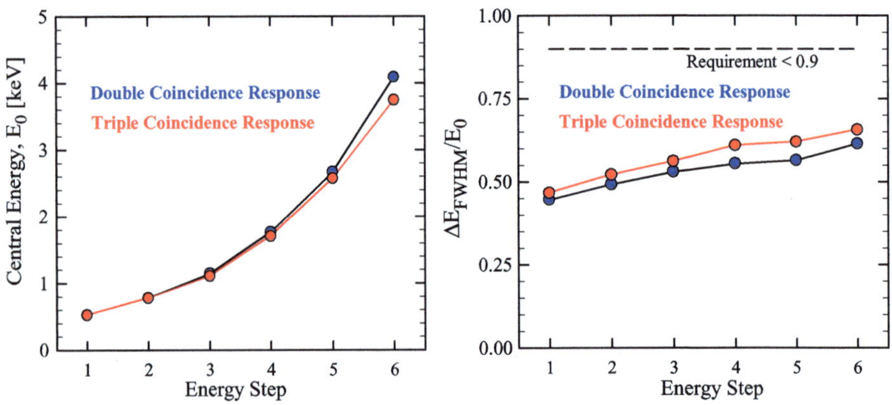

Fig. 15 The central energies (*left panel*) and energy resolutions (*right panel*) of the combined charge conversion and energy analysis subsystems are shown for each energy step as measured during Cal 1. Double coincidences are the sum of Long AB, Long BC, and Qual(Not_C) AC counts. Triple coincidences are Qual(Not_C) ABC counts

are located on one side of each chamber. However, because ionized ENAs are accelerated by 6 keV into the detector subsystem and because of the scalability of electro-optics, we expect that (1) the response for each energy passband can be generally scaled according to E/E_0 where E_0 is the central energy of the passband and E is the initial ENA energy and (2) the detection efficiency for the coincidence combinations are generally independent of energy. Figure 16 shows the detection efficiency for summed double coincidences (defined in the figure caption) as a function of E/E_0 for all passbands and all energies used during Cal 1 and Cal 3. The data for each coincidence combination were fit to $\varepsilon(E/E_0) = c_1 + c_2((E/E_0 + c_3) - c_4)^2$ where c_1, c_2, c_3, and c_4 are fit parameters, and this empirical equation is used as the basis for the detector response in the end-to-end sensor model.

The Cal 3 results show a slightly higher efficiency of the detector subsystem, which likely resulted from new foils installed between Chambers A and B and Chambers B and C. These new foils were mounted using an improved technique and had a higher fraction of coverage over the support grid. To reflect this enhanced detection efficiency, the upper curve in Fig. 16

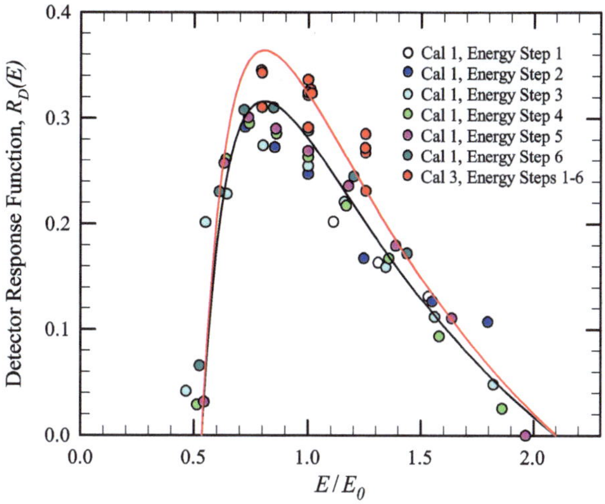

Fig. 16 The empirical detector subsystem response for all energy passbands for double coincidence events is approximately represented by a single equation. Cal 1 results are the sum of Long AB, Long BC and Long AC, whereas the Cal 3 results are the sum of Long AB, Long BC, and Qual(Not_C) AC. The *black line* is an empirical fit to the Cal 1 results, and the *red line* is the Cal 1 empirical equation scaled by a factor of 1.15 to fit the Cal 3 results

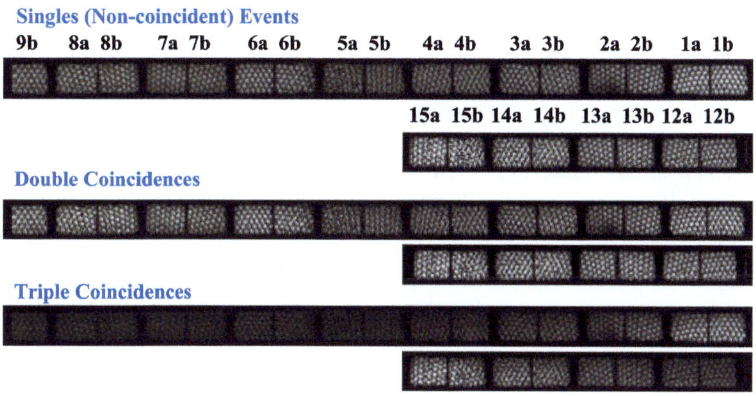

Fig. 17 Response of the IBEX-Hi sensor to a beam of 2.8 keV H^0, where the arc-shaped foil apertures have been flattened such that the horizontal direction corresponds to azimuth. Double coincidences are the sum of Long AB, Long BC, and Qual(Not_C) AC counts. Triple coincidences are Qual(Not_C) ABC counts

was derived by linear scaling of the empirical Cal 1 equation to the Cal 3 results, which were incorporated into the end-to-end sensor response function.

During Cal 3, a 2.8 keV H^0 beam was directed along a snake-like path at a steady rate of 1 mm/sec along a radial line across a foil and through 0.36° azimuthal steps between the radial lines. The incident beam flux was intermittently re-measured using the absolute beam current monitor to monitor beam flux drift. The relative singles (non-coincidence) rates, double coincidence rates, and triple coincidence rates over multiple foils is shown in Fig. 17, in which brighter pixels correspond to higher count rates. The response is observed to change as a function of azimuthal angle (horizontal direction in the figure). This response variation is the direct result of the azimuthal-dependent trajectories through Chambers A, B, and C and, in particular, the location of the CEM within each chamber. The axis of symmetry of the response observed in Fig. 17 and throughout IBEX-Hi calibrations lies along the line between CEM B (corresponding to the middle of Foil 8) and CEMs A and C (located between Foils 1 and 15). Table 5 shows the relative fraction of double and triple coincidence

Table 5 Relative probabilities of occurrence for different double and triple coincidence combinations. Note that coincidence combinations abcABC, acAC, cAC are not included because they are not likely produced by an ENA in the detector subsystem. CEM D is the detector in the IBaM

Triple coincidence event set	Combination	Event probability
Qual(Not_C) ABC	aABC	0.64
	bABC	0.14
	abABC	0.22
Double coincidence event set	**Combination**	**Event probability**
Long AB	aAB	0.18
	bAB	0.03
	abAB	0.06
Qual(Not_C) AC	aAC	0.08
Long BC	bBC	0.49
	cBC	0.07
	bcBC	0.09

events averaged over the energy passbands. The most probable triple and double coincidence events are aABC and bBC, respectively.

Calibration results were used to refine and validate all of the modeled and simulated components of the end-to-end sensor model except for the detector subsystem, which uses the empirical response shown in Fig. 16. The model is then used to interpolate the response for energies, foil locations, and incident angles not measured during calibration and derive the integrated response of the whole sensor. The count rate C_i of IBEX-Hi corresponding to an ENA flux of $J(E,\Omega)$ $\text{cm}^{-2}\,\text{sr}^{-1}\,\text{eV}^{-1}\,\text{s}^{-1}$ at ESA setting i (with central energy E_i) is

$$C_i = \int J(E,\Omega)\, G_i(E,\Omega)\, d\Omega\, dE. \tag{1}$$

The geometric factor G_i is constructed by separating the collimator response $P(\Omega)$ from the comprehensive energy-dependent response $R_i(E)$ of the sensor after the collimator:

$$G_i(E,\Omega) = A T_C\, P(\Omega)\, R_i(E). \tag{2}$$

In this equation, A is the total aperture area of the charge conversion foils (156.6 cm^2) and T_C is the fractional area of the collimator consisting of open apertures (0.67). The point spread function $P(\Omega)$ describes the angular response of IBEX-Hi, which, to a high degree of accuracy, can be represented solely as the energy-independent collimator response shown in Fig. 4. Because the incident ENA flux is expected to be relatively constant over the solid angle of any single azimuthal pixel, the integrated collimator response corresponds to a solid angle of $\int P(\Omega)\, d\Omega = 0.0147$ sr.

The sensor response function $R_i(E)$ describes the cumulative, energy-dependent response to ENAs that exit the collimator and are incident within the total area of the charge conversion foils. $R_i(E)$ therefore includes all factors governing transmission and transport through the charge conversion, energy analysis, and detector subsystems, as well as the probability of detection in each CEM chamber. The response function was generated using the end-to-end sensor model, and the results for double and triple coincident events are shown in Fig. 18. The primary uncertainty of the sensor response function lies in the empirical detector subsystem response function. Using the end-to-end response function, the flight energy passband settings have been defined as shown in Table 6.

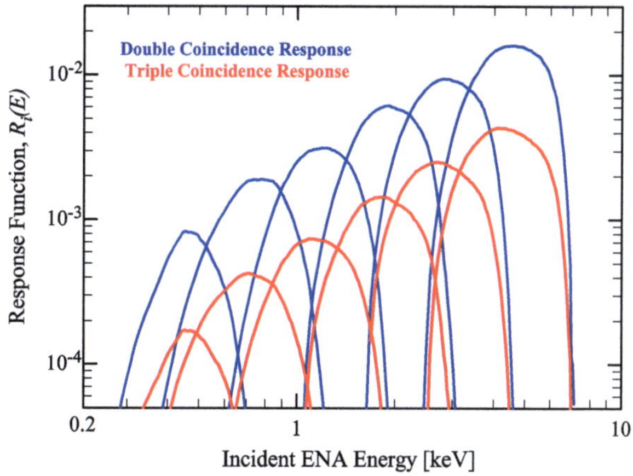

Fig. 18 The response function $R_i(E)$ of the IBEX-Hi sensor is shown for each energy passband i. The double coincidence response includes the sum of Long AB, Long BC, and Qual(Not_C) AC events. The triple coincidence response corresponds to Qual(Not_C) ABC events

Table 6 Flight energy passbands for double and triple coincidence events as derived from the sensor end-to-end model for each of the six IBEX-Hi energy steps. The full-width-at-full-maximum values ($E_{+FW} - E_{-FW}$) represent the 99% level and central energy E_0 represents the energy of the maximum (peak) value of the response function

ESA setting			Double coincidences: Long AB + Long BC + Qual(Not_C) AC					
ESA Step	Outer ESA [kV]	Inner ESA [kV]	E_{-FW} [keV]	E_{-HW} [keV]	E_0 [keV]	E_{+HW} [keV]	E_{-FW} [keV]	$\Delta E/E_0$ [keV]
1	−0.09	−0.78	0.26	0.38	0.45	0.58	0.75	0.20
2	−0.08	−1.28	0.36	0.57	0.78	0.98	1.25	0.41
3	−0.02	−1.85	0.62	0.89	1.21	1.56	1.91	0.68
4	0	−2.88	1.07	1.45	1.92	2.53	3.02	1.08
5	0	−4.46	1.68	2.15	2.80	3.82	4.54	1.67
6	0	−7.00	2.57	3.34	4.51	6.14	6.93	2.80
ESA setting			Triple coincidences: Qual(Not_C) ABC					
ESA Step	Outer ESA [kV]	Inner ESA [kV]	E_{-FW} [keV]	E_{-HW} [keV]	E_0 [keV]	E_{+HW} [keV]	E_{-FW} [keV]	$\Delta E/E_0$ [keV]
1	−0.09	−0.78	0.24	0.38	0.45	0.59	0.75	0.47
2	−0.08	−1.28	0.35	0.52	0.71	0.95	1.23	0.60
3	−0.02	−1.85	0.58	0.84	1.08	1.55	1.93	0.65
4	0	−2.88	1.07	1.36	1.85	2.50	3.02	0.62
5	0	−4.46	1.68	1.99	2.70	3.75	4.54	0.65
6	0	−7.00	2.57	3.13	4.09	6.00	6.93	0.70

Based on simulations and recent estimates of ENA fluxes, the spectral slope of ENAs should sharply decrease with increasing energy above ∼1 keV (e.g., Gruntman et al. 2001; Wurz et al. 2008). Therefore, derivation of the incident ENA flux $J(E,\Omega)$ from the sensor count rate C_i through inversion of (1) should be performed using an iterative, forward-modeling method. Nevertheless, the energy geometric factor, defined as

$$G_i^E = \frac{1}{E_i}\int G_i(E,\Omega)\,d\Omega\,dE, \tag{3}$$

Table 7 Values of the energy geometric factor G^E (cm^2 sr eV/eV) derived using (3) for the flight ESA passbands

ESA step	G^E, double coincidence: Long AB + Long BC + Qual(Not_C) AC	G^E, triple coincidence: Qual(Not_C) ABC
1	0.00060	0.00013
2	0.0016	0.00041
3	0.0028	0.00075
4	0.0053	0.0013
5	0.0086	0.0024
6	0.015	0.0045

can be used to estimate the ENA flux near energy E_i using $J(E_i) = C_i/E_i G_i^E$. The energy geometric factors for double and triple coincidence events are listed in Table 7 for each of the six IBEX-Hi energy passbands.

5 Sensor Operations and Data Handling

IBEX-Hi operates in science mode over the portion of the IBEX orbit when the spacecraft is above 15 R_E, which is about 93% of the ~8-day orbital period. In this mode, all high voltages are at nominal operating levels, and the ESA voltages are stepped once every two spacecraft spins. A complete energy sweep over all six passbands is therefore completed every 12 spins, which corresponds to 3 min based on a nominal spin rate of 4 rpm.

In science mode, IBEX-Hi reports two data products: (1) histograms of accumulated counts as a function of angle, energy step, and coincidence type, and (2) direct event data consisting of individual coincidence events selected using a prioritization scheme and reported at high time (and therefore angular) resolution. The telemetry rate is 42 bps for histogram data, 57 bps for direct event data, and 12 bps for housekeeping that is shared with IBEX-Lo.

Histograms are constructed from accumulated counts from coincident and non-coincident (single) events and sorted by event type. Each histogram is a 6 × 60 array of counts for the six energy passbands and 60 angular bins around the spin axis, each 6° wide. The angular bins of eight primary event type histograms are summed over four ESA cycles (12 minutes); another three secondary histograms are summed over eight ESA cycles. The counts in each angular bin are compressed from 16 bits to 8 bits and stored in the solid-state recorder (SSR) for downloading during the next IBEX perigee pass.

The primary histograms include non-coincident (singles) counts in each detector (Long A, Long B, Long C), four double coincident types (Long AB, Qual AB, Qual AC, and Qual BC), and one triple coincident type (Qual ABC). The secondary histograms include Long AC, Long BC, and Long ABC. The qualification scheme, either Not_C or Not_Equal, can be independently selected for each of the qualified coincidence event types. The histogram data also includes accumulated counts from the IBaM as a function of spin angle.

Direct event data consists of individual coincidence events whose coincidence type is reported along with the spin angle and energy passband at the time that the event was detected. Direct events are recorded at higher time resolution (4.2 ms, or ~0.1° of spin angle) than used for acquisition of the histograms. Because of telemetry constraints, IBEX-Hi is limited to recording ~800 direct events per 12-spin cycle, which corresponds to an average of ~67 events per spin. Therefore, a culling process is used by the CEU to telemeter only the highest priority direct events. The prioritization scheme, which lies in a changeable look up table in the CEU, currently has Qual(Not_C) ABC as the highest priority because these triple coincidences represent the measurement with the lowest noise. These high priority events are stored up to the limit of ~800 events per 12 spins. If space remains, the second priority events, which currently include Long AB, Long BC, and Qual(Not_C) AC, are included in the direct event data. No other event types are stored as direct event data. Because direct events are recorded at greater time resolution and greater coincidence type specificity than the histograms (for example, the qualified-triples histogram includes three different types of direct events), they represent key data for deeper analysis of the global ENA maps.

IBEX-Hi has three other operational modes: high voltage standby mode, gain test mode, and functional test mode. High voltage standby mode is entered every orbit when the spacecraft passes below 15 R_E as it approaches the radiation belts. At these low altitudes, the magnetospheric ENA signal can be significantly greater than that of heliospheric ENAs, and detector background from penetrating radiation may also be substantial. Upon entering this mode, the collimator voltages are turned off, the voltage applied to the detector chambers is reduced from −6 kV to −4 kV, and the voltage across each CEM is decreased by 0.6 kV, which is below the threshold for electron avalanche in the CEM.

Gain tests are run periodically to ensure that the CEMs are operating on the plateaus of their respective gain curves. During a gain test, each CEM voltage is alternately stepped between its current operating voltage and offsets of −200 V, −100 V, +100 V and +200 V from operating voltage. Because the heliospheric ENA signal rate is quite low, gain tests are performed at ~12 R_E to utilize the stronger magnetospheric ENA signal in some of the magnetospheric-viewing azimuthal pixels, resulting in anticipated ENA count rates of ~10 s^{-1} in these pixels. Based on these count rates, the planned integration time per energy passband setting is ~12 minutes. The scheme of returning to the current operating voltage before proceeding to the next voltage offset minimizes the impact of natural fluctuations in the ENA rate on the test results. If a gain test shows that a CEM detector gain has decreased, the gain will be increased by an appropriate increase of the CEM operating voltage.

The electronics functional test is periodically executed to assess the health of the IBEX-Hi amplification and signal processing electronics. This is a shortened and automated version of the electronics checkout performed throughout calibration and during commissioning. Specific diagnostics include determination of the electronic noise floor discriminator setting and the lower level discriminator below which crosstalk between separate detectors is observed.

6 Summary

The IBEX-Hi imager is a sensitive, single pixel ENA imaging camera with a FOV of 6.5° with six contiguous energy passbands spanning an energy range of ~0.38–6 keV. Because of the anticipated tenuous signal at 1 AU of ENAs from the interaction region between the heliosphere and the LISM, the IBEX-Hi sensor design has been driven by maximizing sensitivity and minimizing background and noise, resulting in the most sensitive ENA sensor ever flown over its energy range. An end-to-end IBEX-Hi sensor performance model has been developed with most components validated through extensive testing and calibration. This model allows accurate quantification of the cumulative energy response and sensitivity of the sensor and enables the flexibility to modify instrument operations to optimize our measurement of new discoveries as they arise. The sensor meets the measurement requirements for acquiring global ENA maps that will allow us to answer the key scientific questions regarding the structure and dynamics of the interaction between the heliosphere and the LISM. Although focused primarily on measurement of heliospheric ENAs, IBEX-Hi will also allow us to view and study regions of the terrestrial magnetosphere and possibly other planetary magnetospheres that have never been observed before in ENAs.

Acknowledgements The IBEX-Hi sensor is the result of an outstanding collaboration between scientists, engineers, technicians, and support personnel at multiple institutions, including Jeff Christmas and Lt. Brett Ryder of Air Force Research Lab (AFRL), Kirtland AFB; Charlie Briggs, Greg Ellison, Tracy Gerber, Marty Leahy, Charlotte Ogas, and Dan Wright of ATA Aerospace; Eric Stasiunas and Alex Pimatel at Sandia National Laboratories; Patsy Archuleta, Alan Baca, Sean Apgar, Randy Edwards, John Bernardin, and David Jones at Los Alamos National Laboratory (LANL); Chris Kofoed, Bob Harbaugh, Wendy Mills, Mike Young, Rob Ebert, Joe Westlake, Josh Alquiza, Tom Broiles, Brent Randol, and Roberto Livi at SwRI; and Georg Bodmer and Jürgen Scheer at the University of Bern. Work at Los Alamos National Laboratory was performed under the auspices of the United States Department of Energy.

References

F. Allegrini, R.F. Wimmer-Schweingruber, P. Wurz, P. Bochsler, Measurement of the ion-induced electron yields from thin carbon foils for low energy ions. Nucl. Instr. Meth. B **211**, 487 (2003)

F. Allegrini, D.J. McComas, D.T. Young, J.-J. Berthelier, J. Covinhes, J.-M. Illiano, J.-F. Riou, H.O. Funsten, R.W. Harper, Energy loss of 1–50 keV H, He, C, N, O, Ne, Ar ions transmitted through thin carbon foils. Rev. Sci. Instrum. **77**, 044501 (2006)

F. Allegrini, R.W. Ebert, J. Alquiza, T. Broiles, C. Dunn, D.J. McComas, I. Silva, P. Valek, J. Westlake, A mass analysis technique using coincidence measurements from the IBEX-Hi detector. Rev. Sci. Instrum. **79**, 096107 (2008)

F. Allegrini, G.B. Crew, D. Demkee, H.O. Funsten, D.J. McComas, B. Randol, B. Rodriguez, P. Valek, S. Weidner, The IBEX background monitor. Space Sci. Rev. (2009, this issue)

D.A. Dahl, SIMION for the personal computer in reflection. Int. J. Mass Spectrom. **200**, 3 (2000)

H.O. Funsten, D.J. McComas, B.L. Barraclough, Thickness uniformity and pinhole density analysis of thin carbon foils using keV ions. Nucl. Instrum. Meth. B **66**, 470 (1992)

H.O. Funsten, D.J. McComas, B.L. Barraclough, Ultrathin foils used for low energy neutral atom imaging of planetary magnetospheres. Opt. Eng. **32**, 3090 (1993)

H.O. Funsten, D.J. McComas, M.A. Gruntman, Energetic neutral atom imaging of the outer heliosphere–LIC interaction region, in *The Outer Heliosphere: The Next Frontiers*, ed. by K. Scherer, H. Fichtner, H.-J. Fahr, E. Marsch. COSPAR Colloquia Series, vol. 11 (Pergamon, New York, 2001), pp. 237–244

H.O. Funsten, R.W. Harper, D.J. McComas, Absolute detection efficiency of space-based ion mass spectrometers and neutral atom imagers. Rev. Sci Instrum. **76**, 053301 (2005)

S.A. Fuselier, A.G. Ghielmetti, E. Hertzberg, A. Moore, D. Isaac, J. Hamilton, C. Tillier, E. Moebius, M. Granoff, D. Heirtzler, B. King, H. Kucharek, S. Longworth, J. Nolin, S. Turco, P. Wurz, M. Wieser, J. Scheer, L. Saul, C. Schlemm, D.J. McComas, D. Chornay, J. Lobell, T. Moore, P. Rosmarynowski, R.J. Nemanich, T. Friedmann, H. Funsten, The IBEX-Lo sensor for the IBEX mission. Space Sci. Rev. (2009, this issue)

G. Gloeckler, J. Geiss, Composition of the local interstellar medium as diagnosed with pickup ions. Adv. Space Res. **34**, 53 (2004)

M.A. Gruntman, A.A. Kozochkina, V.B. Leonas, Multielectron secondary emission from thin foils bombarded by accelerated beams of atoms. JETP Lett. **51**, 22 (1990)

M.A. Gruntman, Energetic neutral atom imaging of space plasmas. Rev. Sci. Instrum. **68**, 3617 (1997)

M.A. Gruntman, E.C. Roelof, D.G. Mitchell, H.-J. Fahr, H.O. Funsten, D.J. McComas, Energetic neutral atom imaging of the heliospheric boundary region. J. Geophys. Res. **106**, 15767 (2001)

M.G. Henderson, G.D. Reeves, H.E. Spence, R.B. Sheldon, A.M. Jorgensen, J.B. Blake, J.F. Fennell, First energetic neutral atom images from polar. Geophys. Res. Lett. **24**, 1167 (1997)

D.C. Jones, J.D. Bernardin, Thermal modeling and experimental verification of the interstellar boundary explorer's high energy neutral atom imaging instrument (IBEX-Hi), 2007 AIAA InfoTech at Aerospace Conference (2007), vol. 3, pp. 2364–2375

R. Korde, C. Prince, D. Cunningham, R.E. Vest, E. Gullikson, Present status of radiometric quality silicon photodiodes. Metrologia **40**, S145 (2003)

A.A. Kozochkina, V.B. Leonas, V.E. Fine, Statistics of heavy particle-induced electron emission from a foil, in *Ionization of Solids by Heavy Particles*, ed. by R.A. Baragiola (Plenum, New York, 1993), pp. 223–237

D.J. McComas, F. Allegrini, C.J. Pollock, H.O. Funsten, S. Ritzau, G. Gloeckler, Ultra-thin (~10 nm) carbon foils in space instrumentation. Rev. Sci. Instrum. **75**, 4863 (2004)

D.J. McComas, F. Allegrini, L. Bartolone, P. Bochsler, M. Bzowski, M. Collier, H. Fahr, H. Fichtner, P. Frisch, H. Funsten, S. Fuselier, G. Gloeckler, M. Gruntman, V. Izmodenov, P. Knappenberger, M. Lee, S. Livi, D. Mitchell, E. Möbius, T. Moore, S. Pope, D. Reisenfeld, E. Roelof, H. Runge, J. Scherrer, N. Schwadron, R. Tyler, M. Wieser, M. Witte, P. Wurz, G. Zank, The Interstellar Boundary Explorer (IBEX): Update at the End of Phase B, AIP Conference Proceedings, vol. 858, *Physics of the Inner Heliosheath: Voyager Observations, Theory, and Future Prospects* (2006), pp. 241–250

D.J. McComas, F. Allegrini, J. Baldonado, B. Blake, P.C. Brandt, J. Burch, J. Clemmons, W. Crain, D. Delapp, R. DeMajistre, D. Everett, H. Fahr, L. Friesen, H. Funsten, J. Goldstein, M. Gruntman, R. Harbaugh, R. Harper, H. Henkel, C. Holmlund, G. Lay, D. Mabry, D. Mitchell, U. Nass, C. Pollock, S. Pope, M. Reno, S. Ritzau, E. Roelof, E. Scime, M. Sivjee, R. Skoug, T.S. Sotirelis, M. Thomsen, C. Urdiales, P. Valek, K. Viherkanto, S. Weidner, T. Ylikorpi, M. Young, J. Zoennchen, The Two Wide-angle Imaging Neutral-atom Spectrometers (TWINS) NASA mission-of-opportunity. Space Sci. Rev. **142**, 157 (2009a)

D.J. McComas, F. Allegrini, P. Bochsler, M. Bzowski, M. Collier, H. Fahr, H. Fichtner, P. Frisch, H. Funsten, S. Fuselier, G. Gloeckler, M. Gruntman, V. Izmodenov, P. Knappenberger, M. Lee, S. Livi, D. Mitchell, E. Möbius, T. Moore, S. Pope, D. Reisenfeld, E. Roelof, J. Scherrer, N. Schwadron, R. Tyler, M. Wieser, M. Witte, P. Wurz, G. Zank, IBEX–Interstellar boundary explorer. Space Sci. Rev. (2009b, this issue)

D.G. Mitchell, S.E. Jaskulek, C.E. Schlemm, E.P. Keath, R.E. Thompson, B.E. Tossman, J.D. Boldt, J.R. Hayes, G.B. Andrews, N. Paschalidis, D.C. Hamilton, R.A. Lundgren, E.O. Tums, P. Wilson, H.D. Voss, D. Prentice, K.C. Hsieh, C.C. Curtis, F.R. Powell, High energy neutral atom (HENA) imager for the IMAGE mission. Space Sci. Rev. **91**, 67 (2000)

H. Moestue, The electric field and geometrical factor of an annular curved plate electrostatic analyzer. Rev. Sci. Instrum. **44**, 1709 (1973)

T.E. Moore, D.J. Chornay, M.R. Collier, F.A. Herrero, J. Johnson, M.A. Johnson, J.W. Keller, J.F. Laudadio, J.F. Lobell, K.W. Ogilvie, P. Rozmarynowski, S.A. Fuselier, A.G. Ghielmetti, E. Hertzberg, D.C. Hamilton, R. Lundgren, P. Wilson, P. Walpole, T.M. Stephen, B.L. Peko, B. Van Zyl, P. Wurz, J.M. Quinn, G.R. Wilson, The low-energy neutral atom imager for IMAGE. Space Sci. Rev. **91**, 155 (2000)

G. Paschmann, E.G. Shelley, C.R. Chappell, R.D. Sharp, L.F. Smith, Rev. Sci. Instrum. **41**, 1706 (1970)

C.J. Pollock, K. Asamura, J. Baldonado, M.M. Balkey, P. Barker, J. Burch, E.J. Korpela, J. Cravens, G. Dirks, M.-C. Fok, H.O. Funsten, M. Grande, M. Gruntman, J. Hanley, J.-M. Jahn, M. Jenkins, M. Lampton, M. Marckwordt, D.J. McComas, T. Mukai, G. Penegor, S. Pope, S. Ritzau, M.L. Schattenburg, E. Scime, R. Skoug, W. Spurgeon, T. Stecklein, S. Storms, C. Urdiales, P. Valek, J.T.M. van Beek, S.E. Weidner, M. Wüest, M.K. Young, C. Zinsmeyer, Medium Energy Neutral Atom (MENA) imager for the IMAGE mission. Space Sci. Rev. **91**, 113 (2000)

F.H. Read, N.J. Bowring, P.D. Bullivant, R.R.A. Ward, Penetration of electrostatic fields and potentials through meshes, grids, or gauzes. Rev. Sci. Instrum. **69**, 2000 (1998)

S.M. Ritzau, R.A. Baragiola, Electron emission from carbon foils induced by keV ions. Phys. Rev. B **58**, 2529 (1998)

E.C. Roelof, D.G. Mitchell, D.J. Williams, Energetic neutral atoms ($E \approx 50$ keV) from the ring current: IMP 7/8 and ISEE 1. J. Geophys. Res. **90**, 991 (1985)

E.G. Shelley, A.G. Ghielmetti, H. Balsiger, R.K. Black, J.A. Bowles, R.P. Bowman, O. Bratschi, J.L. Burch, C.W. Carlson, A.J. Coker, J.F. Drake, J. Fischer, J. Geiss, A. Johnstone, D.L. Kloza, O.W. Lennartsson, A.L. Magoncelli, G. Paschmann, W.K. Peterson, H. Rosenbauer, T.C. Sanders, M. Steinacher, D.M. Walton, B.A. Whalen, D.T. Young, The toroidal imaging mass-angle spectrograph (TIMAS) for the Polar mission. Space Sci. Rev. **71**, 497 (1995)

O.L. Vaisberg et al., Complex plasma analyzer SKA-1, in *Interball: Mission and Payload*, ed. by A.A. Galeev (Russian Space Agency–French Space Agency, 1995), pp. 170–177

P. Wurz, A. Galli, S. Barabash, A. Grigoriev, Energetic Neutral Atoms from the Heliosheath. Astrophys. J. **683**, 248 (2008)

P. Wurz, J. Scheer, L. Saul, S.A. Fuselier, A.G. Ghielmetti, E. Hertzberg, E. Möbius, H. Kucharek, H.O. Funsten, P. Janzen, R. Harper, M. Wieser, P.C. Brandt, E. Roelof, D. McComas, F. Allegrini, IBEX Backgrounds and signal to noise ratio. Space Sci. Rev. (2009, this issue)

J.F. Ziegler, J.P. Biersack, U. Littmark, *Stopping and Range of Ions in Solids*, vol. 1 (Pergamon, New York, 1985)

Space Sci Rev (2009) 146: 105–115
DOI 10.1007/s11214-008-9439-8

The IBEX Background Monitor

F. Allegrini · G.B. Crew · D. Demkee · H.O. Funsten · D.J. McComas · B. Randol · B. Rodriguez · N.A. Schwadron · P. Valek · S. Weidner

Received: 30 April 2008 / Accepted: 13 September 2008 / Published online: 10 October 2008

Abstract The IBEX Background Monitor (IBaM) provides a small and lightweight method for independently measuring IBEX's high-energy proton background by integrating the flux of $>\sim 14$ keV protons over a $\sim 7^\circ$ conical FOV. The IBaM is part of the IBEX-Hi sensor and has a co-aligned look direction. This paper describes the principle of the IBaM and details its design and responses. In particular, we show the response of major components to both ions and ultraviolet (UV) light background. We also provide the geometric factor and calculate expected count rates.

Keywords IBEX background · Channel electron multiplier · Carbon foil · Upstream event

1 Introduction

The Background Monitor (IBaM) for the IBEX mission (see McComas et al. 2008) quantifies the background environment by making an integral measurement of $>\sim 14$ keV protons, just above the positive 10 kV potential of the Hi and Lo collimator/entrance subsystems. The IBaM is part of the IBEX-Hi sensor (Funsten et al. 2008) and is located close to the detector section. The IBaM and the IBEX-Hi detector section use identical detectors (channel-electron multiplier, CEM) and share a similar electronics design. The look direction of the IBaM is aligned with that of the IBEX-Hi sensor, with a comparable field-of-view ($\sim 7^\circ$ FWHM).

F. Allegrini (✉) · D. Demkee · D.J. McComas · B. Randol · B. Rodriguez · P. Valek · S. Weidner
Southwest Research Institute, San Antonio, TX, USA
e-mail: fallegrini@swri.edu

G.B. Crew
MIT Kavli Institute for Astrophysics and Space Research, Cambridge, MA, USA

H.O. Funsten
Los Alamos National Laboratory, Los Alamos, NM, USA

N.A. Schwadron
Boston University, Boston, MA, USA

Due to their extreme sensitivity the IBEX sensors will also see numerous sources of noise and background that are described in Wurz et al. (2008). Among these sources, energetic neutral atoms (ENAs) can be created by 1) charge exchange of plasma ions with outgassing species from the spacecraft or 2) by scattering of high energy ions off the edges of a collimator grid; such ions are a particular concern in the magnetosheath and foreshock regions. The foreshock region refers to the region beyond but magnetically connected to the Earth's bow shock. Locally produced ENAs could then masquerade as real ENAs from the termination shock. For both of these sources, the neutral yield is a function of the ambient ion flux, which can be measured with the IBaM.

By design, the IBEX team sought to minimize the effect of the outgassing species from the spacecraft by venting the internal volumes of the spacecraft only along the directions of the spin axis. We will also segregate the observations when the IBEX spacecraft look direction is in the solar wind, magnetosheath, and foreshock regions (see Schwadron et al. 2008). The IBaM will help identify and quantify time periods with potentially high backgrounds from the local energetic protons. This information is valuable for understanding IBEX backgrounds by directly measuring ions with energies in the range that can penetrate through the +10 kV potential on the collimator plates.

Sanderson et al. (1996) show an example of an energetic particle event upstream of the Earth's bow shock. This particular event was observed by instruments on the Wind spacecraft when it was in the solar wind. The spectrum of upstream events usually follows a power law with power index around −4. Such events will be identified by the IBaM, making it straightforward to remove them from the IBEX observations.

2 Background Monitor Principle

The IBaM makes an integral measurement of protons above ~14 keV in a 7° FWHM field-of-view aligned with the IBEX-Hi viewing direction. The principle is illustrated schematically in Fig. 1. With energies above ~14 keV (in red), protons with the proper incident direction will go through the collimator and cross two thick carbon foils (21 and 22 μg/cm^2 for CF1 and CF2, respectively). The protons will be detected with a channel electron multiplier (CEM). The CEM is biased so that the top of the funnel is at about −1700 V, and the anode, where the electron avalanche is collected, is at ground. This bias design allows us to use an almost identical design for the electronics as is used on the IBEX-Hi detector (see Funsten et al. 2008). The grid is attached to the CEM funnel to ensure that the secondary electrons (in green) created in the funnel are accelerated toward the throat of the channel rather than back toward the carbon foils, which are at ground. Most of the protons with less than ~14 keV (in light blue) will not be able to cross both carbon foils because of the energy straggling in the foils. The foils are also thick enough to significantly attenuate the UV radiation (in purple), to which CEM detectors are moderately sensitive. Carbon foils can have pinholes and thickness variations (Funsten et al. 1992a, 1992b). Because of the potential UV background from a pinhole in a single carbon foil, we chose to use two. The likelihood that pinholes in two foils are aligned is extremely low. Thus, the impact of pinholes in the carbon foils is mitigated.

Figure 2 shows a cross section (bottom right) of the IBaM with detailed pictures of the main components. The CEM grid (bottom left) is affixed onto the MCP with conductive epoxy EPO-TEK H21D, which is compatible with CEMs' operation and lifetime (McComas and Bame 1984). The CEM is attached to a ceramic board where the high voltage and signal pickup connections are made.

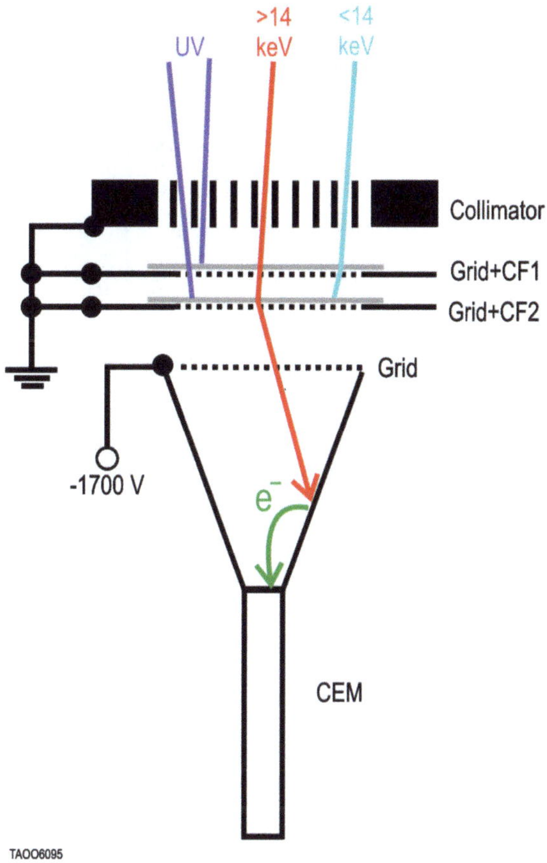

Fig. 1 Principle of the background monitor

Light tightness, required to ensure that stray light does not produce a background in the IBaM, is achieved at the level of the collimator and carbon foils and at the bottom of the IBaM housing. A spacer is placed between the carbon foils with grooves and holes to provide venting for the small volume between them but a tortuous path for UV photons. The bottom part of the IBaM has a removable baffle (dark green) also for venting through the channel on the right of the IBaM housing. This channel also provides a venting path for the electrostatic analyzer and detector subsystems of IBEX-Hi. The channel, removable baffle, and spacer between the carbon foils are blackened using the Ebonal-C process. The most direct access for UV photons to the CEM is through the collimator and the carbon foils.

The carbon foils (dark area on the top left picture) are mounted on electroformed nickel grids that have 70 lines per inch for a transmission of 90%. The grids are formed with a 0.005-inch-thick rim, which gives them enough stiffness to be handled without an additional fixture. The carbon foils were manufactured by Arizona Carbon Foils.

The collimator (yellow part in the cross section image) consists of a 2.5-mm-thick aluminum plate with drilled holes in a close packed pattern. The hole diameter is 0.343 mm, and the spacing between the hole centers is 0.394 mm. There are 2593 holes on a 13-mm × 26-mm area. A detailed electron microscope picture of the collimator holes is shown in the upper right of Fig. 2. The collimator was electro-polished to remove burrs from machining. The collimator, the grids supporting the carbon foils, and the front of the CEM funnel all have the same aperture area: 13 mm × 26 mm.

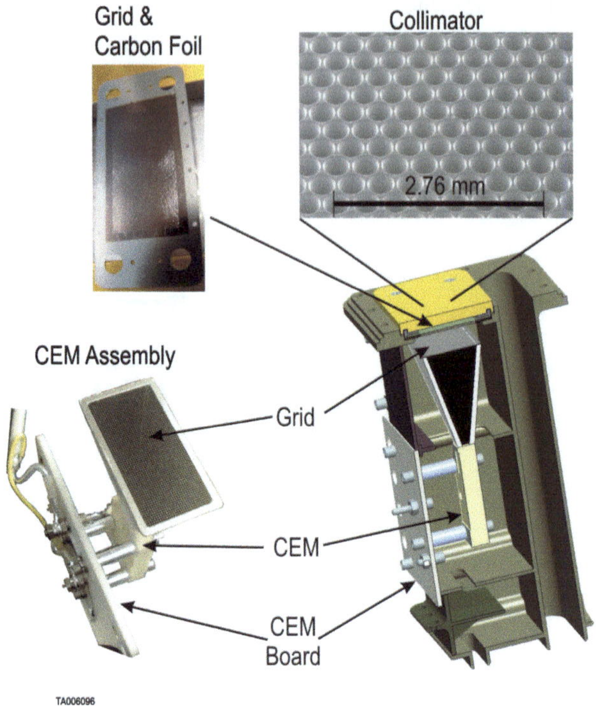

Fig. 2 Engineering cross section of the background monitor and detailed pictures of the major components

3 Background Monitor Response

3.1 Collimator Field-of-View

Measurement of the IBaM field-of-view (FOV) was performed with a 2-mm diameter, 30 keV H^+ beam aimed at the center of the collimator. The IBaM was oriented using two orthogonal rotational actuators (providing rotation angles alpha and beta), and the rate was recorded for each orientation. The result of the angular scan with a resolution (pixel size) of $1° \times 1°$ is illustrated in Fig. 3 where the rate is color-coded and scaled to a maximum of 100. The maximum intensity or the center of the FOV corresponds to normal incidence on the collimator. This direction is found by calculating the weighted average of the rate as a function of angles. By 1) reporting the rate as a function of angular distance between the center of the FOV and each pixel (angle theta) and 2) binning these data into 1° bins, we find the transmission function depicted in Fig. 4. In other words, the data from Fig. 3 of the 3-D angular scan is collapsed and displayed in the form of a histogram in Fig. 4.

The orange curve in Fig. 4 is a fit to the data using the following function

$$T(\theta) = \frac{2}{\pi}\left(\arccos\left(\frac{h}{d}\tan\theta\right) - \frac{h}{d}\tan\theta\sqrt{1-\left(\frac{h}{d}\tan\theta\right)^2}\right) \tag{1}$$

which models the transmission of a cylindrical collimator. The FWHM derived from the fit in Fig. 4 corresponds to a height-to-diameter ratio $h/d = 6.67178$ of the holes in the collimator. The maximum angle for transmission from the center of this FOV is found when the square root in (1) is equal to 0: $\theta_{\max} = \arctan(d/h) \cong 8.52°$. The FOV of the collimator

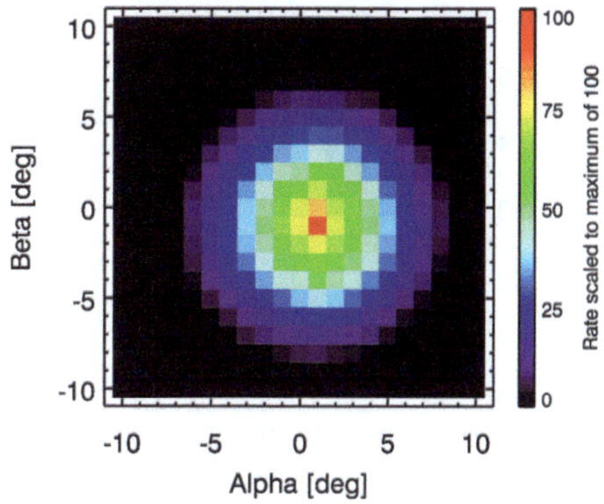

Fig. 3 Collimator transmission as a function of two orthogonal angles

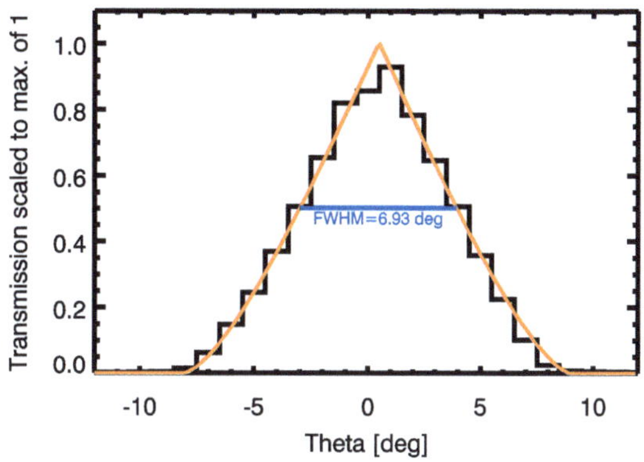

Fig. 4 Collimator transmission as a function of angle with respect to the look direction

is defined by $d\Omega = T(\theta)d\omega$, where $\omega(\theta) = 2\pi(1 - \cos\theta)$ is the solid angle of a cone of apex angle 2θ. Using the fact that $d\omega = 2\pi\sin\theta d\theta$ we find the FOV of the collimator

$$\Omega = \int_0^{\theta_{\max}} T(\theta) 2\pi \sin\theta\, d\theta = 0.0175 \text{ sr.}$$

3.2 Energy Response

We measured the energy response using an 8-mm-diameter proton beam aimed at the center of the IBaM aperture. An absolute beam monitor, based on the principle described in Funsten et al. (2005) was used to infer the beam flux impinging on the IBaM. Hence, we were able to determine the absolute detection efficiency, ε, as a function of energy shown in Fig. 5. Both plots show the same data in linear (on the left) or logarithmic (on the right) representations. The data points and their error bars are in black.

The red curve is a fit to the data points defined by

$$\varepsilon(E) = \begin{cases} 0 & \text{for } E < 9 \text{ keV} \\ 0.125\tanh(0.65(E-16)) + 0.1251 & \text{for } 9 < E < 14.214 \text{ keV} \\ 0.23\arctan(0.155(E-18.023)) + 0.1451 & \text{for } E > 14.214 \text{ keV} \end{cases} \quad (2)$$

where E is in keV. This empirically derived function fits the data quite well. The efficiency asymptotically tends to a maximum of 0.506. This efficiency curve includes the grid and

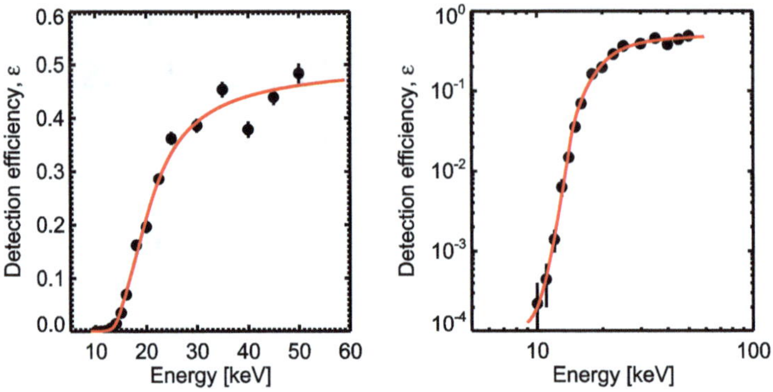

Fig. 5 Detection efficiency as a function of energy in linear (*left*) and logarithmic (*right*) representations

carbon foil transmissions (CF1, CF2), the collimator transmission, and the CEM detection efficiency up to ~50 keV, but it does not include the CEM detection efficiency at higher energies. The CEM detection efficiency for ions above 50 keV decreases very slowly as a function of energy, so neglecting the variation of the CEM detection efficiency has a small impact on the IBaM measurement. At high energies, the flux of protons decreases very rapidly as shown in Sanderson et al. (1996). The contribution of protons with energies larger than 50 keV for a power law distribution with power index of −4 is significantly less than between 10 and 50 keV.

3.3 Active Area

The active area is defined as the surface area of the sensor aperture where particles can be detected. There can be variations of the efficiency across the active area, but it is usually independent of the energy of the incoming particles. In the case of the background monitor, the efficiency across the area depends on the energy of the protons. There are at least two reasons for this energy dependence. First, the protons scatter when they pass through the carbon foils. Those that scatter significantly might miss the CEM funnel and remain undetected. The scattering half-width is inversely proportional to the incident energy of the protons so that scattering increases with decreasing energy. As a consequence, protons entering the IBaM at the center of the active area with a given energy will be more efficiently detected than the ones that enter the IBaM at the edge of the active area. Second, the detection efficiency of the CEM not only depends on the energy of the protons but also on the direction of incidence of the protons and the spatial location where they hit the funnel (Seah and Smith 1991). These characteristics are not quantitatively known for the IBaM CEM.

In order to quantify the spatial dependence of the active area, we measured the relative efficiency variations across the active area at 30 keV with a 2-mm-diameter beam. Figure 6 shows a color-coded map of the IBaM with 1 × 1 mm pixels.

We calculate the active area with the following equation

$$A = 1.3 \cdot 2.6 \sum_i \frac{R_i}{\overline{R_0}} \text{ cm}^2 \tag{3}$$

where 1.3 cm and 2.6 cm are the dimensions of the collimator aperture, $\overline{R_0}$ is the average rate in the center of the area for a ~8-mm-diameter beam, R_i is the rate for pixel i (from the

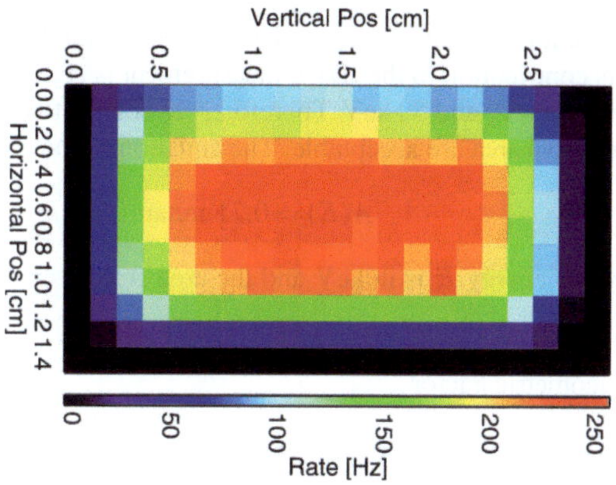

Fig. 6 Map of the IBaM response across the aperture at 30 keV

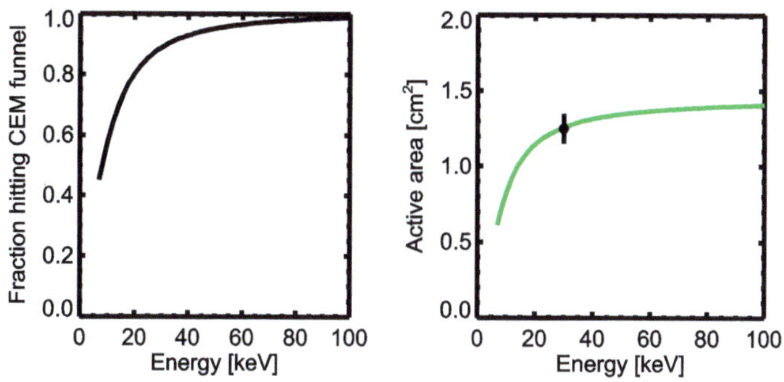

Fig. 7 Fraction of protons hitting the CEM funnel after being scattered in the thick carbon foils

2-mm-diameter beam), and the sum is done over all the pixels. Equation (3) assumes that all the protons at the center of the active area are detected. Even though this assumption is incorrect because of the scattering, if we combine this definition of active area with the energy response of the IBaM and its FOV, the overall result is consistent and leads to the correct estimate of the geometric factor given below. In summary, (3) does not give an area per se, but accounts for the detection efficiency variations across the active area. Applying (3) to our measurements, we find $A = 1.25 \pm 0.10\ \text{cm}^2$ at 30 keV.

We do not have measurements at other energies to determine the energy dependence of the active area defined here. To overcome the lack of measurements and still predict the efficiency variations across the active area, we developed a simple model of the effect of angular scattering of the proton in the two thick carbon foils. In particular, we looked at the fraction of protons that would hit the CEM funnel aperture after being scattered in the foils. This very simple model, based solely on geometrical considerations, lets us understand the basic relationship between the active area and the energy. Figure 7 shows the fraction of protons hitting the CEM funnel as a function of energy (left) and the same function scaled such that the model curve passes through the data point at 30 keV (right). This choice is arguable because we do not have enough data points to verify our predictions. However, an error in the estimate of the active area at low energies (around 10 keV) or at high energies (more than $\sim$50 keV) has a minor impact on the IBaM measurement. At low energies, the driver of the IBaM response (geometric factor) is the cutoff due to the carbon foils as shown in Fig. 5 with the efficiency curve. At high energies, we can use the same argument as for the detection efficiency described above: the proton flux decreases very rapidly with energy and its contribution to the IBaM measurement is significantly less than that from the proton flux between 10 and 50 keV. Therefore, we believe that the assumptions made for the model of the active area are acceptable. The equation of the green curve in Fig. 7 is

$$A(E) = 0.71 \arctan(0.13(E-4)) + 0.35 \tag{4}$$

where the energy E is in keV and the area A is in cm^2, with an asymptotic value of 1.465 cm^2.

3.4 Geometric Factor

The IBaM-reduced geometric factor, G', is defined as the product of the solid angle, Ω, with efficiency, ε, and active area, A, defined above. Since both the efficiency and the active area depend on the energy, the reduced geometric factor also depends on the energy. Thus, $G'(E) = \Omega\varepsilon(E)A(E)$. Plugging in (2) and (4) for ε and A and the numerical value of Ω, we obtain an analytical expression for the reduced geometric factor

$$G'(E) = 1.75 \times 10^{-2} \cdot (0.71 \arctan(0.13(E-4)) + 0.35) \times \begin{cases} 0 & \text{for } E < 9 \text{ keV or } E > 1000 \text{ keV} \\ 0.125 \tanh(0.65(E-16)) + 0.1251 & \text{for } 9 \le E < 14.214 \text{ keV} \\ 0.23 \arctan(0.155(E-18.023)) + 0.1451 & \text{for } 14.214 \text{ keV} \le E \le 1000 \text{ keV} \end{cases} \tag{5}$$

where E is in keV, and G' is in cm^2 sr. The upper limit of 1000 keV is simply set for convenience. It does not influence the expected rate calculations since the flux at these energies is much lower than around 10–50 keV. A more general expression for the geometric factor, G, can be derived such that the units of G are cm^2 sr eV/eV

$$G(E) = G \cdot T(E) \tag{6}$$

where $G = 12.34$ cm^2 sr eV/eV and $\int_9^{1000} T(E)\,dE = 1$

$$T(E) = \frac{1.75 \cdot 10^{-2}}{12.34E} \cdot (0.71 \arctan(0.13(E-4)) + 0.35) \times \begin{cases} 0 & \text{for } E < 9 \text{ keV or } E > 1000 \text{ keV} \\ 0.125 \tanh(0.65(E-16)) + 0.1251 & \text{for } 9 \le E < 14.214 \text{ keV} \\ 0.23 \arctan(0.155(E-18.023)) + 0.1451 & \text{for } 14.214 \text{ keV} \le E \le 1000 \text{ keV.} \end{cases} \tag{7}$$

3.5 Typical Response and Count Rate Estimate

In order to estimate the count rate that the IBaM is likely to observe in orbit, we use the spectrum given in Sanderson et al. (1996). The count rate is given by

$$R = \int_{E_{\min}}^{E_{\max}} G'(E)j(E)\,dE \tag{8}$$

where $E_{\min}$ and $E_{\max}$ are the limits of the integration (for example, $E_{\min} = 9$ keV and $E_{\max} = 1000$ keV), $G'(E)$ is given by (5), $j(E)$ is the differential intensity and the energy is expressed in keV. Figure 8 shows the term $G'(E)j(E)$, which represents the IBaM

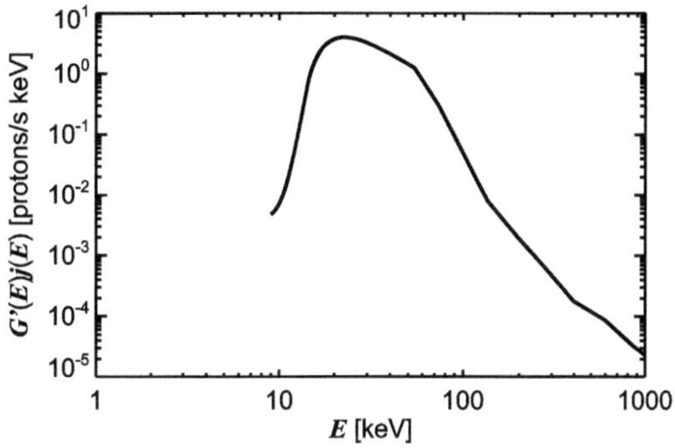

Fig. 8 Background monitor response for the upstream event shown in Sanderson et al. (1996)

response for a typical spectrum, as a function of energy. The response peaks between 20 and 30 keV with a width of ∼60 keV, which confirms that most of the counts that the IBaM will record fall between ∼10 to ∼50 keV. The discontinuities in the curve come directly from the spectrum.

The result of the integration is a rate of ∼120 Hz, which needs to be compared with the "intrinsic" background noise of the IBaM. We measured the background noise of the IBaM over an accumulated period of 19.1 hours during the IBEX payload cross calibrations and found a rate of 11.5 ± 0.4 mHz. The background noise is many orders of magnitude smaller than the expected signal.

As described below, the signal from the IBaM will be integrated over ∼3 minutes and divided into sixty 6-degree pixels in the sky. If the signal hits the same pixel at the estimated rate of 120 Hz, we can expect a total number of counts of the order of 360 events in each of the 60 angular pixels. Clearly these measurements are more than adequate to identify high background intervals for removal from the IBEX ENA observations.

3.6 UV Sensitivity

We tested the response of the IBaM to ultraviolet (UV) light using a krypton line lamp (Resonance Ltd., model KrLM-L) with two main lines at wavelengths of 116.5 and 123.6 nm and a typical intensity $F \geq 3 \times 10^{15}$ photons/(s sr). Taking into account the intensity of the lamp, the distance between the lamp, the different grids and collimator transmissions, and the IBaM aperture, we calculated a rate of $R = 1.2 \times 10^{12}$ photons/s that would reach the CEM if there was no carbon foil. The IBaM measured a rate of ∼650 Hz when the UV lamp was shining straight into the IBaM aperture (also maximum rate). Thus, the attenuation factor (including CEM detection efficiency) is roughly $650/1.2 \times 10^{12} \approx 5.4 \times 10^{-10}$. Assuming a CEM detection efficiency of about 0.02 at the wavelength of the UV source (Timothy and Bybee 1975; Paresce 1975), we obtain a carbon foil attenuation factor of $5.4 \times 10^{-10}/0.02 \approx 2.7 \times 10^{-8}$.

Hsieh et al. (1980) measured the transmittance of UV at wavelengths of 121.6 nm and 58.4 nm through carbon foils between 2 and 14 μg/cm² thick. At 121.6 nm they found that the transmittance is $0.26\ \exp(-0.37x)$ where x is the carbon foil thickness in μg/cm². By extrapolating to the IBaM total carbon foil thickness of $21 + 22 = 43$ μg/cm², the predicted attenuation factor is 3.2×10^{-8}. The estimate based on the extrapolation is fairly close to our measurement given the fact that it is well beyond the thickness range of their measurements. Since the IBaM FOV is co-aligned with the IBEX-Hi FOV, it looks out essentially perpendicular to the spacecraft's Sun-pointing spin axis. In this orientation the IBaM never views toward the Sun and will only see minimal counts when it sweeps past particularly bright UV stars.

4 Operations and Data Products

Upon consideration of the likely background sources detectable by the background monitor, it was deemed prudent to collect its data in phase with the IBEX-Hi direct events. This results in a flight telemetry product approximately every three minutes, which is a reasonable timescale for magnetospheric phenomena. As with IBEX-Hi, the counts are separated in sixty 6-degree azimuthal bins, which may provide adequate discrimination among possible background sources.

These count histograms are then compressed and sent to the ground along with the IBEX-Hi singles rates which are collected on an approximately 12-minute cycle. Thus, 4 cycles of background monitor data may be directly compared with the IBEX-Hi singles rates for any 6-degree patch in the sky. Since we do not expect count rates above $\sim$1 kHz, the compression scheme saturates at approximately 1 kHz. However, above this level the saturated compression still identifies the counts in that pixel as very large (>1 kHz), which is the information needed to cull the interval.

After the telemetry is delivered to the IBEX Science Operations Center (ISOC) (see Schwadron et al. 2008), it is processed to identify periods of high background from which IBEX-Hi and IBEX-Lo data should be excluded from consideration. This identification process will be calibrated on orbit based on the response to known sources, as well as any unanticipated ones as they are discovered. This processing will routinely produce a background monitor intensity curve (counting rate with time and angle) and a list of background time intervals. Additional products are planned on an as-requested basis through a web interface. This would include counting rates with time from selected regions of the sky and sky maps of counting rates over multiple orbits.

Acknowledgements We are grateful to many individuals who contributed at different stages of the design, development, testing, and implementation of the background monitor. In alphabetical order they are Josh Alquiza, Tom Broiles, Rob Ebert, Danny Everett, Mike Gruntman, Ron Harper, Paul Janzen, Chris Kofoed, Roberto Livi, Stefano Livi, Matt Maple, Susan Pope, Mike Young, and the IBEX team in general. We would like to thank the Goddard Space Flight Center for supporting acoustic testing of the flight carbon foils for the background monitor.

References

H.O. Funsten, B.L. Barraclough, D.J. McComas, Pinhole detection in thin foils used in space plasma diagnostic instrumentation. Rev. Sci. Instrum. **63**(10), 4741–4743 (1992a)

H.O. Funsten, D.J. McComas, B.L. Barraclough, Thickness uniformity and pinhole density analysis of thin carbon foils using incident keV ions. Nucl. Instrum. Meth. Phys. Res. B **66**(4), 470–478 (1992b)

H.O. Funsten, R.W. Harper, D.J. McComas, Absolute detection efficiency of space-based ion mass spectrometers and neutral atom imagers. Rev. Sci. Instrum. **76**, 053301 (2005)

H.O. Funsten et al., Space Sci. Rev. (2008, this issue)

K.C. Hsieh, E. Keppler, G. Schmidtke, Extreme ultraviolet induced forward photoemission from thin carbon foils. J. Appl. Phys. **51**(4), 2242–2246 (1980)

D.J. McComas, S.J. Bame, Channel multiplier compatible materials and lifetime tests. Rev. Sci. Instrum. **55**(4), 463–467 (1984)

D.J. McComas et al., Space Sci. Rev. (2008, this issue)

F. Paresce, Quantum efficiency of a channel electron multiplier in the far ultraviolet. Appl. Opt. **14**(12), 2823–2824 (1975)

T.R. Sanderson et al., WIND observations of energetic ions far upstream of the Earth's bow-shock. Geophys. Res. Lett. **23**(10), 1215–1218 (1996)
N.A. Schwadron et al., Space Sci. Rev. (2008, this issue)
M.P. Seah, G.C. Smith, Energy and spatial dependence of the electron detection efficiencies of single channel electron multiplier used in electron spectroscopy. Rev. Sci. Instrum. **62**(1), 62–68 (1991)
J.G. Timothy, R.L. Bybee, One-dimensional photon-counting detector array for use at EUV and soft X-ray wavelengths. Appl. Opt. **14**(7), 1632–1644 (1975)
P. Wurz et al., Space Sci. Rev. (2008, this issue)

Space Sci Rev (2009) 146: 117–147
DOI 10.1007/s11214-009-9495-8

The IBEX-Lo Sensor

S.A. Fuselier · P. Bochsler · D. Chornay · G. Clark · G.B. Crew · G. Dunn · S. Ellis · T. Friedmann · H.O. Funsten · A.G. Ghielmetti · J. Googins · M.S. Granoff · J.W. Hamilton · J. Hanley · D. Heirtzler · E. Hertzberg · D. Isaac · B. King · U. Knauss · H. Kucharek · F. Kudirka · S. Livi · J. Lobell · S. Longworth · K. Mashburn · D.J. McComas · E. Möbius · A.S. Moore · T.E. Moore · R.J. Nemanich · J. Nolin · M. O'Neal · D. Piazza · L. Peterson · S.E. Pope · P. Rosmarynowski · L.A. Saul · J.R. Scherrer · J.A. Scheer · C. Schlemm · N.A. Schwadron · C. Tillier · S. Turco · J. Tyler · M. Vosbury · M. Wieser · P. Wurz · S. Zaffke

Received: 31 July 2008 / Accepted: 20 February 2009 / Published online: 9 May 2009

S.A. Fuselier (✉) · A.G. Ghielmetti · J.W. Hamilton · E. Hertzberg · D. Isaac · A.S. Moore · C. Tillier
Lockheed Martin Advanced Technology Center, 3251 Hanover St, Palo Alto, CA 94304, USA
e-mail: stephen.a.fuselier@lmco.com

A.G. Ghielmetti
e-mail: gmetti@spasci.com

J.W. Hamilton
e-mail: Jon.Hamilton@lmco.com

E. Hertzberg
e-mail: Eric.Hertzberg@lmco.com

D. Isaac
e-mail: Donald.Isaac@lmco.com

C. Tillier
e-mail: Clemons.Tillier@lmco.com

G. Clark · S. Ellis · J. Googins · M.S. Granoff · D. Heirtzler · B. King · U. Knauss · H. Kucharek · F. Kudirka · S. Livi · S. Longworth · E. Möbius · J. Nolin · M. O'Neal · L. Peterson · S. Turco · J. Tyler · M. Vosbury · S. Zaffke
University of New Hampshire, 39 College Road, Morse Hall, Durham, NH 03824, USA

E. Möbius
e-mail: Eberhard.Moebius@unh.edu

P. Bochsler · D. Piazza · L.A. Saul · J.A. Scheer · P. Wurz
Physikalisches Institut, University of Bern, Sidlerstrasse 5, 3012 Bern, Switzerland

P. Wurz
e-mail: peter.wurz@phim.unibe.ch

G. Dunn · J. Hanley · D.J. McComas · S.E. Pope · J.R. Scherrer
Southwest Research Institute, 6220 Culebra Rd., San Antonio, TX 78238, USA

Abstract The IBEX-Lo sensor covers the low-energy heliospheric neutral atom spectrum from 0.01 to 2 keV. It shares significant energy overlap and an overall design philosophy with the IBEX-Hi sensor. Both sensors are large geometric factor, single pixel cameras that maximize the relatively weak heliospheric neutral signal while effectively eliminating ion, electron, and UV background sources. The IBEX-Lo sensor is divided into four major subsystems. The entrance subsystem includes an annular collimator that collimates neutrals to approximately 7° × 7° in three 90° sectors and approximately 3.5° × 3.5° in the fourth 90° sector (called the high angular resolution sector). A fraction of the interstellar neutrals and heliospheric neutrals that pass through the collimator are converted to negative ions in the ENA to ion conversion subsystem. The neutrals are converted on a high yield, inert, diamond-like carbon conversion surface. Negative ions from the conversion surface are accelerated into an electrostatic analyzer (ESA), which sets the energy passband for the sensor. Finally, negative ions exit the ESA, are post-accelerated to 16 kV, and then are analyzed in a time-of-flight (TOF) mass spectrometer. This triple-coincidence, TOF subsystem effectively rejects random background while maintaining high detection efficiency for negative ions. Mass analysis distinguishes heliospheric hydrogen from interstellar helium and oxygen. In normal sensor operations, eight energy steps are sampled on a 2-spin per energy

D. Chornay · J. Lobell · T.E. Moore · P. Rosmarynowski
Goddard Space Flight Center, Greenbelt, MD 20771, USA

T.E. Moore
e-mail: Thomas.e.moore@nasa.gov

M. Wieser
Swedish Institute of Space Physics, Box 812, 98128 Kiruna, Sweden
e-mail: wieser@irf.se

C. Schlemm
Applied Physics Laboratory, Johns Hopkins University, 11100 Johns Hopkins Road, Laurel, MD 20723, USA
e-mail: chuck.schlemm@jhuapl.edu

K. Mashburn
Montana State University, Bozeman, MT, USA
e-mail: kmashburn@swri.edu

H.O. Funsten
ISR Division MS B241, Los Alamos National Laboratory, Los Alamos, NM 87535, USA

T. Friedmann
Sandia Laboratory, Mail Stop 1415, PO Box 5800, Albuquerque, NM 87185, USA
e-mail: tafried@sandia.gov

R.J. Nemanich
University of Arizona, Tuscon, AZ, USA

N.A. Schwadron
Boston University, 725 Commonwealth Ave, Boston, MA 02215, USA
e-mail: nathanas@bu.edu

G.B. Crew
MIT Kavli Institute for Astrophysics and Space Research, 77 Massachusetts Ave, Cambridge, MA 02139, USA
e-mail: gbc@sapce.mit.edu

step cadence so that the full energy range is covered in 16 spacecraft spins. Each year in the spring and fall, the sensor is operated in a special interstellar oxygen and helium mode during part of the spacecraft spin. In the spring, this mode includes electrostatic shutoff of the low resolution (7° × 7°) quadrants of the collimator so that the interstellar neutrals are detected with 3.5° × 3.5° angular resolution. These high angular resolution data are combined with star positions determined from a dedicated star sensor to measure the relative flow difference between filtered and unfiltered interstellar oxygen. At the end of 6 months of operation, full sky maps of heliospheric neutral hydrogen from 0.01 to 2 keV in 8 energy steps are accumulated. These data, similar sky maps from IBEX-Hi, and the first observations of interstellar neutral oxygen will answer the four key science questions of the IBEX mission.

Keywords Neutral atom imaging · Heliosphere · Termination shock · Energetic neutral atoms · Magnetosphere · Surface ionization

Contents

1 Introduction and Basic Science

The IBEX mission achieves its sole objective by answering four fundamental science questions that are described in detail in the IBEX mission overview and the IBEX science introduction (McComas et al. 2004, 2009). These four questions focus on the global interaction between the solar wind and the interstellar medium. They are:

Question 1: What is the global strength and structure of the termination shock?
Question 2: How are energetic protons accelerated at the termination shock?
Question 3: What are the global properties of the solar wind flow beyond the termination shock and in the heliotail?
Question 4: How does the interstellar flow interact with the heliosphere beyond the heliopause?

The science questions are answered using sky maps of heliospheric neutral fluxes from 10 eV to 6 keV and with observations of the interstellar neutral oxygen and helium flows. The sky maps are obtained from the combined IBEX science payload.

The IBEX payload consists of the IBEX-Hi and -Lo sensors and the Combined Electronics Unit (CEU). The payload is designed to operate as a single entity. Thus, there is little distinction between IBEX-Hi and IBEX-Lo sensor science objectives. Two neutral atom sensors are needed for the payload because interstellar and heliospheric neutrals span an energy range that is large enough to require two different detection techniques. In particular, techniques used to convert neutrals to ions (so that the ions can be accelerated and analyzed) change at energies around several hundred electron volts (Wurz 2000). This natural division in conversion techniques results in two sensors with significant energy overlap that provide independent measurements of heliospheric hydrogen neutrals over the entire energy range of interest.

The IBEX-Lo sensor measures the low-energy part of the heliospheric neutral spectrum. The energy range for this sensor is from 10 eV to 2000 eV. The lower bound of the energy range is set by sensor sensitivity and practical interpretation of the observations. A 10 eV neutral from the heliospheric termination shock requires almost 11 years to complete a trip from the shock to the inner heliosphere. Neutrals with energies less than 10 eV would take more than a solar cycle to complete the trip. In addition, the sensitivity decreases relatively rapidly below 10 eV because the neutral-to-ion conversion efficiency decreases at energies below 10–20 eV (see Sect. 3.3 and also see Wurz et al. 2006).

The upper bound of the energy range is set by electric fields in the sensor, voltages on the electrodes, and, ultimately, on sensor mass resources. In total, the IBEX-Lo energy range covers very low energy neutrals expected to survive the journey into the inner heliosphere, through nominal solar wind energy neutrals (and the expected energy peak in heliospheric neutral flux for some heliospheric interaction models) and into the energy range for suprathermal neutrals accelerated in the solar wind and at the heliospheric termination shock. Since the IBEX-Hi sensor measures down to ~ 300 eV and has very good sensitivity at ~ 1000 eV, there is significant energy overlap between the two sensors for important heliospheric measurements near the nominal solar wind energy of 1000 eV.

IBEX-Lo also provides measurements that are used to answer the fourth IBEX mission science question: How does the interstellar flow interact with the heliosphere beyond the heliopause? Because interstellar oxygen and helium neutrals have energies up to several hundred electron volts, contributions to this science question come primarily from the IBEX-Lo sensor. Figure 1 shows locations in the Earth's orbit plane where interstellar oxygen and helium will be measured and also shows predicted energies of the neutrals. The Earth is in the upwind direction in June. Each year in the "northern hemisphere fall" (135° from the upwind direction, with prime viewing from 17 Sept to 21 October), the Earth is moving in the same direction as interstellar neutrals from the upwind direction. Because interstellar neutrals are accelerated by the Sun's gravitational force, they have higher energies than when they first enter the solar system, but they have to "catch up" to the Earth. IBEX-Lo will observe oxygen at a center energy of only 32 eV and helium will be at 8 eV energy with respect to the Earth's motion. Thus, helium will be just below the sensor's 10 eV low energy cutoff. Each year in the "northern hemisphere winter/early spring" (hereafter referred to as "spring") (225° from the upwind direction, with prime viewing from about 10 January to 23 February), the Earth's velocity vector and the interstellar neutral velocity vector are directed nearly opposite one another, so neutrals have considerably higher energies. IBEX-Lo will observe oxygen at 522 eV and helium at 130 eV.

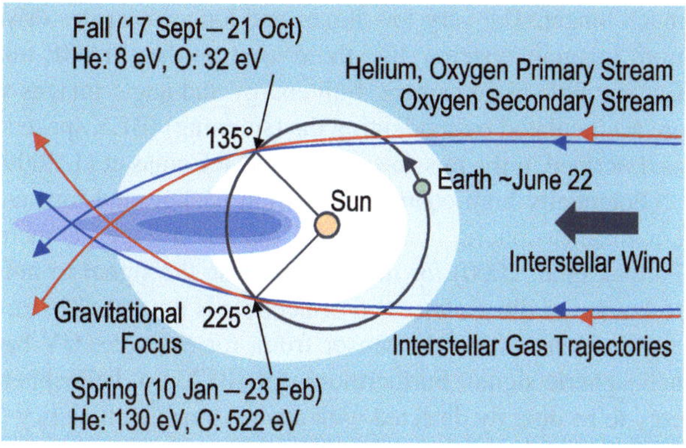

Fig. 1 Interstellar oxygen and helium observations occur in the fall and spring, where the streams intersect the Earth's orbit and the IBEX-Lo sensor FOV. In the fall, the interstellar neutral flux vector is in the same direction as the Earth's orbit velocity, so energies are low. In the spring, the two vectors are nearly oppositely directed, so energies are high

A key observable parameter for science question 4 is the arrival direction of the primary and secondary "filtered" neutral oxygen streams relative to the helium arrival direction. Measuring this arrival direction drives the IBEX-Lo sensor design in several ways discussed later in this paper.

2 Basic Sensor Requirements

Heliospheric and interstellar neutral fluxes are low and potential background contributions are very high. For comparison, 1000–2000 eV neutral fluxes from the Earth's magnetosphere (ring current) are $\sim 4 \times 10^4$ ENAs/(cm^2 s sr) and ~ 50 eV neutral fluxes from high latitude ionospheric outflow range from 6×10^4 to 1×10^6 ENAs/(cm^2 s sr). These neutral fluxes can vary significantly on timescales of tens of minutes. Magnetospheric and ionospheric neutrals are readily imaged with moderately large neutral atom imagers like the ones that were on the IMAGE spacecraft (Pollock et al. 2000; Moore et al. 2000; McComas et al. 2002; Fuselier et al. 2006). These imagers have time resolution of minutes, commensurate with the variability timescales of the source neutrals.

In contrast, ~ 1000 eV neutral fluxes from the outer heliosphere are predicted to range from 10^1 to 10^3 ENAs/(cm^2 s sr) and ~ 50 eV fluxes are predicted to range from 3×10^1 to 3×10^3 ENAs/(cm^2 s sr), depending on several factors including the termination shock strength (Gruntman et al. 2001). Recent measurements of energetic neutral atoms from the heliosheath suggest that these estimates may be somewhat low (Galli et al. 2006; Wurz et al. 2008b). However, heliospheric neutral fluxes are still considerably lower than magnetospheric ENA fluxes.

Variability timescales for ENAs from the outer heliosphere are not known. However, propagation times for neutrals from the vicinity of the termination shock to the inner solar system range from months (for ~ 1 keV ENAs) to years (for ~ 50 eV ENAs). ENAs from the termination shock propagating into the inner heliosphere have a 10–20% probability of charge exchange with solar wind protons, (i.e., only 80–90% of the original ENA signal can be detected at Earth orbit). This charge exchange is localized within the last 10 AU. However, if the ENA path goes through the plasma of a coronal mass ejection (CME), where there can be significantly higher proton densities, complete extinction of the ENA signal is possible. Given the typical CME frequencies with respect to propagation times of the ENAs, signal fluctuations at timescales of the order of days are possible.

Thus, compared to magnetospheric neutral fluxes, heliospheric neutral fluxes are from 10 to more than 1000 times lower, but the required time resolution for the observations is much longer. The very low fluxes and long timescales drive the IBEX sensor design away from imaging systems (like those used on the IMAGE mission) and to a large geometric factor, single pixel camera. Full energy and angle images with appropriate time resolution are accumulated by reorienting the (spinning) IBEX spacecraft over the course of 6 months, as described in the mission overview (McComas et al. 2009).

Potentially high contributions from background ions and UV also drive sensor design. For magnetospheric imagers, one can use a large geometric factor, pinhole camera (e.g., Pollock et al. 2000) for imaging because the signal to noise ratio is relatively large. This technique of direct detection of neutrals is not possible for the IBEX mission because creation of ions within the sensor from, for example, UV background could overwhelm the heliospheric signal. Furthermore, for IBEX-Lo, heliospheric neutrals have too low an energy to be directly detected with any reasonable efficiency (Wurz 2000). These background considerations and detection efficiencies drive the overall sensor design away from a direct detection, pinhole camera concept. Instead, heliospheric neutrals are ionized in the sensor and resulting ions are accelerated (to improve detection efficiency) and deflected away from their incident trajectory (to separate signal ions from potential backgrounds such as UV). Finally, coincidence measurements are used because this technique combines high detection efficiency with very high background rejection.

High throughput of the relatively weak heliospheric signal and very good background rejection are key elements in each of the IBEX-Lo sensor subsystems. In the next section, the sensor and these subsystems are described, with emphasis on contributions each subsystem makes to the overall signal detection and background rejection.

3 IBEX-Lo Sensor Subsystems

3.1 Introduction

The IBEX-Lo sensor is a large geometric factor, single pixel camera. It uses a large annular entrance to collimate the neutral flux. This entrance has positive and negative electrodes that reject incoming ions and electrons. The neutrals pass through the collimator and strike a conversion surface at a shallow angle (nominally 15°) where a fraction of them are converted to negative ions. These negative ions are accelerated in an electrostatic analyzer (ESA) that also selects the sensor energy range and resolution. Upon exiting the ESA, negative ions are further accelerated into a multiple carbon foil time-of-flight (TOF) mass spectrometer that measures ion mass.

Figure 2 shows a cross-section of the sensor (rotationally symmetric about the centerline of the figure with a maximum radius of about 15 cm). Major components are labeled on the right hand side, and representative trajectories of neutrals, ions, and electrons, are shown on the left hand side. A picture of the front entrance of the sensor (in the calibration vacuum chamber) with some of the components labeled is shown in Fig. 3. Table 1 shows a summary of the sensor parameters and resources.

The sensor consists of four major subsystems: the entrance, ENA to ion conversion, energy analysis, and mass (TOF) analysis subsystems. These four subsystems are attached to the "optics deck", which provides the stable mechanical platform for the ion optics and also a stable connection to the IBEX spacecraft. The four subsystems operate together, maximizing sensitivity and minimizing background to produce the highest possible signal to noise measurements over the IBEX-Lo energy range.

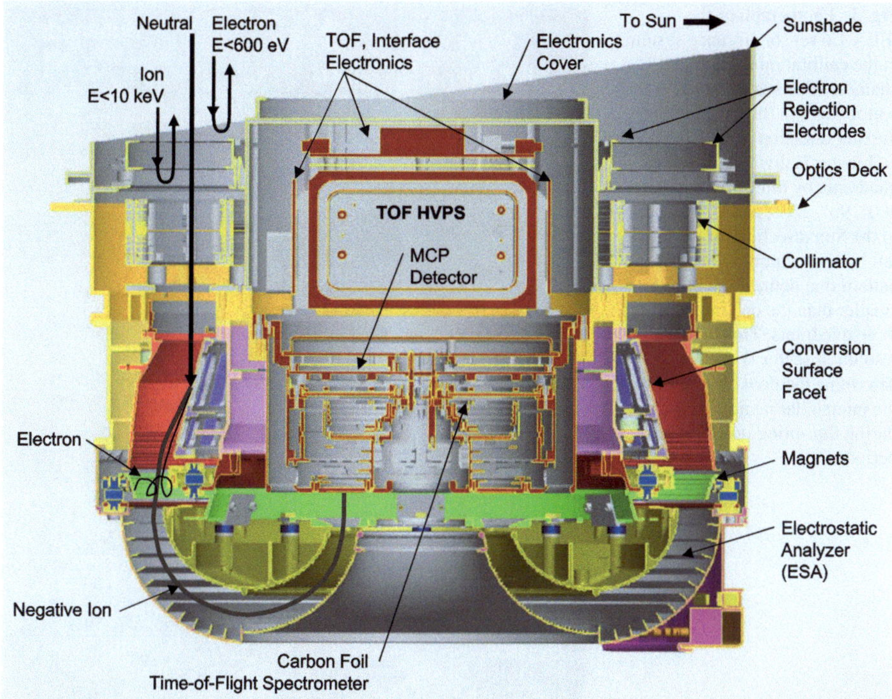

Fig. 2 Cross-section of the IBEX-Lo sensor showing the primary components. The sensor is rotationally symmetric about the centerline axis of the figure. Electrons, neutrals, and ions all enter the sensor through the collimator. Charged particles are rejected by the collimator and electron rejection electrodes. Neutrals pass through the collimator and strike a conversion surface. A fraction of these incident neutrals leave the conversion surface as negative ions and pass through the electrostatic analyzer. Electrons from the conversion surface are deflected by two concentric rings of permanent magnets. Negative ions exit the ESA, are accelerated and enter a triple coincidence time-of-flight (TOF) mass spectrometer. In this subsystem, the ion mass is determined

3.2 Entrance Subsystem

The entrance subsystem consists of the sunshade, electron rejection electrodes, and collimator. The sunshade is cut at an angle so that sunlight cannot reflect off any part of the sensor onto the collimator. Eliminating scattered sunlight from the collimator entrance is critical for background reduction. The sensor views 90° away from the spacecraft spin axis and the spin axis is reoriented towards the Sun every orbit (8 days). By setting the spin vector a few degrees off the Sun and letting it precess through the Sun direction, the spin vector will move a maximum of about ±4° off the Sun direction over the course of an orbit. Thus, with margin, the sunshade was designed to block sunlight for a spin axis that is up to 8° off of the Sun direction. Elimination of sunlight reduces UV flux into the sensor to interstellar levels (~ 800 Rayleighs maximum flux at Lyman Alpha wavelength) during prime science observations. The UV flux will be considerably greater (~ 20 kRayleighs at Lyman Alpha wavelength) when the sensor views the Earth (twice per year), but, at those times, the sensor will be viewing through the Earth's magnetosphere and not making prime heliospheric science measurements.

Since the sensor views 90° from the Sun direction, the cold (temperatures of ~ 10's of eV maximum), flowing (~ 400 km/s away from the Sun) solar wind ion distribution does

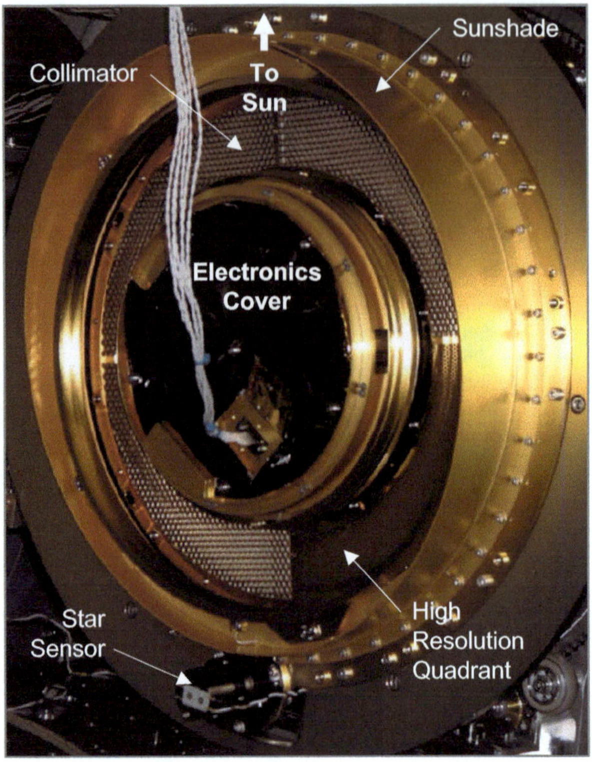

Fig. 3 Photograph of the IBEX-Lo sensor entrance system in the calibration vacuum chamber. The star sensor is at the bottom and, in this orientation, the Sun direction is up. The collimator is divided into 4 quadrants by thin spokes located at 0°, 90°, 180°, and 270° relative to the Sun direction. In the lower left hand quadrant, the hexagon pattern that defined the FOV is smaller than the one in the other three quadrants. This high resolution sector is used to determine the flow direction of the interstellar neutral oxygen during the spring observing period

Table 1 IBEX-Lo sensor parameters and resources

Energy Range (eV)	10–2000
Energy resolution ($\Delta E/E$)	0.8
Mass Range	1–32 amu
Mass Resolution $M/\Delta M$	>4
Field of View—Low Resolution	6.54° × 6.54°
Field of View—High Resolution	3.19° × 3.19°
Geometric Factor* (cm^2 sr)	0.91
Mass (kg)	11.5
Power (W)	3.46
Telemetry (bps)	122

*This geometric factor refers only to the collimator opening and is the sum of the high and low resolution sectors, not including grid transmission factors, neutral-to-ion conversion efficiencies, ESA or TOF efficiencies

not have direct access to the sensor through the collimator. However, the solar wind (halo) electron distribution has sufficient temperature that a significant electron flux in the energy range from 10's to 100's of eV could enter the sensor. These electrons could ionize residual gas inside the sensor through electron impact ionization. The newly created ions would be accelerated into the conversion surface, and resulting negative ions would be indistinguishable from those created by source heliospheric neutrals. To eliminate low-energy electrons, a pair of electron rejection electrodes encircles the entrance to the collimator. The electrodes are charged to −3.1 kV, creating a field that rejects up to 600 eV electrons from the collimator.

The collimator defines the instantaneous fields-of-view (FOVs) of the sensor and is held at a positive high voltage of 10 kV to keep energetic, positive ions (up to 10 keV) out of the sensor. The collimator designs for IBEX-Hi and -Lo are identical except that the annular diameter of the IBEX-Lo collimator is smaller and the IBEX-Lo collimator has two separate FOVs. The two FOVs are a high (angular) resolution FOV (one 90° azimuthal quadrant) and a low resolution FOV (three 90° azimuthal quadrants). The high resolution FOV has approximately one fourth the intrinsic angular FOV of the low resolution quadrant and is used to measure interstellar neutral oxygen in the springtime (see Fig. 1). The full sensor (combined high and low resolution FOVs) is used for heliospheric neutral hydrogen measurements throughout the year and for measurements of interstellar neutral oxygen in the fall.

To provide a well-defined angular FOV with the largest possible collection area for neutral atoms, the collimator has a hexagon shape multi-hole aperture that uses a stack of identical photo-etched plates. A linear version of this type of collimator was used on the ACE/SEPICA instrument (Möbius et al. 1998a).

Figure 3 shows the front of the collimator hexagon pattern. The collimator is divided into four 90° quadrants by 4 spokes at approximately 0°, 90°, 180°, and 270° from the vertical (Sun) direction. The high-resolution quadrant is between the 90° and 180° spokes measured clockwise from vertical in Fig. 3. Optical tests of the IBEX-Lo collimator show that the full-width half-max (FWHM) FOV of the low resolution quadrants is $6.54° \pm 0.23°$ and the FWHM FOV of the high resolution quadrant is $3.19° \pm 0.2°$, or very close to the designed low- and high-resolution FOVs of 7° × 7° and 3.5° × 3.5°, respectively.

The FOV is determined solely by the width w (~ 4 mm for the low resolution sectors) of the hexagon openings and the total height h (~ 25.9 mm) of the collimator stack, i.e., between the entry and exit plate, as shown in Fig. 4. The angular width of the FOV in each direction is calculated from progressive clipping of trajectories through the aperture pair as the angle θ relative to the normal direction of the collimator plates increases. The maximum throughput is at normal incidence and is determined by the transparency T of the collimator. The transparency is dependent solely on the ratio of the line width d and the width of the hexagons w, and is written as:

$$T = 1/(1 + d/w)2. \tag{1}$$

The collimator has a transmission of $68.8 \pm 0.3\%$ for the low resolution quadrants and $61.7 \pm 0.5\%$ for the high resolution quadrant.

In Fig. 4, particles can pass through neighboring channels in a multi-hole collimator as long as their angle θ exceeds $\theta \geq d/h$. To prevent such "leakage" of particle trajectories, identical collimator plates are stacked in a roughly geometric sequence, with the largest plate separation $h_n \leq d \cdot h/w$ and the smallest separation so that $h_1 \leq d \cdot \tan\theta_{Max}$, where

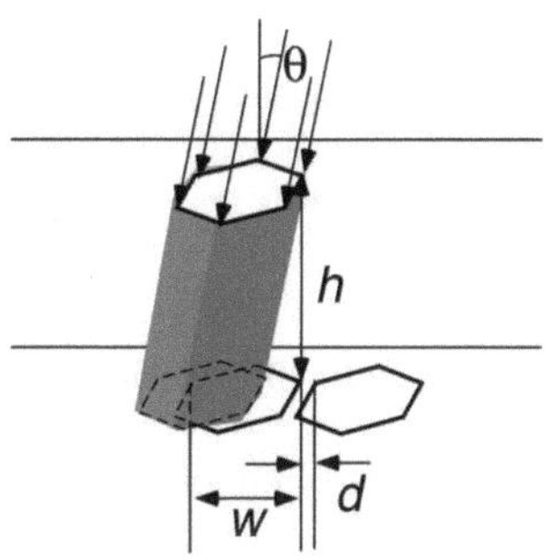

Fig. 4 Pair of entry and exit aperture holes of the IBEX-Lo collimator. As indicated by the shaded hexagon shape, only the fraction of particles pass the collimator that fall into the intersection of the two hexagon areas at the exit plate. The fraction depends on the distance h between the plates, the opening width w, and the separation d between hexagons

θ_{Max} is the largest possible incidence angle for particles. The angle θ_{Max} is limited to $\leq 50°$ by a precision-milled pre-collimator with trapezoid-shaped hexagon ribs, whose width is ≥ 50 μm less than d. The geometric sequence of plate separations starts with the largest one at the exit plate and alternates to smaller and smaller separations from both the top and bottom toward the center so that a series of 6 plates with the smallest identical separations is placed at the center of the collimator. With this pattern, the fraction of particles scattered at the edges of the holes into the FOV (an unavoidable effect) is minimized. Very thin (0.5 mm thick) etched plates with sharp edges are also used to reduce the scattered particle fraction. Finally, plate separations are reduced compared to a strictly geometric progression to account for manufacturing tolerances and deviations from planarity in the collimator plates.

Accounting for collimator transmission, correcting for losses due to spokes, incomplete hexagons at the edges, and shadowing at the edges, and combining the FOVs of 3 low resolution quadrants and one high resolution quadrant, the total geometric factor of the collimator is 0.91 ± 0.04 cm^2 sr.

The collimator floats at +10 kV relative to the spacecraft (and sensor) ground potential and is attached to the optics deck by 16 high voltage insulators. This positive potential keeps up to 10 kV ions out of the sensor. While nominal solar wind ion fluxes directed into the sensor with energies above a few 10's of eV are very low, energetic ion fluxes (up to several keV) in the magnetosheath and in the Earth's foreshock upstream from the bow shock can be high enough to create measureable background in the sensor. The positively biased collimator rejects these ions.

Without the +10 kV collimator voltage, the IBEX-Lo design provides some mitigation against ion background. First, the negative electrodes in front of the collimator act as a defocusing lens for low energy ions. Simulations show that ions below about 200 eV are defocused enough that they hit the collimator plates before entering the sensor. Second, there is a conical-shaped grid between the collimator and the ENA to ion conversion subsystem (described in Sect. 3.3 below) that deflects ions away from the conversion surface. Finally, IBEX-Lo is inherently a negative ion sensor. Any positive ion that enters the sensor must be converted to a negative ion on the conversion surface in order to be detected as background.

3.3 ENA to Ion Conversion Subsystem

By rejecting the majority of high energy ions and nearly all electrons, only UV, a very low flux of solar wind energetic ions with energies > 10 keV, and neutrals exit the back of the collimator. These constituents enter the subsystem where a fraction of the neutrals are converted to negative ions. The key to this subsystem is a diamond-like carbon (DLC), or more accurately described as a tetrahedral amorphous carbon (ta-C), conversion surface (e.g., Wieser et al. 2005).

As stated in the introduction, 10 eV to several hundred eV neutrals do not have sufficient energy to be detected directly with any efficiency using standard detector technology. Furthermore, heliospheric neutrals are accompanied by interstellar UV fluxes that would overwhelm a detector placed directly behind the IBEX-Lo collimator. Thus, neutrals must be converted to ions so that they can be accelerated (thereby raising their detection efficiency) and deflected away from the direct path taken by UV background (Wurz 2000; Wurz et al. 2006).

Ionization by scattering from charge-state conversion surface offers the highest ionization efficiencies in the energy range below 1 keV. This technique was first proposed for space applications by Gruntman (1993) and Wurz et al. (1993). Early low energy neutral atom imager designs were proposed using low work function surfaces for converting neutrals to ions during surface impact and reflection (e.g., Ghielmetti et al. 1994). However, these surfaces must be re-conditioned and regenerated often, placing difficult requirements on sensor resources and operations. Furthermore, changing surface conditions result in variable conversion efficiencies and ultimately result in uncertainties in overall sensitivity of the sensor. This uncertainty creates the need for a separate, accurate monitor of conversion efficiency.

Since these early designs, there has been a concerted search for a stable, inert, high-yield, low-scatter conversion surface. The Low Energy Neutral Atom (LENA) imager on the IMAGE mission was the first to use this type of surface conversion. A highly polished polycrystalline tungsten surface was used for neutral to ion conversion, with ionization facilitated by natural contaminants, most likely adsorbed water (Moore et al. 2000). Surface conversion efficiencies were $\ll 1\%$ for hydrogen.

Since this pioneering mission, several surfaces have been identified that have better negative ion yield. Among these, natural diamond crystals demonstrated reasonably high negative ion production for hydrogen and oxygen (Wurz et al. 1997). The large conversion surface area required for most neutral atom imagers makes the use of natural diamond surfaces impractical from a cost standpoint. Instead, diamond-like carbon surfaces make an excellent substitute (Scheer et al., 2005, 2006; Wieser et al. 2005; Wurz et al. 2006). These diamond-like carbon surfaces are readily grown on large, very smooth silicon substrates, retaining the surface smoothness of the underlying substrate.

For IBEX-Lo, 3 inch silicon wafers were cut into trapezoidal facets 62 mm long and ~ 30 mm wide at the center. The edges of the trapezoids were beveled so that they fit together to form an annular cone that is inclined at 15° from the incident direction of the neutrals that pass straight through the collimator. A 100 nm thick tetrahedral amorphous carbon (ta-C) layer was grown on each trapezoid facet. At the start of this process, the silicon surface smoothness was < 0.1 nm RMS and, at the end, the DLC layer had surface smoothness ~ 0.1 nm RMS. These surfaces were then treated with a hydrogen beam in a vacuum. This process, called hydrogen termination, is used to chemically terminate exposed, non-diamond-like carbon bonds on the surface. It removes oxygen from the surface, making it more inert, and it also lowers the work function of the surface and does not add to surface roughness.

The end result was 28 facets that are inert, slightly conductive, extremely smooth, and have reasonably high negative ion yield for neutral impact at grazing (15°) incidence. A picture of one of the facets is shown in Fig. 5. The negative ion yield properties for hydrogen are shown in Fig. 6. Ionization efficiencies increase with increasing energy, reaching $\sim 5\%$ for hydrogen. Measurements of 4 of the facets are shown at 4 different energies. The empirical curve is based on measurements over the entire energy range using a variety of conversion surfaces and detectors (Wieser 2005; Wurz et al. 1998, 2006; Wieser et al. 2007). Tests of these surfaces over periods of more than several years indicate that the conversion efficiency is stable for many years (Scheer et al. 2005, 2006).

While Fig. 6 shows relatively high ionization efficiencies for the conversion surfaces, the conversion efficiency is the product of the ionization efficiency and the reflection efficiency. That is, the total negative ion yield is the ratio of the number of negative ions off the surface divided by the number of neutral atoms incident on the surface. The reflection efficiency plays an important role in determining the overall conversion efficiency for these diamond-like surfaces. The reflection efficiency is also energy and mass dependent (Scheer et al. 2008). The reflection efficiency is the number of scattered particles (atoms and negative

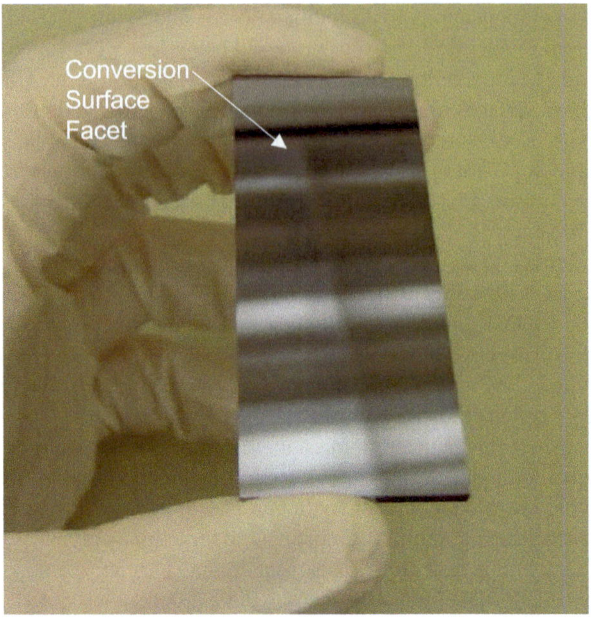

Fig. 5 Photograph of a conversion surface facet. The substrate is a highly polished silicon wafer cut into a trapezoid. A ~100 nm diamond surface is grown on this surface. The resulting conversion surface is smooth to within ~0.1 nm. Twenty eight of these facets are used in a conical configuration in the sensor

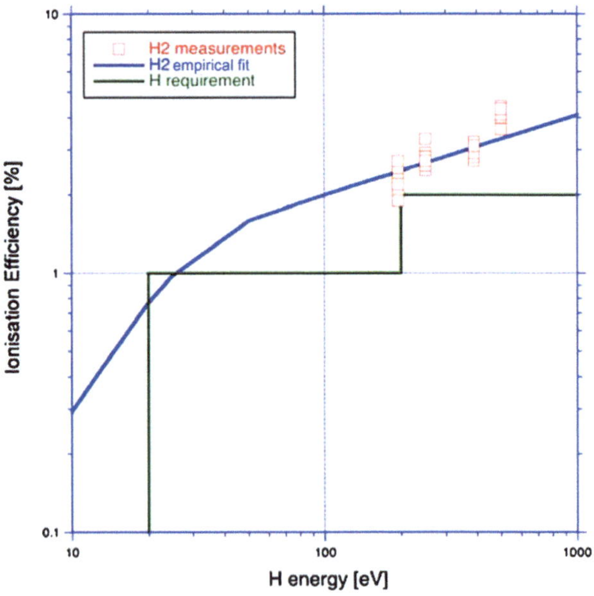

Fig. 6 Measured neutral to negative ion ionization efficiencies of IBEX-Lo conversion surfaces for hydrogen. The *green line* indicates the sensor requirement and the blue line is an empirical fit to measurements from different synthetic diamond coatings on silicon (Wieser 2005). The conversion efficiency is the product of these ionization efficiencies times the (energy dependent) reflection efficiencies

ions) in the specular direction within the ion-optical acceptance angle of the ion optical system. The roughness of the surface determines the fraction of negative ions that scatter in the specular direction. The IBEX-Lo conversion surfaces are smooth on an atomic level, maximizing this fraction of scattered negative ions. However, even these surfaces are corrugated at an atomic level. Therefore, there is always angular scatter away from the specular direction. The number of particles that scatter away from the specular direction increases with particle energy and with angle of incidence because the incoming particles probe deeper into the surface potential well of the conversion surface. Thus, the reflection efficiency decreases with particle energy, since the angular scatter increases (Wahlström et al. 2008), and a lower fraction of the angular scatter width is within the angular acceptance of the ion optical system. In addition to this effect, there is the possibility that incoming particles get stuck in the surface.

Tests using the University of Bern's conversion surface test facility (Wurz et al. 1997; Jans et al. 2000) show that the reflection efficiency for the IBEX-Lo DLC surfaces is of the order of 10% for 200 eV hydrogen and of the order of 4% for 700 eV hydrogen (Scheer et al. 2008). Combined with the ionization efficiency (Fig. 6), these tests result in an overall conversion efficiency for hydrogen neutrals that is $< 1\%$ over the IBEX-Lo energy range. However, the IBEX-Lo sensor has ion optics that are designed to maximize capture of reflected negative ions, even some ions that scatter away from the specular direction, so the DLC surfaces in the sensor have measured conversion efficiencies that approach 1% at the upper energy limit. While this efficiency seems low, it is still one of the highest for inert conversion surfaces like the DLC surface.

Finally, ionization and reflection efficiencies also depend on incidence particle type. Ionization efficiencies are a strong function of the electron affinity of the incident neutral. Oxygen has a higher electron affinity than hydrogen and a much higher negative ion yield ($\sim 30\%$ compared to $\sim 5\%$ for hydrogen). Reflection efficiency is also dependent on mass. Tests conducted at the University of Bern's conversion surface test facility indicate that oxygen has, on average, a lower reflection efficiency than hydrogen. Also, the energy loss upon reflection is greater for oxygen than for hydrogen. Differences in the reflection efficiency and energy loss between oxygen and hydrogen are probably associated with the fact that hydrogen is much lighter than carbon (the conversion surface material) while oxygen and carbon have similar masses. The details of this interaction are still the subject of investigation (Wahlström et al. 2008).

3.4 Energy Analysis Subsystem

Negative ions from the conversion surface facets are accelerated away because the facets are held at a negative potential. Another electrode that faces the annular ring of conversion surface facets is also at a negative potential. In combination, these fields deflect and focus negative ions in the radial direction into the entrance of the Electrostatic Analyzer (ESA) (see the negative ion trajectory in Fig. 2). Large angle scattering off the conversion surface facets in this direction is at least partially compensated for by this focusing effect.

The energy analysis subsystem consists of a toroidal electrostatic analyzer (ESA), two electrodes at the entrance to the ESA that help deflect and focus the negative ions into the ESA, and a third electrode at the ESA exit that helps focus the negative ion beam into the mass analysis subsystem.

The toroidal ESA defines the sensor energy pass band. The analyzer has the shape of a "bundt" baking pan (Moestue 1973), and this geometry was used in the Toroidal Imaging Mass Angle Spectrograph (TIMAS) on the Polar spacecraft (Shelley et al. 1995). Parameters were adjusted so that the annular ring at the ESA exit was the same size as a standard size microchannel plate in the TOF spectrometer. Also, the plate separation between the inner and outer ESA is quite large (see Fig. 2), commensurate with the large passband of the sensor (the ESA $\Delta E/E$ is estimated to be 0.67). With such a large plate separation, there could be considerable UV background. To reduce the background, the ESA outer shell has very large "fins" that are very effective light traps (see Fig. 2). In addition, the inner ESA shell is serrated and both the inner and outer shells were blackened with a porous black coating that further reduces UV reflection. These measures counteract the effect of the large plate gap and maintain an overall approximately 3-bounce system for UV to reach the entrance to the TOF system. Before the design of the IBEX-Lo sensor, a prototype sensor was built and all ion-optical properties were verified (Wieser 2005; Wieser et al. 2007) (see also Sect. 3.8).

In addition to setting the energy passband of the sensor, the ESA folds the ion optics so a smaller, standard size detector can be used. This reduction does not come without a price. Two fundamental ion optics properties of the ESA are that it focuses in the radial direction, but it disperses in the azimuthal direction. A narrow-angle beam entering the collimator at one point will disperse into an arc that is greater than $\sim 180°$ in azimuthal extent at the detector. Since it is not necessary to image in the azimuthal direction, this dispersive property is not an issue for IBEX-Lo. However, ions that disperse to large azimuthal angles will exit the ESA at a very large angle with respect to the normal angle of the TOF entrance. The sensor properties at the ESA exit are designed to help deflect the azimuthal trajectories of negative ions so the ions arrive at the TOF entrance with a small angle relative to the normal. This deflection is done in two ways: by shaping the ESA exit so that it focuses ions and by applying a large acceleration voltage between the ESA exit and the TOF entrance. However, ions with very large azimuthal trajectories exit the ESA with too large an angle from the radial direction. These ions do not pass completely through the TOF system and are lost.

This fundamental property of the ESA feeds back to conversion surface smoothness. If negative ions leave the conversion surface with more than 10° azimuthal angle relative to their incident direction, then their trajectories in the ESA become large spirals and they exit the ESA at very oblique angles with respect to the normal to the TOF entrance and are lost. These ions are lost in the TOF system. Tests of the IBEX-Lo conversion surface facets indicates that they are smooth to ~ 0.1–0.2 nm RMS. Nonetheless, a significant fraction of the ions are lost in this manner and that fraction is energy dependent.

In addition to negative ions, incident neutrals produce electrons from the conversion surface. In fact, many more electrons are produced than negative ions because UV photons also reach the conversion surface. Since the ESA and energy analysis system are designed to accelerate negative ions, electrons could become a serious background in the TOF subsystem. Specifically, if a sufficient number of electrons are accelerated through the ESA and hit the first carbon foil in the TOF system, then they could overwhelm the TOF electronics.

To counter this background, the sensor takes advantage of the azimuthal defocusing property of the ESA. The energy analysis subsystem uses permanent magnets to deflect the electron trajectories so they have a large azimuthal component to their velocities. This electron suppression scheme was used effectively on the IMAGE/LENA imager (Moore et al. 2000) and tested in the IBEX-Lo prototype (Wieser et al. 2007) (see Sect. 3.8). For IMAGE/LENA, the electrons were suppressed after they were accelerated to several keV in the ion optics, and relatively large permanent magnets were needed (Moore et al. 2000). For IBEX-Lo, it was possible to design a magnetic suppression system for the electrons before they were accelerated significantly (electrons leaving the conversion surface have only a few eV energy) (Wieser et al. 2007). Furthermore, instead of deflecting the electrons so that they cannot enter the ESA, all that was required was to add a large azimuthal component to their velocities. Two nearly concentric circles of permanent magnets were used on the inner and outer electrodes that define the entrance to the ESA (see Fig. 2). These magnets (~ 1.5 mm in diameter) face each other across the ESA entrance gap and create a 3 millitesla field directed radially across the gap. Electrons up to a few 10's of eV are effectively deflected in this field and, if they enter the ESA, their trajectories have significant azimuthal components, and therefore they do not reach the TOF entrance. The magnetic field is low enough that trajectories of even the lowest energy negative ions are unaffected.

Finally, the energy analysis subsystem has one more requirement. As discussed in Sect. 3.2 (the entrance subsystem), one of the four quadrants of the collimator has a high resolution, 3.2° × 3.2° FOV. For interstellar neutral oxygen measurements in the springtime, this quadrant must be used and the other three quadrants must be "shut off". Shutoff is achieved electrostatically by applying a large, negative voltage (−2.5 kV) to the inner electrode at the entrance to the ESA for the three quadrants behind the low resolution collimator quadrants. This potential pushes negative ions in the three low resolution quadrants to the outer wall, where they scatter and do not enter the ESA. Tests of the IBEX-Lo sensor demonstrate that this shutoff works very well. The edges of the high resolution sector were of particular concern since fringe fields could affect ion trajectories in the low resolution sector and possibly allow "leakage" of these ions from low resolution sectors to the detector. Field termination electrodes are used to minimize fringe fields and therefore minimize leakage. Tests indicate that leakage is $< 1.5\%$ and, at this level, leakage does not affect high angular resolution measurements.

3.5 Mass (TOF) Analysis Subsystem

The mass (time-of-flight, TOF) analysis subsystem is a triple coincidence carbon foil-based time-of-flight ion mass spectrometer. It is designed to distinguish hydrogen and oxygen negative ions and suppress background random events through triple coincidence measurements. While distinguishing hydrogen and oxygen is the minimum mass resolution that is required, the TOF is designed with the goal to distinguish hydrogen and helium negative ions so that interstellar neutral helium fluxes can be measured separate from the interstellar neutral oxygen. This type of double and triple coincidence TOF system is a novel design that is based on the TOF systems used on FAST, Cluster, Equator-S, and STEREO/PLASTIC (e.g., Möbius et al. 1998b). The triple TOF system has several major advantages over previous designs. Advantages such as the superior background suppression and higher efficiency when single TOF channels are used are important for the IBEX-Lo sensor (Möbius et al. 2007).

The basic TOF operation is shown in Fig. 7. This figure shows a radial cross-sectional cut of the TOF subsystem. It is rotationally symmetric about the left hand side of the figure, so that the microchannel plate (MCP) detector stack (pink in the Fig. 7) is an annular ring. The section of the TOF in Fig. 7 is shown in the inverse orientation compared to Fig. 2. A picture of the entrance end of the flight model of the TOF is shown in Fig. 8. The eight carbon foils that make up the first set are supported on grids that are seen through the ultra-thin foils. Vent holes surround the foils to protect them from perforation by acoustic shock.

Negative ions are accelerated into the first set of foils at the top of Fig. 7 because the entire TOF ion optics section floats at a nominal +16 kV post-acceleration (PAC) high voltage. (The blackened ground cylinder that surrounds the optics is shown in Fig. 8). As discussed in Sect. 3.4, this high post acceleration voltage helps straighten out negative ion trajectories between the ESA exit and TOF entrance foils. More importantly, high post acceleration allows a TOF measurement with high enough resolution after energy loss in the entrance foil.

Upon striking the first set of carbon foils, negative ions knock off secondary electrons. These secondary electrons are focused on the outmost radius of the MCP and constitute the first start pulse (start 1 or "a" in Fig. 7). As the negative ions pass through the first

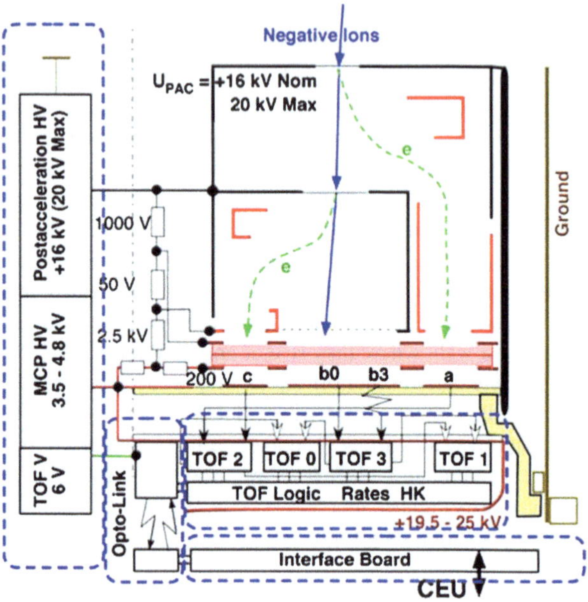

Fig. 7 Schematic of the TOF mass spectrometer. The TOF is rotationally symmetric about the right hand side of the figure. Negative ions from the ESA strike the first foil at the top. These ions pass through the foil (some become neutral) and knock off electrons that are accelerated and steered to the outer edge of the annular microchannel plate stack. The signal from these ions (a, on the anode below the pink MCP stack) is the start 1 signal. Ions and neutrals pass through a second, interior foil. Electrons from this foil are accelerated to the inner edge of the MCP stack and create the start 2 signal. Finally, ions and neutrals strike the MCP stack at position b0/b3 and create the stop signal. By combining the starts and stops, the mass of the incident negative ion is determined

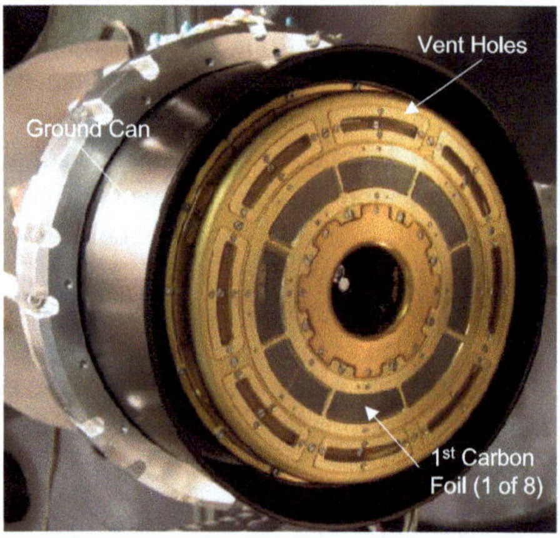

Fig. 8 Photograph of the front entrance of the TOF mass spectrometer in the test vacuum chamber. The 8 ultra-thin carbon foils are transparent and are mounted on high transmission grids. Vent holes around the grids and a general open design minimize possible acoustic damage to the foils. The entire TOF floats at 16 kV and is surrounded by the blackened aluminum ground can

foil, a fraction of them become neutral again. The ions and neutrals strike a second foil and knock off secondary electrons. These electrons are focused on the innermost radius of the MCP and constitute the second start pulse (start 2, or "c" in Fig. 7). Finally, the ions and neutrals pass through the second foil and strike the MCP in the center radius. The signal from these ions and neutrals constitutes the stop pulse (stop, or "b0" in Fig. 7). The stop anode is segmented into 4 quadrants, labeled b0, b1, b2, and b3 in Fig. 9. Delay lines are placed between anodes b0 and b1, b1 and b2, and b2 and b3. Using the delay between the signals from anodes b0 and b3, the arrival quadrant of the signal is determined. Although it is relatively crude angular information, this sectoring of the signal provides important additional background rejection when the high resolution mode is used. Since three of the four collimator quadrants are shut off in this mode, there should be minimum signal from quadrant b1, the quadrant opposite the high resolution quadrant. Tests conducted

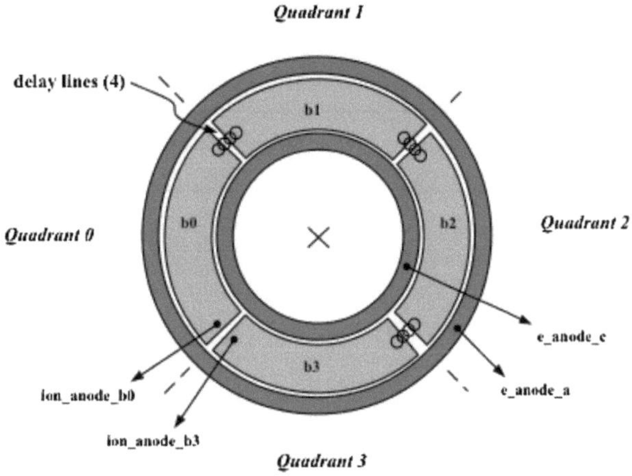

Fig. 9 The TOF anode is divided into four sectors with delay lines between three sectors. By analyzing the signal delay between anode b0 and b3, the quadrant for the stop signal is determined

during the sensor calibration show that the ratio of the quadrant that has maximum counts to the opposite (background) quadrant is ~ 2000. This high ratio indicates that the ion optics is behaving as designed and that the delay line detection of the ion arrival location is an effective additional background suppression technique for the interstellar neutral oxygen measurements.

3.6 TOF and Other Electronics

Electron avalanches from the back of the MCP are collected on the start 1 (a), start 2 (c), and stop (b0, b1, b2, and b3) anodes. These anodes are biased 200 V positive relative to the MCP back to accelerate the electrons from the MCP back (see Fig. 7). Four TOF ASIC chips (Paschalidis et al. 2002, 2003) combine the signals to give the ion TOF over the entire path from the first foil to the stop MCP, the half path from the second foil to the stop MCP, and between the stop anodes b0 through b3. A valid double coincidence event requires a start (two possibilities) and stop. A valid triple coincidence event requires both starts and a stop. Furthermore, a valid triple event meets the criterion that the TOF over the full 60 mm distance from the first foil to the stop MCP is equal to the sum of the TOF over the 30 mm distance from the first to the second foil and the TOF over the 30 mm distance from the second foil to the MCP. Checking that this criterion is met greatly reduces background due to random double coincidences. In particular, it eliminates events where one TOF is very near zero.

The TOF board located directly behind the MCP anode (Fig. 7) performs these logic timing determinations, monitors overall rates, and performs other housekeeping duties. Signals from this board (at the MCP high voltage) are transferred to the interface board (at sensor ground) through a pair of optical links (one for signals into the board and the other for signals out). The interface board controls TOF and PAC high voltages and is connected to the IBEX Combined Electronics Unit (CEU) through a serial port. The CEU provides conditioned, low-voltage power (± 12 V, $+5$ V) to the interface board, and this board distributes power that is used to create the PAC, MCP, and TOF digital voltages in the TOF HV supply (see Fig. 2).

Voltages for other parts of the IBEX-Lo sensor come from the CEU. These include the entrance subsystem voltages (high voltages for the electron repeller ($CO- = -4.1$ kV) and the collimator ($CO+ = +11$ kV)), optics voltages (U+, 4.8 kV and U−, −2.1 kV) and

Table 2 Optics voltage settings for the IBEX-Lo sensor for the normal and special oxygen science modes

Heliospheric hydrogen science mode			
Energy Step	U+ Voltage (volts)	U− Voltage (volts)	Center energy of incident H neutrals (eV)
1	42.1	−16.8	14
2	81.2	−32.3	27
3	156	−62.2	52
4	307	−122	102
5	592	−236	197
6	1212	−482	451
7	2346	−934	908
8	4511	−1795	1903
Interstellar oxygen mode			Center energy of incident O neutrals (eV)
Jan–Feb	1035	−414	534
Oct–Nov	81	−32	33
Interstellar helium mode			Center energy of incident He neutrals (eV)
Jan–Feb	381	−152	134
Oct–Nov	41	−16	8.3

the voltage used to shut off low-resolution quadrants when in the high-resolution mode (Uso $= -2.5$ kV). All of these voltages are commandable to several levels, even voltages that are "fixed" at a specific voltage (e.g., Uso) in nominal science operations. Ion optics voltages on various electrodes are determined by the set point of the U+ and U− voltages and two high precision, high resistance resistor strings, one for each voltage. The CEU sets the five high voltages, controls their changes, and commands the TOF interface board for a particular science or engineering mode. The CEU is described in the IBEX flight segment description (Scherrer et al. 2009).

Set points for the optics voltages are shown in Table 2. For the eight energy channels in the normal (heliospheric hydrogen) science mode, the voltages fix center energies of the ESA passband that are 15% lower than the center energies of the incident neutral hydrogen. These settings assume that negative ions leave the conversion surface with 15% less energy than the incident neutrals. In the calibration, it was discovered that this energy loss is energy and mass dependent. Therefore, the highest two energy channels are not evenly logarithmically spaced from the first six channels. Also, for the special oxygen and helium modes in the spring and fall, the energy loss for oxygen off the conversion surface is much larger than that for hydrogen, so the voltages are set for correspondingly lower negative ions off the conversion surface.

3.7 Star Sensor

Accurate, absolute directional determination of interstellar neutral oxygen is critical for IBEX science closure. Therefore, a star sensor is co-aligned with the IBEX-Lo sensor to determine the absolute neutral oxygen arrival direction with respect to several stars. The star

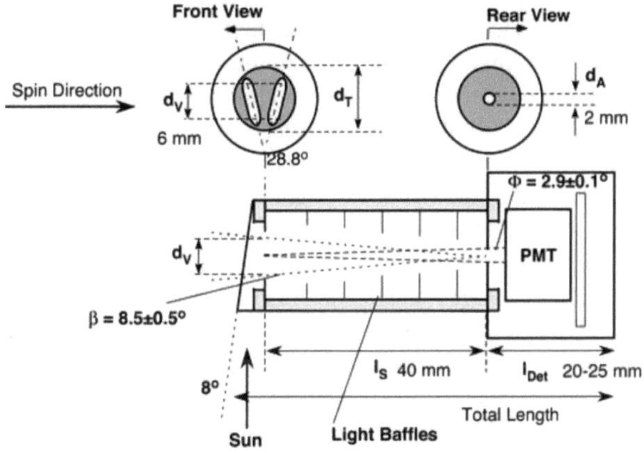

Fig. 10 Schematic of the IBEX-Lo star sensor. The front aperture is shaped into a "v" so that as a star passes in front of the spinning aperture, a double pulse is produced in the photomultiplier tube (PMT). The time between the two pulses is used to determine the elevation angle (up/down in the figure) of a star. The azimuthal (within the spin plane) direction is determined from the center time of the two pulses and spacecraft attitude information

sensor provides data for determining positions of as many stars as possible to an accuracy of ±0.1° relative to the IBEX-Lo collimator bore sight (after ground processing).

The star sensor basic design is shown in Fig. 10, and a picture of the star sensor attached to the optics deck (with a protective cover over the aperture) is shown in Fig. 3. The star sensor operates similarly to a Sun sensor on a spinning spacecraft. It consists of an entrance aperture and collimation tube, exit pinhole, and photomultiplier tube (PMT). The entrance aperture has two slits in the shape of a "V". As the spacecraft spins, the light curve from a star in the FOV generates two 3° wide triangular shaped peaks with full-width half-maximum separation equal to the angular aperture width. The time difference between peaks determines the elevation angle of the star with respect to the star sensor bore sight. The IBEX spin period is planned to be 4 ± 0.5 rpm. Calculations show that an integration period of 11 ms is equivalent to 1° in the spin direction. Given a FWHM of 3° for the triangular shaped peak for a star, this resolution is adequate for determining star directions within the accuracy requirements. The star sensor is sensitive to between 50 and 100 stars brighter than magnitude 2.5 in the visible part of the spectrum.

Star sensor signals are accumulated in CEU memory over multiple spins (typically ∼ 64) in 720, 0.5° bins to form a 360° histogram. The absolute reference of this histogram is the spacecraft spin pulse, which is provided by the spacecraft to the CEU. These data are processed on the ground to determine the absolute direction of the star from the azimuth location of the two peaks in the spin plane and the time separation of the two peaks.

3.8 Prototype Tests Prior to Sensor Development

Prior to the IBEX-Lo design phase, a prototype IBEX-Lo sensor was developed and tested using both ions and neutrals (Wieser 2005; Wieser et al. 2007). This prototype had the same basic geometry and design as the flight sensor except that the conversion surface was placed on the outside circumference of the prototype. Also, for initial tests, a single microchannel plate detector was used in place of the TOF mass spectrometer. Test and calibration were performed at the University of Bern MEFISTO calibration facility (Marti et al. 2001). This facility provides a calibrated neutral beam in the energy range from 10 eV to 3 keV. The same facility was used to test and calibrate the IBEX-Lo flight sensor.

The prototype design was somewhat different because one of the tests used a positive ion beam injected at the position and angle of specularly reflected negative ions off the conversion surface. These tests verified the ion optics properties of the ESA. In particular, it verified radial focusing and azimuthal defocusing properties of the ion optics system. Positive ions were used (with the appropriate reversal of ion optics voltages) because ion beam angular width, energy spread, and flux are much better controlled than neutral beam parameters. After these tests, the aperture was replaced with a DLC conversion surface like the one used in the IBEX-Lo flight sensor, ion optics voltages were reversed, and a neutral beam was used to complete testing.

The neutral beam was produced by surface neutralization (Wieser and Wurz 2005): a 3 kV ion beam (H^+ and O^+ beams were used for the prototype testing, and H^+, O^+, He^+ and C^+ were used in the IBEX-Lo flight sensor calibration) with narrow (about 3 eV) energy spread was injected into the neutralizer unit (Wieser and Wurz 2005). In the neutralizer unit, the ion beam was slowed in an ion deceleration stage by retarding potentials to select a beam energy from 10 eV to a maximum of 3 keV (i.e., no deceleration of the initial beam). The decelerated ion beam was directed onto a highly polished, mono-crystalline tungsten surface at a very shallow angle (10°) where it is very efficiently neutralized. Residual ions in the resulting neutral beam were deflected away from the neutralizer exit slit using a set of electrostatic deflection plates. The neutral beam had a large energy and angle spread caused by the reflection/neutralization process. This large energy and angle spread complicates analysis of the test data. The current off the surface was calibrated prior to prototype tests (and also prior to and after the flight sensor calibration) so that absolute neutral fluxes were known to ~ 20–30%.

Prototype tests using the neutral beam confirmed the energy resolution of the system, verified the ESA transmission function ($= 0.4$, approximately independent of energy), and demonstrated the overall geometric factor of the sensor. Later, the sensor was upgraded to include "fins" on the outer ESA (like the ones in Fig. 2) and to include magnets. The upgrades verified the importance of both the fins in reducing the scattered ion background and the magnets in reducing electron transmission through the ESA.

Upon completion of the tests of the upgraded prototype, the flight sensor design was developed by starting with the prototype geometry and adjusting and optimizing voltages and geometries of the electrodes. This optimization was done in an iterative process using a computer code (Wieser et al. 2008) to maximize sensor throughout and add features that simplify manufacturing and reduce the number of high voltage supplies needed to control the ion optics.

3.9 Flight Sensor Calibration and Performance

For the flight sensor, the entrance and mass analysis subsystems were tested separately. Tests of the entrance system were done to verify collimator performance including transparency, energetic particle rejection, and off axis leakage. In all of these tests, the collimator performed within the specifications. Figure 11 shows the solid angle FOV of the low-resolution quadrants of IBEX-Lo as obtained by a Monte-Carlo simulation with maximum manufacturing tolerances on the etched plates that make up the collimator. The simulated leakage over the entire accessible angle space was less than 10^{-6} of the total FOV, (i.e. two orders of magnitude better than the requirement). Combined effects of leakage through neighboring channels of the collimator and scattering off edges of the collimator plates were investigated using the collimator, a detector, and an intense Argon ion beam. An angular scan across the collimator FOV, with the ion beam intensity increased by a factor of 100 for large angles θ, is shown in Fig. 12. The observed particle rate outside the collimator FOV is typically a factor of 10 below the required suppression of 10^{-4}.

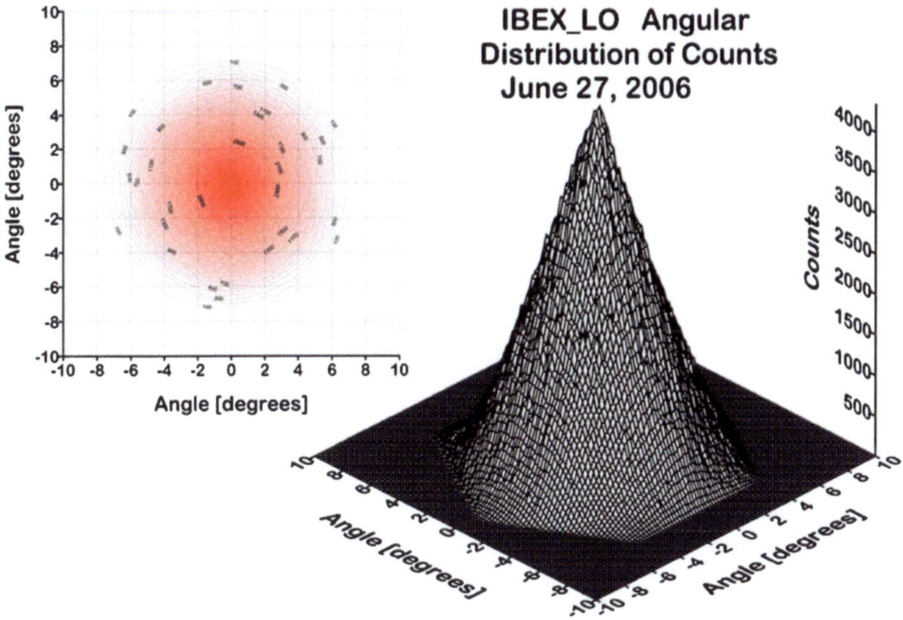

Fig. 11 Calibration results. Transmission function of high resolution sectors of the IBEX-Lo collimator. The transmission is nearly symmetric with a FWHM of $\sim 7^\circ \times 7^\circ$

Fig. 12 On- and off-axis performance of the IBEX-Lo collimator. The on-axis profile (*blue curve*) shows the near-gaussian response of the collimator. The off-axis profile (*red curve*, note the change in scale by a factor of 100), is well below the requirement for leakage outside of $\sim 14^\circ$ yaw angle

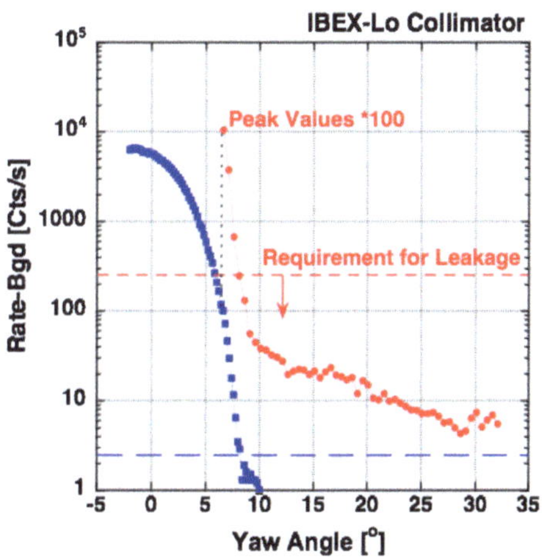

A positive biased collimator collects plasma electrons from the environment and photo-electrons emitted from the sun-lit side of the spacecraft. In significant numbers, these electrons could be responsible for a substantial background. Thus, electron suppression, especially at low energies, is a critical requirement for the IBEX-Lo sensor. The suppression factor for electrons is shown as a function of electron energy in Fig. 13. These data are taken from tests with the IBEX-Lo entrance subsystem and an electron beam.

The mass analysis subsystem was sufficiently complicated that it required separate testing. In addition, a significant simplifying feature of the mass analysis subsystem was that

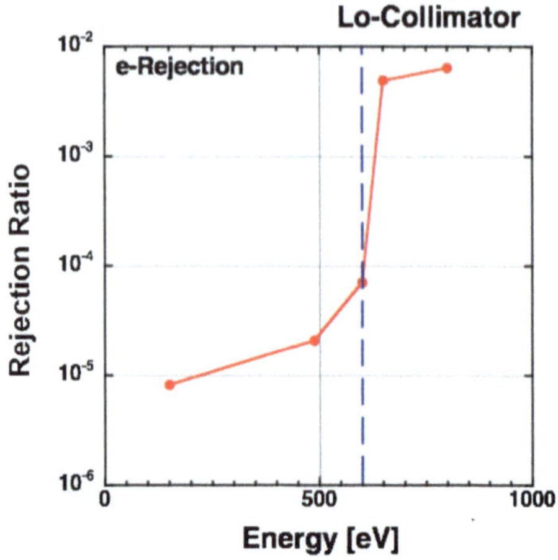

Fig. 13 Measured electron rejection properties of the IBEX-Lo collimator. With −3.1 kV on the electron rejection rings (see Fig. 2), electron fluxes below 600 eV are reduced by almost 3 orders of magnitude

voltages could be reversed and positive ions could be used to verify basic performance, mass resolution, and overall efficiency. These positive ion beam tests were performed at the University of New Hampshire. Similar to the prototype tests, positive ions have the advantage that ion beam angular width, energy spread, and flux are much better controlled than similar neutral beam parameters.

The mass analysis subsystem performed within specifications. Because singles rates and all double coincidence rates are monitored in the TOF and because the triple coincidence rate is determined from these TOF events, the absolute TOF efficiency can be determined independent of whether the TOF is tested with the sensor or tested alone. In particular, because the triple and double coincidence rates can be used to derive detector efficiencies, the IBEX-Lo TOF subsystem is fully self-calibrating, even in flight, without need to correct for any background or sensor inefficiencies. Figure 14 shows TOF double and triple coincidence efficiencies for hydrogen and oxygen as a function of MCP voltage. These tests were done during final calibration, but they confirm measurements from the University of New Hampshire ion beam tests prior to sensor assembly.

The sensor was assembled and tested in two stages. First, the ENA to ion conversion subsystem, energy analysis subsystem, and mass analysis subsystems were tested (without the entrance subsystem) using the neutral beam. These initial "pre-cal 1" tests verified basic sensor performance in a configuration similar to the prototype tests. In particular, the calibration verified that the energy subsystem had an overall throughput of 0.4, essentially independent of energy. After these first tests, the complete sensor was tested and calibrated with the neutral beams. Figures 15 and 16 show the installation of the complete IBEX-Lo sensor into the calibration vacuum chamber and the sensor as installed for the final calibration tests. Figure 15 provides a good perspective of the sensor size compared to a person. The sensor was installed in a rotation stage that was mounted on a 5-axis motion table. The rotation stage allowed tests of the azimuthal response of the sensor while the motion table was used to test the radial and radial angle response. In Fig. 16, the neutral beam source is at the left. (The ion beam that feeds this neutralizer source is in a separate beam line that enters the chamber from the left of the picture.)

Figure 17 shows sample results from the IBEX-Lo calibration. The eight peaks in Fig. 17 correspond to the 8 energy bins of the sensor and show the energy response. These data were obtained by setting the neutral beam to the center energy of a particular energy bin and scanning the sensor energy acceptance over the beam energy. The flux measured at each

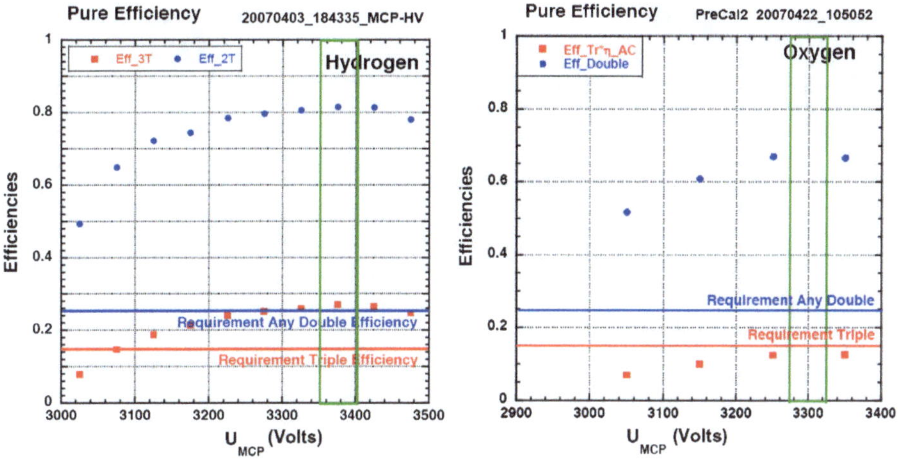

Fig. 14 TOF detection efficiencies for hydrogen (*left panel*) and oxygen (*right panel*) as a function of the MCP voltage. The MCP voltage will be set so that the detector is run in saturation mode. In this mode, the double and triple coincidence efficiencies are well above the requirements

Fig. 15 Installation of the IBEX sensor into the calibration vacuum chamber at the University of Bern. The sensor is installed in a rotation stage so that the neutral beam can be directed into different parts of the collimator. The rotation stage is mounted on a motion table that allows vertical and horizontal motion to investigate sensor radial and radial angle responses

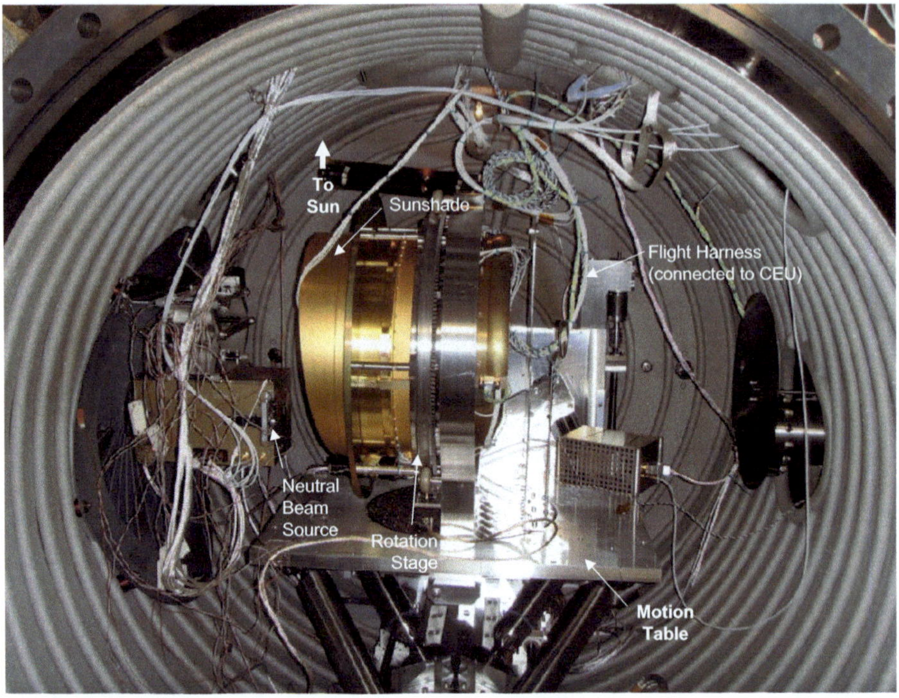

Fig. 16 IBEX-Lo installed in the calibration chamber. The neutral beam is mounted on the left and directs neutrals into the lower part of the collimator

sensor energy setting was normalized to the flux measured at the nominal center energy of the beam. Although some energy spread is due to the broad energy spread of the incident neutral beam, most of the energy spread in Fig. 17 is due to the broad $\Delta E/E$ of the energy analysis subsystem. These data were used to derive the $\Delta E/E$ of the sensor. The sensor $\Delta E/E = 0.8$ and is the combined passband of the ESA alone (which is predicted to be ~ 0.67) and the extraction system of the conversion surface. This $\Delta E/E$ is constant over the energy range.

Analysis of the calibration data indicated that the peak countrate does not occur when the sensor energy step is the same as the center energy of the beam. At low energies (below several hundred eV), this discrepancy is explained by the difficulty to produce a neutral beam with energies between ten and several hundred eV in the calibration facility. At all energies, the center energy of the beam is difficult to predict because neutrals lose energy off the tungsten neutralization surface and then lose more energy when they interact with the conversion surface in the sensor. However, measured center energies of the first 5 energy steps correspond reasonably well to the predicted energies. For the last 3 energy steps, energy loss off the IBEX-Lo conversion surface is greater than the predicted value of 15%. Figure 18 shows the energy loss of hydrogen and oxygen off the conversion surface as a function of incident neutral energy. Since voltages on the ESA were designed to pass negative ions with 15% less than the incident neutral energy, greater loss off the conversion surface translates into a higher incident neutral energy. Thus, in Table 2, the center energy of energy steps 6, 7, and 8 are separated from one another and from lower energy steps by greater than a logarithmic spacing (but still without any gaps). For oxygen, energy losses off the conversion surface are

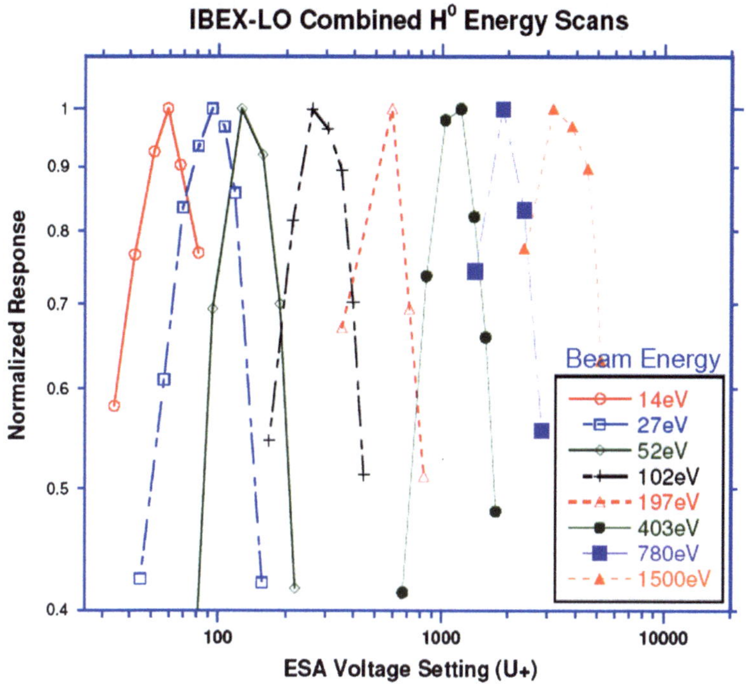

Fig. 17 Measured hydrogen fluxes in each of the 8 energy bins as a function of the ESA voltage. In this test, the neutral beam energy was fixed at the values shown in the legend. Energy bins have significant overlap with a $\Delta E/E$ of 0.8

Fig. 18 Negative ion energy loss as a function of neutral beam energy. Negative ions lose energy off the conversion surface and this energy loss is species and beam energy dependent. The ESA bandpasses are designed to account for this energy loss

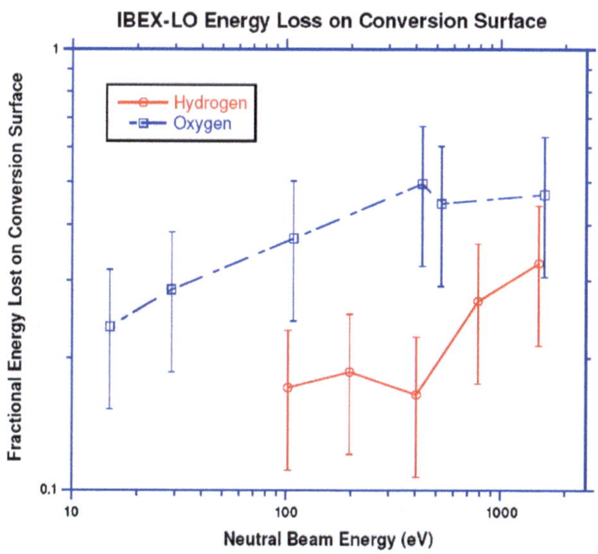

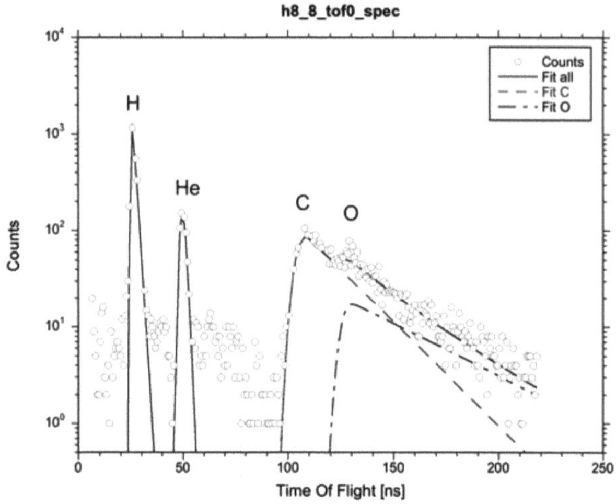

Fig. 19 TOF mass spectrum from the IBEX-Lo flight sensor. A neutral helium beam was used in this test. The masses observed include helium (converted to He- at the conversion surface) and H, C, and O sputtered from the conversion surface and from the breakup of water on the conversion surface

even higher as are the corresponding center energies of incident oxygen neutrals. However, the science objectives of the sensor focus on detection of heliospheric hydrogen neutrals. Thus, for normal science operations, ESA voltages and corresponding energy steps are designed to produce a quasi-logarithmically spaced set of energy channels for hydrogen from 10 eV to ~ 2 keV. The energy steps for oxygen are determined from these voltages, but the voltages are set for hydrogen. For the special operations to detect interstellar neutral oxygen in October and January, the sensor energy step is fixed for the center energy of arriving oxygen neutrals. These voltages, using the calibrated higher energy loss off the conversion surface, are shown in Table 2.

Figure 19 shows a mass spectrum measured by the IBEX-Lo sensor. This figure illustrates important properties of low energy neutral detection using the conversion surface ionization technique. For this test, a 1.5 keV neutral helium beam was directed into the IBEX-Lo sensor. The sensor was set to detect neutrals centered at 1.5 keV, the same energy as the beam. There are several mass peaks in the spectrum. The H, C, and O mass peaks are caused by recoil sputtering of negative ions from the DLC conversion surface by the neutral helium beam. Only the mass peak identified as He is produced by true conversion of neutral helium into He^- on the conversion surface. Unlike neutrals with high electron affinity, He^- is not stable and survives only because of the relatively short flight time from the conversion surface to the TOF entrance (Wurz et al. 2008a). The ionization efficiency for helium is very low ($\sim 10^{-5}$), thus the peak is considerably lower than the recoil sputtered products. For other neutral beams with high electron affinity (i.e., neutral hydrogen or oxygen), nearly 100% of the signal observed at beam energies ~ 1 keV is true conversion to a negative ion. However, all neutrals produce sputtered products at low energies ($\sim$ 10's of eV and greater). Thus, analysis of the IBEX-Lo signal at low energies requires knowledge of the flux of high energy (~ 1 keV) neutrals on the surface so that low-energy sputtered products from these neutrals can be subtracted from the observed total flux.

Table 3 shows sensor geometric factors for each energy step for double and triple coincidence hydrogen. These factors were determined from the calibration and include the sensor $\Delta E/E$, collimator solid angle FOV, all of the efficiencies of transmission through the collimator, internal grids transmission, effects of the spokes that separate each azimuthal quadrant, the energy dependent conversion efficiency, and TOF efficiencies.

Table 3 $G\Delta E/E$ for the IBEX-Lo sensor, determined from calibration

Energy Step	Center Energy (eV)	$G\Delta E/E$ for any double coincidence hydrogen (cm^2 sr eV/eV)	$G\Delta E/E$ for triple coincidence hydrogen (cm^2 sr eV/eV)
1	14	7.5×10^{-5}	2.7×10^{-5}
2	27	1.5×10^{-4}	5.3×10^{-5}
3	52	2.2×10^{-4}	8.1×10^{-5}
4	102	2.5×10^{-4}	9.1×10^{-5}
5	197	2.5×10^{-4}	9.0×10^{-5}
6	451	2.9×10^{-4}	1.0×10^{-4}
7	908	5.4×10^{-4}	1.9×10^{-4}
8	1903	7.6×10^{-4}	2.7×10^{-4}

Background suppression is a critical element of the IBEX-Lo design. Suppression of any background that can masquerade as signal neutrals is particularly important because the heliospheric neutral source strength is low. In a separate paper in this volume (Wurz et al. 2009), background sources are discussed in detail. In nearly all instances, background consists of positive ions produced at or behind the collimator exit that are accelerated to high voltage by the collimator positive voltage. This background flux depends on the residual gas pressure in this part of the sensor. To reduce this gas pressure, the sensor has significant vent paths that bypass this critical region (see Fig. 2), sensor electronics are vented separately from sensor optics, there are no vent paths to the spacecraft interior, and materials in the optics path were carefully chosen for their low outgassing properties. Based on expected on-orbit electron, ion, and photon fluxes that produce background, a residual gas pressure of $\sim 10^{-8}$–10^{-7} mbar is needed to keep signal to noise > 10 over the full energy range of the sensor.

Estimating residual gas pressure in the region behind the collimator is very difficult. The internal pressure depends strongly on the pumping speed of the sensor, which is determined by the ratio of interior to exterior gas pressures and the amount of pumping area available.

Because the sensor internal pressure drives most of the important background levels, this pressure was measured and compared to the pressure inside the calibration vacuum chamber. The internal pressure was measured using a nude ion gauge installed in the access port at the bottom center of the sensor (see Fig. 2). Because this location is deep inside the sensor, the pressure in this region is probably higher there than in the region just in back of the collimator. Table 4 shows results of this pressure test. The pressure was measured as the vacuum chamber pressure was decreasing by over an order of magnitude over a few hours. The internal to external pressure ratio is only a factor of ~ 2, presumably because of the extensive measures used to vent the sensor. It is doubtful that this internal to external pressure ratio will remain a factor of two over many more orders of magnitude in pressure. However, these results are encouraging for on-orbit performance, where external pressures are expected to be 10^{-10}–10^{-14} mbar. With these low pressures, the pressure in the critical region behind the collimator is likely to be $\ll 10^{-8}$ mbar and the background will be correspondingly low.

3.10 Sensor Operation

IBEX-Lo sensor operations on orbit are relatively simple. After initial on-orbit checkout and high voltage turn-on, the sensor operates in a single science mode during most spacecraft

Table 4 Internal sensor and external vacuum chamber pressures measured during the IBEX-Lo calibration

External (vacuum chamber) pressure (mbar)	Internal (sensor) pressure (mbar)	Ratio internal/external pressure
1.2×10^{-6}	2.5×10^{-6}	2.1
6.0×10^{-7}	1.4×10^{-7}	2.3
1.6×10^{-7}	3.8×10^{-7}	2.4

Table 5

Product	Bits per second (bps)	Comments
Quaternions	57	Spacecraft attitude
Payload	12	Sensor housekeeping
IBEX-Hi total	98	Direct events and histograms
IBEX-Lo total	115	Direct events and histograms
IBEX-Lo Telemetry Detail		
Direct events	95	About 3 events per second
H histogram	7	Various start/stop rates and additional monitor rates
O histogram	7	
Star sensor	6	Collected with 0.5° resolution

orbits. At low altitudes (< 10 Earth radii, R_E), the sensor is switched from science mode to a standby mode. In this mode, high voltages are turned down or off to eliminate high countrates in the Earth's radiation belts. Since science operations are performed above this altitude, there is no loss of science in these standard operations.

In science mode, the sensor is set at a fixed energy step for 2 spins, so that the entire energy range is sampled in 16 spins. These measurements are repeated without interruption over the entire science operations part of the spacecraft orbit.

Twice per year, the sensor is switched into a special interstellar neutral oxygen and helium mode in the science operations part of the spacecraft orbit. These times are shown in Fig. 1 and this mode is described in more detail in Möbius et al. (2009). During the fall interstellar neutral observing period, only oxygen is observed (helium is below the energy range of the sensor). In the part of the spacecraft spin when the IBEX-Lo sensor is viewing ±30° around the ecliptic, the standard 2-spin energy step sequence is interrupted and the sensor is set at a fixed energy corresponding to the expected energy of interstellar neutral oxygen.

In the spring interstellar neutral observing period, there is a similar interruption of the standard 2-spin energy stepping sequence. This time, low resolution sectors are electrostatically switched off in the region ±30° around the ecliptic and the sensor is set at a fixed energy corresponding to the expected energy of the interstellar neutral oxygen. This sequence is repeated for 7 spins. Every 8th spin, the sensor is set at a fixed energy corresponding to the expected energy of interstellar neutral helium.

3.11 Data Products

Regardless of science mode, data products and telemetry from the IBEX-Lo sensor are the same. Table 5 shows the data products that are accumulated over 64 spins (about 16 min) and transmitted to ground during spacecraft perigee passes. The telemetry budget allows an average of 282 bps of payload telemetry (115 bps for IBEX-Lo) for most of the 8-day orbit.

Although IBEX-Lo is a triple coincidence TOF mass spectrometer, the triple coincidence is composed of double coincidence measurements. Therefore, considerably more information at higher sensitivity is available from the sensor. In particular, there are 4 double coincidence times of flight available, TOF 0, 1, 2, and 3. TOF0 is the time between the first start and the stop signals, TOF1 between the first start and the second start, TOF2 between the second start and the stop, and TOF3 is the delay line signal. TOF3 (anode 3) determines the quadrant where the stop signal originated. Each TOF is determined separately and encoded to 11 bits (0.16 ns resolution). A triple event has all 4 TOFs valid and a "golden" triple satisfies the following equation:

$$\text{Checksum} = \text{TOF0} + \text{TOF3} - \text{TOF2} - \text{TOF1} \sim 0. \tag{2}$$

The golden triples are the most interesting events because they represent the best, lowest noise signal. A full description of these events is available for transmission to ground provided the event rate is ~ 1 event/second (which is about a factor of 10 higher than the current expected rate for heliospheric neutrals). Consistent with the primary science objectives, the IBEX CEU must avoid using all of the telemetry allocation on a bright source (like the magnetosphere) and to give priority to the heliospheric measurements (which are distributed somewhat evenly over the spin). This prioritization is done in the CEU, using an algorithm that distributes the number of golden events allowed approximately uniformly over the sky. The essential component of the algorithm is that direct events are collected over an interval that is short enough that the heliosphere contributes only a count or two in any sky bin, and these are all sent to ground. Anomalously bright regions (e.g., the magnetosphere) will saturate and are counted in histograms, but all direct events from these regions are not telemetered.

Using these data, the number of triple events that could not be included in the telemetry can be reconstructed. In this manner, the actual strength of bright sources such as the magnetosphere or possibly a planetary source (e.g., Jupiter) can be determined.

4 Summary

The IBEX-Lo sensor is a single pixel, large geometric factor camera. It detects 10 eV to 2 keV heliospheric neutral hydrogen in 8, rather broad energy bands. The sensor uses surface conversion to convert neutrals into negative ions and then accelerates the ions so that they can be deflected away from UV background and can be analyzed with higher efficiency. The sensor uses a triple coincidence TOF mass spectrometer to distinguish hydrogen, oxygen, and helium from the heliosphere and from the interplanetary medium. The sensor has undergone significant test and calibration that have demonstrated that it meets requirements for the IBEX mission. Together with the IBEX-Hi sensor, this sensor will achieve the science objectives of the mission.

Acknowledgements The IBEX-Lo sensor is the result of efforts from a large number of scientists and engineers located at many institutions around the world. From conceptual design to final cross-calibration, this sensor has been a collaborative effort. All who contributed to this sensor share in its success. The success of this sensor would not be possible without the strong support of the IBEX project. This support went well beyond standard contractual and financial support and included creative solutions to procurement, processing, and testing issues that occurred during the course of the design, development, and test of the sensor.

References

S.A. Fuselier, E.S. Claflin, S.B. Mende, C.W. Carlson, T.E. Moore, Combined in situ and remote sensing of ionospheric ion outflow. Geophys. Res. Lett. **33**, L04103 (2006). doi:10.1029/2005GL024055

A. Galli, P. Wurz, S. Barabash, A. Grigoriev, R. Lundin, Y. Futaana, H. Gunell, M. Holström, E.C. Roelof, C.C. Curtis, K.C. Hsieh, A. Fedorov, D. Winningham, R.A. Fram, R. Cerulli-Irelli, P. Bochsler, N. Krupp, J. Woch, M. Fraenz, Direct measurements of energetic neutral hydrogen in the interplanetary medium. Astrophys. J. **644**, 1317 (2006).

A.G. Ghielmetti, E.G. Shelley, S.A. Fuselier, F. Herrero, M.F. Smith, P. Wurz, P. Bochsler, T. Stephen, Mass spectrograph for imaging low energy neutral atoms, in *Instrumentation for Magnetospheric Imaging II*, ed. by S. Chakrabarti, Proc. SPIE 2008, pp. 105–112, 1993; Opt. Eng. **33**, 362 (1994)

Gruntman, Magnetospheric imaging (1993)

M. Gruntman, E.C. Roelof, D.G. Mitchell, H.J. Fahr, H.O. Funsten, D.J. McComas, Energetic neutral atom imaging of the heliospheric boundary region. J. Geophys. Res. **106**, 15767 (2001)

S. Jans, P. Wurz, R. Schletti, T. Fröhlich, E. Hertzberg, S. Fuselier, Negative ion production by surface ionization using aluminum-nitride surfaces. J. Appl. Phys. **85**, 2587 (2000)

A. Marti, R. Schletti, P. Wurz, P. Bochsler, Calibration facility for solar wind plasma instruments. Rev. Sci. Instrum. **72**, 1354 (2001)

D.J. McComas, P. Valek, J.L. Burch, C. Pollock, R.M. Skoug, M.F. Thomsen, Filling and emptying of the plasma sheet: Remote observations with 1–70 keV energetic neutral atoms. Geophys. Res. Lett. **29**, 2079 (2002). doi:10.1029/2002GL016153

D.J. McComas et al., The interstellar boundary explorer (IBEX), in *Physics of the Outer Heliosphere, Third Annual IGPP Conference*, ed. by V. Florinski, N.V. Pogorelov, G.P. Zank, AIP CP719 (2004), p. 162

D.J. McComas et al., Space Sci. Rev. (2009, this issue)

H. Moestue, The electric field and geometrical factor of an annular curved plate electrostatic analyzer. Rev. Sci. Instrum. **44**, 1709 (1973)

T.E. Moore, D.J. Chornay, M.R. Collier, F.A. Herrero, J. Johnson, M.A. Johnson, J.W. Keller, J.F. Laudadio, J.F. Lobell, K.W. Ogilvie, P. Rozmarynowsky, S.A. Fuselier, A.G. Ghielmetti, E. Hertzberg, D.C. Hamilton, R. Lundgren, P. Wilson, P. Walpole, T.M. Stephen, B.L. Peko, B. Zyl, P. Wurz, J.M. Quinn, G.R. Wilson, The low-energy neutral atom imager for IMAGE, in *The IMAGE Mission*, ed. by J.L. Burch. (Kluwer, Dordrecht, 2000). Space Sci. Rev. **91** (2000), pp. 155–195

E. Möbius, D. Hovestadt, B. Klecker, L.M. Kistler, M.A. Popecki, K.N. Crocker, F. Gliem, M. Granoff, S. Turco, A. Anderson, H. Arbinger, S. Battell, J. Cravens, P. Demain, J. Distelbrink, I. Dors, P. Dunphy, J. Gaidos, J. Googins, A. Harasim, R. Hayes, G. Humphrey, H. Kästle, E. Künneth, J. Lavasseur, E.J. Lund, R. Miller, G. Murphy, E. Pfeffermann, K.-U. Reiche, E. Sartori, J. Schimpfle, E. Seidenschwang, M. Shappirio, K. Stöckner, S.C. Taylor, M. Vosbury, W. Wiewesiek, V. Ye, The solar energetic particle ionic charge analyzer (SEPICA) and the data processing unit (S3DPU) for SWICS, SWIMS and SEPICA. Space Sci. Rev. **86**, 447 (1998a)

E. Möbius, L.M. Kistler, M. Popecki, K. Crocker, M. Granoff, Y. Jiang, E. Sartori, V. Ye, H. Rème, J.A. Sauvaud, A. Cros, C. Aoustin, T. Camus, J.L. Médale, J. Rouzaud, C.W. Carlson, J.P. McFadden, D.W. Curtis, H. Heetderks, J. Croyle, C. Ingraham, E.G. Shelley, D. Klumpar, E. Hertzberg, B. Klecker, M. Ertl, F. Eberl, H. Kästle, E. Künneth, P. Laeverenz, E. Seidenschwang, G.K. Parks, M. McCarthy, A. Korth, B. Gräwe, H. Balsiger, U. Schwab, M. Steinacher, The 3-D plasma distribution function analyzers with time-of-flight mass discrimination for CLUSTER, FAST and Equator-S, measurement techniques in space plasmas, ed. by R. Pfaff, J. Borowski, D. Young. Geophys. Monograph 102 (1998b), p. 243

E. Möbius, S. Fuselier, M. Granoff, E. Hertzberg, B. King, H. Kucharek, S. Livi, S. Longworth, N. Paschalidis, L. Saul, J. Scheer, C. Schlemm, M. Wieser, P. Wurz, Time-of-flight detector system of the IBEX-Lo sensor with low background performance for heliospheric ENA detection, Proc. of the 30th Int. Cosmic Ray Conf., on CD (2007)

E. Möbius et al., Space Sci. Rev. (2009, this issue)

N.P. Paschalidis et al., A CMOS time of flight system on a chip for spacecraft instrumentation. IEEE Trans. Nucl. Sci. **49**, 1156–1163 (2002)

N.P. Paschalidis, Advanced system on a chip microelectronics for spacecraft and science instruments. Acta Astronaut. **52**(2–6), 411–420 (2003)

Pollock et al., Medium energy neutral atom (MENA) imager for the IMAGE mission, in *The IMAGE Mission*, ed. by J.L. Burch (Kluwer, Dordrecht, 2000), pp. 113–154

J.A. Scheer, M. Wieser, P. Wurz, P. Bochsler, E. Hertzberg, S.A. Fuselier, F.A. Koeck, R.J. Nemanich, M. Schleberger, High negative ion yield from light molecule scattering. Nucl. Instr. Meth. Phys. Res. B **230**(1–4), 330–339 (2005). doi:10.1016/j.nimb.2004.12.063

J.A. Scheer, M. Wieser, P. Wurz, P. Bochsler, E. Hertzberg, S.A. Fuselier, F.A. Koeck, R.J. Nemanich, M. Schleberger, Conversion surfaces for neutral particle imaging detectors. Adv. Space Res. **38**(4), 664–671 (2006)

J. Scheer, P. Wahlström, P. Wurz, E. Hertzberg, S. Fuselier, Scattering properties of hydrogen and oxygen on artificial diamond surfaces using in space flight, Nucl. Instr. Meth. B (2008, in preparation)
J. Scherrer et al., Space Sci. Rev. (2009, this issue)
X. Shelley et al., The toroidal imaging mass-angle spectrography (TIMSA) for the polar mission. Space Sci. Rev. **71**, 497 (1995)
P. Wahlström, J. Scheer, P. Wurz, E. Hertzberg, S. Fuselier, Calibration of charge state conversion surfaces for neutral particle detectors. J. Appl. Phys. **104**, 034503-1–034503-6 (2008). doi:10.1063/1.2957064
M. Wieser, Detection of energetic neutral atoms and its application to heliospheric science. PhD Thesis, University of Bern (2005)
M. Wieser, P. Wurz, Production of a 10 eV—1000 eV neutral particle beam using surface neutralization. Meas. Sci. Technol. **16**, 2511–2516 (2005)
M. Wieser, P. Wurz, R.J. Nemanich, S.A. Fuselier, Secondary electron emission of chemical-vapor-deposited diamond by impact of slow H^+, D^+, H_2^+, C^+, O^+, and O_2^+ ions. J. Appl. Phys. **98**, 034906 (2005). doi:10.1063/1.1996855
M. Wieser, P. Wurz, E. Möbius, S.A. Fuselier, E. Hertzberg, D.J. McComas, The ion-optical prototype of the low energy neutral atom sensor of the interstellar boundary explorer mission (IBEX). Rev. Sci. Instr. **78**, 124502-1–124502-14 (2007)
M. Wieser, P. Wurz, E. Möbius, S.A. Fuselier, E. Hertzberg, D.J. McComas, Development of the low energy neutral atom sensor of the Interstellar Boundary Explorer Mission (IBEX). Rev. Sci. Instrum. (2008, in press)
P. Wurz, Detection of energetic neutral particles, in *The Outer Heliosphere: Beyond the Planets*, ed. by K. Scherer, H. Fichtner, E. Marsch (Copernicus Gesellschaft e.V., Katlenburg-Lindau, 2000), pp. 251–288
P. Wurz, P. Bochsler, A.G. Ghielmetti, E.G. Shelley, F. Herrero, M.F. Smith, Concept for the Hi-LITE neutral atom imaging instrument, in ed. by P. Varga and G. Betz, *Proceedings of Symposium on Surface Science*, Kaprun, Austria (1993), p. 225
P. Wurz, R. Schletti, M.R. Aellig, Hydrogen and oxygen negative ion production by surface ionization using diamond surfaces. Surf. Sci. **373**, 56–66 (1997)
P. Wurz et al., Formation of negative ions by scattering from a diamond (111) surface, in *Proc. of the Week of Doctoral Students*, ed. by J. Safrankova, A. Koruka (Charles University, Prague, 1998), p. 257
P. Wurz, J. Scheer, M. Wieser, Particle scattering off surfaces: application in space science. e-J. Surf. Sci. Nanotechnol., **4**, 394–400 (2006)
P. Wurz, L. Saul, J.A. Scheer, E. Möbius, H. Kucharek, S.A. Fuselier, Negative helium generation upon surface scattering: Application in space science. J. Appl. Phys. **103**, 054905 (2008a)
P. Wurz, A. Galli, S. Barabash, A. Grigoriev, Energetic neutral atoms from the heliosheath. Astrophys. J. **683**, 248 (2008b) doi:10.1086/589854
P. Wurz, Space Sci. Rev. (2009, this issue)

Space Sci Rev (2009) 146: 149–172
DOI 10.1007/s11214-009-9498-5

Diagnosing the Neutral Interstellar Gas Flow at 1 AU with IBEX-Lo

E. Möbius · H. Kucharek · G. Clark · M. O'Neill · L. Petersen · M. Bzowski · L. Saul · P. Wurz · S.A. Fuselier · V.V. Izmodenov · D.J. McComas · H.R. Müller · D.B. Alexashov

Received: 17 September 2008 / Accepted: 16 March 2009 / Published online: 22 April 2009

Abstract Every year in fall and spring the Interstellar Boundary Explorer (IBEX) will observe directly the interstellar gas flow at 1 AU over periods of several months. The IBEX-Lo sensor employs a powerful triple time-of-flight mass spectrometer. It can distinguish and image the O and He flow distributions in the northern fall and spring, making use of sensor viewing perpendicular to the Sun-pointing spin axis. To effectively image the narrow flow distributions IBEX-Lo has a high angular resolution quadrant in its collimator. This quadrant is employed selectively for the interstellar gas flow viewing in the spring by electrostatically shutting off the remainder of the aperture. The operational scenarios, the expected data,

E. Möbius (✉) · H. Kucharek · G. Clark · M. O'Neill · L. Petersen
Space Science Center & Department of Physics, University of New Hampshire, Durham, NH 03824, USA
e-mail: Eberhard.Moebius@unh.edu

M. Bzowski
Space Research Centre, Polish Academy of Sciences, Warsaw, Poland

L. Saul · P. Wurz
Physikalisches Institut, Universität Bern, 3012 Bern, Switzerland

S.A. Fuselier
Lockheed Martin Advanced Technology Lab, Palo Alto, CA, USA

V.V. Izmodenov
Moscow State University and Space Research Institute, Russian Academy of Sciences, Moscow, Russia

D.J. McComas
Southwest Research Institute, San Antonio, TX, USA

H.R. Müller
Department of Physics and Astronomy, Dartmouth College, Hanover, NH, USA

D.B. Alexashov
Space Research Institute (IKI) and Institute for Problems in Mechanics, Russian Academy of Sciences, Moscow, Russia

and the necessary modeling to extract the interstellar parameters and the conditions in the heliospheric boundary are described.

The combination of two key interstellar species will facilitate a direct comparison of the pristine interstellar flow, represented by He, which has not been altered in the heliospheric boundary region, with a flow that is processed in the outer heliosheath, represented by O. The O flow distribution consists of a depleted pristine component and decelerated and heated neutrals. Extracting the latter so-called secondary component of interstellar neutrals will provide quantitative constraints for several important parameters of the heliosheath interaction in current global heliospheric models. Finding the fraction and width of the secondary component yields an independent value for the global filtration factor of species, such as O and H. Thus far filtration can only be inferred, barring observations in the local interstellar cloud proper. The direction of the secondary component will provide independent information on the interstellar magnetic field strength and orientation, which has been inferred from SOHO SWAN Ly-α backscattering observations and the two Voyager crossings of the termination shock.

Keywords Interstellar gas · Heliosphere · Instrumentation

1 Introduction and Context

The local galactic environment of the Sun consists of a warm, relatively dilute, partially ionized, and quite structured interstellar gas cloud (e.g. reviews by Cox and Reynolds 1987; Frisch 1995). Apparently, the Sun finds itself close to a cloud boundary, possibly with a significant gradient in the ionization fraction of He (e.g. Cheng and Bruhweiler 1990; Wolff et al. 1999; Slavin and Frisch 2002). The environment and structure of the local interstellar cloud, including integral densities and relative speeds, has been studied on scales of several parsec through UV line absorption by the surrounding medium in the light of nearby stars (e.g. McClintock et al. 1978; Frisch 1981; Crutcher 1982; Lallement and Bertin 1992; Linsky et al. 1993). In a recent workshop "From the Heliosphere to the Local Bubble" at the International Space Science Institute in Bern, Switzerland, the heliosphere and its surroundings have been discussed in the context of their more extended environment of the Local Bubble (Möbius 2009, and references therein). It was also suggested in a consensus decision to call the interstellar medium immediately outside the heliosphere "Circum-Heliospheric InterStellar Medium (CHISM)", which we use henceforth. The influence of the interstellar gas reaches deep into the heliosphere, for example, with the generation of pickup ions (e.g. Möbius et al. 1985; Gloeckler and Geiss 1998) and of anomalous cosmic rays (e.g. Klecker 1995; Jokipii 1998) as well as a slow down of the solar wind (Richardson et al. 1995).

The conditions in the surrounding interstellar medium and their consequences have changed dramatically over the history of the solar system (for a recent comprehensive compilation see Frisch 2006). In particular, the inventory of neutral interstellar gas, its spatial distribution, and its products in the inner heliosphere change substantially with external conditions, as does the filtering at the interface (Möbius et al. 2006; Müller and Zank 2004; Zank et al. 2006). Within the hot and very dilute plasma of the Local Bubble the heliosphere is moving through a succession of warm clouds that are similar to the one it is in currently. In fact, the heliosphere may be at the edge of one cloud while in transition to another (Redfield 2008).

The flow of several neutral gas species through the inner solar system is a very important tool for detailed diagnostics of the local cloud conditions. In this paper we will discuss how the Interstellar Boundary Explorer (IBEX, McComas et al. 2009, this issue) will further this toolset by observing simultaneously at least interstellar oxygen and helium at two different locations in the Earth's orbit. We will begin by laying out the current knowledge and important questions that arose from recent observations, followed by an overview of neutral gas measurements of the interstellar flow, how modeling has to be combined with observations, and some anticipated results. Section 4 will contain a discussion of the IBEX capabilities in terms of the mission design and the instrumentation, followed by the planning of the interstellar flow observation campaigns in Sect. 5, and some conclusions with a look to the future.

2 Current Knowledge of the CHISM and Its Interaction with the Heliosphere

Together with the strength and variations of the solar wind, the very local conditions in the CHISM control the size and shape of the heliosphere. They also are responsible for the processes that control the heliospheric boundary regions. The basic understanding of the heliosphere and its interaction with the interstellar medium has been summarized in early reviews (Axford 1972; Fahr 1974; Holzer 1977; Thomas 1978). Since then substantial progress has been made in the global heliospheric modeling (e.g. Baranov and Malama 1993; Pauls et al. 1995; Zank et al. 1996; Linde et al. 1998; Fahr et al. 2000; Zank and Müller 2003; Malama et al. 2006; Izmodenov et al. 2008; and reviews by Zank 1999; Izmodenov and Kallenbach 2006). However, fixing the parameters in these models requires detailed knowledge of the very local boundary conditions. In a concurrent contribution Frisch et al. (2009, this issue) describe in detail the current knowledge of the interstellar medium that surrounds the heliosphere and how this information has been gathered. Therefore, we will only discuss the observations of interstellar neutrals inside the heliosphere here and how they pertain to the task at hand, i.e. the direct observation of neutral gas velocity distributions from a 1 AU vantage point.

Interstellar neutral gas penetrates into the inner heliosphere as a wind due to the relative motion between the Sun and the local interstellar medium. Through the interplay between this wind, the ionization of the neutrals upon their approach to the Sun, and the Sun's gravitational field (distinctly modified by radiation pressure for low energy H) a characteristic flow pattern and density structure is formed with a cavity close to the Sun and gravitational focusing on the downwind side (for all species except H). Starting with the analysis of backscattered solar Lyman α intensity sky maps (Bertaux and Blamont 1971; Thomas and Krassa 1971), the parameters of H became accessible to observations. Using high-resolution line profiles of Ly-α with the Copernicus spacecraft, Adams and Frisch (1977) obtained first reasonable values for the H bulk speed and constrained its temperature. This method to determine the kinetic H parameters was substantially improved with the use of hydrogen absorption cells (Bertaux et al. 1985). Through the gravitational focusing of the interstellar He flow on the downwind side of the Sun, maps of the backscattered solar He I line at 58.4 nm that became available a few years after the H observations (Weller and Meier 1974) were used to obtain the bulk flow velocity vector, density, and temperature of He (see also Fahr et al. 1978; Wu and Judge 1979). Chassefière et al. (1986) compiled a critical evaluation of all interstellar density measurements, with values for n_{H} from 0.02 to 0.068 cm^{-3}, including their latest value of 0.065 ± 0.01, and n_{He} from 0.0035 to 0.032 cm^{-3}, including theirs of 0.01 ± 0.0045. Puzzling at that time was a counterintuitive difference between H and He in temperature, or depending on the parameter choice in their speed.

The discovery of interstellar He pickup ions at 1 AU (Möbius et al. 1985) introduced a first in-situ method for probing interstellar gas. This enabled an independent determination of the interstellar He flow parameters (Möbius et al. 1995), but the method was hampered in its accuracy by strong variations in pickup ion fluxes (Möbius et al. 1998a) and the discontinuous data set of an Earth orbiting spacecraft. Continuous coverage in interplanetary space out to 5.4 AU and access to H^+ and He^{2+} pickup ions with Ulysses SWICS provided a more precise determination of the H (0.11 cm^{-3}) and He (0.015 cm^{-3}) densities referenced to the termination shock (Gloeckler et al. 1997), and a more direct evaluation of the abundance of minor species, such as N, O and Ne, (Gloeckler and Geiss 2001) than had been obtained through anomalous cosmic rays (Cummings et al. 2002). Finally, direct observations of the neutral gas velocity distribution (Witte et al. 1996) have become available for He, with the most complete information on this key CHISM species yet.

Combining the aforementioned three in-situ observation methods for interstellar He and adding new observations of the relevant ionization rates in the inner heliosphere, Möbius et al. (2004) consolidated the physical parameters known for He from the CHISM, with consensus values of $n_{\rm He} = 0.015 \pm 0.002$ cm^{-3}, $v_{\rm He} = 26.3 \pm 0.4$ km/s, and $T_{\rm He} = 6300 \pm 390$ K. A benchmark set of parameters of the CHISM proper was established because the heliosphere is essentially transparent for He. It became also quite clear that the direct neutral imaging observations provided by Ulysses GAS (Witte 2004) contributed with unambiguous and most detailed information about the kinetics of the CHISM and its flow through the heliosphere. This achievement leaves the total H density, the ionization environment coupled with the filtration of species that show a much stronger charge exchange interaction with the protons of the surrounding interstellar plasma, and the interstellar magnetic field to be determined.

The observation of H pickup ions (Gloeckler and Geiss 2001; Bzowski et al. 2008) and of the solar wind slowdown due to massloading with implanted interstellar H pickup ions (Richardson et al. 2008) provide the means to determine the neutral H density at the termination shock, while still leaving the neutral density in the CHISM up to modeling. Recently, both Voyager spacecraft have reached the termination shock. These observations constrain the overall size of the heliosphere and thus the CHISM pressure in the ram direction of the heliosphere. They also place strong constraints on the neutral and ion density of the surrounding interstellar medium (Izmodenov et al. 2004).

For the typical plasma densities in the surrounding interstellar cloud the mean free path for charge exchange of He exceeds several 1000 AU. The comparable number for H and O is of the order of 200 AU, i.e. comparable to the size of the heliosphere. Therefore, H and O, as well as a number of other species are subject to a substantial filtration between the pristine CHISM and the inner heliosphere (for reviews see Zank 1999; Müller and Zank 2004; Izmodenov et al. 2004; Izmodenov 2007). A noticeable fraction of the original neutral gas density of these species becomes part of the interstellar plasma that is forced to flow around the heliosphere, and thus the observed particle distribution of these species is depleted in the inner heliosphere. This is the definition of filtration. In turn, some of the ions of the outer heliosheath are converted into neutral atoms, a fraction of which has velocity vectors that bring them into the inner heliosphere. These atoms retain the velocity distribution of the outer heliosheath with a higher temperature than the original interstellar neutrals and a slower bulk velocity towards the inner heliosphere. This new distribution of neutrals is referred to as secondary neutrals. It has a distinctly different velocity distribution from the primary neutral gas population. A measurement of the velocity distribution function of all incoming neutrals (e.g., with a technique employed by Ulysses GAS) will provide much more direct diagnostics of the filtration than inferring the depletion factor from combined modeling of the neutral and ion densities inside and outside the heliosphere. Of course, this is true only if both the primary and secondary components can be identified in the combined velocity distribution.

The kinematics of the interstellar neutrals, including their secondaries, also carries the imprint of the interstellar magnetic field in the CHISM. Observations of an offset of the flow direction of interstellar H relative to that of the pristine He (Lallement et al. 2005) have provided insight into the direction and strength of the interstellar magnetic field (Izmodenov et al. 2005). In addition, the Voyager encounters with the termination shock have provided evidence of a distinct asymmetry of the heliosphere, which has also been interpreted as a specific tilt direction of the interstellar magnetic field, which appears to be consistent with what can be derived from the H flow observations (Izmodenov et al. 2005; Opher et al. 2006; Pogorelov et al. 2008). However, the direction determination still contains large uncertainties, and many questions about the strength of the influence of neutrals on the resulting deflection in the necessary global modeling remain (Pogorelov et al. 2006). Furthermore, the heliosphere is a rather dynamic entity, and temporal variations of the solar wind alter the location of the boundary structures in time and space, thus injecting into the observed asymmetry of the termination shock through the two consecutive encounters with the two Voyager spacecraft an additional uncertainty. Although the combined temporal and spatial variations of the shock location in an asymmetric heliosphere in three dimensions have been addressed in some of the global modeling attempts (Washimi et al. 2007; Izmodenov et al. 2008), the combined ambiguities cannot be resolved based on just the two Voyager observations.

Therefore, direct observations of the interstellar flow distribution of a species that is affected by the heliospheric interface will be key to resolving these issues. IBEX will provide observations of O, which will be complementary to the available information from the H backscattering measurements because the strength of the interaction and the kinematics of O are quite different since O is a minor constituent and has a much larger mass than H. Furthermore, this will be yet another independent measurement of the heliospheric asymmetry with a different technique. Also, IBEX will provide such data over the course of several years (in particular if the mission can be successfully extended beyond the primary mission of two years) so that uncertainties arising from temporal variations can be separated.

3 Direct Neutral Gas Observations and Their Inferences

As discussed above, the least model-dependent and most detailed information about the interstellar neutral gas flow through the heliosphere can be obtained from observations of the velocity distribution of the neutral gas under investigation close to the Sun. Keplerian trajectories provide a unique transformation of the original velocity distribution, after the passage through the heliospheric boundary, into an angular distribution as viewed by sensors between 1 and 2 AU from the Sun. After inclusion of all pertinent effects along the neutral trajectories, such as ionization and deflection by the Sun's gravitation and radiation pressure, the distribution at the boundary can be constructed from observations of directional maps of the flow at the peak energy of each species.

This enabling characteristic of the Sun's influence is very important because current observation methods for neutral atoms can provide precise angular images of the flow, but do not retain adequate energy resolution. This is true for the observation of interstellar helium through sputtering off a lithium-fluoride (LiF) surface as used in the Ulysses GAS instrument (Witte et al. 1996) and for negative ion conversion instruments for species, such as O and H (Fuselier et al. 2009, this issue). Therefore, the differential deflection of the interstellar neutral gas flow in the gravitational field of the Sun is utilized to convert speed differentials into an angular image. The analysis of such images through a deconvolution technique has been described in detail by Banaszkiewicz et al. (1996). The very precise temperature and bulk flow values for CHISM He have been derived in this way (Witte 2004).

If the velocity distribution of H and O were studied with the same scrutiny, this would return decisive information about the filtration of these CHISM components, which is needed to derive exact neutral densities and the ionization-state of the CHISM. We will start the discussion with the velocity distribution of interstellar neutrals as they enter the heliosphere at the termination shock after having penetrated the boundary regions. This distribution contains unique information about the related processes, because filtration is accompanied by an effective slowdown and heating of the gases compared with their original state (Izmodenov et al. 1997), and about any deflection of the plasma due to the symmetry breaking role of the interstellar magnetic field (Izmodenov et al. 2005; Opher et al. 2006; Pogorelov et al. 2008). In detail, neutrals are first lost to the plasma flow by charge exchange and now contribute to a rather hot plasma distribution that is diverted around the heliosphere. In turn ions of this hot flow are converted to neutrals, of which a good fraction is directed towards the inner heliosphere. These neutrals constitute an additional secondary and rather hot gas distribution that is added to the original depleted flow. Because the fraction of these secondaries and their distribution scale with the column density of the ions in the outer heliosheath, the observation of the secondary neutral distribution holds a key to understanding the filtration processes in these regions and to constraining the overall geometry.

To estimate the observable effect and to define the parameters of the needed instrumentation we have simulated the neutral O distribution in the inner heliosphere and the respective observations at various positions of a spacecraft. It was assumed that the original O distribution in the CHISM resembles that of He with $v_{\text{Bulk}} = 26.3$ km/s and $T = 6400$ K as obtained by Witte (2004). According to Izmodenov et al. (1997), the filtration was adjusted so that the resulting neutral gas flow contains 50% of the primary distribution. Twenty percent of a secondary distribution with a reduced speed of $v_{\text{Sec}} = 21$ km/s and $T_{\text{Sec}} = 10\,000$ K were added. Figure 1 shows such a simulated velocity distribution of interstellar O as it enters the heliosphere at the termination shock. It is seen that the resulting distribution is asymmetric, with a substantial contribution from the secondary component on the low velocity side of

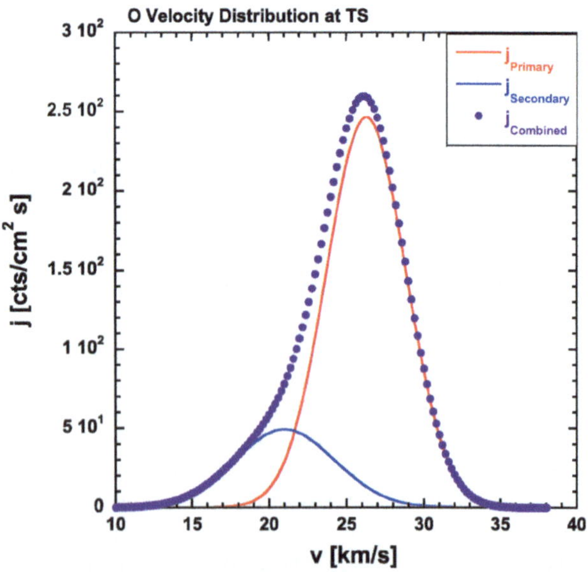

Fig. 1 Simulated velocity distribution of the interstellar O-flow, as it enters the heliosphere at the termination shock. In addition to the primary O component with velocity and temperature as obtained for CHISM He, a secondary component of 20% of the overall distribution is added, produced by charge exchange in the heliospheric interface according to Izmodenov et al. (1997)

the distribution. If the distribution of the primary component is known, for example, from a simultaneous measurement of the He flow distribution, which resembles the pristine interstellar flow, then the secondary component can be reconstructed. In addition, a deviation of the flow vector direction from the original He CHISM flow for any species that are affected by the interface will also provide constraints on the local interstellar magnetic field.

To illustrate the combination of these effects, Fig. 2 shows the density distributions of neutral H and O. These distributions are normalized to the respective density values in the CHISM and are shown in a color-coded representation in a heliospheric cut that contains the interstellar magnetic field vector. Shown are the results from a global heliospheric simulation performed with a 3D kinetic magnetohydrodynamics model by the Moscow simulation group. For the simulations shown here, an interstellar magnetic field strength of 4.4 μG (a somewhat higher field strength than favored previously) and a direction that is inclined by 20° relative to the interstellar flow vector have been adopted. This parameter set produces an approximately 10 AU difference in the distance of the termination shock from the Sun for the times and locations of the Voyager 1 and 2 crossings, as actually observed (Izmodenov et al. 2008). Because at least part of the observed termination shock distance difference could also be attributed to asymmetries arising from the interplanetary magnetic field configuration and to dynamic variations of the solar wind (Pogorelov et al. 2006; Washimi et al. 2007) it is important to note that other important observables result from this asymmetry that will be captured by IBEX.

Both the H and the O densities increase over their respective values in the CHISM when approaching the heliopause. The production of a slow secondary neutral component from the interstellar plasma that is forced to divert around the heliosphere overcompensates the loss of primary H and O from the original interstellar flow. Due to the inclined interstellar magnetic field the shapes of the heliospheric boundaries and the H and O wall are strongly asymmetric. A result of this asymmetry is a change in the direction of the mean velocity of the neutral atoms, which is imprinted on the secondary components of H and O. Comparing these predicted flow directions directly with observed interstellar neutral flow observations using IBEX-Lo will enable us to eliminate the persisting ambiguities in the evaluation of the Voyager asymmetry observations. The upper panel of Fig. 3 shows the flow deflection angles for interstellar H and O. For O the results are also broken out into the primary and secondary component. The overall deflection of H and O is that of the weighted combination of both components. The deflection of the combined O flow close to the Sun is still 1.2°, i.e. approximately half the deflection of H. Using predicted IBEX-Lo observations, it will be demonstrated below that both the overall deflection and the secondary component will be well visible and can be compared directly with interstellar He. The lower panel of Fig. 3 shows the normalized total number densities of neutral H and O as a function of distance from the Sun for the four cardinal directions within the plane that contains the interstellar magnetic field. The density increase in the H and O wall outside the heliopause is obvious for the upwind and side-wind directions. The depletion in the downwind direction is less prominent for O than for H. At 1 AU, where the observations with IBEX will be performed, the density of O is down to $\approx 7\%$ of the CHISM value in the $+$ and $-90°$ locations, which leads to neutral atom fluxes that are comfortably in IBEX-Lo's sensitivity range.

It should be noted that the filtration level and the resulting ratio of the primary and secondary components, which represents an observable quantity for O with IBEX-Lo, depend on the local ionization state of the CHISM. These parameters have been studied for H, O, and N by Izmodenov et al. (2004). In another sensitivity study, Müller and Zank (2004) modeled the processing of interstellar O and O^+ against three different hydrogen backgrounds. The primary O is depleted while passing through the various heliospheric regions.

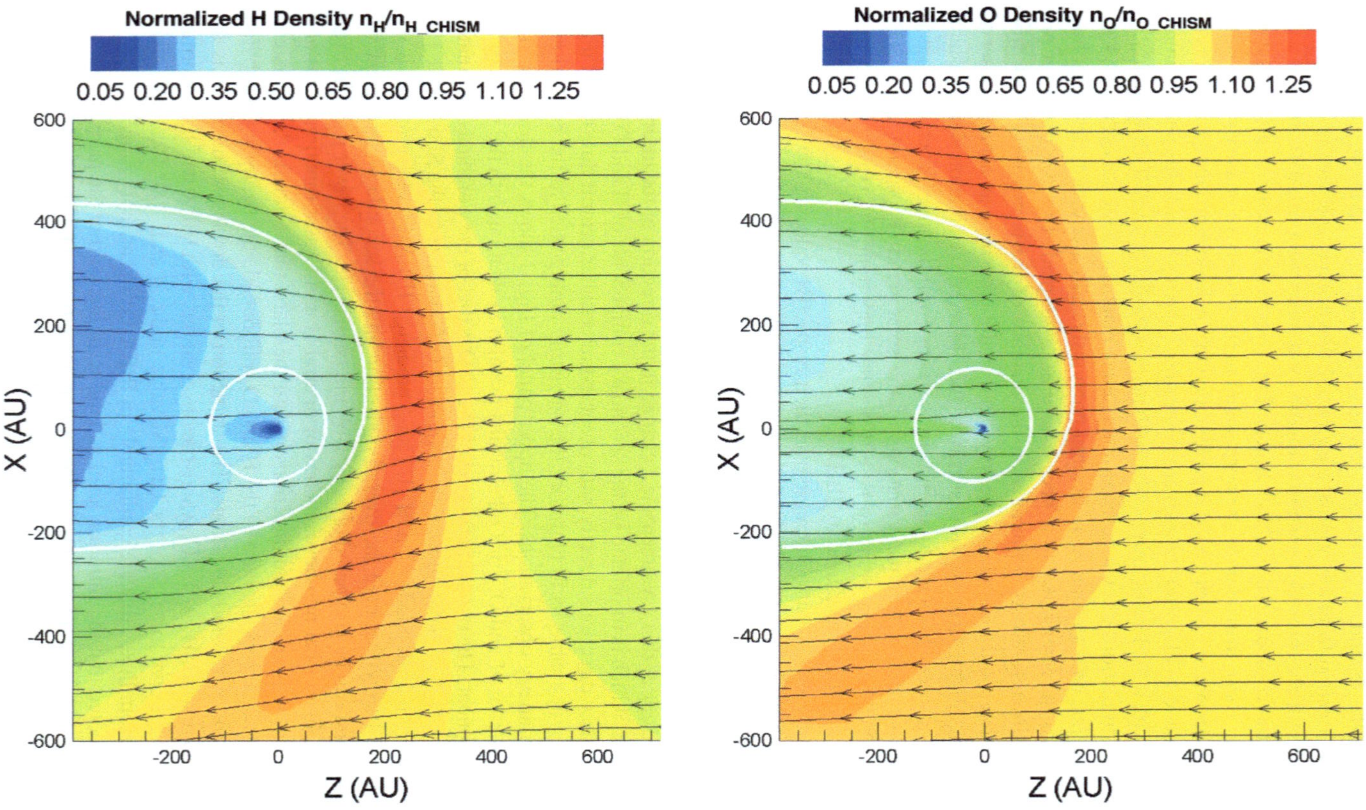

Fig. 2 Color-coded representation of the H (*left*) and O (*right*) densities $n_{\mathrm{H}}/n_{\mathrm{H\_CHISM}}$ and $n_{\mathrm{O}}/n_{\mathrm{O\_CHISM}}$ in a heliospheric cut that contains the interstellar magnetic field vector ($B = 4.4\ \mu\mathrm{G}$, vector 20° relative to the CHISM flow vector)

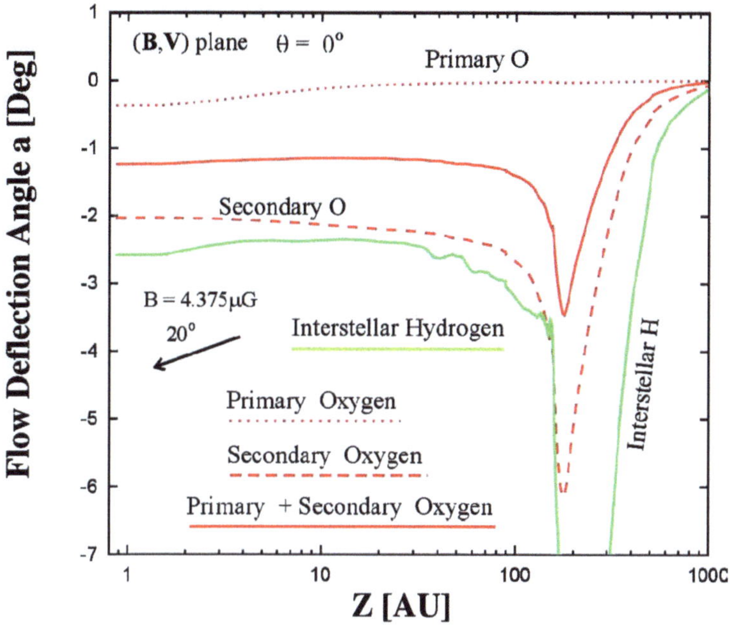

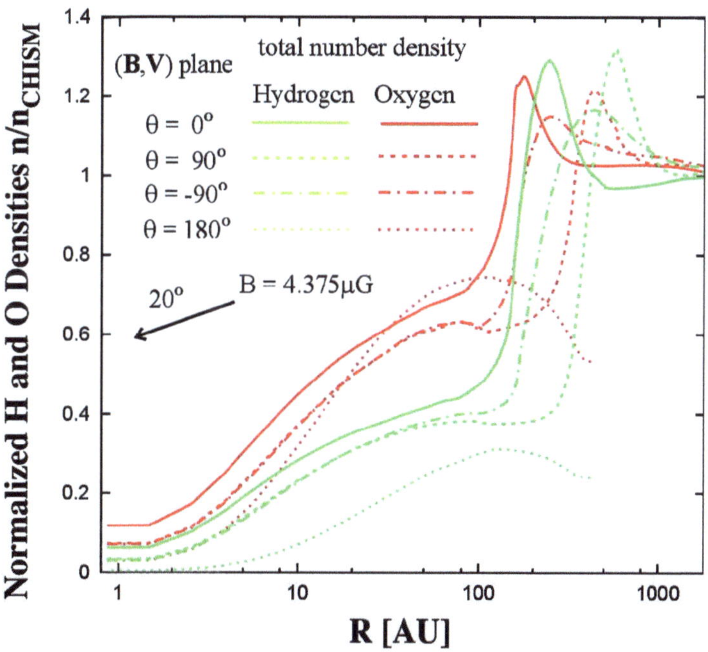

Fig. 3 Deflection of the neutral gas flow from the original CHISM flow as a function of distance from the Sun along the inflow direction (*upper panel*) and normalized neutral densities as a function of radial distance from the Sun for four directions (*lower panel*). Shown are weighted combinations of the primary and secondary components for H and O. For the O flow directions in the *upper panel* also the primary and secondary component are shown separately

However, a substantial amount of secondary O is generated, especially when interstellar O^+ encounters the hydrogen wall. Because of the lower speeds upwind of the heliopause, characteristic mean-free-paths for charge exchange are quite low in this region, in the range of 10–100 AU. At the termination shock, the combined primary and secondary neutrals reach densities comparable or even exceeding the pristine neutral O CHISM density. This result was shown to be relatively insensitive to the different hydrogen backgrounds, which represented a range of hydrogen filtration values (Müller and Zank 2004). This interplay between O neutrals and ions with the hydrogen in the inner and outer heliosheath, which in turn depends on the strength of the interstellar bow shock (Müller et al. 2008), deserves further detailed study. In comparison with IBEX observations such simulations may therefore place important constraints on the existence and parameters of a heliospheric bow shock.

In the inner heliosphere, i.e. at <2 AU, the Sun's gravitational field can be employed to deduce flow velocities, from the observed image of the flow in the sky, at locations downstream of the Sun in the spring and fall of each year. A comparison of the He and O flow characteristics in the inner heliosphere will provide a powerful tool to deduce the amount of filtration and its related processes quantitatively from observations in the inner heliosphere. To turn the velocity distributions at large distances in a quantitative way into the angular distributions observed at 1 AU, the velocity distributions are handed over to the ray tracing program developed by the group at the Space Research Centre in Warsaw. This ray tracing of individual atoms includes all relevant effects, such as the Sun's gravitation and radiation pressure, as well as ionization on the way (Bzowski 2008).

Ideally one would also like to obtain the flow pattern of H in the inner heliosphere, which is being observed indirectly through Ly α backscattering (Quémerais et al. 1999; Lallement et al. 2005). While IBEX-Lo is also sensitive to interstellar H in the ram direction at the low end of the sensor energy range, H is very strongly affected by radiation pressure so that predictions for its observation are rather uncertain. In turn, imaging neutral H from various locations may provide the most accurate account of this effect and will allow unambiguous separation of interface and radiation pressure related effects. In addition, radiation pressure may also serve as a discriminator between H and D, because the strength of the effect depends on the particle mass. As a result the D image will be separated from the H image in angle, thus giving a neutral imaging mass spectrograph an advantage in the observation of D in the CHISM. The D/H ratio in the interstellar medium has been considered an important data point for the evolution of matter (Linsky 1998; Mullan and Linsky 1998; Prantzos 1998) because D is the only isotope that is clearly being depleted by reactions inside stars over the primordial amount. Whereas local observations of ^{3}He have been obtained, this task appears to be much more difficult for D (Gloeckler and Geiss 1996). Recent estimates by Tarnopolski and Bzowski (2008) have shown that IBEX-Lo may be able to obtain direct D observations. Potential observations of H and D have substantial discovery potential, but their expected count rate is much more uncertain than the projected O observations because of the strong sensitivity of H and D flows in the inner heliosphere to the Sun's radiation pressure. Therefore, we will restrict the remainder of the discussion to O.

4 IBEX Capabilities to Measure the Interstellar Gas Flow

The IBEX Mission has been implemented to focus on a single broad-scope scientific objective, the investigation of the global interaction of the heliosphere with the surrounding CHISM that occurs in the boundary regions. To achieve this goal experimentally, IBEX will take the first full-sky images of the heliospheric boundary in the "light" of energetic neutral atoms (ENAs) in the energy range 10–6000 eV that emerge from the plasma and suprathermal ion populations in the interaction processes at the termination shock and in the heliosheath (McComas et al. 2009, this issue). Of course, the flow of the pristine interstellar neutral gas and of their secondary products will be a natural part of these images, which will stand out because of their much higher directed fluxes. Because of the journey of these ENAs through the boundary regions into the inner heliosphere they are probes for the interaction in the outer heliosheath beyond the heliopause, thus addressing the fourth specific science question of the IBEX mission: What is the interstellar flow beyond the heliopause?

This section describes how the IBEX-Lo sensor will be used to probe the interstellar neutral gas flow distribution as it arrives at 1 AU.

4.1 Influence of the IBEX Mission Design

To provide a series of full-sky images with ENAs, IBEX observations will be made with two large geometric factor single-pixel ENA cameras, which, in combination, cover the entire energy range with a comfortable overlap. These cameras are flown on a Sun-pointing spinning spacecraft in a high elliptical orbit around Earth. To keep the spin axis aligned with the Sun within $\pm4°$, the spacecraft is re-oriented so that it points $\approx4°$ east of the Sun at the beginning of each orbit (with a period of about 8 days). After completion of the orbit the spin axis will point approximately 4° west of the Sun. The resulting IBEX viewing conditions are shown in Fig. 4 for two positions of the Earth on its orbit over the course of one year when the satellite is oriented such that the interstellar gas flow passes the field-of-view (FoV) of the IBEX-Lo sensor over a period of several weeks.

During a satellite spin, the sensors scan the angular flow distribution in ecliptic latitude. The two panels at the bottom of Fig. 4 show scans of the angular flow distribution in ecliptic longitude, as they would be accumulated by a spacecraft that keeps its spin axis exactly pointing at the Sun while following the Earth around the Sun. It is a combination of the changing orientation of the sensor FoV and the varying gravitational deflection of the flow along the Earth's orbit. Shown are the expected oxygen counts accumulated each day in an operation that keeps the IBEX-Lo sensor at the peak energy for each scan of the flow distribution, separately for the primary and secondary O component along with their combination. The latter is the actual observable in this measurement. As can be seen in Fig. 4, both combined distributions appear asymmetric. The asymmetry is due to the addition of the secondary component, which arrives at slightly larger deflection angles. The difference in deflection angle between these two components is shown as a function of ecliptic longitude, referenced to the inflow direction of the interstellar gas outside the heliosphere in Fig. 5. The angle difference is substantially larger in fall when compared with spring. This is due to the fact that in fall the observer is receding from the flow, while the observer is moving into flow in spring, following the Earth in its orbit. This behavior is also reflected in the overall much narrower angular distribution in spring. As will be discussed in connection with the sensor capabilities below, a smaller FoV will be employed to cope with the narrow distribution in spring. The width of the FoV must be significantly smaller (approximately by a factor of four) than the expected angular width of the distribution of about 13° FWHM (full width at half maximum) to avoid degradation of the asymmetry in the distribution (shown in the right hand panel of Fig. 4), which is key to the evaluation of the secondary neutral O component.

Taking the O flow distributions and demonstrating that they are asymmetric will satisfy the discovery level of the IBEX mission. However, to extract quantitative information on the interaction of the interstellar O with the heliospheric boundary at the exploration and the investigation level, the O distribution must be compared with the interstellar He distribution, which is unaffected by the interface. Ideally, the primary O distribution (shown in blue

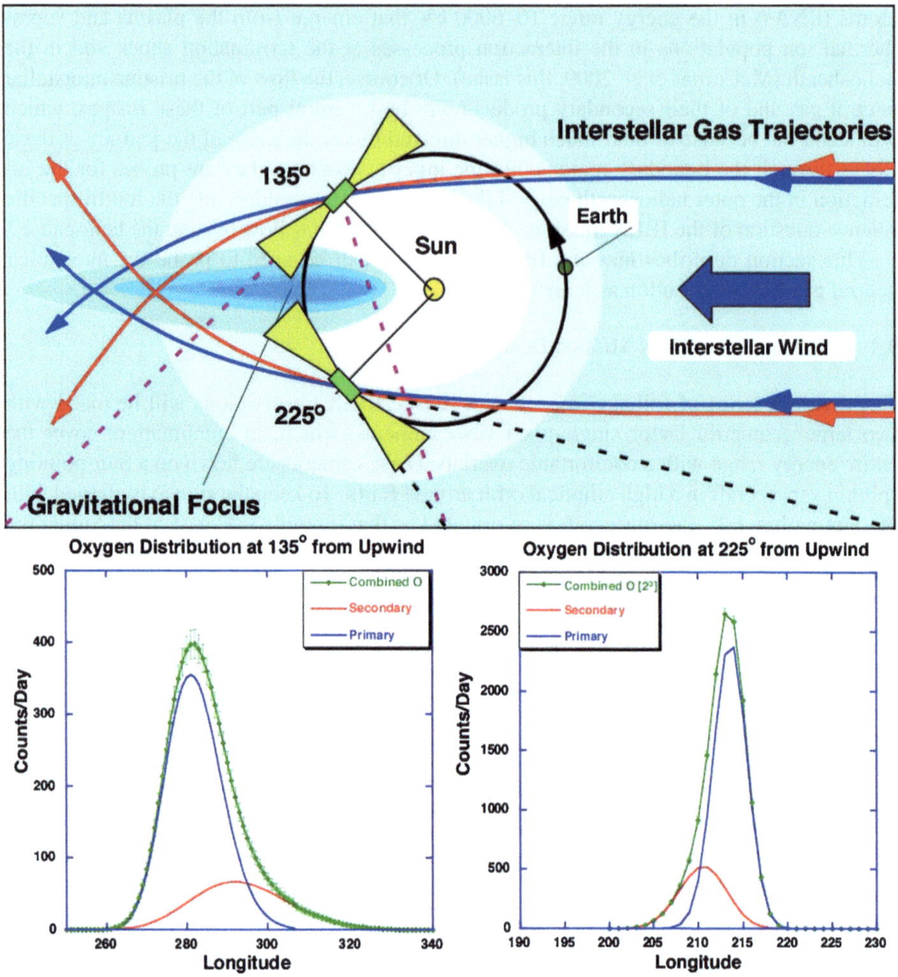

Fig. 4 Schematic view of the interstellar gas flow through the inner heliosphere with the Earth's orbit and the IBEX-FoV for the two time periods of interstellar gas flow observations. The *insets* show the expected angular distributions of the gas flow with the primary and secondary components separate

in Fig. 4) should be subtracted from the combined distribution to reveal the unknown secondary O distribution. This can be achieved either by constructing the primary distribution from the accurately measured interstellar He flow based on observation with Ulysses GAS (Witte 2004) or by simultaneous observation of the He flow with IBEX-Lo. The former requires knowledge of the pointing of IBEX-Lo to better than $\pm 0.2°$ in absolute terms, i.e. the accuracy with which the He flow vector is known from the Ulysses observations; the latter requires a detection capability for He with IBEX-Lo in addition to observing O.

To facilitate accurate pointing of the IBEX-Lo sensor without undue impact on spacecraft resources and without setting stringent requirements on sensor pointing accuracy relative to the spacecraft attitude system, a small star sensor has been implemented as part of the IBEX-Lo sensor. By directly mounting the star sensor to the common baseplate of the IBEX-Lo sensor, which also holds the collimator, or the sole optical element, the impact of mechanical

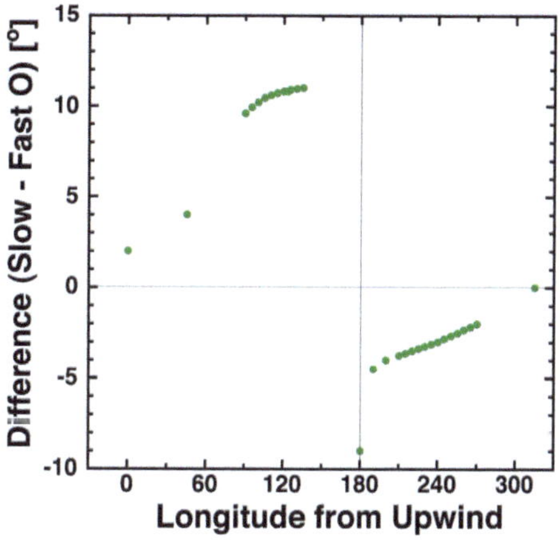

Fig. 5 Angle difference of the slow (secondary) O component relative to the fast (primary) O component as a function of ecliptic longitude relative to the interstellar upwind direction

alignment tolerances has been minimized. In pointing tests of the star sensor it has been established that the boresight of the IBEX-Lo collimator can be referenced in flight to stars in the star sensor FoV to $\leq \pm 0.2°$. This capability will allow a direct comparison of IBEX flow vector measurements to those obtained by Ulysses GAS (Witte 2004) with the same accuracy as achieved with that instrument. A brief description of the star sensor and its calibration is provided in the paper by Fuselier et al. (2009, this issue).

As will be discussed below, IBEX-Lo is indeed capable of observing simultaneously O and He neutral distributions. Therefore, a direct comparison of the interstellar O flow distribution, which is modified by the heliospheric interface, and that of He, which reflects the pristine CHISM conditions, will be performed with the same sensor. The precise absolute pointing capability provided by the star sensor will enable a triangulation of the He flow from different vantage points by combining the previous Ulysses observations and the future IBEX observations, with the prospects of improving the current knowledge.

4.2 The IBEX-Lo Sensor

IBEX-Lo is one of the two large geometric factor single-pixel neutral atom cameras that together cover the energy range 10–6000 eV with considerable overlap from 250 to 2000 eV. Key requirements for both the IBEX-Lo and Hi cameras are large collecting power; excellent suppression of any unwanted background to unambiguously detect the low fluxes of the termination shock ENAs of 5×10^3–1.5×10^5 $cm^{-2}\,s^1\,sr^{-1}$ in the energy range from 200 eV to 6 keV; and angular resolution of $\leq 7°$ FWHM (McComas et al. 2009, this issue).

IBEX-Lo is dedicated to the low energy portion from 10 to 2000 eV, and, because it covers the range of the interstellar flow components, it is outfitted as a mass spectrometer that can distinguish between H, He, and O. All requirements, functions, and implementation of the IBEX-Lo sensor are described in detail in the paper by Fuselier et al. (2009, this issue). Therefore, only the basic principles of the sensor and the characteristics most relevant to the interstellar flow observations will be presented here. Figure 6 shows a schematic cross-sectional view for one radial section of IBEX-Lo. The sensor is a large area single-pixel camera with a mechanical collimator as the key optical element, which accepts incoming neutral atoms within a solid angle FoV of $\leq 7°$ FWHM in its normal operational mode. To adequately observe the O flow distribution during the spring period when the angular

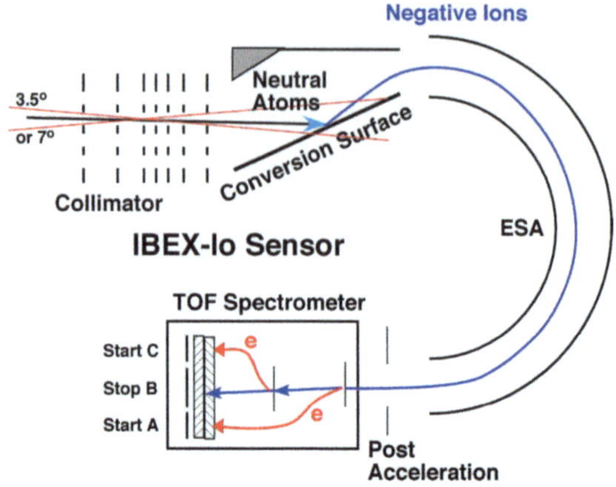

Fig. 6 Schematic cross-sectional view of the IBEX-Lo sensor. In sequence incoming particles pass through a collimator, bounce off a neutral to negative ion conversion surface, pass an electrostatic analyzer, are post-accelerated, and then analyzed in a triple TOF system

distribution is about 7° in FWHM (see Fig. 4), IBEX-Lo has also been equipped with a high-resolution section of 3.3° FWHM angular acceptance that can be selected electrostatically by command. It is important to emphasize that the peak of a narrow flow distribution or its center of mass can be determined to much better accuracy than the angular resolution defined by the FoV, in this case to significantly better than 1°, mainly limited by counting statistics. Also, the mechanical collimator determines both the FoV and the pointing of the sensor, and the strong angular scattering at the conversion surface (CS) behind the collimator is largely irrelevant because (for the purpose of defining the angular acceptance) the CS is the detector element.

While bouncing off a CS a certain fraction of the neutrals become negative ions (Wurz 2000). The CS used is a highly polished Si wafer with a coating of tetrahedral amorphous carbon (ta-C) as described in Fuselier et al. (2009, this issue). This surface is inclined by 15° relative to the sensor boresight. The conversion is rather efficient for species that form stable negative ions, such as H, C, and O. However, it is rather inefficient for those that can only produce meta-stable negative ions, such as He. For the other noble gases a negative ion does not exist (Smirnov 1982). Also for N a stable negative ion does not exist (Hotop and Lineberger 1975), and the binding energies for the meta-stable negative N ion are so low (Mazeau et al. 1978) that useful negative ion yields are not expected for the IBEX investigation.

This behavior makes the CS a technique of choice for key neutral interstellar species, such as H and O, but it left the question open whether He could be detected directly with high enough efficiency. It should be pointed out that neutral atoms with energies above a few tens of eV will also release atoms that are present on the CS, of which a certain fraction take off as negative ions via sputtering and knock-on processes. For a relatively strong directional source of neutral atoms, such as the interstellar He flow at 1 AU, this process can serve as a valuable proxy for the direct detection of surface converted negative ions. In fact, this resembles the successful method used with the Ulysses GAS experiment (Witte 2004). GAS collects positive Li ions that are generated by the impact of incoming He atoms on a LiF surface. To evaluate how effective IBEX-Lo, with its diamond-like C conversion surface, is in the generation and collection of sputtered negative H, C, and O ions (from the diamond and always present water), and to find out whether a significant portion of negative He is produced, the sensor was also calibrated with a neutral He beam, as discussed below.

Negative ions are then collected electrostatically and guided through a toroidal electrostatic analyzer (ESA) with $\approx 80\%$ energy resolution (right part of Fig. 6). For our purposes this energy band choice allows a very effective collection of ions from the entire energy range of the interstellar flow distribution, while the analysis of the velocity distribution is left to the angular resolution of the entrance system and the gravitational deflection of the flow by the Sun, as described in Sect. 4.1. After post-acceleration by a voltage $U_{\mathrm{Acc}} = 16\,\mathrm{kV}$ the ions are discriminated against remaining noise and background and analyzed for their mass in a triple time-of-flight (TOF) subsystem.

The TOF unit has been adapted from Cluster CODIF, similar sensors on FAST and Equator-S (Möbius et al. 1998b), and the more recent version STEREO PLASTIC (Galvin et al. 2007). Because of the extreme importance of background suppression for the IBEX mission, two similar TOF sections are used in series to achieve a genuine triple coincidence for the incoming ions (bottom left of Fig. 6). The incoming ions generate secondary electrons at the first carbon-foil that are guided to the outermost section of the micro-channel plate (MCP) detector. Similarly, electrons are generated at the second foil and guided to the innermost section of the MCP. The ions are detected in the center section, which is divided into 4 quadrants to allow for a separate background characterization. A single MCP pair is used. Three TOF values are determined between each of the foils and the MCP (TOF0 and TOF1) and between the two foils (TOF2). Each of these three TOF measurements is equivalent to the observation with a CODIF-like sensor. The ability to use any of the three measurements provides for very high efficiency. The quadrant position is derived from a delay line TOF measurement (TOF3). Ion detection with a maximum of background suppression is achieved when all 4 TOF signals are registered, but already a single TOF measurement provides adequate background suppression for the directed interstellar gas flow.

Combining the energy selected by the ESA E_{ESA} with the post-acceleration and the measured TOF τ the species mass M is determined as follows:

$$M = 2(E_{\mathrm{ESA}} + e \cdot U_{\mathrm{Acc}}) \cdot \alpha \cdot \tau^2/d^2, \tag{1}$$

where d is the length of the TOF unit and α describes the reduction in energy due to losses in the entrance foils, which is energy and species dependent. This energy loss is determined during sensor calibration and in simulations. With a mass resolution $M/\Delta M \geq 4$ as established during calibration, IBEX-Lo can unambiguously distinguish between H, He, and O and can even differentiate C and O contributions by fitting the respective TOF distributions of these two species. In its left half Fig. 7 shows a TOF spectrum taken during tests with IBEX-Lo using a neutral He beam. Shown are two TOF0 spectra, i.e. with the TOF measured between the first C-foil and the MCP, for which the separation between H and He is the largest, for energy step 8 (1500 eV incoming and detected particles; top) and for energy step 4 (130 eV particles, bottom). Shown are spectra compiled with only such events that contain all time-of-flight measurements, i.e. with the best noise and background suppression that the sensor offers. The peaks of H, He, and a combination of C and O are clearly separated, where a marked double-peak distribution also enables a quantitative separation of C and O in a two-peak fit. In step 8 converted He shows a peak that is clearly standing out (containing 53 events over the entire run). In step 4, at an energy commensurate with the interstellar He flow encountered by IBEX in spring, He counts are still seen, but with only 4 events the statistical significance for a positive identification of He is marginal. However, these measurements have established that He^- is indeed produced so that interstellar He may be directly identified with IBEX-Lo at 1 AU (Wurz et al. 2008). In addition, its presence can be inferred from the much more prominent sputtered products (H, C, and O). The H, C, and O present in the TOF spectra consists of negative ions created via sputtering and knock-off on the diamond-like carbon of the CS that typically contains a mono-layer

Fig. 7 Time-of-flight spectra taken between the first C-foil and the MCP (TOF0, see Fig. 6) for a 1500 eV (*top*, Step 8) and 130 eV (*bottom*, Step 4 He beam (*right*)), taken for triple coincidence events. The TOF range for accepting He is indicated by the two *vertical lines*

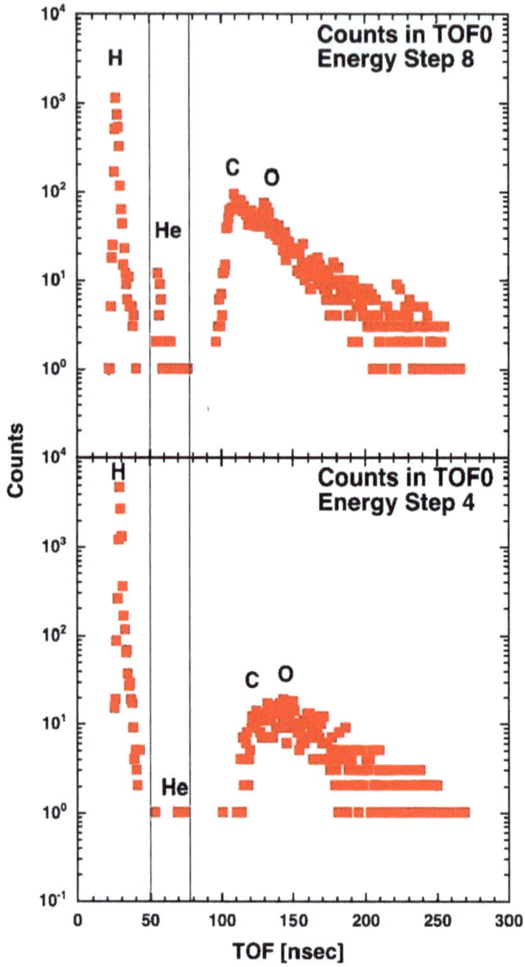

of water. Hence the mass spectrum has the products of water and of carbon from the surface. The multi-species TOF spectra shown here also indicate that interstellar O is clearly distinguished from other species by IBEX-Lo. It should be noted here that the O flow will be detected at energies that are a factor of two, or two energy steps, above those of He. In the following we will see that sputtering leads to a very high yield so that the interstellar He flow can be observed with superb counting statistics in spring and in fall.

Figure 8 shows both the direct conversion efficiency of He and the efficiency for producing sputtered H as a function of energy over the entire energy range of IBEX-Lo, with the two semi-transparent rectangles indicating the energy ranges where interstellar He is observed in the spring and fall. For these energy ranges the combined sputter and sensor collection efficiency, which is relevant for the detection of incoming neutral atoms, is close to 10^{-3}, i.e. only a factor of 20 below the combined conversion and collection efficiency for O, while the He flux density at 1 AU is approximately a factor of 1500 above that of O. Thus He will be observed with much more ease than O. The derived He conversion and collection efficiencies are much lower (by a factor of 5000), and it should be pointed out that at present only measurements with less than 10 events (4 triple events each that fall into

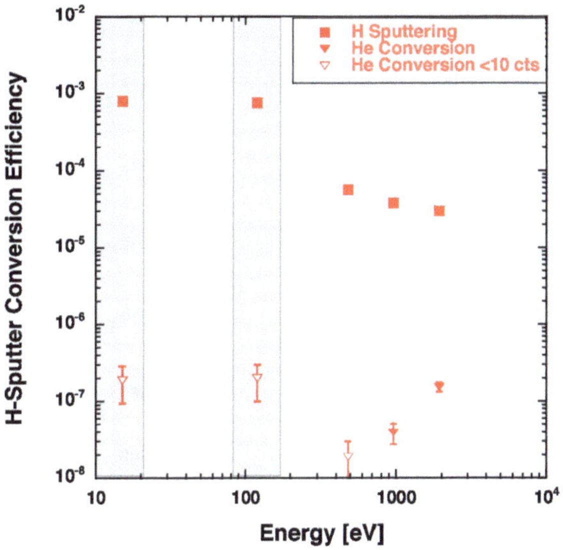

Fig. 8 He conversion efficiency and efficiency for the combined production and collection of sputtered H versus energy. The two energy passbands relevant for interstellar He flow observations in spring (higher energy) and fall (lower energy) are highlighted. *Error bars* indicate statistical uncertainties of the measurements

the He TOF range) have been collected. The error bars indicate the statistical uncertainty of each data point. With so few events (while absolutely excluding any background) it cannot be excluded that some events from the far tail of the H TOF distribution with orders of magnitude higher intensity contribute, and thus the final He conversion efficiency is even lower. For the purpose of this paper we are adopting the efficiency values shown here when we estimate the expected count rates for the detection of interstellar neutrals in spring and fall. It should be noted that the efficiency for the production of sputtered negative ions found in the energy range where the original He atoms arrive is surprisingly high. While the total yield of sputtered products is usually highest at energies significantly below the energy band of incoming neutral atoms, the yields shown here for the same energy band as the incoming particles produce rather high count rates as will be shown below. Using the energy steps of the incoming He atoms for the observation will avoid potential confusion with the possible collection of interstellar H atoms that will arrive with energies in the lowest energy step in spring. Also, this choice enables the potential direct detection of He through conversion, which will further validate the observation through sputtered particles.

5 Observation Campaign Planning

With the detection capability of IBEX-Lo for the key interstellar neutral species He and O clearly established in the sensor calibration, it remains the task of the IBEX mission planning to optimize a combined observation of the interstellar He and O flow distributions in the spring and fall when the bulk flow will be within the sensor FoV. Figure 9 shows expected accumulated counts per day for He and O in its upper panel and the direction of the flow relative to the Sun in the lower panel. The counts per day and the flow direction are plotted as a function of the position of the Earth (and thus IBEX) in ecliptic longitude, referenced to the upwind direction of the interstellar gas flow at large distances from the Sun. In the simulation, the geometric factors for the full collimator (lo-resolution) and the hi-resolution section only as established in the IBEX-Lo calibrations (Fuselier et al. 2009, this issue) were used to determine the expected count rates, and operation in the interstellar flow mode as

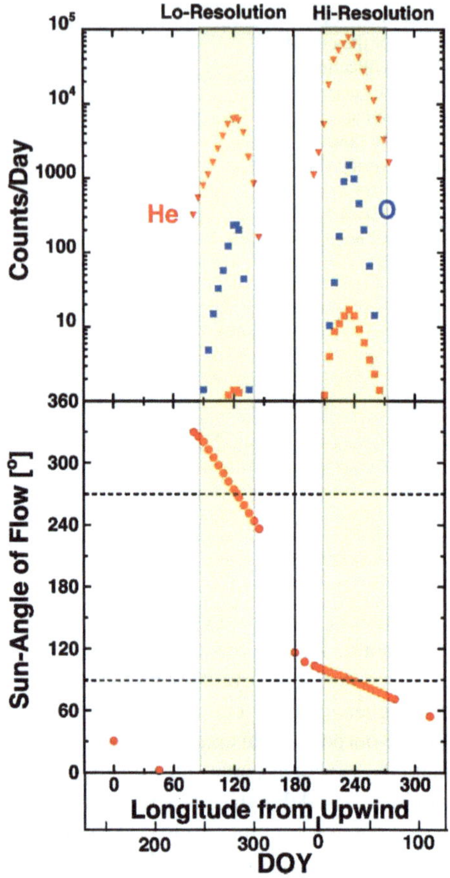

Fig. 9 Expected number of counts/day integrated in spin direction for interstellar O (*blue*) and He (*red*) during the fall and spring passage of the flow with the IBEX-Lo FoV, with the sensor switched into its flow-sensing mode (*top panel*). Shown for He are the expected counts from surface conversion of He (*squares*) and those from sputtering of H off the surface (*triangles*), which are more abundant. Peak flow direction relative to the Sun as a function of ecliptic longitude (*bottom panel*). Both are shown as a function of ecliptic longitude from the interstellar medium upwind direction and day of the year (DOY) over two consecutive years. The two observation campaigns are highlighted

described below was assumed. In the representation in Fig. 9 we assumed for simplicity that the spin axis of the IBEX spacecraft would always point perfectly towards the Sun.

The viewing of both species is optimized and the necessary angular resolution of the directed flow measurement is achieved by switching into a flow sensing mode during two ≈1.5-month periods in spring and fall. For ±30° elevation above and below the ecliptic in the arrival direction of the flow, the electrostatic analyzer is set to the bulk flow energy of O for 7 steps and to that of He for 1 step of the 8 step energy cycle. In fall the regular collimator FoV of 6.5° FWHM is adequate to distinguish the pattern of the two O streams, but in spring the hi-resolution quadrant with a FoV of 3.3° FWHM is needed. In the spring, the remaining three quadrants with lo-resolution are switched off electrostatically at the CS. In this way the fact that the full collimator can be used for the broader angular distribution in fall compensates conveniently for the much lower neutral gas flux because the Earth recedes from the flow in the fall, while it moves into the ram direction in the spring. Figure 9 shows the number of counts per day collected for O and He (integrated in spin direction, when operated in this way) in the upper panel and the flow arrival direction relative to the Sun in the bottom panel.

The use of the interstellar gas flow mode leads to reduced coverage for the normal ENA H maps as a function of energy with IBEX-Lo for these observation periods. However, IBEX-Lo and IBEX-Hi have a substantial overlap in energy coverage from 250 eV to 2000 eV so that IBEX-Hi still covers most of the energy range, and six months before and after each interstellar flow observation the same region of the sky is covered. Thus the interstellar flow mode operation only leads to reduced coverage for energies outside the energy band of the interstellar gas flow in its arrival directions in the sky. Here, the coverage of O, which arrives in the spacecraft rest frame at energies of 540 eV in the spring and 32 eV in the fall, is substantially increased by switching to the O flow energy for 7 out of 8 cycles. The remaining 8th cycle is used to collect a sample of the He flow for comparison with O.

As can be seen in the lower panel of Fig. 9, the peak of the gravitationally deflected interstellar flow distribution passes the sensor FoV at 90° and at 270° relative to the Sun from day of the year (DOY) 1 through 70 and from DOY 250 through 300 each year. The angle-scan results in the variation in the number of O and He atoms detected per day shown in the upper panel. The cut-off in the O viewing is used as the criterion for the two observation campaigns, which are indicated by the highlighted rectangles in Fig. 9. The campaigns will be run between January 5 and March 5 and between September 15 and October 25. In spring and in fall O and He will be seen with reasonable counting statistics. Because the sputtering efficiency is substantially higher than the direct conversion efficiency for He and the local flux at 1 AU is approximately 1500 times that of O, sputtered H, due to the impact of interstellar He, will be by far the strongest signal at 135 eV in spring and at 10 eV in fall. The count rates shown in Fig. 9 reflect the results from the IBEX-Lo calibration. The expected total number of interstellar O atoms will be $\approx 2.2 \times 10^4$ in the spring and ≈ 3400 in the fall. Accounted for in the same way, the total of directly converted He atoms would potentially be ≈ 470 in the spring and only 40 in the fall, if the He conversion efficiencies shown in Fig. 6 do not contain events from the tail of the H distribution, which can be clearly assessed with observations that contain a better counting statistics than the available calibration measurements. Both the O and He conversion numbers are dwarfed, however, by the collection of sputtered H, with $\approx 2.1 \times 10^6$ in the spring and $\approx 3 \times 10^5$ in the fall. Both the converted He and the sputtered H will be collected simultaneously at the same energy setting for 135 eV in the spring and 10 eV in the fall, thus enabling a direct validation of the sputtered H as a product of the interstellar He. However, the huge number of sputtered particles from He will not result in additional background for the interstellar O, which is seen at 540 eV and 32 eV, respectively. Sputtering always creates products at energies lower than that of the incoming particles.

After the launch on October 19, 2008, the first spring passage of the interstellar flow falls into the time period when the first global ENA maps are taken with IBEX (McComas et al. 2009, this issue), for which a contiguous sky coverage with regular energy cycles and the full aperture of IBEX-Lo was anticipated. While the normal IBEX-Lo operation means visiting the peak energy of the interstellar O flow once every cycle of eight steps, the sensor geometric factor is now larger by a factor of 8.6, which more than compensates the duty cycle loss of a factor of 8. Using the entire collimator gains a factor of 4 in area and a somewhat higher transparency for the low-resolution section. Another factor of 2 is gained with the larger FoV angle in longitude, while a change in spin angle has no effect for a beam distribution when using integrated counts. Therefore, the total number of counts collected in this mode will be approximately the same for O and increased by a factor of 8.6 for He. Only the angular resolution of the measurement is not optimal for a quantitative evaluation of the O distribution whose angular half width is 13° in spring. Therefore, the normal operation mode will yield a first survey of the interstellar O and He flow distributions and at least yield an upper limit on the flow difference between the primary and secondary O streams.

In the anticipated operation scenario, IBEX will be in a high elliptical orbit with a typical orbital period of 8 days. At the beginning of each orbit the spin axis will be re-oriented to

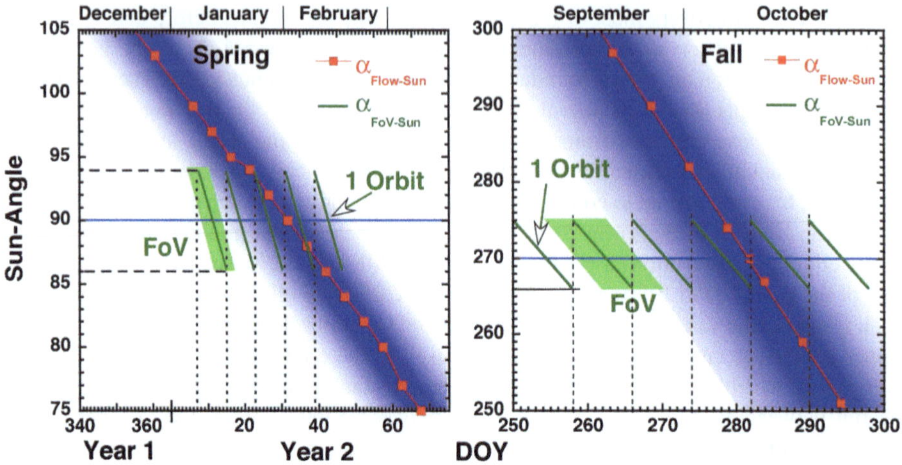

Fig. 10 Variation of the CHISM flow arrival direction relative to the Sun (*red line*) along with indication of the angular distribution (*blue shading*) as a function of DOY. The change in the orientation of the FoV for a typical orbit with a period of 8 days is shown in *green*. (spring left and fall right)

about 4° behind or east of the Sun. Thus over the course of one orbit the FoV will drift across the direction 90° from the Sun until the spin axis points about 4° ahead or west of the Sun at the end of the orbit. At this point the spin axis will be reoriented so that it points again 4° east of the Sun. This behavior of the FoV is shown in Fig. 10 along with the change in the flow arrival direction for spring (left) and fall (right) as a function of the DOY. In spring this "ratcheting" motion of the FoV scans the full angular distribution over the course of several orbits with considerable overlap between two consecutive orbits, while in fall the drift of the FoV is almost parallel to the variation of the arrival direction. However, in fall the wider FoV is used and the distribution is wider. In this way the angular distribution of interstellar gas is still covered without gaps by consecutive FoV scans taking into account the FWHM of the FoV (green area in Fig. 10).

Figure 11 shows the resulting pattern of expected counts accumulated per day for O and He for the sputtered H counts over one pass of the interstellar flow distribution in spring. The periodicity of the shift of the spacecraft orientation with each orbit across the flow is clearly seen as a saw tooth pattern in the count rate. While this figure has been constructed from the simulation shown in Fig. 10 by transformation into the fixed orientation of the IBEX spacecraft for an entire orbit, the idealized observation can be reconstructed from the actual observations by reversing this step in the modeling. To compare model distributions of the interstellar flow with the observations both complete forward modeling up to the observations as simulated in Fig. 10 or reverse modeling of the observations into the expected interstellar flow distributions will be employed.

Making use of the spacecraft spin and timing information tagged onto each recorded particle, the flow distribution can also be evaluated in ecliptic latitude out of the ecliptic. Figure 12 shows the expected counts of interstellar O for the spring passage as shown in Figs. 10 and 11 in a color-coded 2-dimensional representation. The accumulated counts have been distributed by 6° pixels in latitude (commensurate with the 6° pixels of the all-sky ENA maps) and over 12 hours each to cover helio-longitude. The ratcheting of the FoV across the sky at the cadence of the spacecraft orbit is visible again in a similar pattern as in Fig. 11. The interstellar flow direction into −5.3° in helio-latitude is evident from the offset

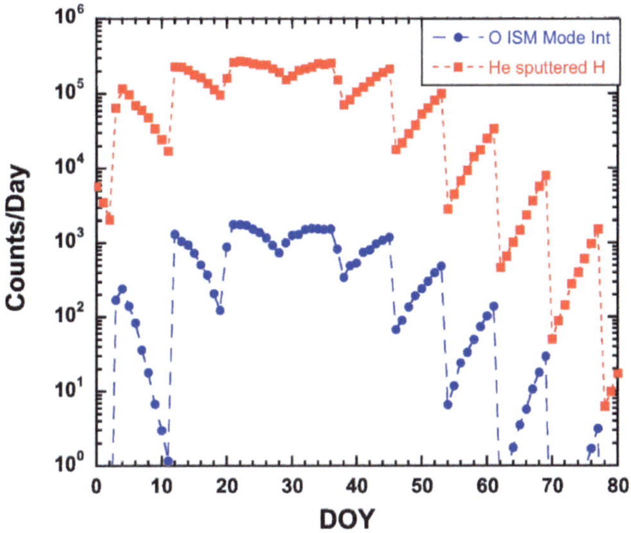

Fig. 11 Simulation of expected counts accumulated per day for O and He over one pass of IBEX through the interstellar flow distribution in spring 2009. An eight-day orbit and re-orientation of the spin axis at the start of each orbit are assumed. Shown for He are the counts expected from sputtering of H off the CS

Fig. 12 2-dimensional view of the incoming interstellar O distribution in a color-coded representation. 6° pixels are shown in latitude, and a 12-hour cadence is chosen for the variation of the viewing direction relative to the peak of the flow distribution with spacecraft position in helio-longitude

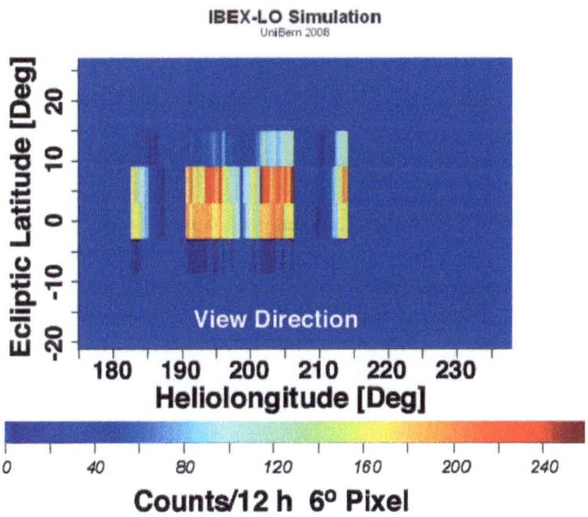

of the maximum of counts in ecliptic latitude in Fig. 12. Because the timing of the incoming particles is recorded with an accuracy that is equivalent to better than 0.05°, the peak flow direction in latitude can be determined from the center of mass of the observed distribution with very high accuracy. It should be pointed out here that the expected deflection of the secondary neutrals due to the orientation of the interstellar magnetic field is mainly helio-latitude and thus can be identified in a deviation of the peak in the spin phase.

6 Conclusions

Interstellar neutral gas enters the inner heliosphere in the form of a wind due to the relative motion of the Sun. While the pristine He population enters the heliosphere unimpeded, the charge exchange interaction with the interstellar plasma that is diverted around the heliopause produces a substantial secondary flow component for H and O with slower speed and higher temperature. Comparing the flow distributions of any of these two species with the undisturbed He flow will yield invaluable information to further constrain the interaction processes in the outer heliosheath and thus the ionization state in the surrounding CHISM. In addition, an interstellar magnetic field that is inclined relative to the interstellar gas flow vector distorts the shape of the heliosphere and as a consequence leads to a deflection of the secondary neutral gas flow component. Therefore, the combined velocity distributions of primary and secondary interstellar H and O entering the heliosphere are distinctly different from that of He. Any deflection of these flows relative to the pristine flow vector of the CHISM due to magnetic field effects becomes directly visible in ENA maps of these species as a deviation from the He flow vector. In comparison with 3-dimensional and time-dependent global heliospheric modeling, these observations will be used to further constrain the poorly known interstellar magnetic field strength and direction and to remove persisting ambiguities due to multiple causes of observed asymmetry effects in the heliosphere.

Close to the Sun the gas flow is subject to substantial gravitational deflection, which maps the original velocity distribution that is present at the termination shock into a related directional distribution. In turn, the velocity distributions can be obtained from directional distributions of interstellar neutrals.

By scanning through the angular distributions of interstellar neutrals with the single-pixel IBEX-Lo sensor, the Sun-pointed spinning IBEX satellite will obtain concurrent images of O and He in spring and fall when the deflected interstellar flow falls into the sensor FoV. Results from the calibration of the IBEX-Lo flight sensor demonstrate that the instrumentation will provide sufficient sensitivity and angular resolution for both O and He so that a direct comparison of the two different angular distributions becomes possible.

Acknowledgements The authors are grateful to the many individuals at the IBEX co-investigator institutions and the NASA Small Explorer Office who have made this program possible. Several of us (E.M., M.B., and V.I.) are indebted to the International Space Science Institute (ISSI) in Bern for the support of several workshops and scientific teams, which have helped the idea behind this paper from conception to implementation. This work is supported by NASA's Explorer Program under Grant NNG05EC85C, by NASA SR&T Grant NNG06GD55G, by the Swiss National Science Foundation, PRODEX, the Space Research Centre of the Polish Academy of Sciences, and the Russian Federation.

References

T.F. Adams, P.C. Frisch, Astrophys. J. **212**, 300 (1977)
W.I. Axford, in: *Solar Wind*. NASA SP-308, ed. by E.P. Sonnett, P.J. Coleman, J.M. Wilcox (1972), p. 609
M. Banaszkiewicz, M. Witte, H. Rosenbauer, Astron. Astrophys. Suppl. Ser. **120**, 587 (1996)
V.B. Baranov, Yu.G. Malama, J. Geophys. Res. **98**, 15157 (1993)
J.L. Bertaux, J.E. Blamont, Astron. Astrophys. **11**, 200 (1971)
J.L. Bertaux, R. Lallement, V.G. Kurt, E.N. Mironova, Astron. Astrophys. **150**, 1 (1985)
M. Bzowski, Astron. Astrophys. **483**, 1155 (2008)
M. Bzowski, E. Möbius, S. Tarnopolski, V. Izmodenov, G. Gloeckler, Astron. Astrophys. **491**, 7 (2008)
E. Chassefière, J.L. Bertaux, R. Lallement, V.G. Kurt, Astron. Astrophys. **160**, 229 (1986)
K.-P. Cheng, F.C. Bruhweiler, Astrophys. J. **364**, 573 (1990)
D.P. Cox, R.J. Reynolds, Ann. Rev. Astron. Astrophys. **25**, 303 (1987)
R.M. Crutcher, Astrophys. J. **254**, 82 (1982)
A.C. Cummings, E.C. Stone, C. Steenberg, Astrophys. J. **578**, 194 (2002)
H.J. Fahr, Space Sci. Rev. **15**, 483 (1974)
H.J. Fahr, G. Lay, C. Wulf-Mathies, COSPAR Space Res. **18**, 393 (1978)
H.J. Fahr, T. Kausch, H. Scherer, Astron. Astrophys. **357**, 268 (2000)
P.C. Frisch, Nature **293**, 377 (1981)

P.C. Frisch, Space Sci. Rev. **72**, 499 (1995)
P.C. Frisch, *Solar Journey: The Significance of our Galactic Environment for the Heliosphere and Earth*. Astrophysics and Space Science Library, vol. 338 (Springer, Berlin, 2006)
P.C. Frisch et al., Space Sci. Rev. (2009, this issue)
S. Fuselier et al., Space Sci. Rev. (2009, this issue)
A.B. Galvin et al., Space Sci. Rev. (2007). doi:10.1007/s11214-007-9296-x
G. Gloeckler, J. Geiss, Nature **381**, 210 (1996)
G. Gloeckler, L.A. Fisk, J. Geiss, Nature **386**, 374 (1997)
G. Gloeckler, J. Geiss, Space Sci. Rev. **86**, 127 (1998)
G. Gloeckler, J. Geiss, Space Sci. Rev. **97**, 169 (2001)
T.E. Holzer, Rev. Geophys. Space Phys. **15**, 467 (1977)
H. Hotop, W.C. Lineberger, J. Phys. Chem. Ref. Data **4**, 539 (1975)
V.V. Izmodenov, Space Sci. Rev. **130**, 377 (2007)
V.V. Izmodenov, Yu. Malama, R. Lallement, Astron. Astrophys. **317**, 193 (1997)
V.V. Izmodenov, R. Kallenbach, eds., *The Physics of the Heliospheric Boundaries*, ISSI Scientific Report Series, vol. 5 (ESA, 2006). ISBN 1608-280X
V.V. Izmodenov et al., Astron. Astrophys. **414**, L29 (2004)
V.V. Izmodenov, D.B. Alexashov, A.V. Myasnikov, Astron. Astrophys. **437**, L35 (2005)
V.V. Izmodenov, Y.G. Malama, M.S. Ruderman, J. Adv. Space Res. **41**, 318 (2008)
J.R. Jokipii, Space Sci. Rev. **86**, 161 (1998)
B. Klecker, Space Sci. Rev. **72**, 419 (1995)
R. Lallement, P. Bertin, Astron. Astrophys. **266**, 79 (1992)
R. Lallement, E. Quemerais, J.L. Bertaux et al., Science **307**, 1447 (2005)
T.J. Linde, T.I. Gombosi, P.L. Roe, K.G. Powell, D.L. DeZeeuw, J. Geophys. Res. **103**, 1889 (1998)
J.L. Linsky, Space Sci. Rev. **84**, 285 (1998)
J.L. Linsky, A. Brown, Gayley et al., Astrophys. J. **402**, 694 (1993)
Yu.G. Malama, V.V. Izmodenov, S.V. Chalov, Astron. Astrophys. **445**, 693 (2006)
J. Mazeau, F. Grestau, R.I. Hall, A. Huetz, J. Phys. B **11**, L557 (1978)
W. McClintock, R.C. Henry, J.L. Linsky, W.H. Moos, Astrophys. J. **225**, 465 (1978)
D. McComas et al., Space Sci. Rev. (2009, this issue)
E. Möbius, Space Sci. Rev. **143**, 465 (2009)
E. Möbius, D. Hovestadt, B. Klecker et al., Nature **318**, 426 (1985)
E. Möbius, D. Ruciñski, D. Hovestadt, B. Klecker, Astron. Astrophys. **304**, 505 (1995)
E. Möbius, D. Rucinski, M.A. Lee, P.A. Isenberg, J. Geophys. Res. **103**, 257 (1998a)
E. Möbius et al., in: *Measurement Techniques in Space Plasmas*, ed. by R. Pfaff, J. Borowski, D. Young, Geophys. Monograph, vol. 102 (1998b), p. 243
E. Möbius et al., Astron. Astrophys. **426**, 897 (2004)
E. Möbius, M. Bzowski, H.-R. Müller, P. Wurz, in *Solar Journey: The Significance of our Galactic Environment for the Heliosphere and Earth*, ed. by P.C. Frisch (Springer, Berlin, 2006), pp. 209–258
H.-R. Müller, G.P. Zank, J. Geophys. Res. **109**, A07104 (2004)
H.-R. Müller, V. Florinski, J. Heerikhuisen et al., Astron. Astrophys. **491**, 43 (2008)
D.J. Mullan, J.L. Linsky, Astrophys. J. **511**, 502 (1998)
M. Opher, E.C. Stone, P.C. Liewer, Astrophys. J. **640**, L71 (2006)
H.L. Pauls, G.P. Zank, L.L. Williams, J. Geophys. Res. **100**, 21595 (1995)
N.V. Pogorelov, G.P. Zank, T. Ogino, Astrophys. J. **644**, 1299 (2006)
N.V. Pogorelov, J. Heerikhuisen, G.P. Zank, Astrophys. J. **675**, L41 (2008)
N. Prantzos, Space Sci. Rev. **84**, 225 (1998)
E. Quémerais, J.L. Bertaux, R. Lallement et al., J. Geophys. Res. **104**, 12585 (1999)
S. Redfield, Space Sci. Rev. (2008). doi:10.1007/s11214-008-9422-4
J.D. Richardson, Y. Liu, C. Wang, D.J. McComas, Astron. Astrophys. **491**, 1 (2008)
J.D. Richardson, K.I. Paularena, A.J. Lazarus, J.W. Belcher, Geophys. Res. Lett. **22**, 1469 (1995)
J. Slavin, P.C. Frisch, Astrophys. J. **565**, 364 (2002)
B.M. Smirnov, *Negative Ions* (McGraw-Hill, New York, 1982)
S. Tarnopolski, M. Bzowski, Astron. Astrophys. **483**, L35 (2008)
G.E. Thomas, R.F. Krassa, Astron. Astrophys. **11**, 218 (1971)
G.E. Thomas, Ann. Rev. Earth Planet. Sci. **6**, 173 (1978)
H. Washimi, G.P. Zank, Q. Hu, T. Tanaka, in: *Turbulence and Non-Linear Processes in Astrophysical Plasmas*. AIP Conf. Proc., vol. 932 (2007), p. 153
C.S. Weller, R.R. Meier, Astrophys. J. **193**, 471 (1974)
M. Witte, Astron. Astrophys. **426**, 835 (2004)
M. Witte, M. Banaszkiewicz, M. Rosenbauer, Space Sci. Rev. **78**, 289 (1996)

B. Wolff, D. Koester, R. Lallement, Astron. Astrophys. **346**, 969 (1999)
F.M. Wu, D.L. Judge, Astrophys. J. **231**, 594 (1979)
P. Wurz, in *The Outer Heliosphere: Beyond the Planets*, ed. by K. Scherer, H. Fichtner, E. Marsch (Copernicus Gesellschaft e.V., Katlenburg-Lindau, 2000), p. 251
P. Wurz, L. Saul, J. Scheer et al., J. Appl. Phys. **103** (2008). doi:10.1063/1.2842398
G.P. Zank, Space Sci. Rev. **89**, 413 (1999)
G.P. Zank, H.-R. Müller, J. Geophys. Res. **108**, 1240 (2003). doi:10.1029/2002JA009689
G.P. Zank, H.L. Pauls, L.L. Williams, D.T. Hall, J. Geophys. Res. **101**, 21639 (1996)
G.P. Zank, H.-R. Müller, V. Florinski, P.C. Frisch, in *Solar Journey: The Significance of our Galactic Environment for the Heliosphere and Earth*, ed. by P.C. Frisch (Springer, Berlin, 2006), p. 23

Space Sci Rev (2009) 146: 173–206
DOI 10.1007/s11214-009-9515-8

IBEX Backgrounds and Signal-to-Noise Ratio

P. Wurz · S.A. Fuselier · E. Möbius · H.O. Funsten · P.C. Brandt · F. Allegrini · A.G. Ghielmetti · R. Harper · E. Hertzberg · P. Janzen · H. Kucharek · D.J. McComas · E.C. Roelof · L. Saul · J. Scheer · M. Wieser · Y. Zheng

Received: 16 January 2009 / Accepted: 21 April 2009 / Published online: 9 June 2009

Abstract The Interstellar Boundary Explorer (IBEX) mission will provide maps of energetic neutral atoms (ENAs) originating from the boundary region of our heliosphere. On IBEX there are two sensors, IBEX-Lo and IBEX-Hi, covering the energy ranges from 10 to 2000 eV and from 300 to 6000 eV, respectively. The expected ENA signals at 1 AU are low, therefore both sensors feature large geometric factors. In addition, special attention has to be paid to the various sources of background that may interfere with our measurement. Because IBEX orbits the Earth, ion, electron, and ENA populations of the Earth's magnetosphere are

P. Wurz (✉) · L. Saul · J. Scheer
Physics Institute, University of Bern, Sidlerstrasse 5, 3012 Bern, Switzerland
e-mail: peter.wurz@space.unibe.ch

S.A. Fuselier · A.G. Ghielmetti · E. Hertzberg
Space Physics Department, Lockheed Martin Advanced Technology Center, 3251 Hanover Street, Palo Alto, CA 94304, USA

E. Möbius · H. Kucharek
Space Science Center and Department of Physics, University of New Hampshire, 39 College Road, Durham, NH 03824, USA

H.O. Funsten · R. Harper · P. Janzen
Los Alamos National Laboratory, Los Alamos, NM 87545, USA

M. Wieser
Swedish Institute of Space Physics, Box 812, 98128 Kiruna, Sweden

P.C. Brandt · E.C. Roelof · Y. Zheng
Applied Physics Laboratory, 11100 Johns Hopkins Road, Laurel, MD 20723, USA

F. Allegrini · D.J. McComas
Space Science and Engineering Division, Southwest Research Institute, San Antonio, TX 78228-0510, USA

Present address:
P. Janzen
Department of Physics & Astronomy, University of Montana, 32 Campus Drive, Missoula, MT 59812, USA

prime background sources. Another potential background source is the magnetosheath and the solar wind plasma when the spacecraft is outside the magnetosphere. UV light from the night sky and the geocorona have to be considered as background sources as well. Finally background sources within each of the sensors must be examined.

Keywords Interstellar boundary · Energetic Neutral Atom · ENA · Instrumentation

1 Introduction

The Interstellar Boundary Explorer (IBEX) mission will explore in detail the boundaries of our heliosphere, which is where the solar wind interacts with the interstellar medium (see McComas et al. 2009, this issue). For that purpose, IBEX produces full-sky maps of energetic neutral atoms (ENAs) originating from the boundary region of our heliosphere. There are two sensors, IBEX-Lo and IBEX-Hi, described in detail by Fuselier et al. (2009, this issue) and Funsten et al. (2009, this issue), respectively. IBEX-Lo and IBEX-Hi cover the energy ranges from 10 to 2000 eV and from 300 to 6000 eV, in eight and six energy steps, respectively.

The ENA signals from the boundaries of the heliosphere are expected to be low (Wurz et al. 2008). Thus both sensors feature large geometric factors. In addition, special attention has to be paid to the various sources of background that may interfere with our measurements. As with most applications of ENA instruments in space the fluxes of unwanted signal sources are much larger than the ENA flux, and their efficient suppression is crucial for successful ENA measurements (Wurz 2000). IBEX is in a highly elliptical Earth orbit of about $3 \times 50\ R_E$ (Earth radii). Thus, ion and electron populations of the Earth's magnetosphere and in the solar wind when the spacecraft is outside the magnetosphere are prime sources for background particles. Moreover, ENAs emitted from the terrestrial magnetosphere are of much higher intensity than the expected interstellar signal. Additional background sources include cosmic rays and diffuse UV light backscattered from the interstellar gas and geocorona. Finally, there are also some internal background sources typical for each sensor that need careful consideration.

1.1 Expected Signal

Because of the obvious observational difficulties and the lack of dedicated missions, there are, at present, only a few measurements available on ENAs from beyond the termination shock region. The existing ENA measurements from the termination shock region have been summarised in Fig. 1. They range from 200 eV upwards to almost 100 keV (Hilchenbach et al. 1998; Kallenbach et al. 2005; Wurz et al. 2008). When integrated over energy the ENA fluxes are found to be low with typical values of the order of $10^4\ \mathrm{cm^{-2}\ s^{-1}\ sr^{-1}}$ (Wurz et al. 2008). Most recently, Wang et al. (2008) reported such ENA measurements above a few keV from the STEREO mission. However, these measurements have later been reinterpreted to be X-rays from stars near the galactic centre (Hsieh et al. 2009).

At present, there are no ENA measurements reported below 200 eV. Wood et al. (2007) extrapolated the ENA measurements to lower energies by comparing them with Lyman-α absorption measurements. Based on these extrapolations it was concluded that a straight continuation of the ENA spectrum to lower energies is in conflict with the profiles of the Lyman-α absorption. At energies at and below about 100 eV some sort of flattening of the slope or even a roll over is expected, as suggested by several model calculations (see also Fig. 1).

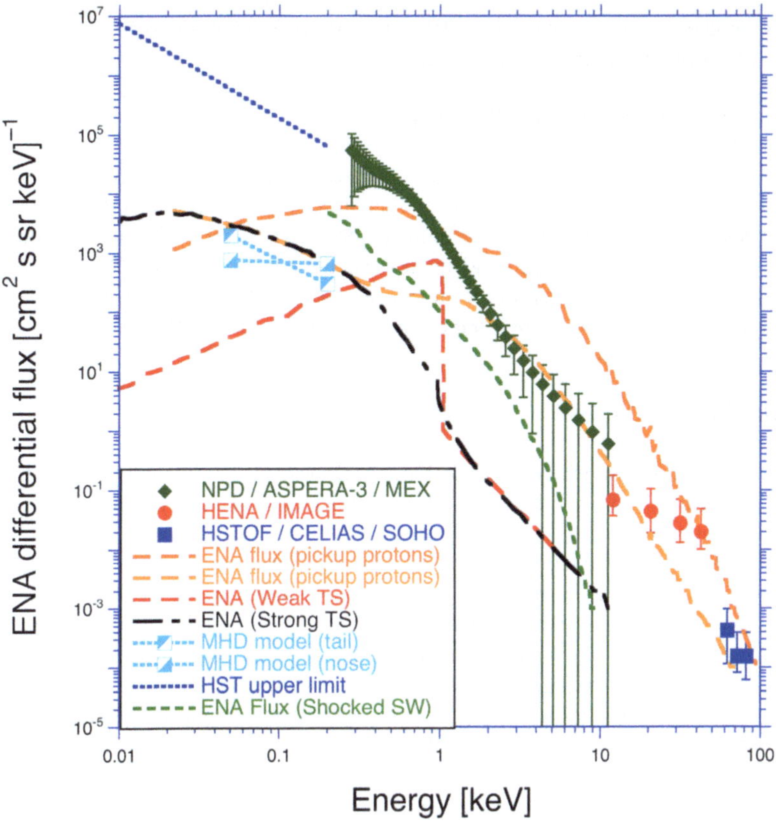

Fig. 1 Points with error bars: Compilation of experimental data of ENAs from the termination shock region. The *red circles* are not actual ENAs from the termination shock region, but represent an upper limit to these fluxes. The *lines* give model fluxes. For details see main text

Our present understanding of the global properties of the region beyond the termination shock relies mostly on detailed model calculations. These calculations not only predict the shape and the physical parameters in the various regimes of the heliospheric interface, but they also predict energy spectra of ENAs produced in that region and travelling inward to Earth orbit. Several calculated ENA energy spectra for an observer at Earth orbit are reproduced in Fig. 1 as well (Wurz et al. 2008, and references therein). The black dashed and red dashed-dotted lines in Fig. 1 (McComas et al. 2004) represent a strong and a weak termination shock, respectively. The light blue lines represent MHD model nose and tail data from a self-consistent axisymmetric MHD plasma/kinetic neutral model (Heerikhuisen et al. 2007). The green short-dashed line shows ENA flux for shocked solar wind data from Fahr and Scherer (2004); the orange long-dashed curves represent two cases of ENA flux for pickup protons from the calculation by Chalov et al. (2003). The HST measurements are indicated by the dotted line that represents an upper limit for the ENA fluxes derived from the observations by Wood et al. (2007).

Currently, there is still is a large range in the predictions of ENA intensities by the various models as can be seen in Fig. 1. For our discussions of the background levels we will use the ENA predictions for the weak and strong termination shocks shown in Fig. 1, which are from the model by Gruntman et al. (2001), since they represent conservative assumptions for the evaluation of the signal-to-noise ratio.

1.2 Background and Suppression Requirements

Based on careful analysis of the expected ENA signal from the termination shock region and the corresponding count rates in the sensors, requirements were developed for the suppression of background and of unwanted stimuli and noise. These requirements were developed to make sure that resulting background and noise count rates will be well below the expected signal rates. Thus requirements for the suppression were defined based on the evaluation of the charged particle and UV environment at the location of the IBEX spacecraft. These requirements are listed below.

1.2.1 IBEX-Lo Requirements

- Noise (excluding cosmic rays) for IBEX-Lo shall be $\leq 8 \cdot 10^{-3}$ counts/sec for all acceptable "any double" coincidence events. Acceptable means that the measured time-of-flight is consistent with the detection of an ion.
- The goal for noise (excluding cosmic rays) for IBEX-Lo is $\leq 8 \cdot 10^{-4}$ counts/sec for all acceptable "any double" coincidence rate.
- Average background for IBEX-Lo shall be $< 2 \cdot 10^{-1}$ ENA/sec/time pixel in energy range between 200 eV and 1 keV incident at the entrance to the energy analysis subsystem.
- The goal for average background is $< 2 \cdot 10^{-2}$ ENA/sec/time pixel in energy range between 200 eV and 1 keV incident at the entrance to the energy analysis subsystem.
- Signal-to-noise ratio for IBEX-Lo shall be > 1.
- The total singles rate from photons and electrons should be $<$ 1000/sec.
- Maximum illegitimate triple coincidence count rate (all other particles masquerading as signal) should be 10 per day when IBEX is in the solar wind.
- The IBEX-Lo collimator shall reject ions of energy $\leq$ 10 keV. The reduction in integral flux shall be > 100 (goal $> 1 \cdot 10^4$) for energies $<$ 10 keV.
- The IBEX-Lo collimator shall reject electrons of energy less than 600 eV. The reduction in integral flux shall be $> 1 \cdot 10^4$ for energies $<$ 150 eV, and > 100 for energies between 150 and 600 eV.
- IBEX-Lo shall deflect and separate the ionised ENAs from ambient UV light and ions of energy greater than 10 keV.
- IBEX-Lo shall use low outgassing materials with strict control of organic volatiles.
- The IBEX-Lo TOF electronics shall not vent through or into the optical or TOF detector subsystems.
- The IBEX-Lo optics and TOF subsystems shall not vent through or into the interior of the spacecraft.

1.2.2 IBEX-Hi Requirements

- The electrostatic energy analyser shall incorporate anti-reflection measures including serrations and two-bounce pass minimum for background particles to enter detector subsystem.
- IBEX-Hi shall have a coincidence subsystem that provides coincidence detection of ionised ENAs that transit the energy analysis subsystem.
- The IBEX-Hi individual channel electron multiplier (CEM) background count rate shall be less than 1 count/sec.

- The IBEX-Hi triple coincidence count rate of the detector subsystem in the solar wind shall be less than $2 \cdot 10^{-4}$ counts/sec. This corresponds to a background flux of atoms or ions at the conversion foil of $3.8 \cdot 10^{-2}$ particles/sec.
- The intrinsic IBEX-Hi triple coincidence count rate of the detector subsystem in the solar wind should be less than $1 \cdot 10^{-5}$ counts/sec.
- The IBEX-Hi detector subsystem shall reside in a light-tight enclosure having minimum 2-corner flange pockets for all edges of enclosure parts.
- IBEX-Hi shall use low outgassing materials with strict control of organic volatiles per the IBEX Contamination Control Plan.
- The IBEX-Hi detector electronics shall not vent through the collimator subsystem.
- The IBEX-Hi ESA and detector subsystems shall not vent through or into the interior of the spacecraft.
- In the solar wind IBEX-Hi shall have a background count level $< 2 \cdot 10^{-2}$ ENAs/sec at the entrance to the energy analysis subsystem in the energy ranges from 1 keV to 3 keV.
- The above background rate should be less than $2 \cdot 10^{-4}$ counts/sec in the energy ranges from 1 keV to 3 keV and at the entrance to the energy analysis subsystem in the solar wind.

1.3 Principles of Background Suppression

Because the expected signal levels of ENAs arriving from the termination shock region are extremely low compared with other diffuse particle and photon sources, any sources of background must be efficiently suppressed to enable a sufficient signal-to-noise ratio to answer the four fundamental science questions underlying the IBEX mission. Major sources of background include charged particles in the local space environment, i.e., electrons and ions, as well as UV photons and penetrating energetic particles. To cope with such demanding requirements background suppression was a key driver for the subsystems of each sensor. In this first section, we start with a brief overview of the general measurement techniques used by both sensors. Section 2 follows with a quantitative evaluation of all identified external background sources. Section 3 will then describe background sources specific to the sensor designs and how they have been minimised.

Both sensors employ the same basic principles to image and detect ENAs and suppress background. They use a succession of mechanical collimation for imaging, conversion of neutrals into ions for further analysis, electrostatic collection and energy analysis, post-acceleration, and coincidence detector techniques (double and triple) for unambiguous signal identification and to achieve the lowest possible background levels. IBEX-Lo converts ENAs to negative ions, which is most efficient at low energies (Wurz 2000), and IBEX-Hi converts them into positive ions, which is most efficient at higher energies. This difference leads to background considerations that are specific to each sensor. Therefore, we start with a brief compilation of background suppression in each of the four subsystems, followed by some notes on specific features in IBEX-Lo and IBEX-Hi.

1.3.1 Common Background Suppression Concepts

Charged Particle Suppression: As will be shown in more detail below, charged particles are many orders of magnitude more abundant than ENAs and are electrostatically rejected at the aperture, before even reaching any sensor structure. This rejection is implemented with negatively biased annular electrodes in front of the collimator aperture but outside of the sensor field-of-view (FOV) to repel ambient electrons with energies up to 600 eV. Electron rejection is followed by electrostatic rejection of positive ions by applying a +10 kV bias directly to the collimator stack, resulting in rejection of ions with energies up to 10 keV. Because the charged particle suppression introduces secondary background sources within the collimator that need to be understood when discussing the external background environment, the charged particle rejection is described in more detail at the end of this section.

Electrostatic Analyser (ESA): After conversion of the ENAs into ions (negative ions for IBEX-Lo and positive ions for IBEX-Hi), the ESA selects the energy passbands for observation, and ions with energies outside of the passband are prevented from transiting the ESA and entering the detector subsystem. Ions outside the selected energy passband will impact an ESA plate and be absorbed in the structure of the ESA. The curvature of the ESA plates likewise suppresses the propagation of UV photons, which impact the outer ESA plate first. Furthermore, the ESA design requires a minimum two-bounce path for any scattered photon to enter the detector subsystem, further suppressing the small fraction of photons that reflect within the ESA subsystem. For IBEX-Lo, the suppression is increased through the use of copper sulphide blackening of all surfaces and serrations of the outer ESA electrode. For IBEX-Hi, the suppression is enhanced by serrations and a dendritic Ebonol-C coating.

Coincidence Measurements: For the successful identification of an ion, at least two detections (at a start and a stop detector) are required in both sensors. A coincident detection event is considered valid if these sequential detections lie within a prescribed time window. The coincidence technique substantially reduces background from random events, such as those arising from penetrating radiation, UV, or detector dark counts. To reduce the background even further, a triple coincidence measurement has been implemented that requires three detections each in a different detector with a specific temporal sequence. Triple coincidence events have the largest background suppression and thus comprise the highest signal-to-noise measurement of ENAs.

Zero Energy Bin: In addition to the eight ENA passbands for IBEX-Lo and six energy passbands IBEX-Hi, there is a zero energy bin where the ESA voltages are set near zero, which prevents ions from passing through the ESA toward the detector subsystem. For this ESA voltage setting, the coincident and non-coincident count rates from UV, cosmic ray, and internal background sources can be measured in orbit for each sensor. This zero energy bin will be used periodically throughout the mission to assess background levels.

Background Detector: To assess the actual energetic ion environment of IBEX a background ion detector was added to IBEX-Hi that measures the integral flux of ions with energies above 14 keV (see Allegrini et al. 2009, this issue).

1.3.2 Specific IBEX-Lo Background Suppression Concepts

IBEX-Lo works at the low end of the ENA energy range and, after ENAs traverse the collimator, converts them to negative ions. These characteristics provide two additional challenges. Ultraviolet light that enters the collimator unimpeded and strikes either collimator structure or the conversion surface can generate copious photo-electrons. Electrons (i.e., photo-electrons and secondary electrons) released from the conversion surface are accelerated towards the ESA. Because of their negative charge, electrons can pass through the ESA provided their energy is inside the selected ESA energy passband. Therefore, magnets were implemented as an additional suppression stage that is not needed in IBEX-Hi.

Magnets: Between the conversion surface unit and the entrance to the ESA there are two nearly concentric rings of permanent magnets that suppress the passage of secondary electrons (from UV or particle induced) into the ESA (see Wieser et al. 2007; Fuselier et al. 2009, this issue).

Time-of-Flight (TOF) Measurement: IBEX-Lo is also required to distinguish different species, such as H and O. Therefore, the detector system is a TOF mass spectrometer (see Fuselier et al. 2009, this issue). With the triple coincidence measurement IBEX-Lo even provides three TOF measurements, which need to be consistent for actual particles, thus providing unprecedented background suppression.

1.3.3 Specific IBEX-Hi Background Suppression Concepts

In IBEX-Hi, ENAs that pass through the collimator transit an ultrathin charge conversion foil. Ultraviolet can likewise pass through the collimator and generate photo-electrons at the entrance surface of the foil and exposed parts of the foil frame. Because the collimator is nominally biased to +10 kV, these photo-electrons would be accelerated towards the collimator and ionise any residual gas between the foil and the collimator. These ions are accelerated towards the foil at an energy between nearly 0 V (if ionisation occurs near the foil surface) and 10 keV (if ionisation occurs in the collimator). Ions impacting the foil at energies $\lesssim 7$ keV can masquerade as heliospheric ENAs. To mitigate this ion production by photo-electrons, a high (95%) transmission grid biased to −300 V was placed immediately above the foil to suppress photo-electrons from the foil and foil frame by a factor of 20.

Coincidence System: IBEX-Hi employs a simpler coincidence detection system than IBEX-Lo because mass analysis of the ENAs is not necessary. To provide superior background suppression, specific coincidence timing windows have been implemented. Different combinations of such windows provide a range of signal-to-noise values (Funsten et al. 2009, this issue).

1.3.4 The Collimator Subsystem

The collimator system plays a crucial role in the suppression of incoming charged particles that would interfere with the performance of the two sensors. The collimator subsystems for IBEX-Lo and IBEX-Hi are very similar, and the subsystem performance is explained in detail in the two instrument papers (Fuselier et al. 2009, this issue; Funsten et al. 2009, this issue). Here we will only address the suppression of unwanted particle fluxes and possible background signals that have their origin in the collimator system. The necessary sequence of electrodes at negative and positive bias potentials has to be considered in detail in the analysis.

Figure 2 shows a schematic diagram of the IBEX collimator (top panel) and the voltages applied to the grids (bottom panel). In the entrance aperture, a negative high voltage, −HV, is applied to reject external electrons with energies up to 600 eV. The mechanical collimator itself is at a high positive voltage, +HV, to reject ions in the local space environment with energies up to 10 keV/q. Charged particles with higher energies can pass through the collimator into the charge conversion subsystem, but their transmission is largely reduced by the geometrical collimation of the hexagonal channels and the electrostatic defocusing property of the pre-collimator. To assist in understanding the background from ions with energy larger than 10 keV, the IBEX Background Monitor (IBaM, Allegrini et al. 2009, this

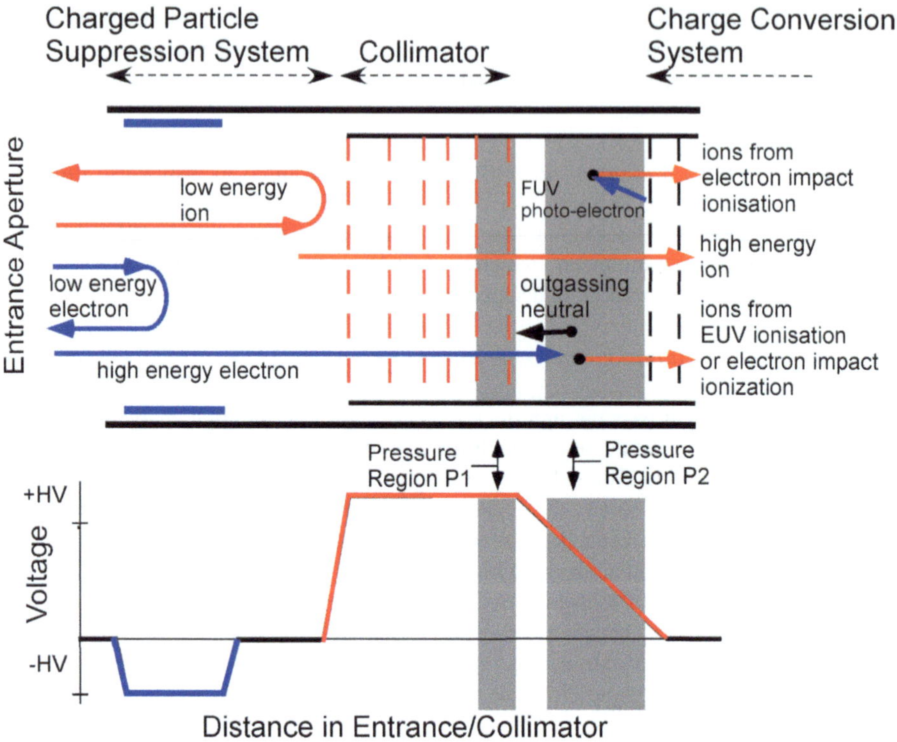

Fig. 2 Schematic of the IBEX collimator (*top*) and voltages applied to the grids (*bottom*). Source neutrals, ions, and electrons enter from the entrance aperture on the *left hand side*

issue), which is incorporated within and boresighted with the IBEX-Hi sensor, will measure the integral fluxes of ions in the local space environment with energies $\gtrsim 14$ keV within a FOV identical to IBEX-Hi and IBEX-Lo.

Even after employing the most careful cleanliness procedures in the sensor design, sensor components will continuously outgas, and a residual gas will be present within the collimator system. The partial pressure of the gas species depends on the area of the outgassing components, the outgassing rate of these components, and the conductance through the collimator subsystem. The volume between the last collimator grid and the charge conversion subsystems of both sensors is of particular concern because of the ion production in which secondary electrons, photo-electrons, and UV photons ionise the residual gas atoms and molecules, which are then accelerated into the charge conversion subsystem. The sensitive volume for the generation of background ions that can masquerade in the sensor as legitimate ENAs are the two pressure regions P1, which lies within the collimator as shown in Fig. 2, and P2, which is between the collimator and the charge conversion subsystem. Because the composition of the residual gas in space is expected to be dominated by water vapour we focus the discussion on water vapour in the following analysis.

There are two external sources for the generation of background ions near the exit region of the collimator (pressure region P1 shown schematically in Fig. 2) and between the collimator and charge conversion subsystem (pressure region P2). First, UV and X-rays from several sources (including short wavelength UV backscattered from the interstellar helium and X-rays from astrophysical sources) can directly photo-ionise neutral gas. Second, these photons (including longer wavelength UV such as hydrogen Lyman-α) can produce photo-electrons at the field shaping grids at the entrance to IBEX-Lo and the charge conversion foil on IBEX-Hi, and these photo-electrons can ionise residual gas atoms and molecules as they are accelerated across region P2 and through region P1. In pressure region P2, these ions are created in an electric field that accelerates them towards the conversion subsystem. A fraction of these ions can enter the charge conversion subsystem within the energy passband of the ESA subsystem and can therefore reach the detector subsystem. In pressure region P1, the ions are created nearly at rest at the potential of the collimator (+HV). A fraction of these ions have a thermal velocity toward the collimator exit, and a larger fraction are accelerated out of the collimator exit by the fringe fields at the exit of the collimator's hexagonal channels. Therefore, these ions can escape into region P2. They are then accelerated by +HV and strike the conversion surface (IBEX-Lo) or the conversion foil (IBEX-Hi). These ions are composed almost entirely of water molecules. Although the molecules strike the conversion surface (IBEX-Lo) or transit the conversion foil (IBEX-Hi) at an energy greater than the highest energy passband of the sensors, the molecule is dissociated during charge conversion, and the molecular energy is partitioned to the atomic constituents based on the atomic mass. Thus, the energy of H after charge conversion is 1/18 of the collimator voltage +HV; if this voltage is 10 kV, then H will have an energy at the charge conversion surface or foil of 550 eV, which lies directly in the energy range of both sensors. Thus, such ions represent a significant background, and every effort has been made to minimize their generation and transmission further into the sensor. These suppression efforts will be discussed in detail in the following sections.

2 External Backgrounds

External background sources include all sources that result in an ion at the exit of the conversion subsystem that could masquerade as a signal ENA from the termination shock region. For IBEX-Lo a set of conversion surfaces where impacting atoms or ions are converted to negative ions upon reflection is used as conversion subsystem (Fuselier et al. 2009, this issue). IBEX-Hi uses a set of ultra-thin carbon foils, the charge conversion foils, through which atoms or ions pass, then exit positively ionised (Funsten et al. 2009, this issue).

2.1 Ultraviolet Photon Background

The Sun and stars are sources of UV photons, to which the detectors used in the IBEX sensors are sensitive. The onset of photo emission occurs at photon energies corresponding to the work function of the material and impurities at the material surface, but should be approximately 4 to 7 eV. For reference, the detection subsystems of both sensors utilise electron emission from carbon foils as an initial step to register passage of an ENA, with the electrons subsequently being registered by a microchannel plate (MCP) or CEM detector in the case of IBEX-Lo or IBEX-Hi, respectively. The work function of graphite is approximately 4.6 eV, which corresponds to an UV wavelength of $\lambda \approx 270$ nm and defines the limit of sensitivity. Toward shorter wavelengths, the solar UV is a continuum spectrum with rapidly declining photon fluxes. At wavelengths below about 150 nm the solar spectrum is mostly a line spectrum, with large temporal fluctuations (Fröhlich and Lean 2004). Hydrogen Lyman-α (121.59 nm, corresponding to approximately 10.2 eV) is a prominent UV line and can be detected directly by typical MCP or CEM detectors used in IBEX-Lo and IBEX-Hi, respectively, thus forward scattering of these photons to the detector subsystem must be efficiently suppressed.

The interaction of molecular water with UV light results in molecular ions, fragment ions and neutrals (Huebner et al. 1992). The ionisation potential for the formation of H_2O^+ is about 12.62 eV, which corresponds to $\lambda = 98.27$ nm (Tonkyn et al. 1991; Merkt et al. 1998), and the thresholds for ion fragments resulting from the interaction of H_2O molecules with UV photons are between 18.1 and 18.7 eV (Huebner et al. 1992).

For a solar spectrum, the photo-ionisation rate of H_2O^+ at 1 AU is in the range (3.31–8.28) $\cdot 10^{-7}$ s^{-1} for quiet and active Sun, respectively (Huebner et al. 1992). Using this solar photo-ionisation rate for a pressure of 10^{-8} mbar (assumed to be mostly H_2O) in the pressure region P2 results in about 60,000 ions s^{-1} being produced in region P2 if the sensor would be pointed directly at the Sun. Since IBEX is not viewing the Sun, this number is prorated to the expected UV fluxes of the sky, which results in the formation of about 0.1 ions/s, which are accelerated up to 10 kV into the sensor onto the conversion surface. This is an overestimation, because the interstellar UV spectrum is composed mostly of the Lyman-α and He I lines. In the following we will calculate the ionisation rates for the He I line for the background evaluation.

Fortunately, the H Lyman-α line is not energetic enough to ionise most of the atoms and molecules in a typical residual gas (i.e., in the collimator system at the sensor entrance), thus it does not create a direct ion background. However, there are ion background sources arising from the Lyman-α, which will be discussed below. Another prominent UV line is the He I line at $\lambda = 58.433$ nm (corresponding to an energy of 21.22 eV), which is energetic enough to ionise most of the atomic and molecular species that might be present in the residual gas. Thus, He I photon fluxes have to be examined in more detail.

A coincident event measured in the detector subsystem requires two detections within a prescribed time window. Because photons scattered into the detector subsystem have sufficient energy to generate only one photo-electron, a coincident event in the detector subsystem requires the random occurrence of multiple photons generating photo-electrons that are detected by multiple detectors within the time window. This detection is called accidental coincidence, because the photons are not related to each other. A recent example of such an accidental coincidence background by UV photons has been discussed by Galli et al. (2006) in detail. In addition to the ion background from photo-ionisation, which may create an ion that is detected directly, there is photon stimulated desorption (PSD) in which UV photons will release atoms, molecules, and ions from exposed surfaces, causing a small particle background. This release of neutral particles adds to the local residual gas, which can be photo-ionised as a second step. Ions generated by PSD may be detected directly. Photon stimulated desorption is highly effective for releasing adsorbed water layers on surfaces for UV wavelengths ≤ 250 nm and is used in vacuum technology to clean high vacuum systems.

Resonant scattering of solar photons on interstellar hydrogen and helium causes the entire sky to be the source of Lyman-α and He I photons. Also, the geocorona is a source of these UV photons with about 1% of the solar flux, which has to be considered when assessing the UV background. Therefore, efficient UV suppression was needed for IBEX, although the IBEX sensors are never pointed toward the Sun.

The Lyman-α background from solar photons scattered at the interstellar hydrogen in interplanetary space is in the range from 400 to 900 R ($1\ R = 1/4\pi \cdot 10^6$ photons cm^{-2} sr^{-1} s^{-1}, Baker and Romick 1976), depending on solar activity and look direction (Quémerais et al. 2006). The brightness of UV bright stars, which are clustered around the galactic plane, reaches up to 3000 R. Figure 3 shows a map of the UV sky in galactic coordinates, recorded with the SWAN instrument on SOHO (Bertaux et al. 1995). Since UV bright stars are localised sources they can be identified in the recorded ENA maps and eliminated if necessary.

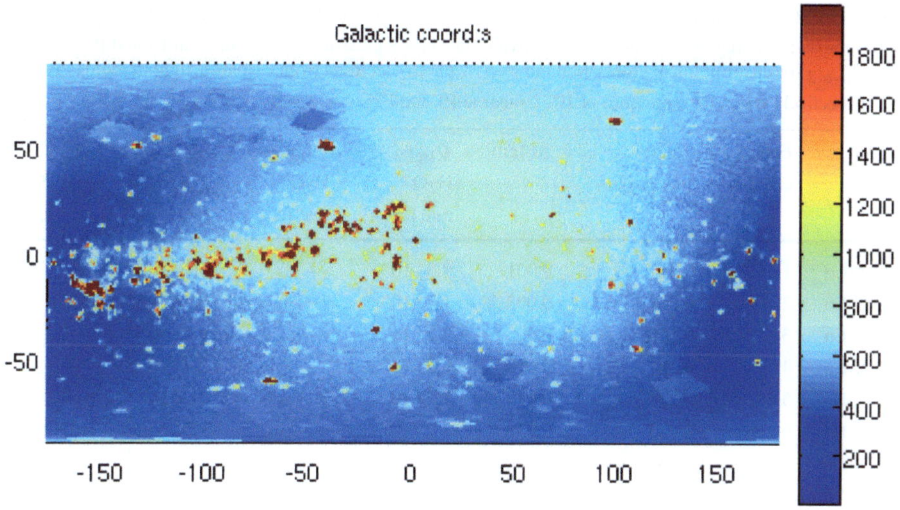

Fig. 3 SOHO/SWAN FUV fluxes (from 115 to 180 nm, including Lyman-α) for the full sky in galactic longitude and latitude (Bertaux et al. 1995). The bright regions (≥ 2 kR) near zero degrees galactic latitude are UV stars in the galactic plane in the direction of the galactic centre

The flux of He I ($\lambda = 58.433$ nm) at 1 AU is known from observations of interstellar He. According to Vallerga et al. (2004) the flux at 1 AU exactly downwind along the He focusing cone is about $\Phi_{\mathrm{HeI}} = 7\ R$ during solar minimum. Outside the focussing cone the typical He I flux is about $\Phi_{\mathrm{HeI}} = 2\ R$. In the following we will use $\Phi_{\mathrm{HeI}} = 10\ R$ as maximum illumination to obtain an upper limit. Moreover, IBEX viewing will at best be perpendicular to the focussing cone at 1 AU.

The rate of ions produced via photo-ionisation by the He I line in region P2 as shown in Fig. 2 is given by

$$Q_{\mathrm{HeI},2} = A \cdot \Delta l \cdot \frac{p_2}{k_B T} \cdot \Phi_{\mathrm{HeI}} \cdot \Delta\Omega \cdot T_{\mathrm{col}} \cdot \sigma_{\mathrm{HeI}} \tag{1}$$

where A is the aperture area, Δl is the effective length over which ions are generated with the correct energy for the energy passband of the actual ESA step, p_2 is the residual gas pressure in region P2 (see Fig. 2), k_B is the Boltzmann constant, and T is the gas temperature, i.e., the sensor temperature during laboratory testing. For the aperture area we take a value of $A = 150\ \mathrm{cm}^2$; the actual aperture areas of IBEX-Lo and IBEX-Hi are a little bit different, which is not important for the calculations performed here. Φ_{HeI} is the ionising He I photon flux, $\Delta\Omega$ is the FOV from the region P2 to the outside ($7^\circ \times 7^\circ$, which gives $\Delta\Omega = 0.0149$ sr), T_{col} is the collimator transmission ($T_{\mathrm{col}} = 0.69$), and σ_{HeI} is the photo-ionisation cross section. For water the ionisation cross section is $2 \cdot 10^{-17}\ \mathrm{cm}^2$ at 58.4 nm (Berkowitz 2002). Using a pressure of $p_2 = 10^{-8}$ mbar in the pressure region P2 (a conservative estimate for the flight situation) and a He I photon flux of $\Phi_{\mathrm{HeI}} = 10\ R$ (again a conservatively large flux), the background rate $Q_{\mathrm{HeI},2}$ was evaluated. Table 1 shows these results for all energy steps of IBEX-Lo and IBEX-Hi. These background fluxes are H_2O^+ ions at the front of the conversion surface (IBEX-Lo) or the conversion foil (IBEX-Hi), which eventually masquerade as ENA signals. Fortunately, the estimated fluxes are significantly lower than the requirement and therefore they are of no concern.

Table 1 Calculated background rate for each of the IBEX energy channels: $Q_{\mathrm{HeI},1}$ and $Q_{\mathrm{HeI},2}$ for He I photo-ionisation in regions P1 and P2, respectively (rates for a pressure of 10^{-8} mbar at P1 and P2); $Q_{\mathrm{Ly}\alpha,1}$ and $Q_{\mathrm{Ly}\alpha,2}$ for electron impact ionisation induced by photo-electrons from Lyman-α light in regions P1 and P2, respectively (rates for a pressure of 10^{-8} mbar at P1 and P2)

Sensor	Energy channel	Centre energy [keV]	Δl [cm]	$Q_{\mathrm{HeI},2}$ H_2O^+ rate [cnt/s]	$Q_{\mathrm{HeI},1}$ H_2O^+ rate [cts/s]	$Q_{\mathrm{Ly}\alpha,2}$ H_2O^+ rate [cts/s]	$Q_{\mathrm{Ly}\alpha,1}$ H_2O^+ rate [cts/s]
IBEX-Lo	1	0.015	0.001	$8.3 \cdot 10^{-6}$	—	$3.2 \cdot 10^{-6}$	—
IBEX-Lo	2	0.029	0.003	$1.7 \cdot 10^{-5}$	—	$6.4 \cdot 10^{-6}$	—
IBEX-Lo	3	0.056	0.005	$3.2 \cdot 10^{-5}$	—	$1.2 \cdot 10^{-5}$	—
IBEX-Lo	4	0.107	0.010	$6.3 \cdot 10^{-5}$	—	$2.4 \cdot 10^{-5}$	—
IBEX-Lo	5	0.208	0.019	$1.2 \cdot 10^{-4}$	—	$4.5 \cdot 10^{-5}$	—
IBEX-Lo	6	0.403	0.036	$2.4 \cdot 10^{-4}$	$6.5 \cdot 10^{-3}$	$9.0 \cdot 10^{-5}$	$4.9 \cdot 10^{-3}$
IBEX-Lo	7	0.781	0.070	$4.7 \cdot 10^{-4}$	—	$1.7 \cdot 10^{-4}$	—
IBEX-Lo	8	1.515	0.136	$9.0 \cdot 10^{-4}$	—	$3.3 \cdot 10^{-4}$	—
IBEX-Hi	1	0.425	0.038	$2.8 \cdot 10^{-4}$	$1.3 \cdot 10^{-2}$	$9.4 \cdot 10^{-5}$	$2.4 \cdot 10^{-3}$
IBEX-Hi	2	0.815	0.073	$4.8 \cdot 10^{-4}$	—	$1.8 \cdot 10^{-4}$	—
IBEX-Hi	3	1.375	0.124	$8.3 \cdot 10^{-4}$	—	$3.0 \cdot 10^{-4}$	—
IBEX-Hi	4	2.250	0.203	$1.4 \cdot 10^{-3}$	—	$4.9 \cdot 10^{-4}$	—
IBEX-Hi	5	3.075	0.277	$1.8 \cdot 10^{-3}$	—	$6.8 \cdot 10^{-4}$	—
IBEX-Hi	6	4.950	0.446	$2.9 \cdot 10^{-3}$	—	$1.1 \cdot 10^{-3}$	—

The effective length Δl is computed from the length of the electric field region, the total voltage across the gap, and the energy pass band of the analyser as

$$\Delta l = h \frac{E}{+\mathrm{HV}} \cdot \left(\frac{\Delta E}{E}\right)_{\mathrm{ESA}} \tag{2}$$

where h is the minimum gap between +HV and ground ($h = 10.0$ mm), +HV is the collimator bias (+HV = 10.0 kV, nominally), E is the centre energy of the energy passband of interest, and $\Delta E/E$ is the energy resolution of the ESA (for IBEX-Lo $\Delta E/E \approx 0.8$, for IBEX-Hi $\Delta E/E \approx 0.45$–0.65 depending on energy). Δl linearly scales with the centre energy of the ESA passband and reaches its maximum at the highest energy ($E = 4.9$ keV for IBEX-Hi, energy step 6). For this energy $\Delta l = 0.45$ cm. For comparison, $\Delta l = 0.14$ cm for the highest energy step of IBEX-Lo ($E = 1.5$ keV, energy step 8). Table 1 also shows the effective length Δl for all energy steps of IBEX-Lo and IBEX-Hi. The background fluxes $Q_{\mathrm{SW,E}}$ are H_2O^+ at the front of the conversion surface (IBEX-Lo) or the conversion foil (IBEX-Hi) for each energy step of IBEX-Lo and IBEX-Hi.

Photo-electrons can be emitted in region P2 by photon impact on exposed surfaces of the charge conversion subsystem (see Fig. 2). These photo-electrons are then accelerated back towards the collimator that lies at +10 kV, and in their transit, they can ionise the neutrals in the P2 region. Of particular concern in IBEX-Hi is the production of photo-electrons off the large area consisting of conversion foils and foil frames, and their acceleration back into the P2 region. For this reason, a high transmission Ni grid is located between the conversion foils and the collimator and is biased to −300 V to suppress photo-electrons back to the foils and foil frames. This grid reduces the flux of ions generated by photo-electrons from the foils and foil frames by a factor of approximately 20. There is no similar issue for IBEX-Lo because there are only high transmission grids between the region P2 and the conversion surface (electrons produced on the conversion surface are suppressed by the magnets in that region).

As a result of these preventative measures, the only photo-electrons that can ionise ions in region P2 are generated at the grids between the collimator and charge conversion subsystem. Since these grids are designed to have high transmission ($T_{\text{grid}} > 95\%$), the amount of surface area available for photo-electron production is minimised, although they span the same area as the aperture. We can calculate the rate of ions produced by photo-electron impact using

$$Q_{Ly\alpha,2} = A \cdot \Delta l \cdot \frac{p_2}{k_B T} \cdot \Phi_{\text{ele}} \cdot \Delta\Omega \cdot T_{\text{col}} \cdot \sigma_2 \cdot (1 - T_{\text{grid}}) \quad (3)$$

where Φ_{ele} is the ionising flux of photo-electrons off the high transmission grids behind the collimator (see Fuselier et al. 2009, this issue for details of the sensor design), T_{grid} is the grid transmission, and σ_e is the electron ionisation cross section ($\sigma_e = 1.25 \cdot 10^{-16}$ cm^2, the average between 100 eV and 10 keV electrons), and the other symbols as before. For the IBEX observations we only need to consider the Lyman-α flux of the night sky, for which we take $\Phi_{Ly\alpha} = 1200\ R$ (as a conservative value). For the photo-electron flux we get $\Phi_{\text{ele}} = \Phi_{Ly\alpha} \cdot \varepsilon$, with ε the efficiency of photo-electron emission. Using $\varepsilon = 0.01$ (Krolikowski and Spicer 1970), we get $\Phi_{\text{ele}} = 1.9 \cdot 10^6$ electrons/(cm^2 s sr). Again, these background fluxes are much lower than the requirement and Table 1 shows the calculated results for this background.

As with region P2, there is photo-ionisation in region P1 within the collimator (see Fig. 2), which can be expressed as

$$Q_{\text{HeI},1} = A \cdot \Delta l_{P1} \cdot \frac{p_1}{k_B T} \cdot \Phi_{\text{HeI}} \cdot \Delta\Omega \cdot T_{\text{col}} \cdot \sigma_{\text{HeI}} \cdot 2P_{\text{Hit}} \quad (4)$$

where Δl_{P1} is the size of region P1 ($\Delta l_{P1} = 1$ cm), p_1 is the residual gas pressure in region P1, Φ_{HeI} is the ionising flux of He I photons, and the other symbols as before. The factor of 2 in (4) accounts for the 2 hydrogen atoms in a water molecule, which have an equal probability of exiting the conversion foil or leaving the conversion surface with a positive or negative charge, respectively. P_{Hit} is the probability that the ion created in region P1 will hit the conversion surface (IBEX-Lo: $P_{\text{Hit}} = 0.5$) or the conversion foil (IBEX-Hi: $P_{\text{Hit}} = 1.0$).

Using $p_1 = 10^{-8}$ mbar pressure in the pressure region P1 and a He I photon flux of $\Phi_{\text{HeI}} = 10\ R$, the background rate $Q_{\text{HeI},1}$ was evaluated. Ions created in region P1 may drift further into the sensor and are accelerated by the collimator potential to 10 keV, nominally. Water ions impinging on the conversion surface or at the conversion foil at 10 keV energy will disintegrate into its atomic constituents. The created atomic constituents, H and O, will leave the surface with an energy proportional to their mass (and a small energy loss). Thus, only hydrogen with about 550 eV will be inside the energy range of IBEX-Lo and IBEX-Hi. Table 1 shows these two results for the pertaining energy steps of IBEX-Lo and IBEX-Hi. The expected background is smaller than the value in the requirement. Given that we used a rather high He I flux for our estimates, the closeness of this background to the requirement is not a concern.

As with region P2, there is ionisation by photo-electrons in region P1 (see Fig. 2), $Q_{Ly\alpha,1}$, for which (3) also applies with $\Delta l = \Delta l_{P1}$. Again, Table 1 shows the calculated results for this background, which is found to be significantly below the requirement and thus of no concern.

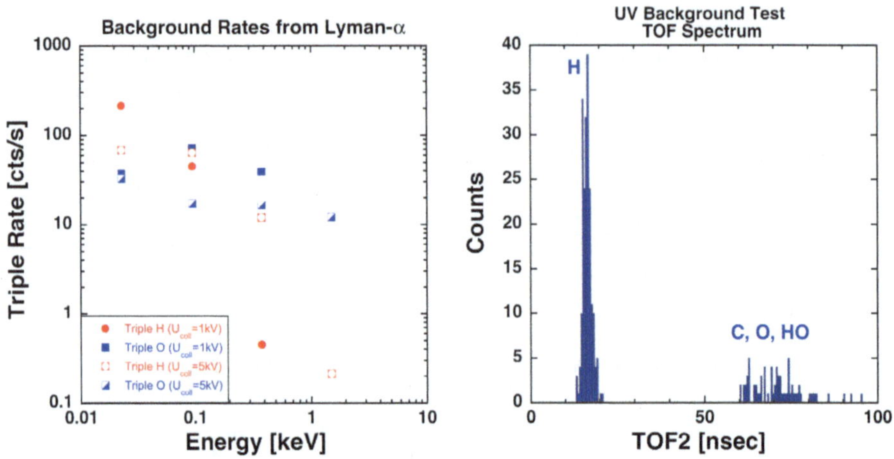

Fig. 4 Results of the UV illumination of IBEX-Lo. *Left*: Triple rates observed for H, C, and O for two collimator voltages. *Right*: Typical mass spectrum recorded during UV tests. Further details are given in the text

2.1.1 The IBEX-Lo UV Background

During the IBEX-Lo calibration campaign a UV background test was performed. For that test an excimer VUV lamp (Ushio, Japan) operating at 126 nm was used. A section of the IBEX-Lo entrance was illuminated with $7.2 \cdot 10^{13}$ photons/s, which corresponds to $1.2 \cdot 10^{6}$ times the Lyman-α background rate expected in flight ($5.8 \cdot 10^{7}$ photons/s) entering IBEX-Lo. The wavelength of 126 nm (close to Lyman-α) used during the UV background test is too long to cause direct ionisation of residual gas molecules. As discussed above, released photo-electrons will cause ionisation resulting in an ion flux of $Q_{Ly\alpha,2}$ from region P2, and photon stimulated desorption of neutrals and ions will take place as well. Figure 4, left panel, shows the triple TOF rates of the detected species as a function of particle energy and collimator voltage recorded during the UV background test. We find that the UV induced background rates depend on the ESA step and the collimator potential, as expected for the $Q_{Ly\alpha,2}$ background ion flux. The mass spectrum given in Fig. 4, right panel, shows a prominent hydrogen peak indicative of water ions, and a broader peak, which is a mixture of C, O, and HO. The ions (mostly water) produced by ionisation through photo-electrons are accelerated toward the charge conversion system to substantial energies and hit the conversion surface where they break up and also may cause sputtering of surface adsorbates. This fully explains the observed mass spectrum shown in Fig. 4.

However, the observed dependence on energy and collimator voltage does not agree with the prediction, since (3) shows an increase in this background signal (see also Table 1) and we observe a decrease with energy. The measured energy dependence of the background fluxes suggests sputtering from the exit structures of the collimator as the main source rather than electron impact ionisation. In addition, a test was run during which the residual gas pressure in the test chamber was deliberately increased by a factor of 2 to see whether most of the UV generated positive ions come from the residual gas in the pressure region P2 or from the exit structures of the collimator. No significant change of the background flux with the ambient gas pressure was seen. This is consistent with the collimator exit structure as the main source, i.e., from electron- and photon-induced desorption of positive ions from the last collimator grids. Therefore, the observed particle rate can be taken as a measure of the expected background arising from the Lyman-α.

Taking the measured UV background rates (Fig. 4, left panel) prorated to the situation during the measurements in space we can calculate backwards to derive the "equivalent" neutral particle fluxes on the conversion surface. Using a TOF efficiency of 0.2 (for triples), an ESA transmission of 0.5, an ionisation efficiency of 0.04 for hydrogen (Wurz et al. 2006) and a reflection probability of 0.06, we get a total efficiency of $2.7 \cdot 10^{-4}$ at 400 eV to convert the triple rates to a flux on the conversion surface. Thus, the measured triple rate of H for energy bin #6 (about 400 eV, see Fig. 4) converts to an equivalent flux of 0.030 s^{-1}, taking into account the ratio of UV fluxes. A comparison of the expected signal with the measured background rates is given in Fig. 5. For energy bin #4 (about 100 eV) we get 0.24 s^{-1} hydrogen ENAs background flux, compared to 10 to 100 hydrogen ENAs per second of the ENA signal from the termination shock region. For most of the energy range the expected background resulting from Lyman-α illumination is significantly below the expected ENA signal, and the signal-to-noise ratio is comfortably large (see Fig. 5). Depending on the actual shape of the energy spectrum of the observed ENAs the signal-to-noise ratio becomes very small only at the lowest two energy bins.

The background due to Lyman-α photons attributed to photo-electron release and subsequent electron impact ionisation of molecules in the pressure regions P1 and P2 has been calculated above and is given in Table 1. These values are much lower than the measured ones, even when considering the higher pressure of 10^{-7} mbar during the measurements. Thus, the background attributed to photo-electron release and subsequent electron impact ionisation can be neglected in further analysis, and only the measured rates given in Fig. 5 need to be considered.

2.1.2 The IBEX-Hi Ultraviolet Background

IBEX-Hi was also tested for response to UV light, particularly to quantify the background created by Lyman-α light. During the calibration campaign, an Ar-purged deuterium lamp, followed by two notch filters used to maximize the fraction of Hydrogen Lyman-α (121.6 nm), and a MgF_2 window directly illuminated the conversion foils. The photon rate at the foils was $\sim 4 \cdot 10^{10}\ s^{-1}$ as measured using a calibrated UV photodiode (Korde et al. 2003). The photon-induced ion production was observed when the collimator was biased to +10 kV, yielding the following information: individual count rates in CEMs A, B, and C of 17, 12, and 1.6 s^{-1} respectively; a total double coincidence rate of about 0.1 s^{-1}; and a total triple coincidence rate below 0.05 s^{-1}. During these tests no voltage was applied to the photo-electron suppression grid to maximise the effect. Unfortunately, experimental difficulties during the calibration campaign prevented us from measuring this background at increasing voltages on the suppression grid, therefore the present background numbers are an upper limit of what is expected in space. As expected, the double and triple coincident rates dropped to the internal background levels when the collimator voltage was switched from +10 kV to 0 V. Adjusted to the interstellar Lyman-α photon flux, the total double coincidence rate becomes $1.5 \cdot 10^{-4}\ s^{-1}$ and the total triple coincidence rate becomes $7.5 \cdot 10^{-5}$ s^{-1}, which are in reasonable agreement with the expected values (see Table 1).

We also tested the response of the IBaM to UV light using a UV krypton line lamp (Resonance Ltd., model KrLM-L, serial number R399) with 116.5 and 123.6 nm wavelengths and intensity close to 1 solar unit of H Lyman-α at the aperture. We varied the angle between the centre of the IBaM FOV and the direction to the lamp. The lamp then shines straight into the IBaM when this angle is equal to 0°. Figure 6 shows the count rate as a function of this

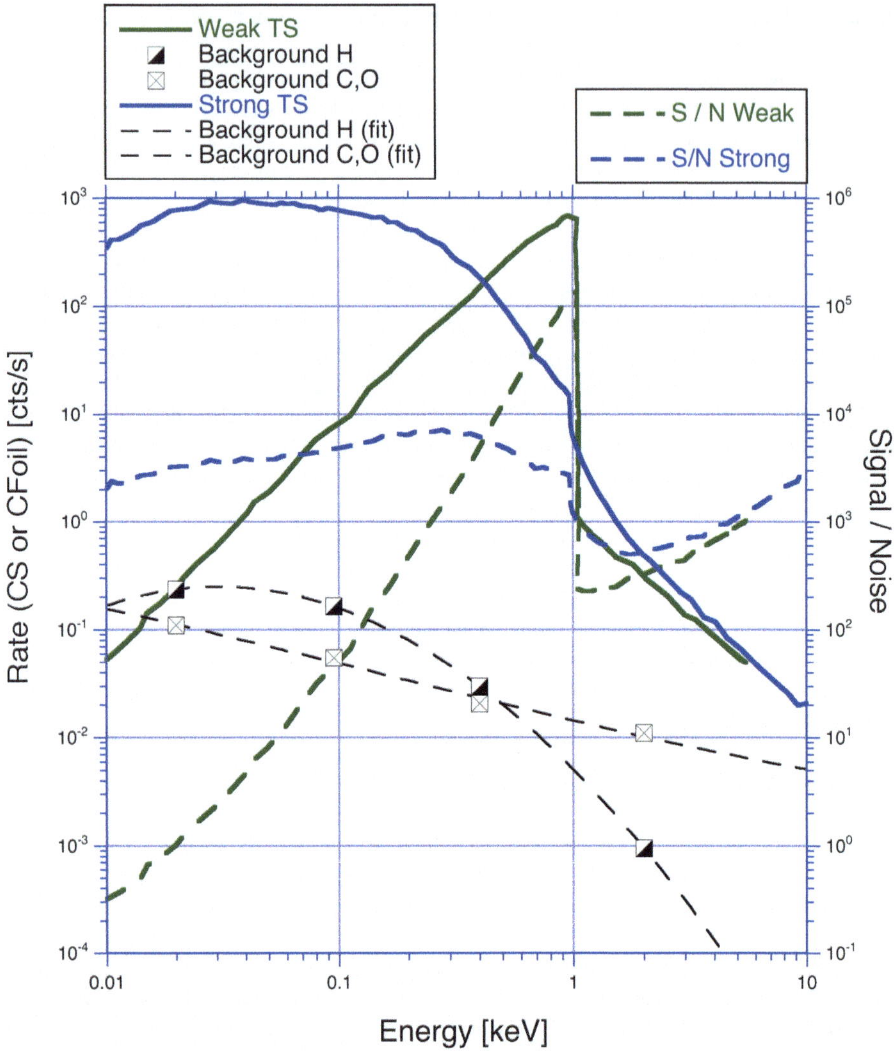

Fig. 5 Measured IBEX-Lo particle background (triples) on the conversion surface as a result of Lyman-α flux compared to the expected signal on the conversion foil or the conversion surface. The expected ENA signal is based on the model by Gruntman et al. (2000)

angle, revealing the expected peak at 0° look direction with a count rate of about 600 s^{-1} when the UV radiation is shining straight into the IBaM's collimator aperture. For this test, the IBaM was placed in the engineering ground support equipment. An additional light leak test was performed during the IBEX cross calibration when the IBEX-Hi sensor was in its flight configuration. We did not find any light leaks for the beam monitor, which verified that the light tight design worked as expected.

Prorating the maximum background rate of about 600 s^{-1}, when UV goes straight through the collimator, from the solar UV fluxes to the interstellar UV fluxes gives a background rate of $9 \cdot 10^{-4}$ s^{-1}. Assuming that the UV intensity reflected by the Earth is roughly 1/100 of the solar H Lyman-α intensity at 1 AU ($\sim 3 \cdot 10^{11}$ photons cm^{-2} s^{-1}), the max-

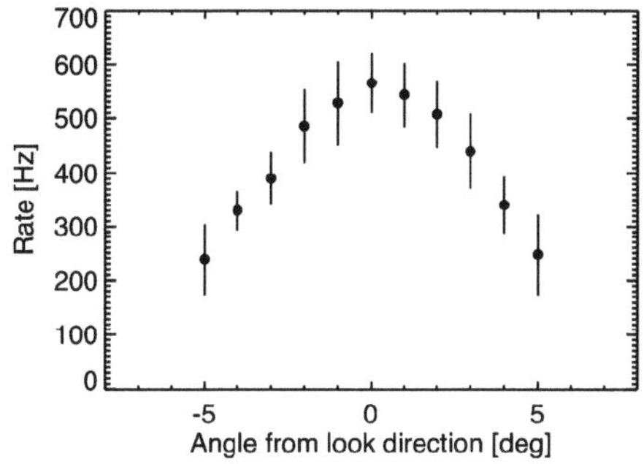

Fig. 6 UV induced background in the IBEX Background Monitor as a function of the angle from the look direction, which is at 0°

imum H Lyman-α intensity when the IBaM will look straight at the Earth will be about $3 \cdot 10^9$ photons $\text{cm}^{-2}\,\text{s}^{-1}$. Thus, we can expect a background count rate of a maximum of approximately 5.5 s^{-1} from the geocorona.

2.2 Local Ion Populations

Background ENAs are generated from the charge exchange of ambient plasma ions with outgassed atoms and molecules from the spacecraft. Ambient ions, if not screened off by the collimator bias, could strike the collimator grids, charge exchange into neutrals, and then continue on to the conversion subsystem masquerading as ENAs. These background signals are indistinguishable from the signal. The fluxes of background ions depend on the location of the IBEX spacecraft. The fluxes are highest within the magnetosheath and foreshock regions with substantial omni-directional populations of energetic ions. Since both sensor FOVs are always pointed about 90° away from the Sun, and thus from the bulk solar wind flow direction, the solar wind ion population does not play a role for this background, which will be shown below.

Aside from the generation of neutrals on the edges of the grids in the collimator, the production of neutrals locally near the spacecraft that masquerade as ENAs from the distant heliosphere depends on the outgassing pressure from the spacecraft and the ion flux. Outgassing products from the sensor typically consist of water and associated molecules. The conservative estimate of this background ENA differential flux, j_0, is

$$j_0 = \int_{r_0}^{\infty} \mathrm{d}r \cdot n(r) \cdot \sigma_{\mathrm{H}} \cdot J_I \tag{5}$$

where n is the number density of outgassing species (that decreases as r^{-2} with distance r from the spacecraft), σ_{H} is the charge exchange cross section of H^+ on water and water related species, and j_I is the ambient ion differential flux. The integral is performed from the spacecraft surface, r_0, to infinity. The neutral density is expressed as $n(r) = n_{S/C}(r_{S/C}/r)^2$ where $r_{S/C}$ is the spacecraft radius and $n_{S/C}$ is the density at the spacecraft surface, so the integral equation yields

$$j_0 = n_{S/C} \cdot r_{S/C} \cdot \sigma_{\mathrm{H}} \cdot j_I = \frac{p_{S/C}}{k_B T} \cdot r_{S/C} \cdot \sigma_{\mathrm{H}} \cdot j_I \tag{6}$$

where $p_{S/C}$ is the pressure at the spacecraft surface, σ_{H} is the energy dependent charge exchange cross section of H^+ on H_2, O_2, H_2O, $r_{S/C}$ is the radius of the spacecraft ($r_{S/C} \approx$ 50 cm), and j_I is the ion flux directed into the collimator. This estimate assumes that the

Table 2 Neutral fluxes on the conversion surface or conversion foil due to charge exchange with spacecraft outgassed neutral gas for the IBEX energy channels: Q_{CX} for magnetosheath ions (rates for a pressure of 10^{-11} mbar near the spacecraft), $Q_{SW,I}$ for solar wind ions (rates for a pressure of 10^{-11} mbar near the spacecraft), $Q_{SW,E}$ for solar wind electrons from regions P2 and P1 (rates for a pressure of 10^{-8} mbar at P1 and P2)

Sensor	Energy channel	Centre energy [keV]	Q_{CX} [cnt/s]	$Q_{SW,I}$ [cnt/s]	P2: $Q_{SW,E}$ [cnt/s]	P1: $Q_{SW,E}$ [cnt/s]
IBEX-Lo	1	0.015	$3.8 \cdot 10^{-6}$	$9.0 \cdot 10^{-3}$	$3.4 \cdot 10^{-6}$	—
IBEX-Lo	2	0.029	$1.4 \cdot 10^{-5}$	$1.8 \cdot 10^{-3}$	$6.6 \cdot 10^{-6}$	—
IBEX-Lo	3	0.056	$3.1 \cdot 10^{-5}$	$1.0 \cdot 10^{-4}$	$1.3 \cdot 10^{-5}$	—
IBEX-Lo	4	0.107	$7.5 \cdot 10^{-5}$	$8.2 \cdot 10^{-7}$	$2.5 \cdot 10^{-5}$	—
IBEX-Lo	5	0.208	$1.9 \cdot 10^{-4}$	$1.2 \cdot 10^{-10}$	$4.8 \cdot 10^{-5}$	—
IBEX-Lo	6	0.403	$1.7 \cdot 10^{-4}$	$6.4 \cdot 10^{-18}$	$9.0 \cdot 10^{-5}$	$5.1 \cdot 10^{-3}$
IBEX-Lo	7	0.781	$1.1 \cdot 10^{-4}$	0	$1.8 \cdot 10^{-4}$	—
IBEX-Lo	8	1.515	$5.4 \cdot 10^{-5}$	0	$3.5 \cdot 10^{-4}$	—
IBEX-Hi	1	0.425	$1.7 \cdot 10^{-4}$	0	$1.0 \cdot 10^{-4}$	$2.6 \cdot 10^{-3}$
IBEX-Hi	2	0.815	$1.1 \cdot 10^{-4}$	0	$1.9 \cdot 10^{-4}$	—
IBEX-Hi	3	1.375	$5.4 \cdot 10^{-5}$	0	$3.2 \cdot 10^{-4}$	—
IBEX-Hi	4	2.250	$4.7 \cdot 10^{-5}$	0	$5.2 \cdot 10^{-4}$	—
IBEX-Hi	5	3.075	$3.5 \cdot 10^{-5}$	0	$7.1 \cdot 10^{-4}$	—
IBEX-Hi	6	4.950	$6.7 \cdot 10^{-6}$	0	$1.1 \cdot 10^{-3}$	—

charge exchange process does not change the flight direction significantly, and the neutralised ion continues to propagate in the initial direction. This assumption is valid for the ion energies of concern for this calculation (see Fig. 7), but at ion energies of about 10 eV and below this assumption is increasingly violated (Hodges and Breig 1991). The neutral flux j_0 represents a background into the collimator. Thus, the neutral flux onto the conversion surface or conversion foil arising from charge exchange is expressed as

$$Q_{CX} = A \cdot \Delta\Omega \cdot T_{col} \cdot J_0 = A \cdot \Delta\Omega \cdot T_{col} \cdot \frac{p_{S/C}}{k_B T} \cdot r_{S/C} \cdot \sigma_H \cdot j_I. \tag{7}$$

After several months in space, we can assume the pressure of the spacecraft surface is $p_{S/C} = 10^{-11}$ mbar, based on Rosetta measurements (Graf et al. 2008) and because IBEX is a much smaller spacecraft that has undergone stringent cleanliness control. For the magnetosheath (the worst case) we used the measured ion fluxes from the CODIF instrument on the Cluster mission for our analysis (shown in Fig. 7), which are similar to magnetosheath measurements reported earlier (Williams et al. 1988; Paschalidis et al. 1994). Table 2 shows the background fluxes on the conversion surface or conversion foil for the IBEX energy steps under these assumptions and using the cross sections of H^+ on water (Greenwood et al. 2000). These background fluxes are lower than the requirements (see Sect. 1.2) and are therefore no concern for the mission.

The strongest background contribution would originate from ambient ions striking the edges of the front collimator grid. Assuming that the entire edge area of the 50-µm-thick grid is the cross-section for interaction, the resulting neutral-atom fluxes, after charge exchange on the surface from the expected ambient fluxes, would have presented an overwhelming background. This assessment led to the conclusion that all ions up to 10 keV had to be

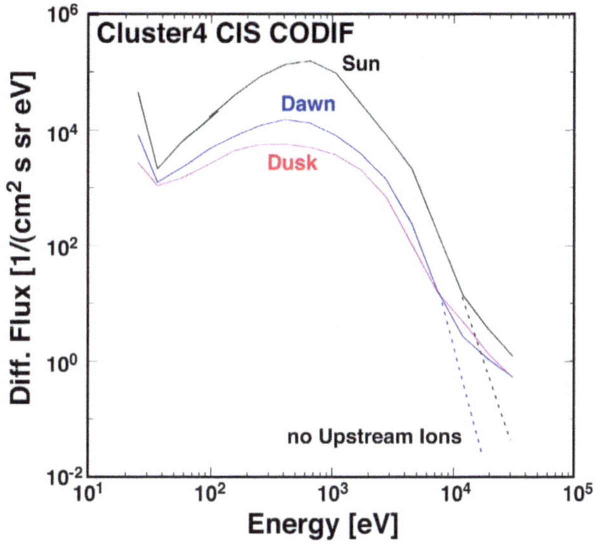

Fig. 7 Spectra of magnetosheath ions at three different locations in the magnetosphere. Data are from CODIF/Cluster

electrostatically rejected. The passage of ions between 10 and 16 keV (the energy range of the sensors after decelerating these ions by the collimator potential) through the collimator had to be minimised. As a result background count rates could be reduced to well below the expected signal for most of the sensor energy range even in the magnetosheath.

Solar wind ions are a supersonic flow of mostly protons, about 4% alpha particles, and the total of heavier ions at the permil level. Solar wind velocities are in the range between 300 and 800 km/s and Mach numbers range between 10 and 20. Since the respective FOVs of the IBEX sensors are pointed roughly perpendicular to the direction of the solar wind flow, there is very little ion flow toward the sensor apertures, therefore we expect only a small flux of neutralised solar wind background, $Q_{\mathrm{SW,I}}$, directed toward the apertures of the IBEX sensors. We estimated this neutral background, $Q_{\mathrm{SW,I}}$, using a Maxwellian-Boltzmann description of the solar wind velocity distribution in three dimensions. The evaluation of the distribution for the energy (i.e., velocity) intervals of the IBEX sensors gives the solar wind ion fluxes j_I toward the entrance apertures, and applying (7) gives the neutral fluxes entering the IBEX sensors. We considered two cases: slow solar wind (with a solar wind speed of $v_{SW} = 400$ km/s, a proton density of $n_{SW} = 8\ \mathrm{cm}^{-3}$, and a kinetic temperature of $v_{th} = 20$ km/s, i.e., Mach number $M = 10$) and fast solar wind ($v_{SW} = 800$ km/s, a density of $n_{SW} = 3\ \mathrm{cm}^{-3}$, and a kinetic temperature of $v_{th} = 40$ km/s, i.e., $M = 20$), but found that the results for the neutral background were very similar for these two cases, with the fast solar wind case resulting in a somewhat smaller neutral background. We found that only the lower energy channels of IBEX-Lo are affected by solar wind plasma and the background is considerably lower than the requirement. The estimates for this background for the slow solar wind case are given in Table 2.

2.2.1 IBEX-Hi Background Monitor

Ions with energies higher than the positive high voltage applied to the collimator, +HV, have direct access to the ENA to ion conversion section. These ions will have the wrong energy for transmission through the ESA, but can cause release of particles by sputtering negative ions from the conversion surfaces or positive ions from the conversion foils. The other way for these ions to create a background signal (as discussed above) is if they scatter off the edge of a collimator grid, become neutral, and pass through the collimator to the ENA to ion conversion subsystem with energy less than their original energy, i.e., reduced by the collimator potential and by some energy loss from the scattering.

In the solar wind the proton flux above 10 keV is very low and of no concern for the ENA measurements, as discussed above. However, if a spacecraft like IBEX is connected magnetically to the certain parts of the Earth's bow shock, creating a so-called upstream event, energetic particles from the bow shock can be observed far upstream in the solar wind (Sanderson et al. 1996). Figure 8 shows an example of an energy spectrum of an upstream event.

There are considerable proton fluxes above energies of 10 keV. Therefore, the IBaM was introduced to measure the ion fluxes above about 14 keV to provide real-time information for this background source. The IBaM reports counts as a function of spin angle with the same resolution as IBEX-Hi and IBEX-Lo. This provides key information about the nature of the local ion environment (e.g., solar wind ions, magnetospheric ions, or ion fluxes through connection to the bow shock) and how to interpret this as a potential background in the ENA spectra and possibly cull the ENA data during these times.

2.3 Local Electron Populations

Unlike solar wind ions, solar wind electrons are subsonic, i.e., their mean thermal speed considerably exceeds the solar wind (ion) bulk speed. From observations it is known that the solar wind electron population is composed of three main components: a cold and almost isotropic collisional core; a hot, variably-skewed halo population, and, in fast solar wind, often a narrow strahl aligned with the magnetic field (Pilipp et al. 1987).

Electrons from the high-energy tail of the distribution of solar wind electrons penetrate through the collimator and enter region P2 (see Fig. 2). There, they will act as an additional ionisation source that creates ions in the P2 region. The rate of ions produced in this manner is

$$Q_{SW,E} = A \cdot \Delta l \cdot \frac{p_2}{k_B T} \cdot \Phi_{SW,E} \cdot \Delta\Omega \cdot T_{\mathrm{col}} \cdot \sigma_e \tag{8}$$

where $\Phi_{\mathrm{SW,E}}$ is the high energy electron flux into the charge exchange region, and σ_e is the electron ionisation cross section. The electron flux $\Phi_{\mathrm{SW,E}}$ is region time dependent. As an estimate, we take an average flux from the solar wind as measured by the Ulysses solar wind electron spectrometer (McComas et al. 1992). The solar wind sensor measured up to 511 eV and these electron fluxes were extrapolated to 600 eV. The differential energy fluxes represent the "average" flux observed over the mission. These fluxes were integrated to determine the electron number flux above 600 eV, which gives $1.5 \cdot 10^4$ electrons/(cm^2 s sr) above 700 eV. The electron impact ionisation cross section of water is $\sigma_e = 2 \cdot 10^{-16}$ cm^2 at 100 eV and $\sigma_e = 0.5 \cdot 10^{-16}$ cm^2 at 10 keV. For the estimation of $Q_{\mathrm{SW,E}}$ we used an average value of $\sigma_e = 1.25 \cdot 10^{-16}$ cm^2. The product $A \cdot \Delta l$ gives the volume in which the ions are created.

As with region P2, there is electron impact ionisation in region P1 (see Fig. 2), which can be expressed as

$$Q_{SW,E} = A \cdot \Delta l_{P1} \cdot \frac{p_1}{k_B T} \cdot \Phi_{SW,E} \cdot \Delta\Omega \cdot T_{\mathrm{col}} \cdot \sigma_e \cdot P_{\mathrm{Hit}}. \tag{9}$$

Again, the factor of 2 in (9) accounts for the 2 hydrogen atoms in a water molecule, which exhibit equal probability to leave the conversion foil with a positive or the conversion surface

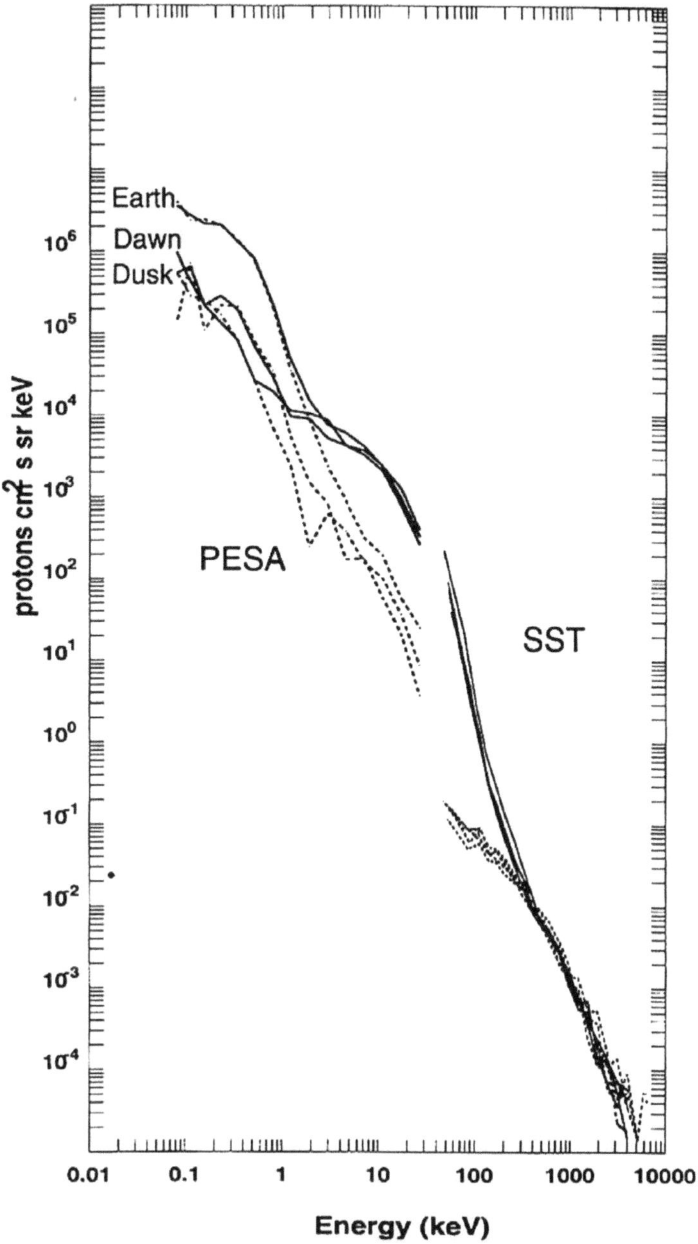

Fig. 8 Composite proton spectrum of upstream event measured by PESA and SST on WIND on 13 August 1995 (Sanderson et al. 1996)

with a negative charge. P_{Hit} is the probability that the ion created in region P1 will hit the conversion surface (IBEX-Lo: $P_{\text{Hit}} = 0.5$) or the conversion foil (IBEX-Hi: $P_{\text{Hit}} = 1.0$). Again, Table 2 shows the results for this background, which are found to be considerably below the requirement.

2.4 Penetrating Radiation

Particles with very high energy (exceeding 1 MeV/nucleon) can penetrate the thin walls of space instrumentation and cause background signals when hitting a detector. The terrestrial radiation belts are a source of such particles. However, the IBEX orbit is designed to have a low radiation dose from radiation belt particles (McComas et al. 2009, this issue), with the IBEX orbit being far from the near-Earth radiation environment during data acquisition. Data acquired while the IBEX spacecraft is inside the magnetosphere are not analysed.

Another source of highly energetic particles is cosmic rays, which consist of approximately 83% protons, 13% alpha particles, 1% heavier nuclei, and 3% electrons. The cosmic ray energy spectrum extends from a few 100 MeV to energies in excess of 10^{20} eV (McDonald and Ptuskin 2001). In the ecliptic plane the intensity of cosmic ray particles depends on the solar cycle, with the modulation being anti-correlated with solar activity. The isotropic flux of galactic cosmic rays at 1 AU is 4 protons/(cm^2 s) at sunspot minimum and 2 protons/(cm^2 s) at sunspot maximum. The cosmic ray flux is measured by a variety of spacecraft, thus measurements are available if needed. In the energy range between 1 MeV/nucl and ≤ 70 MeV/nucleon there are, in addition to the galactic cosmic rays, the anomalous cosmic rays, which are quite variable.

These cosmic ray particles present a constant source of background. They are so energetic that they cannot be effectively shielded in an instrument on a spacecraft, since the necessary mass would be prohibitive. For both IBEX-Lo and IBEX-Hi, the particle detectors are deep inside the sensor, which gives some shielding. However, the best reduction of this background is the use of a triple coincidence technique for the particle detection, together with TOF identification (IBEX-Lo) or TOF gates (IBEX-Hi).

During quiescent periods, measurements from the X-ray satellite Chandra show that cosmic rays produce a count rate of about 1.6 s^{-1} cm^{-2} in the MCP detector (Juda et al. 2003). This results in a count rate of 40 s^{-1} for each of the four IBEX-Lo MCP detectors. Using a TOF window of 200 ns this gives a double rate of $3.2 \cdot 10^{-4}$ s^{-1} and a triple rate of $2.6 \cdot 10^{-9}$ s^{-1}. With the false ion detections randomly distributed in the mass spectra their separation from the signal will be feasible.

Measurements from the Genesis Ion Monitor using energy channels significantly above the solar wind energy show total CEM background count rates of approximately 0.5 s^{-1} (Steinberg 2007). Similarly, we anticipate that the IBEX-Hi CEM count rate from penetrating radiation will be 0.5 s^{-1}. Using a TOF window of 100 ns (corresponding to the long coincidence window) this gives a double rate of $2.5 \cdot 10^{-8}$ s^{-1} and a triple rate of approximately $1.2 \cdot 10^{-15}$ s^{-1}. Although these estimated background rates are very low, they will add directly to the signal rates since there is no mass analysis (i.e., TOF analysis) in IBEX-Hi. To minimize the background from penetrating radiation in the IBEX-Hi sensor, the CEMs are positioned such that no straight penetrating particle trajectory can go through all three CEMs. Thus, a triple coincidence is limited to penetrating particles actually crossing the two carbon foils. Backgrounds due to penetrating radiation are discussed in more detail in Sect. 3.1.2.

Solar Energetic Particle (SEP) events are another source of penetrating radiation. However, their total duration of events lasts less than 2% of the mission. Data during this 2% of the mission are removed from analysis. Solar Energetic Particle events are easily identified for removal by the increase in non-coincident rates of all detectors, and these rates are independent of energy passband and incident angle. Data from other spacecraft also provide the arrival time and duration of these SEP events.

2.5 Neutral Atom Background

Competing ENA signals for the IBEX objective originate from the terrestrial ring current and plasma sheet, which we will refer to as foreground. Heliospheric and interstellar neutral fluxes are relatively low, so care must be taken to identify look directions where the terrestrial ENA intensities dominate. For comparison, typical strong storm-time neutral fluxes from the Earth's magnetosphere (ring current) in the energy range of 1000–2000 eV are about $4 \cdot 10^4$ ENAs/(cm^2 s sr), which exceeds even the highest predictions of ENAs from the termination shock region by a factor of 10 (see Fig. 1). ENAs from high latitude ionospheric outflow at energies around 50 eV range from $6 \cdot 10^4$ to $1 \cdot 10^6$ ENAs/(cm^2 s sr). These neutral fluxes can vary significantly on timescales of tens of minutes.

2.5.1 Ring current

To estimate the extent of terrestrial ENA intensities we simulate the ENA intensities from the expected apogee (40 R_E) and the minimum science operations altitude (10 R_E). Our goal is to provide upper limits on the expected background. Most of the time the terrestrial ENA background will be orders of magnitude lower than the upper limits given here. Energetic neutral atom simulations are produced from runs of the Comprehensive Ring Current Model (CRCM) (Fok et al. 2001, 2003). The boundary conditions of this model run were taken from an event with an unusually dense plasma sheet leading to a strong ring current. Only protons have been assumed since the O^+ intensities (not energy densities!) and oxygen ENA intensities are often less than those of protons. Typical storm-time, ring-current ion spectra have been compiled by Kistler et al. (1989), and the ring current and plasma sheet ion distribution have been reported by several authors (Hamilton et al. 1988).

Estimated ring-current-hydrogen ENA intensities are shown in Fig. 9 together with skymaps of the ENA emissions from the vantage points $y = 40\ R_E$ (panel a) and $y = 10\ R_E$ (panel b). These simulated images have a high angular resolution and are meant to be used as reference ENA distributions representing the ENA foreground outside the instrument. The line plot (solid lines in panel c) in Fig. 9 shows the ENA intensity in the sensor's 7°-wide FOV as a function of look angle for various vantage points. The two strong peaks visible near zero degree spin angle result from the low-altitude emission, which is produced by the charge exchange between the magnetospheric ions and the upper atmosphere. To correctly model the low-altitude emission, multiple charge exchange and charge stripping interactions must be taken into account. In the present simulation we have approximated the low-altitude interaction by taking a "hard" shell at 350 km altitude below which we assume all ions to be lost to the atmosphere. This is acceptable for the purpose of this paper, since it is already implemented in the science operations that the IBEX sensors will not take any measurements when inside 10 Earth radii (McComas et al. 2009, this issue).

It is important to note that the ENA emissions from the plasma sheet are clearly visible in both IMAGE/HENA (Brandt et al. 2002) and IMAGE/MENA (McComas et al. 2002) measurements. Therefore, they probably represent the most extended foreground signal relevant to the IBEX observations. From its near-equatorial orbit it is anticipated that IBEX may detect ENA emissions from the plasma sheet as far down the tail as $x = -20\ R_E$ or more, in the Earthward hemisphere. These ENA emissions are easily avoided by excluding the Earthward hemisphere when IBEX is outside the plasma sheet region. In fact, the vertical scans of the plasma sheet present a very valuable diagnostic tool for the plasma sheet itself.

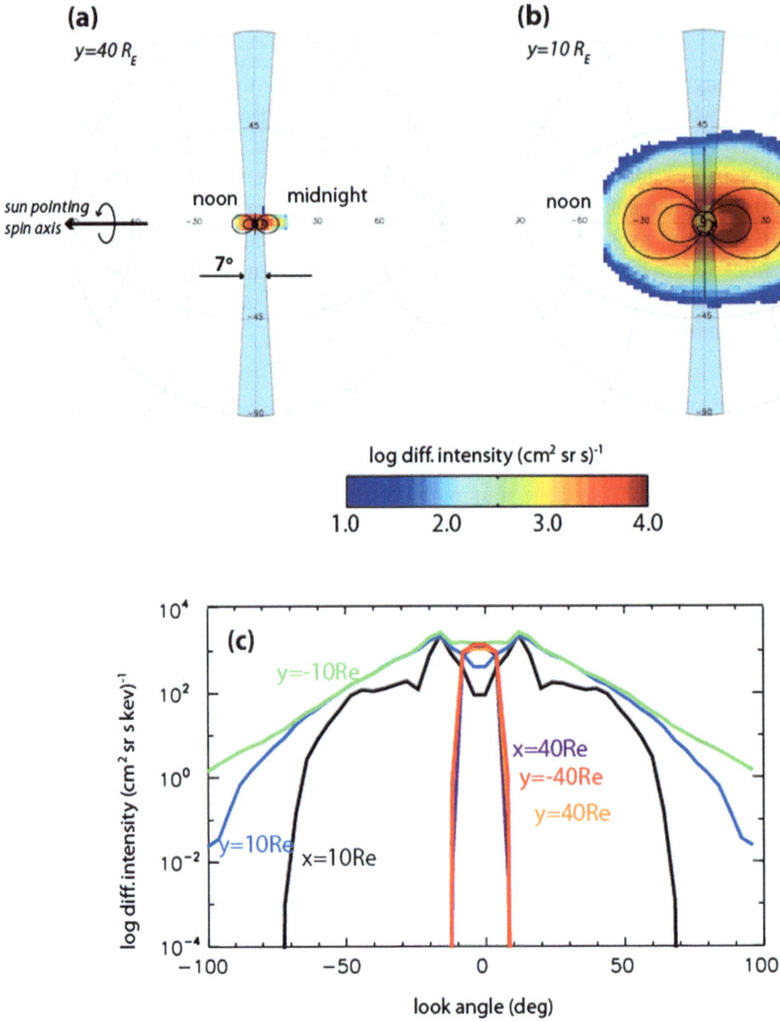

Fig. 9 Simulated worst-case terrestrial ENA intensities in the 1–6 keV range for the ring current and the plasma sheet. (**a**) Full sky map of ENA emissions taken from $y = 40\ R_E$. The *blue swath* represents the angular width of the IBEX telescope. (**b**) Same as (**a**) but from $y = 10\ R_E$. (**c**) *Lines* represent the ENA intensities measured by a spinning 7° wide telescope from various vantage points as indicated

2.5.2 IMAGE/HENA Backgrounds

Even though the IMAGE/HENA instrument was designed to image magnetospheric ENAs (Mitchell et al. 2000), it scanned almost the entire sky like the two IBEX sensors. So there were portions of the scans that could be used to search for ENA sources beyond the heliospheric termination shock. No clear sources of heliospheric emission were detected above the various background sources. There were three significant sources of background in the HENA data, and they were carefully quantified as part of the search for heliospheric sources: 1) ultraviolet radiation (of solar, terrestrial, and heliospheric origin) falling on the front foil, 2) penetrating particles and particles passing through the electrostatic deflection system of the collimators and 3) ENAs from magnetospheric sources. Note that these background sources were not significant for the imaging of magnetospheric ENAs, which have intensities several orders of magnitude larger than the expected intensities of ENAs from the heliosheath.

The coincidence logic for HENA required a pulse on a back foil following a pulse on the front foil within a set of TOF windows. There was a finite count rate on both the front and back foils. The counts on the front foil were primarily due to ultraviolet radiation, and they tended to change with solar activity. The counts on the back foil generally remained nearly constant and are attributed to cosmic rays penetrating the instrument. The accidental coincidence rates estimated from the measured start (front foil) and stop (back foil) counts within the TOF windows were comparable to (but less than) the actual total rates. There were also intermittent periods when there were enhanced background count rates that were closely correlated to increases in solar energetic protons measured by the ACE/EPAM instrument. These periods were easy to identify and eliminate from the data set. Finally, very conservative geometric criteria (based on the geometry of geomagnetic field lines in the particle trapping regions) were set to separate ENAs produced in the magnetosphere from those produced in the heliosheath. In the end, IMAGE/HENA did detect regions of the sky where the actual counts (surviving the culling based on the selection criteria) exceeded the estimated accidental counts. Although it could be a candidate source of ENAs from the heliosheath, it is better ordered in the sky in GSE (Sun-Earth centered) coordinates than it is in ECI (inertial ecliptic) coordinates, thus suggesting a residual magnetospheric contribution from the direction of the geotail.

2.5.3 Neutral Solar Wind

Neutral solar wind (NSW) originates from solar wind ions, which become neutralised during their travel from the Sun through interplanetary space. Consequently, NSW moves in the anti-Sunward direction with solar wind velocity, i.e., 300–800 $\mathrm{km\,s^{-1}}$. Observational evidence of NSW thus far is very scarce because of the difficulty in measuring neutral particle fluxes several orders of magnitude lower than the solar wind flux, while simultaneously being exposed to full sunlight. The only observation thus far was made with the Low Energy Neutral Atom (LENA) imager on IMAGE, from which a NSW fraction of about 10^{-4} was derived (Collier et al. 2001). Enhancements in the NSW flux due to charge exchange in the geocorona were observed, with the NSW fraction increasing to a few times 10^{-4} (Collier et al. 2001). Since NSW flow direction is roughly perpendicular to the FOVs of the sensors it is of little concern for the background discussion.

A special case of the neutral solar wind is produced just upstream of the subsolar region of the magnetopause. The solar wind slows and is heated across the bow shock and as it approaches the magnetopause. The hot ($\sim$1 keV thermal energy), slow (several 10 to $\sim$100 km/s) solar wind is charge exchanged by the Earth's geocorona in the region. Like the ring current and plasma sheet neutrals, this background will be removed from the ENA data when they are in the FOV.

2.5.4 Pre-termination Shock ENAs

Proton distributions in the inner heliosheath are a function of the distance to the termination shock (TS) since pickup ions increase the random thermal energy of the solar wind and are more abundant for larger TS distances. These protons can be accelerated (e.g., at co-rotating interaction regions) and form suprathermal tails on the solar wind and pickup ion distributions. Goeckler (2003) has summarized the behaviour of the ubiquitous suprathermal tails.

They are characterised by power-laws in velocity space in the solar wind frame. When these ions have charge-exchange collisions with the interstellar neutral atoms that flow through the heliosphere, ENAs will be generated, which may be observed in Earth orbit. These ENAs are created inside the TS, thus they are called pre-TS ENAs and will interfere with the heliospheric ENAs of interest. However, it has been concluded earlier (McComas et al. 2004) that these pre-TS ENAs generated from accelerated protons inside the TS are not a significant background.

2.5.5 Summary

In summary, the strongest terrestrial ENA signal comes from the ring current, but the most extended consists of the ENA emissions coming from the plasma sheet during storms. Therefore, to be conservative, any scans across the Earthward hemisphere should be treated with extreme caution even at down tail distances at $x = -20\ R_E$. Similarly, any scans at $x = 10\ R_E$ should be treated with extreme caution. However, such scans will also present valuable opportunities for investigating global plasma sheet and subsolar magnetosheath properties.

3 Internal Backgrounds

Internally generated signals that compete with the expected ENA observations include all sources that randomly trigger the detectors without external stimulus, including the generation of ions internally through knock-off from grids in the particle path. Some additional background sources for IBEX-Lo and IBEX-Hi were identified during the extensive testing and calibration of the sensors and are discussed in the following subsections.

3.1 Random Background Sources

Random MCP or CEM pulses arising from radionuclide decay (mostly from the ^{40}K decay) fall into this category. Typical rates are for MCPs in the range of 0.1–1 counts per second per cm^2. The resulting background is lower than what is expected from penetrating cosmic rays.

Also electronic noise on the signal lines may appear as valid particle detection in a channel. These events occur in a random fashion, with a finite chance for a double coincidence, and a smaller chance for a triple coincidence, which contributes to the background in the particle identification.

Other sources of background come from the high voltage power supplies needed for the ion optical system of each sensor. The generation of high voltages employs oscillators, which may contribute electronic noise on the signal lines if not shielded carefully. Moreover, tiny discharges below the detection level of the partial discharge measurements may create an ion signal in the detector. The typical detection limit of partial discharge measurements is at a charge level of pC, which means a discharge involving 10^6 electrons can happen unnoticed. If some of the electrons propagate to the detector, or create ions on their way (by electron impact ionisation or electron stimulated desorption) such a discharge will contribute to the background.

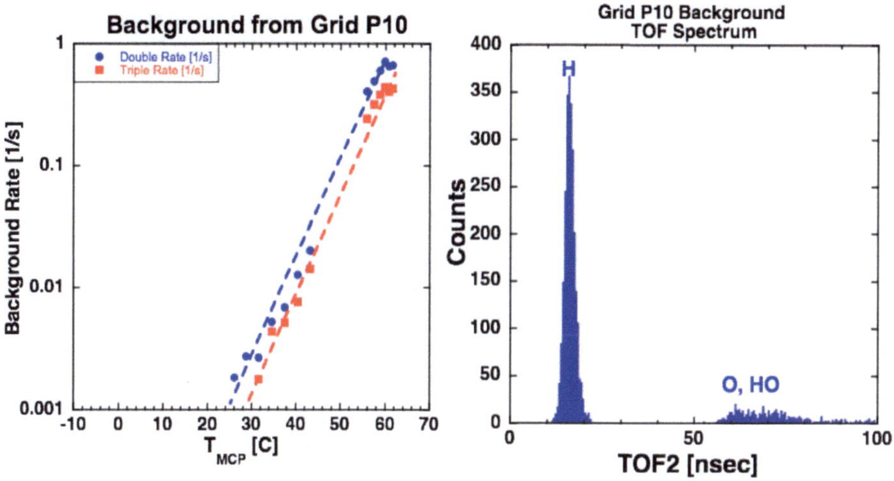

Fig. 10 Internal background of IBEX-Lo during thermal test. *Left*: Background rate as a function of temperature. *Right*: Mass spectrum of background signal

3.1.1 IBEX-Lo Internal Background

To identify any possible internal background sources, IBEX-Lo was operated long term in a mode that prevented any entry of external particles during thermal vacuum testing. No outside stimulus was active, e.g. there was no neutral particle beam. Also, all but one of the ion gauges of the vacuum system were turned off to minimise ambient ions and electrons. The operating ion gauge was of Penning-type and was baffled to largely suppress ion and electron emission into the test chamber. Figure 10 shows a result of the background determination performed during thermal vacuum test, where events with valid TOF combinations for H, O, and OH were indeed observed. The observed background is strongly temperature dependent (Fig. 10, left panel), which indicates that this background involves outgassing (most likely water) at least in an indirect manner. Figure 10, right panel shows a mass spectrum of this background, which is dominated by hydrogen, a fragment from water. The expected operating temperature of IBEX-Lo in space is in the range from 0°C to +18°C, which results in background rates of $2.6 \cdot 10^{-4}\ \text{s}^{-1}$ and $1.25 \cdot 10^{-4}\ \text{s}^{-1}$, for double and triple coincidence detection, respectively, at the upper end of the expected temperature range. When calculating back to the equivalent flux on the conversion surface, this internal background corresponds to about $0.5\ \text{s}^{-1}$ hydrogen ENAs arriving on the conversion surface.

Remember, this background is observed without any external stimulus. Inspection of background location on the four azimuth quadrants in the TOF unit showed that the most likely origin of these ions is the last electrode of the ESA subsystem, P10 (located just before the post-acceleration path to the TOF section), because this background is mostly observed in one quadrant. Figure 11, left panel, shows a schematic view of the ion-optical system near the post-acceleration region. As the first step in the creation of this background, electrons are emitted from the grids of electrode P10, probably via field emission from the edges of the grid wires (red arrow in Fig. 11). The electrons are accelerated toward the TOF unit and impinge the TOF structures with 10 keV and more. If a positive ion is released as a result of the electron impact, this positive ion will be accelerated toward electrode P10 (black arrow in Fig. 11). Upon impact on the grid and other structures there is the possibility that a negative ion is sputtered that in turn is accelerated by the post-acceleration potential toward the TOF

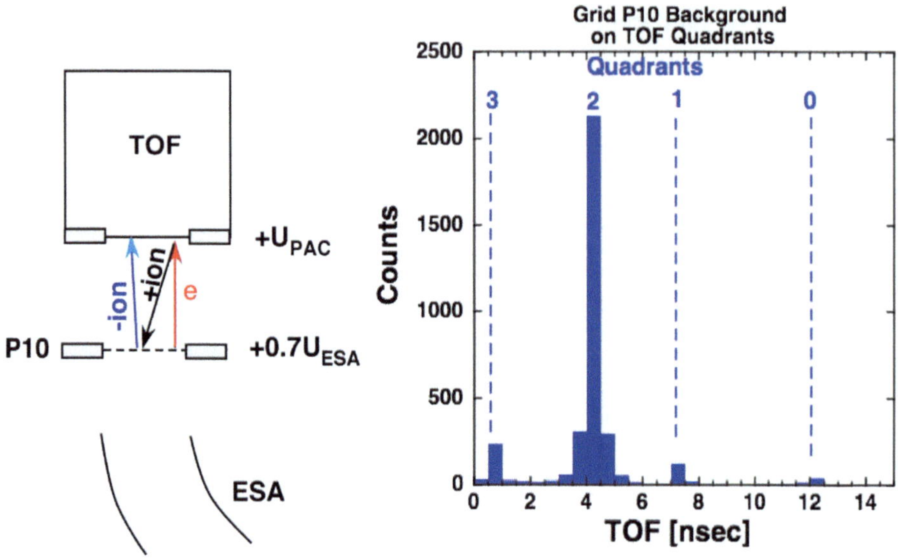

Fig. 11 *Left*: Schematic representation of the ion optical system near the post-acceleration region. *Right*: Histogram of the azimuth sector of background ions

(blue arrow in Fig. 11). If the ion passes the carbon foil it will be detected. An energetic electron impinging onto a metal surface can only remove adsorbates, but not sputter, and apparently is temperature dependent. Therefore, upon better evacuation of the sensor during the mission this background is expected to become smaller. Further support for this concept of background generation arises from the measurement of the azimuth angle of the registered ion in the TOF unit (see Fuselier et al. 2009, this issue, for a detailed sensor description). The registered background ions are mostly clustered in one azimuth quadrant, as can been seen in Fig. 11, right panel. Such a narrow distribution in azimuth can only happen if they are created after ESA, close to the TOF unit. Ions originating on the conversion surface will spread into at least two azimuth sectors even if they start out at a point source on the conversion surface (Wieser et al. 2007; Fuselier et al. 2009, this issue).

This observation also presents an opportunity to further reduce the background in the data analysis. Events from the MCP quadrant with the highest internal background rate can be culled out because the MCP quadrant information is retained in the event data.

3.1.2 IBEX-Hi Internal Background

Internal backgrounds in IBEX-Hi were measured during extended periods of operation when the sensor was fully operational but without external stimuli, i.e., no incident ion or ENA beam.

As a first test, the IBEX-Hi detector subsystem was continuously operated for about 26 h for the determination of internal backgrounds. The background rate of a single CEM detector was in the range (4.0–8.6) $\cdot 10^{-2}$ s^{-1}. For the double rates, i.e., the registration of a signal in two CEM detectors within a short time interval, we obtained rates of about $1.6 \cdot 10^{-2}$ s^{-1}. For the individual coincidence combinations of CEM detectors the background rates were as follows: $5.4 \cdot 10^{-3}$ s^{-1} for coincidences between detectors A and B; $9.2 \cdot 10^{-3}$ s^{-1} for the BC coincidences; and $1.0 \cdot 10^{-3}$ s^{-1} for the AC coincidences. The

Table 3 IBEX-Hi average background count rates for exclusive (non-coincident) single events and a subset of coincidence types measured over 19.6 hours of quiescent operation during cross calibration

Measurement		Background rate [cnt/s]
Singles	CEM A	$89 \cdot 10^{-3}$
	CEM B	$115 \cdot 10^{-3}$
	CEM C	$69 \cdot 10^{-3}$
	CEM D	$12 \cdot 10^{-3}$
Double coincidences	Long AB	$2.8 \cdot 10^{-3}$
	Long BC	$4.2 \cdot 10^{-3}$
	Qual(Not_C) AC	$0.13 \cdot 10^{-3}$
Triple coincidences	Qual(Not_C) ABC	$0.58 \cdot 10^{-3}$
	Long ABC	$1.04 \cdot 10^{-3}$

background rate for triple coincidence (ABC) events were $2.9 \cdot 10^{-3}$ s^{-1}. The background measurement for the single events, the double coincidence rates, and the triple coincidence rate is shown in Fig. 12.

For the full IBEX-Hi sensor the average background count rates were measured over 19.6 hours during the IBEX sensor cross calibration campaign (IBEX-Hi Cal 4 campaign), and the results are listed in Table 3. The singles count rates in each CEM detector were < 0.2 s^{-1}. While the coincident count rates are higher than expected based on random coincidence of the background singles rates in the detectors, we have found that ambient gamma rays are associated with a majority of the background coincidence events observed in the detector subsystem.

The IBaM was also continuously operated for 19.1 h during cross-calibration to determine its internal background. Its background rate was found to be $1.15 \cdot 10^{-2}$ s^{-1} (see panel "D" in Fig. 12 for the actual measurement).

During tests with the engineering model (EM) the sensitivity of the IBEX-Hi detector subsystem to gamma rays was noticed. We used laboratory TOF electronics to measure the TOF between detected events in all coincidence combinations. The EM detector subsystem measures coincidence events from ambient background, likely to be γ-rays from the surrounding building concrete because of accidental coincidences. Possible γ-ray emitters in concrete are ^{40}K, ^{232}Th and daughters, and $^{235/238}$U and daughters. We subsequently measured the background γ-ray environment in the LANL calibration facilities, showing a background flux 6.3 γ cm^{-2} s^{-1} between 0.2–3 MeV. Analysis of the measured background γ-ray spectrum indicated that ^{40}K and ^{226}Ra were the dominant γ-ray line emitters on top of a broad γ-ray continuum that increased with decreasing energy as shown in Fig. 13.

Figure 14 shows a comparison of TOF histograms using laboratory TOF electronics for an incident 3.5 keV proton beam (top panel), a background measurement (middle panel) and a background measurement with an additional ^{137}Cs γ-ray source placed next to the sensor (bottom panel). The left, centre, and right histograms for each of these cases represent triple coincidences in which the first event was recorded in CEMs A, B, and C, respectively. The background run spanning $2.3 \cdot 10^5$ sec (during which the triple coincidence rate was $6 \cdot 10^{-3}$ s^{-1}) and the run with the ^{137}Cs source over $2.5 \cdot 10^4$ sec are very similar. In particular, for both runs the number of triple coincidences in which CEM C was the first detector to register an event is comparable to the number of triple coincidences in which CEM A was the first to register a pulse. In contrast, the proton beam never caused a triple coincidence in which CEM

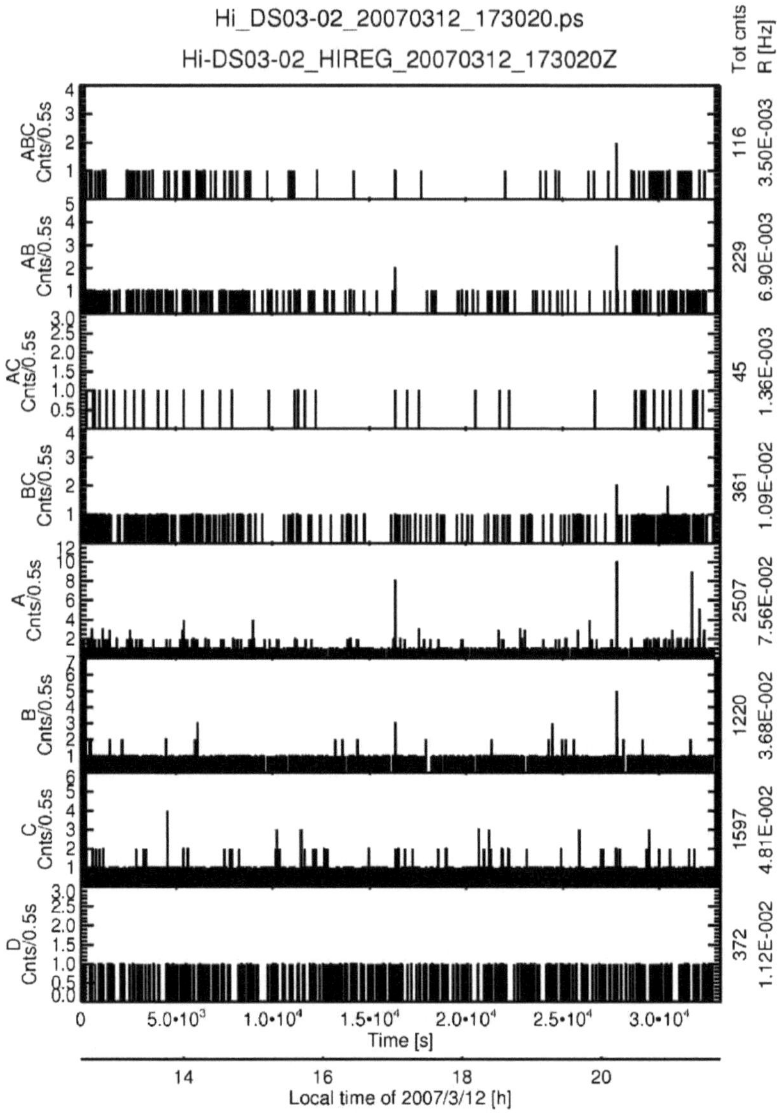

Fig. 12 Background measurement of the IBEX-Hi detector section, showing the single events, the double rates, and the triple rate

C registered the first pulse, which is the direct result of H sequentially transiting detector chambers A, B, and finally C. We can therefore conclude that the measured background during testing in the LANL facility is likely the result of the penetrating γ-ray radiation with its spectrum shown in Fig. 13. Also, the IBaM showed some susceptibility to γ-rays; its background rate went up to $1.22 \cdot 10^{-1}$ s^{-1} when exposed to γ-rays.

We note that ^{137}Cs also emits 1.175 MeV electrons that have a range of about 2.3 mm in Aluminium, but these electrons were completely blocked by the Al walls of minimum thickness > 2.5 mm surrounding the interior of the detector chambers and CEM detectors.

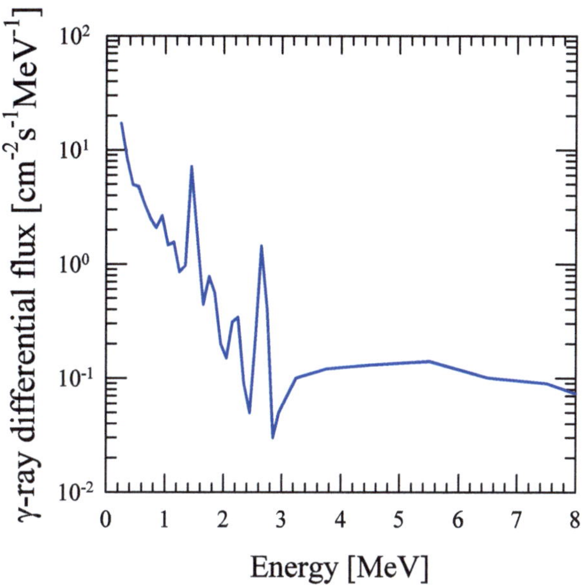

Fig. 13 γ-ray spectrum as measured at the LANL calibration facility

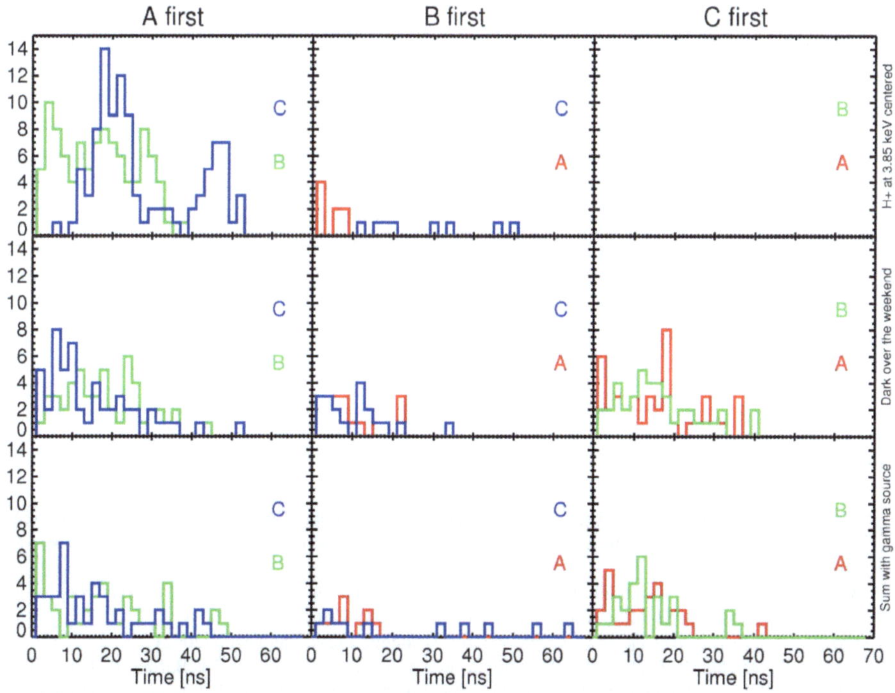

Fig. 14 Triple coincidence TOF distributions for a 3.85 keV hydrogen beam (*top*), a background ("Dark" run) distribution, and a background distribution with a ^{137}Cs γ-ray source was placed next to the sensor

The majority of coincident counts listed in Table 3 are a result of the laboratory γ-ray environment. From the estimates of cosmic ray induced background presented in Sect. 2.4 we expect that in space the measured background count rates are significantly lower, and the present numbers represent overestimates of the background rates. In any case, the triple coincidence count rate in which CEM C generates the first pulse will directly monitor the presence and intensity of penetrating radiation in IBEX-Hi. From the laboratory measurements (see Fig. 14), we know how to relate the background count rate of C-first to the background count rates A-first and B-first, therefore we can subtract the contribution of penetrating radiation in the latter two rates in the data analysis.

4 Conclusions

The IBEX sensors were each designed to maximize the signal-to-noise ratio by utilising a large geometric factor, single-pixel camera design with extensive background rejection. Natural background sources in the IBEX sensors include ENAs from the Earth's magnetosphere and magnetosheath, which actually represent a localised foreground for the planned heliospheric observations. Intensities of these sources are much larger than the ENA signal from the outer heliosphere. This natural foreground will be culled from the data, and the high altitude of the IBEX orbit helps reduce the effect of this foreground on the overall sky maps. Energetic neutral atoms can also be produced in the immediate vicinity of the spacecraft by charge exchange of the solar wind or magnetosheath with the outgassing products (mostly water) from the spacecraft. These backgrounds have been estimated and are found to be orders of magnitude lower than the expected signal. Of course, the outgassing from the spacecraft and the sensors have been minimised by stringent cleanliness control.

Energetic ions and photons interact with gas inside two regions in the IBEX collimators to produce ions that can masquerade as ENAs. This background source represents the most complex and highest background in the sensors. Background from Lyman-α was measured in the sensors during calibration and a background level was determined. Finally, additional background specific to the sensors includes negative ion production deep inside the sensor (IBEX-Lo) and sensitivity of the detector system to penetrating radiation.

The background sources for the IBEX mission have been extensively investigated through simulation, analysis using up-to-date information from other missions, and extensive testing during sensor calibration. Based on this extensive investigation, we conclude that there is ample signal-to-noise margin for measurements over the energy range of the IBEX mission.

References

F. Allegrini, G.B. Crew, D. Demkee, H.O. Funsten, D.J. McComas, B. Randol, B. Rodriguez, N.A. Schwadron, P. Valek, S. Weidner, The IBEX-Hi background monitor, Space Sci. Rev. (2009, this issue)

D.J. Baker, G.J. Romick, The Raleigh: Interpretation of the unit in terms of column emission rate or apparent radiance expressed in SI units. Appl. Optics **15**, 1966–1968 (1976)

J. Berkowitz, *Atomic and Molecular Photoabsorption* (Academic Press, San Diego, 2002)

J.L. Bertaux, E. Kyrölä, E. Quémerais, R. Pellinen, R. Lallement, W. Schmidt, M. Berthé, E. Dimarellis, J.P. Goutail, C. Taulemesse, C. Bernard, G. Leppelmeier, T. Summanen, H. Hannula, H. Huomo, V. Kehlä, S. Korpela, K. KLeppälä, E. Strömmer, J. Torsti, K. Viherkanto, J.F. Hochedez, G. Chretiennot, R. Peyroux, T. Holzer, SWAN: A study of solar wind anisotropies on SOHO with Lyman alpha sky mapping. Sol. Phys. **162**, 403–439 (1995)

P.C. Brandt, R. DeMajistre, E.C. Roelof, D.G. Mitchell, S. Mende, IMAGE/HENA: Global ENA imaging of the plasmasheet and ring current during substorms. J. Geophys. Res. **107**, 1454 (2002). doi:10.1029/2002JA009307

S.V. Chalov, H.J. Fahr, V.V. Izmodenov, Evolution of pickup proton spectra in the inner heliosheath and their diagnostics by energetic neutral atom fluxes. J. Geophys. Res. **108**(A6), 1266 (2003). doi:10.1029/2002JA009492

M.R. Collier, T.E. Moore, K.W. Ogilvie, J.D. Chornay, J.W. Keller, S. Boardsen, J.L. Burch, B. El Marji, M.-C. Fok, S.A. Fuselier, A.G. Ghielmetti, B.L. Giles, D.C. Hamilton, B.L. Peko, J.M. Quinn, T.M. Stephen, G.R. Wilson, P. Wurz, Observations of neutral atoms from the solar wind. J. Geophys. Res. **106**, 24893–24906 (2001)

H.-J. Fahr, K. Scherer, Energetic neutral atom fluxes from the heliosheath varying with the activity phase of the solar cycle. ASTRA Astrophys. Space Sci. Trans. **1**, 3–15 (2004)

M.-C. Fok, R.A. Wolf, R.W. Spiro, T.E. Moore, Comprehensive computational model of Earth's ring current. J. Geophys. Res. **106**, 8417–8424 (2001)

M.-C. Fok, T.E. Moore, G.R. Wilson, J.D. Perez, X.X. Zhang, P.C. son Brandt, D.G. Mitchell, E.C. Roelof, J.-M. Jahn, C.J. Pollock, R.A. Wolf, Global ENA image simulations. Space Sci. Rev. **109**, 77–103 (2003)

C. Fröhlich, J. Lean, Solar radiative output and its variability: evidence and mechanisms. Astron. Astrophys. **12**, 273–320 (2004)

H.O. Funsten, A.A. Guthrie, R.W. Harper, K.H. Kihara, M.P. Manzo, M.J. Fagan, D.J. McComas, S. Weidner, F. Allegrini, D. Everett, B. Rodriguez, G. Dunn, S. Pope, J. Hanley, P. Valek, E. Moebius, J. Nolin, S. Ellis, D. Heirtzler, B. King, H. Kucharek, S. Turco, S. Zaffke, D. Reisenfeld, P. Janzen, S.A. Fuselier, M. Gruntman, E. Roelof, P. Wurz, D. Piazza, L. Saul, P. Bochsler, The interstellar boundary explorer high energy (IBEX-Hi) neutral atom imager. Space Sci. Rev. (2009, this issue)

S.A. Fuselier, A.G. Ghielmetti, E. Hertzberg, A.S. Moore, D. Isaac, J.W. Hamilton, C. Tillier, E. Moebius, M.S. Granoff, D. Heirtzler, B. King, H. Kucharek, S. Longworth, J. Nolin, S. Turco, S. Ellis, J. Googins, F. Kudirka, J. Tyler, M. Vosbury, G. Clark, M. O'Neal, P. Wurz, J.A. Scheer, L.A. Saul, D. Piazza, P. Bochsler, M. Wieser, C. Schlemm, D.J. McComas, J. Scherrer, S. Pope, H.O. Funsten, D. Chornay, J. Lobell, T.E. Moore, P. Rosmarynowski, T. Friedmann, R.J. Nemanich, The IBEX-Lo Sensor. Space Sci. Rev. (2009, this issue)

S. Graf, K. Altwegg, H. Balsiger, P. Bochsler, B. Fiethe, E. Montagnon, Thruster plumes—a source for high pressure and contamination at the payload location. J. Spacecr. Rockets **45**, 57–64 (2008)

A. Galli, P. Wurz, H. Lammer, H.I.M. Lichtenegger, R. Lundin, S. Barabash, A. Grigoriev, M. Holmström, H. Gunell, The hydrogen exospheric density profile measured with ASPERA-3/NPD. Space Sci. Rev. **126**, 447–467 (2006)

G. Goeckler, Ubiquitous suprathermal tails on the solar wind and pickup ion distributions, in *Proc. Solar Wind X*, Am. Inst. Phys., CP679 (2003), pp. 583–588

J.B. Greenwood, I.A. Chutjian, S.J. Smith, Measurements of the absolute, single charge-echange cross section of H^+, He^+, and He^{2+} with H_2O and CO_2. Astrophys. J. **529**, 605–609 (2000)

M. Gruntman, E.C. Roelof, D.G. Mitchell, H.J. Fahr, H.O. Funsten, D.J. McComas, Energetic neutral atom imaging of the heliospheric boundary region. J. Geophys. Res. **106**, 15767–15782 (2001)

D.C. Hamilton, G. Gloeckler, F.M. Ipavich, W. Stüdemann, B. Wilken, G. Kremser, Ring current development during the great geomagnetic storm of February 1986. J. Geophys. Res. **93**, 14343–14355 (1988)

J. Heerikhuisen, N.V. Pogorelov, G.P. Zank, V. Florinski, The effects of global heliospheric asymmetries on energetic neutral atom sky maps. Astrophys. J. **655**, L53–L56 (2007)

M. Hilchenbach, K.C. Hsieh, D. Hovestadt, B. Klecker, H. Grünwaldt, P. Bochsler, F.M. Ipavich, F. Gliem, W.I. Axford, H. Balsiger, W. Bornemann, A. Bürgi, M.A. Coplan, A.B. Galvin, J. Geiss, G. Gloeckler, S. Hefti, D.L. Judge, R. Kallenbach, P. Laeverenz, M.A. Lee, S. Livi, G.G. Managadze, E. Marsch, E. Möbius, M. Neugebauer, H.S. Ogawa, K.U. Reiche, M. Scholer, M.I. Verigin, B. Wilken, P. Wurz, Detection of 55–80 keV hydrogen atoms of heliospheric origin by CELIAS/HSTOF on SOHO. Astrophys. J. **503**, 916–921 (1998)

R.R. Hodges, E.L. Breig, Ionosphere–Exosphere coupling through charge exchange and momentum transfer in hydrogen-proton collisions. J. Geophys. Res. **96**, 7697–7708 (1991)

K.C. Hsieh, P.C. Frisch, J. Giacalone, J.R. Jokipii, J. Kota, D.E. Larson, R.P. Lin, J.G. Luhmann, L. Wang, A re-interpretation of the STERE/STE observations and its consequences. Astrophys. J. **694**, L79–L82 (2009)

W.F. Huebner, J.J. Keady, S.P. Lyon, Solar photo rates for planetary atmospheres and atmospheric pollutants. Astrophys. Space Sci. **195**, 1–294 (1992)

M. Juda, R.H. Donnelly, K.T. Hole, A.T. Kenter, R.P. Kraft, S.S. Murray, D.O. Pease, C.R.A. Wilton, M.V. Zombeck, Characteristics of the on-orbit background of the Chandra X-ray observatory high resolution camera. Proc. SPIE **4851**, 112–123 (2003)

R. Kallenbach, M. Hilchenbach, S.V. Chalov, J.A. Le Roux, K. Bamert, On the "injection problem" at the solar wind termination shock. Astron. Astrophys. **439**, 1–22 (2005)

L.M. Kistler, F.M. Ipavich, D.C. Hamilton, G. Gloeckler, B. Wilken, G. Kremser, W. Stüdemann, Energy spectra of the major ion species in the ring current during geomagnetic storms. J. Geophys. Res. **94**, 3579–3599 (1989)

R. Korde, C. Prince, D. Cunningham, R.E. Vest, E. Gullikson, Present status of radiometric quality silicon photodiodes. Metrologia **40**, S145–S149 (2003)

W.F. Krolikowski, W.E. Spicer, Photoemission studies of the noble metals: II. Gold. Phys. Rev. B **1**, 478–487 (1970)

D. McComas, S.J. Bame, W.C. Feldman, J.T. Gosling, J.L. Phillips, Solar wind Halo electrons from 1–4 AU. Geophys. Res. Lett. **19**, 1291–1294 (1992)

D.J. McComas, P. Valek, J.L. Burch, C.J. Pollock, R.M. Skoug, M.F. Thomsen, Filling and emptying of the plasma sheet: Remote observations with 1–70 keV energetic neutral atoms. Geophys. Res. Lett. **29**, 2079 (2002). doi:10.1029/2002GL016153

D. McComas, F. Allegrini, P. Bochsler, M. Bzowski, M. Collier, H. Fahr, H. Fichtner, P. Frisch, H. Funsten, S. Fuselier, G. Gloeckler, M. Gruntman, V. Izmodenov, P. Knappenberger, M. Lee, S. Livi, D. Mitchell, E. Moebius, T. Moore, D. Reisenfeld, E. Roelof, N. Schwadron, M. Wieser, M. Witte, P. Wurz, G. Zank, The interstellar boundary explorer (IBEX). AIP Conf. Proc. **719**, 162–181 (2004)

D.J. McComas, F. Allegrini, P. Bochsler, M. Bzowski, M. Collier, H. Fahr, H. Fichtner, P. Frisch, H. Funsten, S. Fuselier, G. Gloeckler, M. Gruntman, V. Izmodenov, P. Knappenberger, M. Lee, S. Livi, D. Mitchell, E. Möbius, T. Moore, S. Pope, D. Reisenfeld, E. Roelof, J. Scherrer, N. Schwadron, R. Tyler, M. Wieser, M. Witte, P. Wurz, G. Zank, IBEX – interstellar boundary explorer. Space Sci. Rev. (2009, this issue)

F.B. McDonald, V.S. Ptuskin, Galactic cosmic rays, in *The Century of Space Science* (Kluwer, Dordrecht, 2001), pp. 677–697

F. Merkt, R. Signorell, H. Plam, A. Osterwalder, M. Sommavilla, Towards resolving the hyperfine structure in ions by photoelectron spectroscopy. Mol. Phys. **95**, 1045–1054 (1998)

D.G. Mitchell, S.E. Jaskulek, C.E. Schlemm, E.P. Keath, R.E. Thompson, B.E. Tossman, J.D. Boldt, J.R. Hayes, G.B. Andrews, N. Paschalidis, D.C. Hamilton, R.A. Lundgren, E.O. Tums, P. Wilson IV, H.D. Voss, D. Prentice, K.C. Hsieh, C.C. Curtis, F.R. Powell, High energy neutral atom (HENA) imager for the IMAGE mission. Space Sci. Rev. **91**, 67–112 (2000)

N.P. Paschalidis, E.T. Sarris, S.M. Krimigis, R.W. McEntire, M.D. Levine, I.A. Daglis, G.C. Anagnostopoulos, Energetic ion distributions on both sides of the Earth's magnetopause. J. Geophys. Res. **99**, 8687–8703 (1994)

W.G. Pilipp, H. Miggenrieder, M.D. Montgomery, K.-H. Mühlhäuser, H. Rosenbauer, R. Schwenn, Characteristics of electron velocity distribution functions in the solar wind derived from the HELIOS plasma experiment. J. Geophys. Res. **92**, 1075–1092 (1987)

E. Quémerais, R. Lallement, S. Ferron, D. Koutroumpa, J.-L. Bertaux, R. Kyrölä, W. Schmidt, Interplanetary hydrogen absolute ionization rates: Retrieving the solar wind mass flux latitude and cycle dependence with SWAN/SOHO maps. J. Geophys. Res. **111**, A09114 (2006). doi:10.1029/2006JA011711

T.R. Sanderson, J.P.G. Henrion, K.-P. Wenzel, R.P. Lin, K.A. Anderson, S. Ashford, C.W. Carlson, D. Curtis, R.F. Ergun, D. Larson, J. McFadden, H. Reme, J.M. Bosqued, J. Coutelier, F. Cotin, N. Lormant, C. d'Uston, G.K. Parks, M.P. McCarthy, R.M. Skoug, R.M. Winglee, WIND observations of energetic ions far upstream of the Earth's bow-shock. Geophys. Res. Lett. **23**, 1215–1218 (1996)

J. Steinberg, Private communication. Los Alamos National Laboratory, Los Alamos, NM 87545, USA, 2007

R.G. Tonkyn, R. Wiedmann, E.R. Grant, M.G. White, Rotationally resolved photoionization of H_2O. J. Chem. Phys. **95**, 7033–7040 (1991)

J. Vallerga, R. Lallement, M. Lemoine, F. Dalaudier, D. McMullin, Astron. Astrophys. **426**, 855–865 (2004)

L. Wang, R.P. Lin, D.E. Larson, J.G. Luhmann, Domination of heliosheath pressure by shock-accelerated pickup ions from observations of neutral atoms. Nature **454**, 81–83 (2008)

M. Wieser, P. Wurz, E. Moebius, S.A. Fuselier, E. Hertzberg, D.J. McComas, The ion-optical prototype of the low energy neutral atom sensor of the interstellar boundary explorer mission (IBEX). Rev. Sci. Instr. **78**, 124502-1–124502-14 (2007)

D.J. Williams, D.G. Mitchell, L.A. Frank, T.E. Eastman, Three-dimensional magnetosheath plasma ion distributions from 200 eV to 2 MeV. J. Geophys. Res. **93**, 12783–12794 (1988)

B.E. Wood, V.V. Izmodenov, J.L. Linsky, Y.G. Malama, Lyα Absorption from heliosheath neutrals. Astrophys. J. **657**, 609–617 (2007)

P. Wurz, Detection of energetic neutral particles, in *The Outer Heliosphere: Beyond the Planets*, ed. by K. Scherer, H. Fichtner, E. Marsch (Copernicus Gesellschaft e.V., Katlenburg-Lindau, 2000), pp. 251–288

P. Wurz, J. Scheer, M. Wieser, Particle scattering off surfaces: application in space science. e-J. Surf. Sci. Nanotechnol. **4**, 394–400 (2006)

P. Wurz, A. Galli, S. Barabash, A. Grigoriev, Energetic neutral atoms from the heliosheath. Astrophys. J. **683**, 248–254 (2008)

Space Sci Rev (2009) 146: 207–234
DOI 10.1007/s11214-009-9513-x

The Interstellar Boundary Explorer Science Operations Center

N.A. Schwadron · G. Crew · R. Vanderspek · F. Allegrini · M. Bzowski · R. DeMagistre · G. Dunn · H. Funsten · S.A. Fuselier · K. Goodrich · M. Gruntman · J. Hanley · J. Heerikuisen · D. Heirtlzer · P. Janzen · H. Kucharek · C. Loeffler · K. Mashburn · K. Maynard · D.J. McComas · E. Moebius · C. Prested · B. Randol · D. Reisenfeld · M. Reno · E. Roelof · P. Wu

Received: 8 October 2008 / Accepted: 15 April 2009 / Published online: 5 June 2009

Abstract The Interstellar Boundary Explorer (IBEX) Science Operations Center is responsible for supporting analysis of IBEX data, generating special payload command procedures, delivering the IBEX data products, and building the global heliospheric maps of energetic neutral atoms (ENAs) in collaboration with the IBEX team. We describe here the data products and flow, the sensor responses to ENA fluxes, the heliospheric transmission of ENAs (from 100 AU to 1 AU), and the process of building global maps of the heliosphere. The vast majority of IBEX Science Operations Center (ISOC) tools are complete, and the ISOC is in a remarkable state of readiness due to extensive reviews, tests, rehearsals, long hours, and support from the payload teams. The software has been designed specifically to support considerable flexibility in the process of building global flux maps. Therefore, as we discover the fundamental properties of the interstellar interaction, the ISOC will iteratively improve its pipeline software, and, subsequently, the heliospheric flux maps that will provide

N.A. Schwadron (✉) · C. Prested · K. Goodrich · K. Maynard · P. Wu
Boston University, 725 Commonwealth Ave, Boston, MA 02215, USA
e-mail: nathanas@bu.edu

C. Prested
e-mail: cprested@bu.edu

K. Goodrich
e-mail: kgoodri@bu.edu

K. Maynard
e-mail: maynard@bu.edu

P. Wu
e-mail: pwu@bu.edu

G. Crew · R. Vanderspek
MIT Kavli Institute for Astrophysics and Space Research, 77 Massachusetts Ave, Cambridge, MA 02139, USA

G. Crew
e-mail: gbc@space.mit.edu

R. Vanderspek
e-mail: rvdspek@bu.edu

a keystone for our global understanding of the solar wind's interaction with the interstellar medium. The ISOC looks forward to the next chapter of the IBEX mission, as the tools we

M. Reno
Austin Mission Consulting, Austin, USA
e-mail: mreno@swri.edu

K. Mashburn
Montana State University, Missoula, USA
e-mail: kmashburn@swri.edu

F. Allegrini · J. Hanley · C. Loeffler · D.J. McComas · B. Randol
Southwest Research Institute, 6220 Culebra, San Antonio, TX 78238, USA

F. Allegrini
e-mail: fallegrini@swri.edu

J. Hanley
e-mail: jhanley@swri.edu

C. Loeffler
e-mail: cloeffler@swri.edu

D.J. McComas
e-mail: dmccomas@swri.edu

B. Randol
e-mail: brandol@swri.edu

M. Bzowski
Space Research Centre PAS, Bartycka 18A, 00-716 Warsaw, Poland
e-mail: bzowski@cbk.waw.pl

R. DeMagistre · E. Roelof
JHU/APL, MP3 E132 11100 Johns Hopkins Road, Laurel, MD 20723, USA

R. DeMagistre
e-mail: bob.demajistre@jhuapl.edu

E. Roelof
e-mail: ed.roelof@jhuapl.edu

M. Gruntman
Astronautics and Space Technology Division, USC Viterbi School of Engineering,
University of Southern California, 854 Downey Way, RRB-224, MC-1192, Los Angeles, USA
e-mail: mikeg@usc.edu

J. Heerikuisen
Institute of Geophysics and Planetary Physics, University of California, Riverside, CA 92521, USA
e-mail: jacobh@ucr.edu

H. Funsten
ISR Division, MSB241, Los Alamos National Laboratory, Los Alamos, NM 87545, USA
e-mail: hfunsten@lanl.gov

D. Reisenfeld · P. Janzen
Dept. of Physics and Astronomy, University of Montana, 32 Campus Drive, Missoula, MT 59812, USA

D. Reisenfeld
e-mail: dan.resienfeld@umontana.edu

P. Janzen
e-mail: pjanzen@lanl.gov

have developed will be used in partnership with the IBEX team and the scientific community over the coming years to define our global understanding of the solar wind's interaction with the local interstellar medium.

Keywords Solar wind · Termination shock · Interstellar boundaries · Interstellar medium · Energetic neutral atoms

1 Introduction

IBEX will provide the first global views of the Sun's interstellar boundaries, unveiling the physics of the heliosphere's interstellar interaction, providing a deeper understanding of the heliosphere and thereby astrospheres throughout the galaxy, and creating the opportunity to make even greater unanticipated discoveries. IBEX, which launched on October 19, 2008, will achieve baseline measurements by 2010 shortly after solar minimum, and extended measurements will be achieved in the rise to solar maximum when there will be increased solar activity, from 2010–2012.

The IBEX objective is to discover the global interaction between the solar wind and the interstellar medium by imaging interstellar interactions and interstellar boundaries at the edge of our heliosphere via detection of energetic neutral atoms (ENAs). IBEX achieves its sole objective by answering four fundamental science questions:

- Question I: What are the global strength and structure of the termination shock?
- Question II: How are energetic protons accelerated at the termination shock?
- Question III: What are the global properties of the solar wind flow beyond the termination shock and in the heliotail?
- Question IV: How does the interstellar flow interact with the heliosphere beyond the heliopause?

ENA imaging on several precursor missions and IMAGE (Imager for Magnetopause-to-Aurora Global Exploration) has revealed the global dynamics of Earth's magnetosphere and provided rudimentary measurements of heliospheric neutral atoms. By carrying much more sensitive ENA cameras beyond the region of intense magnetospheric emissions and backgrounds, IBEX globally images ENAs from the outer heliosphere for the first time.

S.A. Fuselier
Space Physics Dept., Lockheed Martin Advanced Technology Center, Dept. ADCS, Bldg. 255, 3251 Hanover St., Palo Alto, CA 93404, USA
e-mail: stephen.a.fusilier@lmco.com

E. Moebius · D. Heirtlzer · H. Kucharek
Space Science Center and Dept. of Physics, University of New Hampshire, Morse Hall, 39 College Road, Durham, NH 03824, USA

E. Moebius
e-mail: eberhard.meobius@unh.edu

D. Heirtlzer
e-mail: dheirtzl@atlas.sr.unh.edu

H. Kucharek
e-mail: kucharek@atlas.sr.unh.edu

G. Dunn
Southwest Research Institute, 6220 Culebra, San Antonio, TX 78238, USA
e-mail: gdunn@swri.edu

Because so little is known about the interstellar interaction, its exploration requires a broadly scoped science strategy. IBEX achieves its objective through three levels of exploration. At the *Discovery Level*, simple raw IBEX images, energy spectra, and fluxes directly reveal the fundamental properties of the interstellar interaction. At the *Exploration Level*, IBEX data products are combined with simple physics-based calculations, theory, and limited 2D and 3D modeling to explore the underlying properties and variations of the interaction. Finally, at the *Understanding Level*, iterative analysis of the IBEX data, in concert with increasingly refined 3D models of the heliosphere, expose the detailed nature of the interstellar interaction.

This paper outlines the science operations for the IBEX mission. Section 2 discusses the IBEX observing strategy, and Section 3 provides an overview of science and mission operations. The operation and design of the IBEX sensors, IBEX-Hi and IBEX-Lo, are detailed in other chapters in this book by Funsten et al. (2009) and Fuselier et al. (2009). Section 4 describes our data products and their flow from the spacecraft to the Mission Operations Center (MOC) at Orbital Sciences Corporation (in Dulles, VA, USA) and to the IBEX Science Operations Center (ISOC) at Boston University and the mirror-ISOC at Southwest Research Institute (SwRi, in San Antonio, TX, USA). Section 5 considers the transmission of neutral atoms from the termination shock to IBEX. Section 6 describes the sensor response functions that will be used for forward modeling of observed distributions. Section 7 outlines the flow of processing from the raw data types into higher level data products of various form. Section 8 summarizes recent results of forward modeling with the ISOC tools, and a summary of the ISOC status is provided in Sect. 9.

2 The IBEX Observing Strategy

The final IBEX orbit was achieved using the Pegasus XL launch vehicle, a STAR-27 Solid Rocket Motor for initial orbit insertion and a Hydrazine Propulsion System (HPS) for apogee and perigee raising burns and routine orbit maintenance. The initial apogee altitude is $\sim 50\ R_e$, which keeps the spacecraft above the Earth's magnetosphere over the majority of each orbit and maximizes the amount of time the sensors view regions of the sky unobscured by the magnetosphere. The initial perigee altitude of $\sim$ 12,000 km is outside the inner radiation belt, reducing the total radiation dose seen by IBEX subsystems. The orbit period is $\sim$ 8 days. Because the final orbit is highly elliptical, it is subject to significant perturbations due to lunar gravity effects.

A principle challenge to the IBEX mission was achieving a highly elliptic orbit with an apogee well beyond the magnetosphere. The $\sim 50\ R_e$ apogee altitude allows good heliospheric viewing beyond the bright ENA emissions of the magnetosphere. Figure 1 shows the orbit orientation relative to the magnetosphere. The orbit and magnetosheath in this figure are scaled up in size for clarity; however, the relative size of the magnetosheath (with a nose near 10 R_e and flanks near 15 R_e) and the orbit are fixed for a 50 R_e apogee. Note that the direction extending from the apogee to perigee (the line of apsides) is oriented about 70° from the tail. The line of apsides is a direction which will be most strongly obscured by the bright magnetospheric ENA emissions. The IBEX mission was designed so that this obscured viewing direction is as close to the tail as possible. For the baseline mission lifetime of 2 years, excluding magnetospheric obstruction, viewing time for each pixel is a minimum of 0.72 hours and an average of 21 hours, as shown in Fig. 2. The orbit orientation changes slightly with time due to lunar pumping.

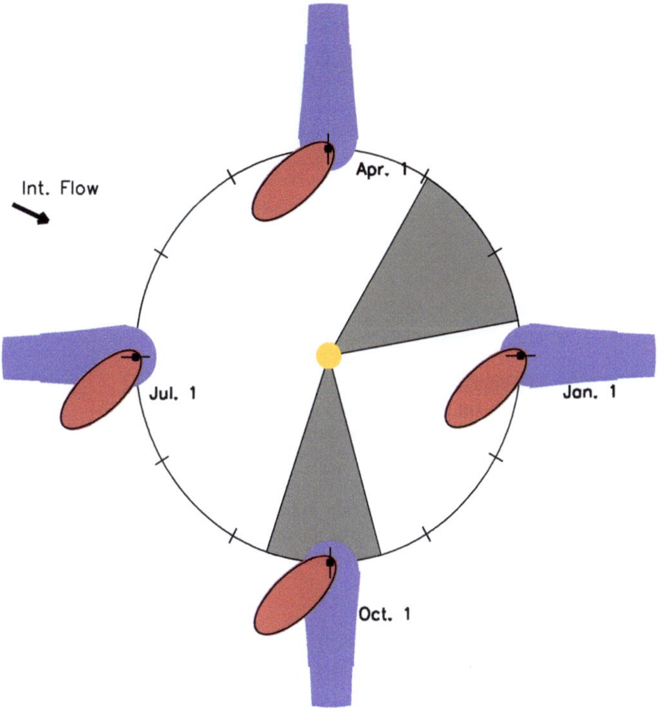

Fig. 1 A cartoon illustrating the IBEX orbit orientation (*red*) and the magnetosheath orientation (*purple*) through the seasons of the year. The sizes of the magnetosheath and the orbit are scaled up for clarity, however the relative sizes are fixed: the magnetosheath has a nose near 10 R_e, flanks near 15 R_e, while the IBEX spacecraft orbits with an apogee of 50 R_e and a perigee altitude between 5,000 and 12,000 km. The *gray shaded regions* indicate periods when the IBEX-Lo sensor will measure inflowing interstellar neutral atoms

The spacecraft spins at about 4 revolutions per minute, and the spin axis points approximately in the direction of the Sun.[1] The IBEX-Lo and IBEX-Hi sensor viewing directions are perpendicular to the spin axis. Over each spin, a great circle on the celestial sphere is viewed. Since the spin direction is fixed inertially during each orbit, the same great circle is viewed on each spin throughout the orbit. However, the repointing maneuver in the next orbit reorients the spin axis of the spacecraft by about 8°, allowing a new great circle to be viewed. The full-width-half-max (FWHM) of the ENA acceptance angle is $\sim 7°$, and the maximum acceptance angle is $\sim 14°$. Hence, the acceptance angles cover the $\sim 8°$ difference between the meridional angle of the great circles viewed on successive orbits. Each spin of the spacecraft is broken into 60 angular bins of about 6° wide sectors, as shown in Fig. 3.

The IBEX-Lo and IBEX-Hi sensors have electrostatic analyzers (ESAs) with voltages that are stepped sequentially. There are 8 voltage steps for IBEX-Lo and 6 voltage steps for IBEX-Hi. Each voltage step is held fixed for 2 spacecraft spins (~ 30 seconds), and the voltages are stepped from low to high. With a spin rate of 4 rpm, an IBEX-Lo cycle takes

[1]There is up to $\sim 7°$ off-pointing since there is only one repointing maneuver during each ~ 8-day orbit. During the repointing maneuver, the spin axis is positioned $\sim 1°$W of the Sun. Through the orbit, the Earth's motion causes the spin vector to move $\sim 1°$ per day in the E direction.

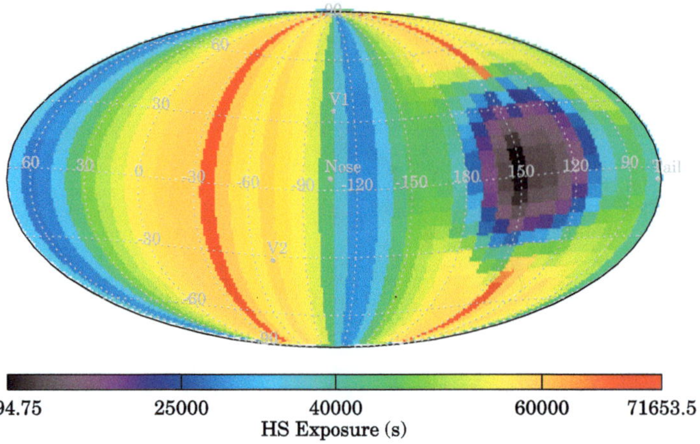

Fig. 2 View time for the baseline 2 year mission including loss of viewing time due to obscuration by the magnetosphere. We show a Hammer-Aitoff projection in heliospheric ecliptic coordinates. Also shown are the positions of Voyager 1, Voyager 2, and the nose of the termination shock (upwind interstellar direction). The heliospheric poles are imaged every spin and therefore are strongly oversampled

8×30 sec $= 4$ minutes, and the IBEX-Hi cycle takes $6 \times 30 = 3$ minutes. Each 48 spins (12 minutes) the stepping sequences of IBEX-Lo and IBEX-Hi fall back in sync at the lowest energy step.

In addition to the nominal voltage stepping sequences, IBEX-Lo has several special modes. There is an "oxygen mode" that is designed for periods of the orbit when IBEX-Lo is well-positioned for sampling interstellar neutral atoms, particularly interstellar oxygen. The times in which we may place IBEX-Lo into oxygen mode are indicated by the gray shaded regions of Fig. 1. In the oxygen mode, the IBEX-Lo sensor will be placed into an energy step tuned for interstellar oxygen sensing when the sensor is between 30° and $-30°$ of the ecliptic plane.

During the spring viewing period (Day-of-Year, DoY, 10–60) in the oxygen mode, the IBEX-Lo sensor may be configured to only accept ENAs from its high-resolution sector; this $\sim 90°$ arc on the detector has a narrower field-of-view (FOV) of $3.5° \times 3.5°$, allowing detection of interstellar neutral atoms with higher angular resolution. This high-resolution oxygen mode is run on a 10-spin cadence. In every 9 out of 10 spins, we use a single energy channel near 522 eV for oxygen detection. In the 10th spin, we change to an energy channel near 120 eV for He detection. The configuration during this period takes into account the Earth's motion into the flow of interstellar neutral atoms, causing a higher flux and energy of measured interstellar atoms. The higher flux, in turn, enables a higher resolution oxygen measurement.

During the fall viewing period (DoY 255–290) in the oxygen mode, the IBEX-Lo sensor accepts ENAs in both the high- and low-resolution sectors of the collimator. The ESA is shifted to a voltage step around 31 eV for interstellar oxygen detection. In this period, Earth's motion causes the IBEX spacecraft to move in the same direction as the interstellar flow, and the oxygen atoms are detected at lower energies. In this fall period, interstellar He is at ~ 8 eV, which makes it more difficult to observe.

In addition to the oxygen mode, there is a background mode in which the ESA is configured so that the ion optics prevent any direct signal from reaching the detection system. This mode is used to infer the level of background counts that are unrelated to ENAs.

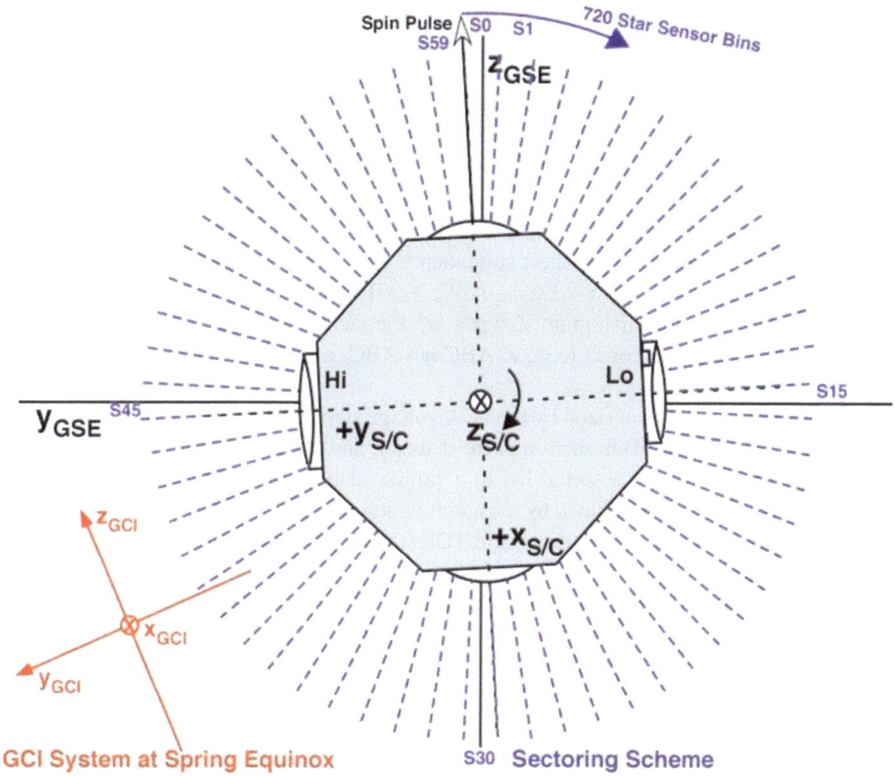

Fig. 3 The angular sectors of a spin during spring equinox. A spin pulse is triggered for each rotation at the position where IBEX-Hi points northward. This corresponds to the spin 0 position. The z-axis in this figure points toward the Sun, with the positive y-axis pointing down the boresight of IBEX-Hi and the negative y-axis pointing down the boresight of IBEX-Lo. Each spin sector is 6° wide, and the spin pulse is issued so that spin-sectors are centered on the poles and the ecliptic

The basis for all IBEX data products are the data elements from the sensors:

- *IBEX-Hi Direct Events, DEs,* are individually detected ENAs. Each DE is characterized by the ESA voltage step, the event arrival time, and the coincidence type (16 possible types defined by the combination of the signals registered in the three channel electron multipliers, CEMs). Direct Events are typically used for creating the global heliospheric maps. The events are packed into a list sorted by spin, coincidence type, and time tag. The 16 basic types are organized into the three categories of single, double and triple coincidence events. For each of the three CEM detectors (A, B, and C), there are long and short coincidences that depend on the timing of the pulses from the detectors. An event-type is defined by the "short" and "long" pulses that are identified. For example, an event of type ab-ABC consists of short pulses identified from detectors A and B, and long pulses from detectors A, B, and C. Events designated as triples include all event-types with long pulses from all three detectors (7 types), while double event-types have long pulses from any two detectors (9 types).

– *IBEX-Hi Histograms* correspond to the number of counts in each 6-deg spin-phase bin. The histograms are partitioned into data-types collected over 12-, 24- and 48-minute periods, depending on the priority of the coincidence-type. For example, triple coincidence

events have all three CEMs triggered and are given the highest priority. The coincidence histograms to be downlinked include the "long" histogram types, L-ABC, L-AB, L-BC, and L-AC, corresponding to event-types with any combination of short coincidences and long coincidences designated by the long histogram type. For example, the L-ABC histogram type contains event type abcABC, ab-ABC, a-cABC, a–ABC, -bcABC, -b-ABC, and –cABC. In addition, depending on the Qualified Count Select register, either the histogram "Y-" types (Y-ABC, Y-AB, Y-BC Y-AC) or histogram "Z-" types (Z-ABC, Z-AB, Z-BC Z-AC) are downlinked as well. Here, the histogram Y-types refer to the subset of long events with the short coincidence combination not equal to the long coincidence combination (e.g., Y-ABC=—ABC, Y-AB=–cAB-, Y-BC=a—BC, and Y-AC=-b-A-C). Similarly, the histogram Z-types are the subset of long events with no short coincidence from detector C (e.g., Z-ABC=ab-ABC, a–ABC, -b-ABC, Z-AB=ab-AB-, a–AB-, -b-AB-, etc.).
- *IBEX-Lo DEs* are characterized by the ESA voltage step, the four values of time-of-flight (TOF) measured in the TOF section of the detector, and the ENA arrival time. The IBEX-Lo DEs are packed into a sorted list in a similar manner to the IBEX-Hi DEs. In this case, coincidence type is defined by the combination of the TOF measurements identified. Triple coincident events have three valid TOF signals, and double coincident events have two. There are 16 different event combinations corresponding to each of the event types.
- *IBEX-Lo histograms* are defined by the number of events per spin-phase bin and have 21 different histogram types depending on the TOF values registered.
- *IBEX Background Monitor (IBaM, Channel D) Histograms* are defined by the number of counts per bin (in each of 720 0.5° bins). The IBaM histograms serve as a basic proxy for backgrounds caused by energetic particles.
- *Star Sensor Histograms* are defined by photon counts in 720 0.5° bins. The star sensor histograms are used to provide accurate pointing for the determination of the interstellar neutral flow direction from IBEX-Lo.

3 Overview of Operations

The operations role of the ISOC is to:

- receive the spacecraft and science data, and generate science products that will be used by the science team to answer the primary science questions;
- evaluate command sequences (Absolute Time Sequences, ATS files) for each orbit, and provide a Command Authorization Report (CAR) back to the MOC;
- generate any needed Science Tasking Files (STFs) for special payload commands.

The operational functions, data flow and interfaces are summarized in Fig. 4.

Routine commanding of IBEX is performed by "rotes", which are rules that determine where in the orbit the sensors are turned on (for example, above 10 R_e), and where the sensor configurations are changed (for example, where in the orbit science data collection should begin). The mission planning software ingests the orbit ephemeris, STFs, and uses the internally stored rotes to determine the ATS of commands that are uploaded to the spacecraft. Specific rotes are triggered when the spacecraft experiences orbital events, such as ascension above 10 R_e. Special operation planning is provided to the MOC in the form of STFs, which are simple ASCII files containing timed commands to be fed to the command system at the MOC. The operational rotes are summarized as follows:

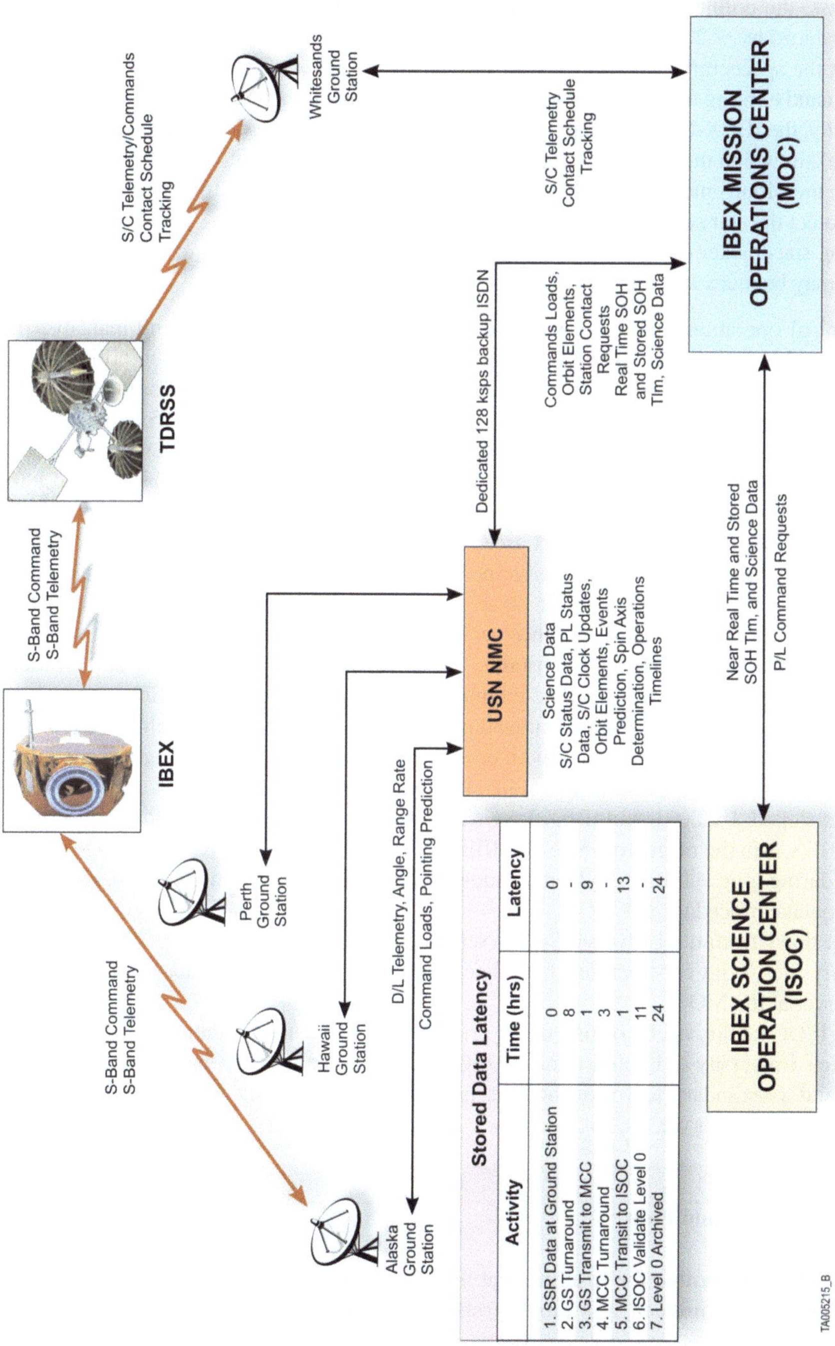

Fig. 4 Ground segment architecture and key elements passed between interfaces

– when the spacecraft descends below 10 R_e,[2] a macro-command in the Combined Electronics Unit (CEU) is triggered to bring the sensors to high-voltage standby. In this process, the collimator voltages are reduced to less than 100 V, and all other voltages are reduced by $\sim 20\%$;
– when the spacecraft ascends above 10 R_e, the CEU triggers the necessary CEU macro-command to bring the sensors from high-voltage standby into nominal science mode. Optionally, the IBEX-Lo sensor can be placed into the "Oxygen-mode" or the "Background-mode", in which the standard stepping of the voltages is modified;
– when the Moon enters the IBEX-Lo star sensor FOV, the photo-multiplier gain is reduced to protect the star sensor;
– during star-tracker outages (for example, during eclipses), the cadence of the star sensor data may be increased.

Nominal operations through a single orbit with its associated passes and maneuver is summarized in Fig. 5. The orbit begins with the 10 R_e ascent; the associated orbital event in the mission planning software generates command sequences that transition the CEU into an appropriate state to collect science data. There is an apogee pass (35 minutes) with low data rate availability (2 kilosymbols/s, ksps; 1 ksps $\approx$ 2 kb/s). The science collection resumes until the spacecraft moves below 10 R_e and the CEU transitions the payload into a "housekeeping state" appropriate for low-altitude operations. In this mode, science data collection is suspended until the next ascent above 10 R_e. Shortly after the descent below 10 R_e, the repointing maneuver is performed, then the science data is downlinked during a high data rate (320 ksps) pass. After the perigee, there will be a tracking pass, and as the spacecraft ascends above 10 R_e, science data collection resumes.

Science Tasking Files provide substantial flexibility in nominal operations. The location of sensor turn-on and turn-off can be modified easily, and the sensors, settings can be adjusted with relative ease. A CEU commanding guide is presently being developed to collect all the possible commands together with operational instructions. Generating STFs is a relatively simple procedure and does not involve special software. The following STFs have been developed: IBEX-Hi and IBEX-Lo Gain Curve Measurements (likely to be performed when IBEX is in the magnetosphere), an IBEX-Lo background test, a STF to change sensor modes during long eclipses, payload shutdown from high-voltage standby mode, and a star sensor outage procedure.

Data flows from the Universal Space Network (USN) to the MOC. The ISOC receives telemetry files, orbit ephemerides, housekeeping data, and additional data from the secure FTP server at the MOC.

The ISOC facility itself consists of the primary IBEX Science Operations Center (ISOC) at Boston University (BU) and a mirror ISOC at SwRI. Each ISOC has a primary server inside and a secondary server outside the firewall, and two RAID Arrays. The BU ISOC also houses a web server.

4 Data Products and Flow

The ISOC is responsible for ensuring that the most complete science results are correctly extracted from the mission in a timely fashion. A relatively small portion of this task is

[2]The actual position in the orbit where the sensors are turned on may change as the science operations are optimized.

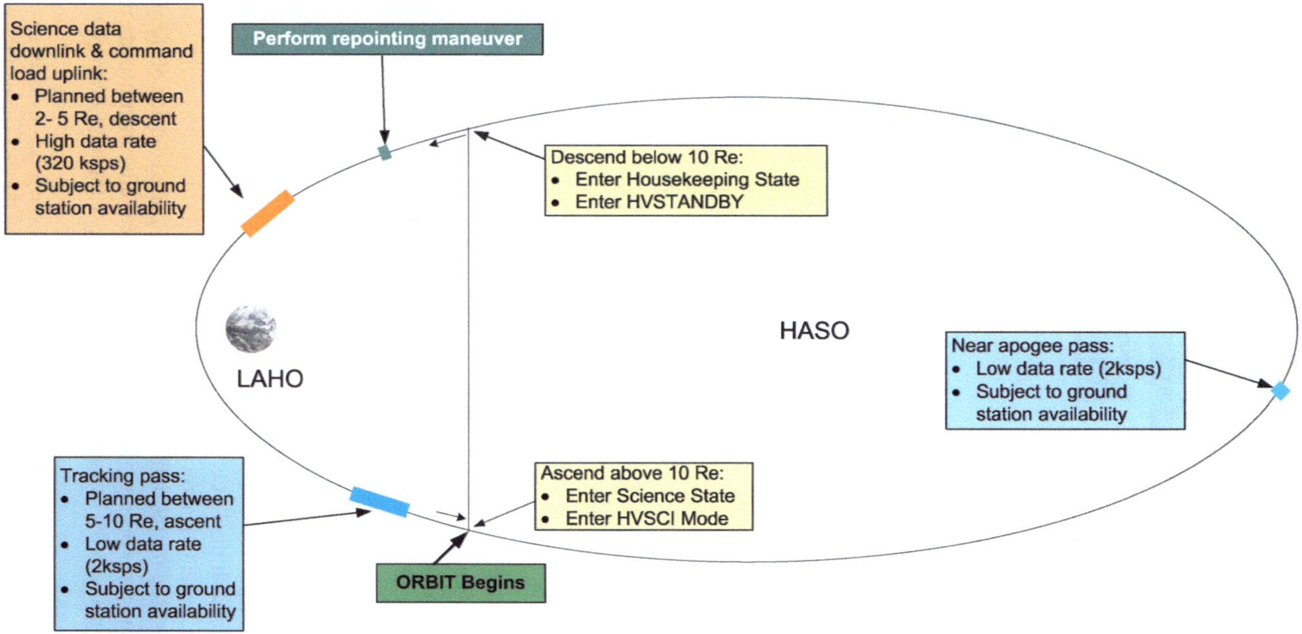

Fig. 5 During nominal operations, there are three passes during each orbit. The high data rate pass (320 ksps) will be used for science data downlinks and command uplinks. The low data rate passes can be used for near-real-time communications with the spacecraft

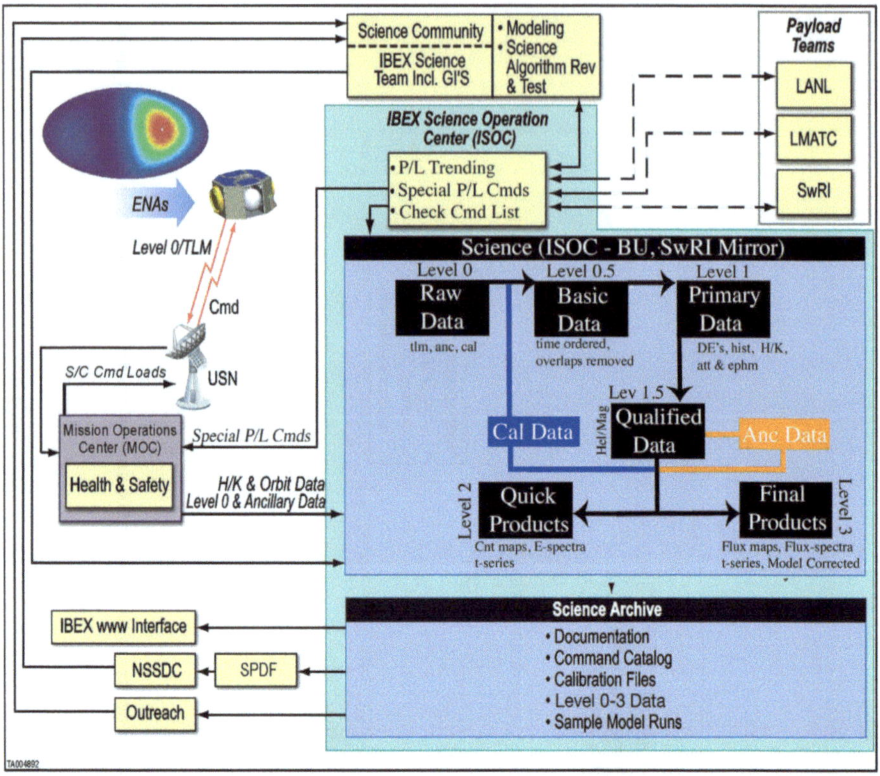

Fig. 6 IBEX Data Flow through the ISOC. The IBEX raw telemetry data and ancillary data are received processed into high level data products, archived and distributed to the scientific community and public

devoted to the commanding of the sensors; they are switched from science mode to standby mode near perigee when there is no useful data to be taken, and switched back into science mode again for the 8-day observing cycle. The bulk of the ISOC effort will go into the analysis of data returned by the sensors themselves.

One way to view this analysis is in terms of a pipeline connecting the principle science data products. The data processing pipeline is a part of a larger framework for receiving telemetry, trending the payload, delivering STFs when needed, archiving data, and distributing data to the scientific community and public (Fig. 6).

Real number designations are used to classify data products in a fashion suitable for IBEX and to distinguish the levels from those used in other NASA missions. (In standard NASA definitions, Level 0 denotes raw telemetry, which is calibrated to yield Level 1, etc.) Integral values are used for those that might be usefully exported outside of the processing pipeline. The others are working products that might appear after intermediate processing steps, but are not necessarily viewable objects. (See Table 1 for brief descriptions of data box contents.)

The raw telemetry (Level 0.0 data) delivered by mission operations may be corrupted, garbled communications, data dropouts or repeats, etc. Thus, the first stage of the processing pipeline is intended to correct for this, producing a "clean copy" of the science data (Level 0.5) as it was stored in the Solid State Recorder (SSR) on IBEX. If necessary, commands can be sent to the spacecraft to reread portions of the recorded data until it has all been received

Table 1 IBEX Data Products. Various data products are created and distributed through the operation of the IBEX data pipeline

Level	Archived	Name	Description
0.0	yes	Raw Data	Telemetry (CCSDS packets) and other basic inputs, including raw calibration data (if delivered to ISOC)
0.5	no	Basic Data	Raw Data reorganized into time order with overlaps removed and fragments reconstructed
1.0	yes	Primary Data	Direct events, event histograms, sensor and spacecraft housekeeping, spacecraft attitude and ephemeris, etc. unpacked and augmented
1.5	no	Qualified Data	A subset of the primary data, qualified for heliospheric (or conversely magnetospheric) work
2.0	yes	Quick Products	Basic high level data products: sky maps of counts, energy spectra, count rate time series, based on selection, but not modelling
3.0	yes	Final Products	Science data products: ENA flux maps, spectra requiring sensor and science model-dependent assessments
n/a	yes	Calibration Data	Individual sensor calibrations, spacecraft geometry, etc.
n/a	no	Ancillary Data	Other useful data, typically data products from other missions

on the ground. There is sufficient space in the onboard SSR to store data for about 4 orbits, so there is roughly a month to request the re-transmission of any missing data. Level 0.0 data products also include other materials (contact logs, ephemerides, etc) received from the MOC.

It should be noted that the payload firmware (PLFW, an 8051 Microcontroller implemented in an Actel FPGA core with 12 MHz clock) has a sophisticated, table-driven algorithm which guides the selection of individual events to be downlinked (whereas all events are included in downlinked histograms). This algorithm has undergone extensive testing in the development phase; and early in the science phase of the mission its performance must be verified, and if necessary, adjustments will be made to its configuration.

The flight algorithm also encodes the data rather compactly, and for subsequent processing, an expanded form of this data (Level 1.0 data products) is created. The most significant change is the correlation of event times (or spin phase) with the corresponding IBEX attitude to determine the arrival direction of each event (or assigned to each histogram bin). The IBEX attitude control system (ACS) provides the PLFW with (1/3 Hz) attitude updates (quaternions), and samples of these (sufficient to reconstruct attitude to the required accuracy) are included in the telemetry.

At this point in the processing pipeline, with a magnetospheric model and knowledge of background sources, one can remove obvious non-signal events and construct ENA count maps and simple spectral estimates for regions of the sky (Level 2.0 data products). However, the analysis path is not completely straightforward, insofar as there are model-dependent choices that play a role at this point. For example, the magnetosphere is a dynamic entity with a rich phenomenology of its own, so the segregation of the magnetospheric ENAs from the heliospheric ENAs may not always be straightforward. Similarly, while a number of noise and background sources have been identified through the mission and sensor development phases, some detective work may be necessary should spurious counts arise. Thus a "toolbox" of methods for the selection and display of results along with some degree of (local or remote) human interaction is implicit at this processing level. Algorithms for this toolbox are provided by the IBEX sensor and science teams. Thus the road to the Level 2.0 data requires intermediate (virtual) data products (shown as Level 1.5 in Fig. 6 and Table 1).

The sensor calibration data also plays a significant role in the processing at this level. The estimation of the true ENA flux into the sensor from the direct event counts is ultimately a model-dependent process. Thus maps and spectra of ENA flux are higher level products (Level 3.0).

At the level of complete understanding, the dissection of the physics of the heliospheric boundary layer is also a model-dependent process (also in product Level 3.0). Some of the modeling requires knowledge of the solar wind and magnetosphere, therefore data products from other missions are desired. Depending on what the IBEX sensors measure, the full science analysis may well extend beyond the life of the mission itself. The final destination of IBEX data products is the SPDF (Space Physics Data Facility) at the NSSDC (National Space Science Data Center) in a form suitable for use by the community. The data products at Levels 0.0, 1.0, and 2.0 are formal deliverables to the SPDF. The Level 3.0 products are essentially publishable results (which might also be archived at the SPDF after publication). Most of the models require solar, geophysical and/or heliospheric data products from other missions. This "Ancillary Data" is produced by other missions and archived by the NSSDC, and therefore does not need to be delivered (back to) the SPDF.

5 Heliospheric Transmission

An important component in the analysis of IBEX data is understanding the transmission of hydrogen ENAs from their points of origin to observation at the IBEX spacecraft (Bzowski et al. 2009; Bzowski 2008). The Level 3 maps represent ENA fluxes from the outer heliosphere, which we take to be a 100 AU boundary (although the results are fairly insensitive to the actual location of the boundary, within $\approx 10\%$). In constructing these maps, we must understand the transmission of ENAs from 100 AU to the spacecraft, and then the response of the sensors to incident ENAs (taken up in the next section). To solve for the ENA transmission, we must take into account loss by ionization (predominantly, photo-ionization and charge-exchange), and the effects of gravitation and radiation pressure. A trajectory solver has been developed that follows ENAs from 100 AU (or some other outer boundary) to 1 AU, or conversely from 1 AU to 100 AU, and takes into account the acceleration exerted by the Sun's gravity and radiation pressure, $\mathbf{a} = -\hat{e}_r GM_\odot(1-\mu)/r^2$; the function μ is the ratio between radiation pressure and gravity, r is the distance from the Sun, and $\hat{e}_r$ is the unit vector in the radial direction. Along each trajectory, we solve also for the survival probability:

$$S_p(t) = \exp\left[-\int_{t_0}^{t} dt'(\sigma_x n_p u + \beta_p)\right], \tag{1}$$

where the charge-exchange rate, $\sigma_x n_p u$, is the charge-exchange cross-section, σ_x, times the proton density, n_p, and the bulk flow speed is u. The photo-ionization rate is β_p. The radiation pressure over gravity term, μ, and the ionization rates are, in general, a function of time, ENA energy, and spatial position from the Sun. A solver for these quantities has been prepared by M. Bzowski and is being used by the ISOC for forward models of ENA count rates, and reverse models that transform incident 1 AU ENA flux maps into ENA

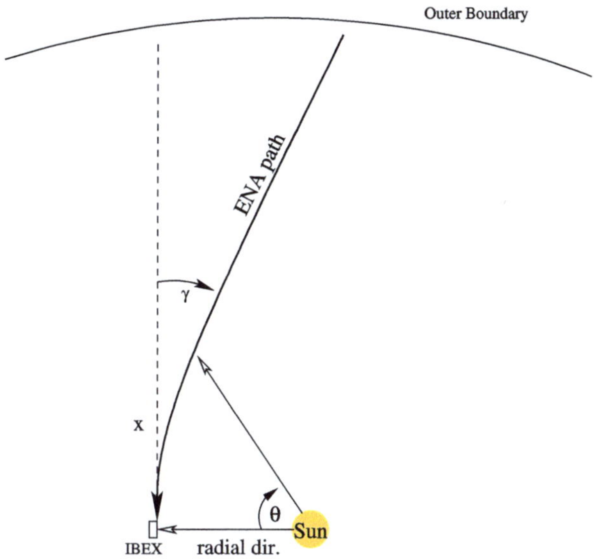

Fig. 7 Schematic showing the deflection of ENAs. The deflection angle, γ, is derived here

emission maps from the outer heliosphere. All terms in (1) are evaluated along the history of the ENA trajectory through time t' from the time of ENA production at t_0 to observation at time t.

The effects leading to ENA loss, deflection and energy loss can be treated analytically using several approximations. We consider an atom that moves inward from 100 AU to 1 AU. We characterize ENA loss in terms of a survival probability, S_p, solved for from the loss rates along the ENA trajectory:

$$v_0 \frac{d \ln S_p}{dx} = \left[\beta_p + n v_x \sigma(E_x)\right] \tag{2}$$

$$E_x = \frac{1}{2} m (u^2 + v_0^2) \tag{3}$$

$$v_x = \sqrt{u^2 + v_0^2} \tag{4}$$

where x is distance along the ENA trajectory (see Fig. 7), v_0 is the speed of the ENA, β_p is the photo-ionization rate, n is the solar wind density, u is the solar wind speed, $\sigma(E)$ is the charge-exchange cross-sections, and E_x is the charge-exchange impact energy (the relative energy between a neutral H atom and a solar wind proton). Note that the impact energy varies slightly with position along the trajectory, and we have taken the value of the impact energy near the end of the ENA trajectory where the solar wind ions move in a direction perpendicular to the ENAs. We take the photo-ionization rate and the solar wind density to fall off with the inverse square of distance from the Sun:

$$\beta_p(r) = \beta_{p1} \left(\frac{r_1^2}{r^2} \right) \tag{5}$$

$$n(r) = n_1 \left(\frac{r_1^2}{r^2} \right) \tag{6}$$

where r_1 is 1 AU, β_{p1} is the photo-ionization rate at 1 AU, and n_1 is the solar wind density at 1 AU. The subtlety in this calculation is that we have considered a straight ENA trajectory moving from the outer heliosphere where the ENA moves inward radially to the point of detection where the ENA moves roughly perpendicular to the solar wind. The approximation of a straight ENA trajectory is justified by the fact that it is near the end of the ENA trajectory where the ionization loss rates are largest. As a result, this approximation turns out to be very accurate. We consider the charge transfer between solar wind protons and the neutral H ENAs, which has a cross-section quantified by Lindsay and Stebbings (2005):

$$\sigma(E) = (a_1 - a_2 \ln E)^2 (1 - \exp^{-a_3/E})^{4.5} \tag{7}$$

where σ is expressed in 10^{-16} cm^2, $a_1 = 4.15$, $a_2 = 0.531$, $a_3 = 67.3$, and E is expressed in keV. We make the further approximations that the speed of the ENA (v_0) and solar wind (u) are constant. In this case, the following analytic solution for the survival probability is obtained:

$$\begin{aligned} S_p(r_1, r_{\text{out}}) &= \exp\left(-\left[\beta_{p1} + n_1 v_x \sigma(E_x)\right][r_1/v_0]\text{atan}(r_{\text{out}}/r_1)\right) \\ &\approx \exp\left(-\left[\beta_{p1} + n_1 v_x \sigma(E_x)\right][r_1\pi/(2v_0)]\right) \end{aligned} \tag{8}$$

where r_{out} is the location of the outer boundary, which can be chosen around 100 AU (the solution is fairly insensitive to the actual choice of the outer boundary, provided that it is many 10's of an AU). The photo-ionization rate at 1 AU varies between 0.8×10^{-8} s^{-1} at solar minimum to 1.5×10^{-7} s^{-1} at solar maximum (Ruciński et al. 1996; Bzowski 2008).

The deflection angle of ENAs from the outer heliosphere to the IBEX spacecraft where they are detected can be approximated based on the solution to the two-body problem. Consider an ENA detected with energy E_0', moving with velocity $\mathbf{v}_0'$. The spacecraft has velocity $\mathbf{v}_{sc}$ so that the velocity of the ENA in the inertial frame of the heliosphere is $\mathbf{v}_0 = \mathbf{v}_{sc} + \mathbf{v}_0'$ and the ENA energy at the point of detection in the heliospheric inertial frame is $E_0 = mv_0^2/2$. Since the look directions of the sensors and the spacecraft motion are primarily perpendicular to the radial direction, the velocity of the ENA at detection is also primarily perpendicular to the radial direction. This means that the observed ENAs are near their perigee. In our analytic approximation, we estimate that the ENAs are *at* their perigee at the point of detection. Deflection of the ENAs from the outer heliosphere to the point of detection (see Fig. 7) takes place in a plane containing the velocity of the ENA and the radial direction. The ENA energy E and the path angle θ are solved for using the gravitational two-body solution. The results are as follows:

$$E = E_0 - E_0 K\left(1 - \frac{r_1}{r}\right) \tag{9}$$

$$\cos\theta = \frac{2(r_1/r) - K}{2 - K} \tag{10}$$

where $K = GM_\odot m(1-\mu)/(r_1 E_0)$, and μ is the ratio of radiation pressure to gravity along the ENA path near 1 AU. In this expression, we have taken μ to be a constant and therefore neglect its dependence on the ENA velocity. Since the distance to the outer heliosphere is large compared to 1 AU, the deflection angle is approximately $\gamma = \theta - 90°$. The case of a straight ENA trajectory at high energies yields $\theta = 90°$ and $\gamma = 0°$.

6 Approach to Analysis: The Sensor Response Functions

A detailed understanding of the sensor responses is critical for both forward models from incident ENA fluxes into sensor count rates and data inversion that converts observed count rates into incident and heliospheric flux maps. The response functions describe the detailed behavior of the sensors as derived from calibration data.

6.1 Functional Form of Response

During a time interval Δt, the counts reported by either sensor in response to a flux of ENAs is:

$$C_{ij}^{x} = \Delta t \sum_{s} \int dE \int d\alpha' \int d\beta' \, \mathcal{R}_{ijs}^{x}(E, \alpha', \beta') \, J_s(E, \alpha', \beta') \tag{11}$$

where $J_s(E, \alpha', \beta')$ is the differential ENA flux, $\mathcal{R}_{ijs}^{x}(E, \alpha', \beta')$ is the response function s is the incident ENA species, E is the incident ENA energy, α' is the incidence spinward angle, β' is the incidence sunward angle, i is the ESA Voltage step, j is the Direct Event type, and x is the detected species.

By including both incident species (s) and detected species (x), the formalism allows for the effects caused by sputtering. This is particularly important for IBEX-Lo since calibration results show that incident H atoms cause sputtered H, O, and C in addition to converted H. These sputtered atoms appear typically at energy steps below that of the incident ENA. In IBEX-Hi, there should be almost no sputtering and, as a result, the detected species and incident species are predominantly the same.

The *Incident Differential ENA Flux* $J_s(E, \alpha', \beta')$ in units of ENAs/(cm^2 s sr keV) is a function of incident energy E and arrival direction. IBEX-Hi is mostly sensitive to H, and expect negligible response from other species. IBEX-Lo is sensitive to both J_H and J_O, and possibly other species that might be present (e.g., J_{He}).

For IBEX-Hi (IBEX-Lo), j refers to one of the 16 coincidence types (14 coincident types for Lo). There are six IBEX-Hi ESA voltage steps labeled $i = 1, \dots, 6$ and eight IBEX-Lo ESA voltage steps labeled $i = 1, \dots, 8$.

In addition to the explicit integrals, we have also integrated over other *calibration variables* that are known only in the calibration setting (e.g., position of arrival on conversion surface or foil). Therefore, this formulation of the response function $\mathcal{R}_{ijs}^{x}(E, \alpha', \beta')$ does not depend on the calibration variables. Dimensionally, the response function is an effective area, expressed in units of cm^2.

The arrival angles sketched in Fig. 8 refer to the direction relative to the sensor boresight. The angle α' to a given fixed location changes with the spin-phase, and the angle β' changes with the repointing of IBEX toward the Sun (i.e., ENAs observed arriving from $\alpha' > 0°$ will be observed a short time later arriving from $\alpha' = 0°$. ENAs outside the FOV with $\beta' \approx 8°$ are observed on the next orbit near $\beta' \approx 0°$.)

6.2 The IBEX-Hi Response Function

A detailed description of the IBEX-Hi sensor is provided by Funsten et al. (2009). We develop here a manageable form of the response function making use of the fact that the ESA transmission function is relatively narrow in energy. This allows us to use functions tuned to each ESA voltage step and consolidate the remaining energy dependence in the transmission function:

$$\mathcal{R}_{ij}(E, \alpha', \beta') = G_i E_i \, P_i(\alpha', \beta') \, T_i(\mathcal{E}(E) - E_i). \tag{12}$$

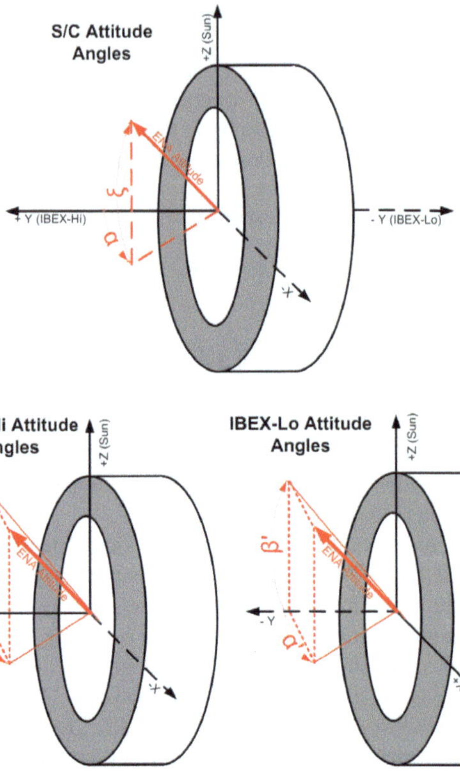

Fig. 8 A cartoon showing the arrival angles, α and ξ in the spacecraft frame (*upper*), and α' and β' in the IBEX-Hi (*bottom-left*) and IBEX-Lo (*bottom-right*) reference frames

Imposing some normalization, and consolidating the units in the geometric factor, we end up with the following components:

- *Geometric Factor*, G_i ($\mathrm{cm}^2\,\mathrm{sr\,keV/keV}$), characterizes the product of effective instrument area, solid-angle acceptance and energy bandwidth ($A\,\Delta\Omega\,\Delta E/E$).
- *Point Spread Function*, $P_i(\alpha', \beta')$, characterizes the distribution of accepted arrival angles. It is a function of the incidence angles (α', β'), and we identify here the incident energy E with ESA voltage step i. This function is normalized:

$$\int d\alpha' \int d\beta'\, P_i(\alpha', \beta') = 1 \tag{13}$$

  so that P_i reflects the relative probability of acceptance of an ENA within the sensor FOV and has units of sr^{-1}. The integrated solid-angle acceptance for the ESA voltage step, $\Delta\Omega_i$, has been transferred to the geometric factor G_i.
- *Conversion Energy*, $\mathcal{E}(E)$, is the energy of a converted ion generated at the foil, resulting from an incident ENA of energy E.
- *Transmission Function*, $T_i(E - E_i)$, characterizes the distribution of accepted energies relative to the central energy of the ESA voltage step, E_i. The transmission function is normalized:

$$\int dE\, T_i(E - E_i) = 1 \tag{14}$$

so that T_i reflects the relative probability of acceptance of an ion within the ESA passband and has units keV^{-1}.

6.3 The IBEX-Lo Response Function

The IBEX-Lo sensor is described in detail in this book by Fuselier et al. (2009). The form of the IBEX-Lo response function is similar to that of IBEX-Hi, however it is necessary to simplify the transmission function because we do not know the detailed energy dependence of the converted, sputtered, and knock-off species from the conversion surface (at an energy resolution below that given intrinsically by IBEX-Lo at its 8 ESA steps). We characterize the response function as follows,

$$\mathcal{R}^x_{ijs}(E,\alpha',\beta') = G^x_{ijs}(E)\,E\,T(E)P_i(\alpha',\beta')Y^x_{is}(E). \tag{15}$$

In this case, the point spread function $P_i(\alpha',\beta')$ is almost identical to the IBEX-Hi point-spread function. The other components in the response function are summarized as follows:

- The geometric factor is inherently species dependent and energy dependent, with different energy-dependencies at each voltage step. The geometric factor also depends on the species generated through conversion, sputtering or knock-off.
- The transmission function, $T(E)$, is the inverse energy width of the ESA acceptance at a given energy, $T(E_k) \approx (\Delta E_k)^{-1}$.
- The *yield matrix*, $Y^x_{is}(E)$, expresses the relative probability of detecting an ion of species x when an ENA of species s is incident at ESA voltage step i. The yield matrix is normalized so that the sum over all generated species is 1:

$$\sum_{\text{all } x} Y^x_{si}(E) = 1. \tag{16}$$

It is useful to simplify the integral form of the response function further. As opposed to integration over all energies, we perform a summation over all energy steps:

$$C^x_{ij} = \Delta t \sum_s \sum_k \int d\alpha' \int d\beta'\,\mathcal{R}^x_{ijsk}(\alpha',\beta')\,\langle J_{sk}(\alpha',\beta')\rangle \tag{17}$$

where $\langle J_{sk}(\alpha',\beta')\rangle$ is the differential energy flux averaged over energy step k. In this case the response function takes the following form:

$$\mathcal{R}^x_{ijsk}(\alpha',\beta') = G^x_{ijsk} E_k\, P_i(\alpha',\beta')\,Y^x_{isk} \tag{18}$$

where the terms G^x_{ijsk} and Y^x_{isk} can be treated as averages over the continuously varying energy-dependent counterparts:

$$G^x_{ijsk} = \int_{\text{step } k} dE\, G^x_{ijs}(E)T(E), \tag{19}$$

$$Y^x_{isk} = \int_{\text{step } k} dE\, Y^x_{is}(E)T(E). \tag{20}$$

The discrete geometric factor, G^x_{ijsk}, comprises the effective area and solid-angle FOV of the sensor. The geometric factor is broken down in terms of the following efficiencies:

$$G^x_{ijsk} = A\Delta\Omega\,\Delta E/E\,\xi_{ijsk}\chi_i\epsilon^g\eta^x_j \tag{21}$$

with the following factors,

- The *absolute geometric factor*, $A\Delta\Omega$, in cm^2 sr characterizes the collimator acceptance area and solid-angle FOV. This factor is independent of species, energy, and time over the mission. It is broken down into three low-resolution quadrants and a high-resolution quadrant.
- The *energy resolution of the ESA*, $\Delta E/E$, is found to be 0.8 keV/keV.
- The *combined conversion and reflection efficiency*, ξ_{ijsk}, is the ratio of converted and reflected ions to incident ENAs on the conversion surface. This factor depends on the incident species, s, the incident energy (in discrete step k), and the ESA-step, i. The branching into different species and conversion and sputtering processes are rolled into the yield matrix.
- The *collection efficiency*, χ_i, is the fraction of incident ions from the conversion surface that pass into the ESA. Currently, we have assumed that the collection efficiency depends only on the observed energy step, but it may depend on incident energy as well.
- The *combined grid transparency*, ϵ^g, is the combined transparency of all grids (the superscript "g" differentiates the combined grid transparency from the transparencies of individual grids) between the collimator and TOF subsystem.
- *TOF efficiency*, η_j^x, includes the grid transparencies of the TOF subsystem and is determined separately for each of the event types, j, and detected species, x. There is no energy dependence in η_j^x because the ion energy in the TOF section is mostly determined by the post-acceleration voltage.

The measured species, x, is identified using the TOF measurements:

$$TOF_k \approx L/\sqrt{2E_{\mathrm{TOF}}/m_x} \tag{22}$$

where L is the same path length for TOF_1 and TOF_2 measurements, m_x is the mass of species x, and the energy in the TOF section is:

$$E_{\mathrm{TOF}} = E_i + E_{\mathrm{PAC}} - \Delta E_x(E_{\mathrm{PAC}}), \tag{23}$$

for the central energy of the ESA, E_i, plus the post-acceleration (PAC) energy, E_{PAC}, minus the energy loss in the C-foils, $\Delta E_x(E_{\mathrm{PAC}})$. TOF_0 provides a triple-coincidence confirmation, and TOF_3 is essentially a diagnostic:

$$\begin{aligned} TOF_0 &\approx TOF_1 + TOF_2 - TOF_3, \\ TOF_3 &\approx 0. \end{aligned}$$

IBEX-Lo has high- and low-resolution collimator sections, so the point spread function is a linear combination of two section-specific functions:

$$P_i(\alpha', \beta') = w_{\mathrm{high}} P_i^{\mathrm{high}}(\alpha', \beta') + (1 - w_{\mathrm{high}}) P_i^{\mathrm{low}}(\alpha', \beta') \tag{24}$$

where $P_i^{\mathrm{low}}(\alpha', \beta')$ is similar to the IBEX-Hi $P_i(\alpha', \beta')$ function. The separate point spread functions in the low- and high-resolution sectors are separately normalized (integration over all angles yields 1). When the low-resolution sections are disabled, only the first term, including the weighting factor, would be used (the overall point spread function would then normalize to w_{high} since the detector has a smaller effective area). The weighting factor for the high-resolution sector, $w_{\mathrm{high}} \approx 0.071$.

7 Construction of Global Heliospheric Maps

We describe here the flow of processing from the raw data types into higher level data products of various forms. The subsections below each describe a component of processing from one level to the next.

7.1 Pixelation of Event Types

One of basic choices for the formation of global maps is how to pixelate the sky. The most obvious technique is to use a grid in co-latitude (θ) and longitude (ϕ). This is the grid that will be used for our initial products, although other options, such as equal-area pixelated maps (Górski 1999, 2005) are being explored.

In each spin of the IBEX spacecraft, one great circle of the sky is viewed by each sensor. The sensors have an angular response with a FWHM width of $\sim 7°$ and a full width of $\sim 14°$. Counts are accumulated in sky pixels based on the fraction of the FWHM angular response exposed to a given pixel. This technique is used to derive an "observed" map, and techniques of map inversion are described in Sect. 7.4.

The method of accumulating counts (and other quantities such as exposure time) is independent of the map resolution. However, the quality of the maps will be somewhat sensitive to the resolution. If a resolution is chosen beneath the intrinsic ($7° \times 7°$) resolution of the sensors and that dictated by the mission measurement strategy, then some pixels may have little or no exposure. In this case, inversion would be needed to extrapolate data to poorly sampled pixels. For our initial map products, we will use a resolution that will give good pixel sampling. In particular, a choice of $\Delta\theta = 6°$ is consistent with the $6°$ bin size for histogram pixels, and $\Delta\phi = 8°$ is consistent with the longitudinal separation of successive sun-pointing angles. For Direct Event maps, the co-latitude resolution may be reduced if count rates are sufficiently high to provide reasonably good statistics in each pixel. The high latitude pixels near the poles have small solid-angles; however, the exposure time per pixel should be comparable to the low-latitude pixels because the great viewing circles converge at the poles of the ecliptic.

Given a nominal resolution of $\Delta\theta = 6°$ and $\Delta\phi = 8°$, pixels are identified by the longitude indices $b = 0, \ldots, 44$ and the latitude indices $a = 0, \ldots, 30$. Note that the double histograms include double types exclusively. The count maps are then denoted by N^d_{abij} where the superscript $d = \text{lo}$ or $d = \text{hi}$ for IBEX-Lo or IBEX-Hi, respectively. Event type is identified by index $j = 0$ for all triples, $j = 1$ for all doubles, $j = 2, \ldots, 17$ for all DE types, $j = 18, \ldots, 25$ for IBEX-Hi histograms, and $j = 18, \ldots, 19$ for IBEX-Lo histograms. The index i denotes the energy channel ($i = 1, \ldots, 8$ for Lo and $i = 1, \ldots, 6$ for Hi).

A similar concept to the count map is developed for viewing time. The exposure map, T^d_{abij}, represents the counting time per pixel for a given energy step i and event type j. The removal of the magnetosphere and other data that needs to be segregated is done by removing both counts and exposure time over the affected pixels and energy steps. The rate map is determined simply by dividing the count map by the exposure map (including only pixels with non-zero exposure):

$$C^d_{abij} = \frac{N^d_{abij}}{T_{abij}}. \tag{25}$$

7.2 Observed Flux Map: IBEX-Hi

We discretize the response function by assuming that the flux is approximately constant over each $6° \times 8°$ pixel and over each energy step. The incident flux at an energy step i and at pixel a, b is $J_{a,b,i}$. The response function simplifies into a matrix multiplication

$$C_{abij} = \sum_{a'=-1}^{1} \sum_{b'=-1}^{1} P_{a'b'} G_{ij} E_i \sum_{i'=0}^{5} T_{i,i'} J_{a+a',b+b',i'} \tag{26}$$

where the response matrices are unitless and defined by the following integrals:

$$P_{a'b'} = \int_{(a'-1/2)\Delta_a}^{(a'+1/2)\Delta_a} d\alpha \int_{(b'-1/2)\Delta_b}^{(b'+1/2)\Delta_b} d\beta\, P(\alpha, \beta) \tag{27}$$

($\Delta_a = 6°$ and $\Delta_b = 8°$) and

$$T_{i,i'} = \int_{E_{i'-a}}^{E_{i'-b}} dE\, T_i(\mathcal{E}(E) - E_i) \tag{28}$$

and the energy step i extends from a lower energy E_{i-a} to a higher energy E_{i-b}.

We generate the full flux map by angle and energy in two steps. The first step is to correct for the energy losses in the sensor, and the second step is to correct for the spread in acceptance angles. The separation in these steps is possible because the response function is divided cleanly between the energy response (transmission function) and the angular response (point spread function). We then separate (26) into two pieces:

$$C_{abij} = G_{ij} E_i \sum_{i'=0}^{5} T_{i,i'} I_{a,b,i'}, \tag{29}$$

$$I_{abi} = \sum_{a'=-1}^{1} \sum_{b'=-1}^{1} P_{a',b'} J_{a+a',b+b',i} \tag{30}$$

where I_{abi} denotes the observed flux map, and J_{abi} denotes the incident flux map.

A useful approximation for the incident flux is to simply divide by the geometric factor of the central energy step:

$$I'_{abij} = \frac{C_{abij}}{G_{ij} E_i}. \tag{31}$$

This provides a first-order estimate for the incident flux. Note that the incident flux does not depend on the event type j; however, the estimates to the incident flux implicitly depend on the event types analyzed.

We generate a better estimate for incident flux map, I_{abij}, by inverting the transmission matrix:

$$I_{abij} = \sum_{i'=0}^{5} T_{i,i'}^{-1} I'_{abij}. \tag{32}$$

The inversion matrices are currently being studied. The assumption that fluxes are constant across each energy step is a concern, and if the incident flux is a steep function of energy, the inversion may yield an incorrect answer for the incident flux map, or even non-sensical results (e.g., negative fluxes in some steps). Therefore, this inversion technique may be viewed as an initial guess for the incident flux map, and forward modeling can be used to iteratively generate improved guesses. There are iterative techniques that can be used employing both forward modeling and inversion; however, these techniques will not converge in all cases. Focused work in this area will be needed once data become available.

7.3 Observed Flux Map: IBEX-Lo

As with IBEX-Hi, we separate the energy and angular inversion. We then break the response function into two pieces:

$$C^H_{abij} \approx \sum_{i'=0}^{7} G^H_{i,j,H,i'} E_{i'} Y^H_{i,H,i'} I_{a,b,i'}, \tag{33}$$

$$I_{abi} = \sum_{a'=-1}^{1} \sum_{b'=-1}^{1} P_{a',b'} J_{a+a',b+b',i}. \tag{34}$$

We denote the I_{abi} as the observed flux map, and J_{abi} as the incident flux map.

Equation (33) can be reduced to a matrix equation:

$$\mathbf{C}[a, b, j, H] = T[j, H, H]\mathbf{I}[a, b] \tag{35}$$

where the **C** and **I** are column vectors with indices running through the energy steps from 1 to 8. The transfer matrix is then defined for element (i, i'):

$$T_{i,i'}[j, H, H] = G^H_{i,j,H,i'} E_{i'} Y^H_{i,H,i'}. \tag{36}$$

This 8×8 matrix can be easily inverted and stored for each event type j. The column vector for the observed flux in pixel (a, b) is:

$$\mathbf{I}[a, b] = T^{-1}[j, H, H]\mathbf{C}[a, b, j, H]. \tag{37}$$

7.4 Incident Flux Map

The next step in the analysis is to infer what the source flux was on the detectors. Since the energy response has already been taken into account, we correct for the angular broadening of the incident fluxes due to the angular response of the detectors. For IBEX-Hi and the low resolution sectors of IBEX-Lo, the response function is cone-shaped, with a FWHM of 7° and a total angular acceptance of 14°.

The first-order approximation is to assume a diagonal matrix, which in this case, means that the observed flux map is approximately equal to the incident flux map, $\mathbf{I}_i \approx \mathbf{J}_i$. A better estimate can be generated using matrix inversion since the transformation from the incident flux to the observed flux is describable as a matrix multiplication:

$$\mathbf{I}_i = P\mathbf{J}_i \tag{38}$$

where $\mathbf{I}_i$ and $\mathbf{J}_i$ are vectors with elements extending through all the pixels on the map. For example, we may take vector index k to be related to the pixel (a, b) through $k = a + n_a b$ where $n_a = 30$ is the number of pixels in latitude. In this case the matrix P is defined by:

$$\begin{aligned} P_{j,j} &= P_{a'=0,b'=0}, \\ P_{j,j\pm1} &= P_{a'=\pm1,b'=0}, \\ P_{j,j\pm n_a} &= P_{a'=0,b'=\pm1}. \end{aligned} \tag{39}$$

Here, care must be taken for the first and final rows and columns of the matrix:

$$\begin{aligned} P_{0,n_an_b-1} &= P_{a'=-1,b'=0}, \\ P_{j=0..n_a-1,n_an_b+j-n_a} &= P_{a'=0,b'=-1}, \\ P_{j=n_an_b-n_a..n_an_b-1,j-n_an_b+n_a} &= P_{a'=0,b'=1}, \\ P_{n_an_b-1,0} &= P_{a'=1,b'=0}. \end{aligned} \tag{40}$$

The solution to this problem boils down to developing a technique to invert the matrix, P. Once the matrix inversion is available, the incident flux is calculated from:

$$\mathbf{J}_i = P^{-1}\mathbf{I}_i. \tag{41}$$

The inverse matrix P^{-1} is still under development.

Figure 9(a) shows an example of an observed distribution overplotted with a source distribution, which was artificially generated. The observed distribution applies the smoothing due to the sensor angular response and the observed distribution is plotted for each 8° pixel. In this case, there is little difference between the observed and source distribution. In Fig. 9(b), we apply the inversion method to recover the source distribution. Because there is so little difference between the observed distribution and the source distribution, the inversion improves the estimate by only a small amount.

Figures 10(a) and 10(b) compare another source distribution with the observed distribution and the inverted distribution. In this case, because the flux changes suddenly within a pixel, the inverted distribution shows both overshoots and undershoots.

In both the examples shown in Figs. 9 and 10, the observed distribution appears to be a more robust estimate of the source distribution. If variations are broader than 7°, the observed distribution provides an accurate estimate. If there are source variations down to and beneath the pixel scale, the inverted distribution may lead to undershoots and overshoots.

The observed map will be used as an initial product. Depending on the results, methods of inversion may be explored, particularly if the observed maps reveal sharp gradients.

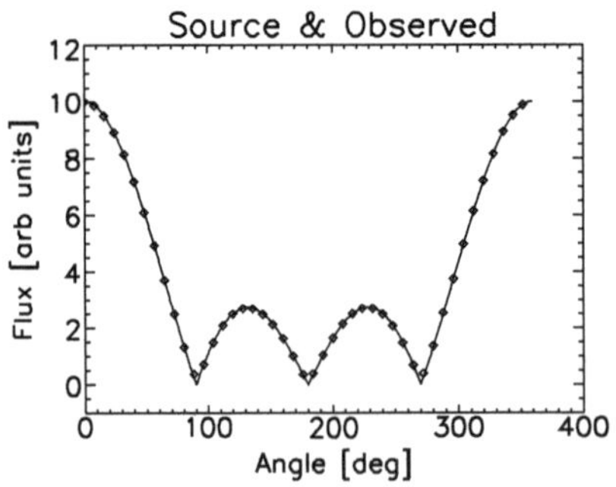

(a) Example observed distribution

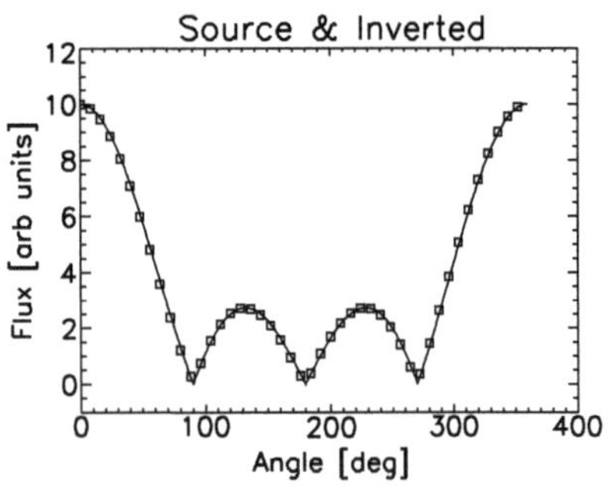

(b) Example inverted distribution

Fig. 9 (*Left*) An artificial source distribution and the observed distribution generated after the angular response of the sensors is applied. (*Right*) The source distribution for the same case plotted against the inverted distribution

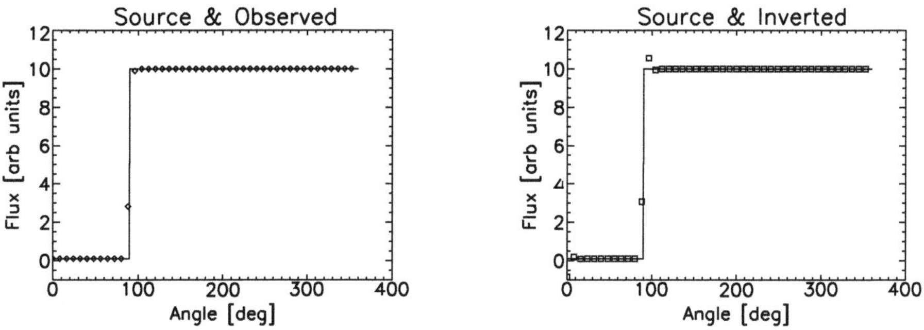

(a) Observed distribution

(b) Inverted distribution

Fig. 10 (*Left*) A second artificial source distribution (a step function) and the observed distribution generated after the angular response of the sensors is applied. (*Right*) The source distribution for the same case (the step function) plotted against the inverted distribution

7.5 Heliospheric Flux Map

The final step is to translate the incident map into a heliospheric map: the ENA fluxes from an external boundary at an outer radius of 100 AU. The main effects that need to be corrected for are the loss of ENAs through the heliosphere due to ionization and the deflection of ENA trajectories due to solar radiation pressure and gravity. We have tested codes in place to correct for both of these effects.

There is also a sampling issue that needs to be managed. Pixels have a solid angle that varies as the sine of co-latitude. One approach is to map rays from each pixel back to the outer boundary, and vary the number of rays per pixel according to the pixel solid area. At the outer boundary, the flux in each pixel would be the average of the loss-corrected fluxes that mapped to the pixel.

8 Forward Modeling Examples

Several examples of forward modeling are considered to demonstrate the effects of heliospheric transmission and response functions on global maps. We emphasize that the nature of the global maps will not be known until observations have been taken. Gruntman et al. (2001) provided three limits for the interstellar interaction from "weak" to "strong". In the weak limit, pickup ions are predominantly transmitted through the shock and simply compressed in the process. In the strong limit, the pickup and solar wind ions form a single population that satisfies the energy, momentum and mass conservation across the shock. The observations from the Voyager 2 satellite (Richardson et al. 2008) show that the core solar wind population lacks sufficient pressure downstream from the shock to satisfy the shock jump conditions. By inference, this implies that some other population, such as pickup protons carries the dominant pressure beyond the shock. This result appears consistent with the weak limit from Gruntman et al. (2001). Recent simulations of the termination shock by Wu et al. (2009) show that a large fraction of solar wind ions are reflected, but the majority of pickup ions are transmitted through the shock without reflection. This also agrees with the weak interaction limit. In contrast, Prested et al. (2008) considered the strong interaction limit using a single kappa distribution to represent the plasma population in the heliosheath.

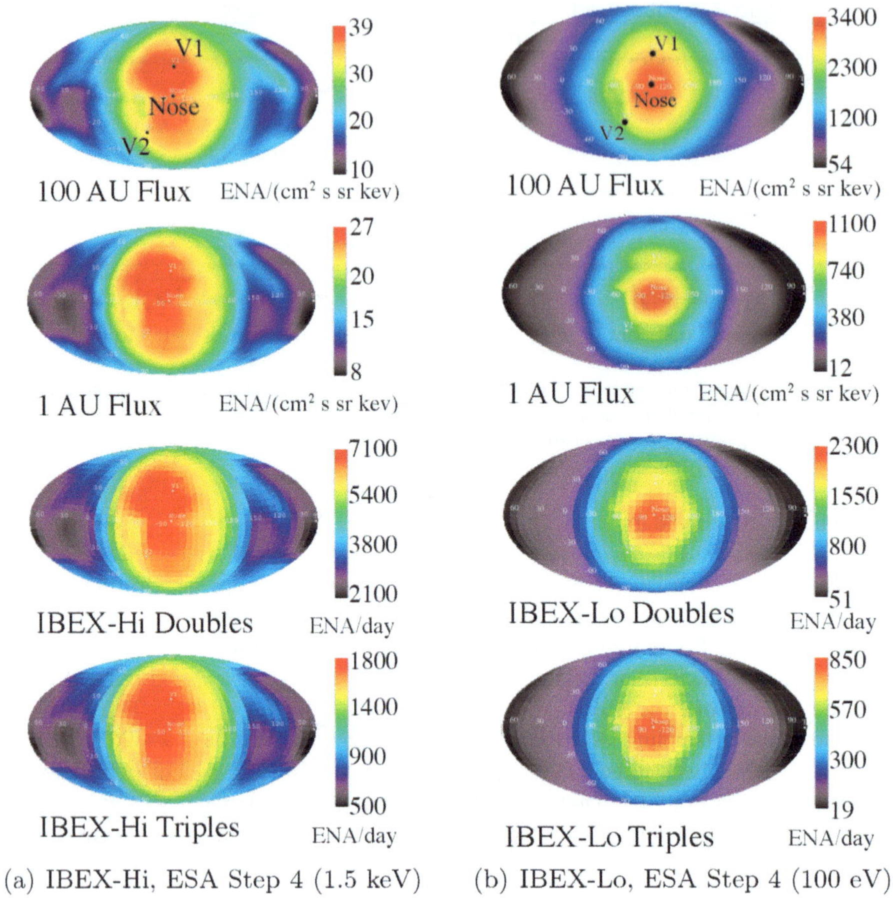

Fig. 11 Forward modeling of IBEX ENA maps at 1.5 keV and 100 eV. (*Top panel*) A global map of the outer heliosphere based on Prested et al. (2008); (*Second panel*) the resulting 1 AU map; (*third* and *bottom panels*) the resulting count rate at IBEX-Hi in double and triple coincidence events

Figure 11 shows the global ENA fluxes in the strong interaction limit (Prested et al. 2008), assuming that the downstream solar wind population has a kappa distribution (a kappa value of 1.6 has been assumed). The top panels show the fluxes at 1.5 keV and 100 eV from the outer heliosphere. The second row of panels shows the incident fluxes at 1 AU. These fluxes are lower due to ENA loss through ionization. The third and fourth rows of panels show the IBEX-Hi double and triple count rate, respectively. These count rates have been generated with the full IBEX-Hi and IBEX-Lo response functions. The results show that the dominant effect of heliospheric transmission is the reduction of flux due to ionization. The deflection of ENAs plays a relatively minor role at these energies with the nominal pixelation ($6° \times 8°$ pixels).

Figure 12 shows a simulated distribution in the weak interaction limit in which pickup ions are transmitted through the termination shock. The modeled distribution includes both pickup ions and solar wind ions (with a downwind temperature of 130,000 K, a density of 0.002 cm^{-3} and a speed of 150 km/s); however pickup ions dominate at all observable ENA energies. The pickup ions are modeled using a Vasyliunas and Siscoe (1976) upstream

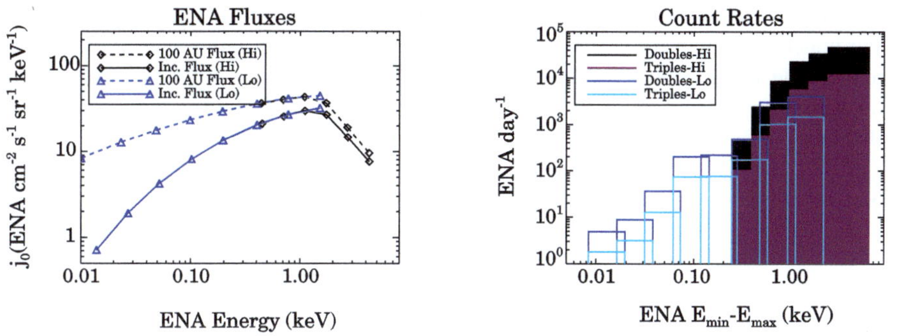

(a) ENA fluxes at 100 and 1 AU (b) Resulting ENA Count Rates

Fig. 12 The incident fluxes (*left*) and resulting ENA count rates at IBEX (*right*). The energy distribution is shown for a weak interaction in which pickup ions dominate the plasma internal energy in the heliosheath

distribution with a neutral H density of 0.1 cm^{-2} and a net ionization rate of 7×10^{-7} s at 1 AU. The pick-up ions are then compressed adiabatically based on a shock compression ratio of 4. This shock compression is large compared to the V2 observations at the shock. However, the net compression between the up and downstream plasma over a broad range shows a transition from $\sim$ 400 km/s to $\sim$ 100 km/s, and is roughly consistent with our assumed compression. The left panel shows both the outer heliospheric flux (at 100 AU, dashed line curve) and the incident flux at 1 AU (solid curves). The right panel shows the resulting ENA double and triple coincident event rates. The large energy dependence of the ENA survival is apparent.

9 Summary

We have described the ISOC's approach to science operations and analysis. The analysis flow proceeds by levels from raw data (Level 0) through final products (Level 3). The ISOC has developed a complete response to ENA fluxes from an outer boundary to the measured counts. This comprehensive response includes the effects of gravitational focusing, loss through ionization, the spacecraft motion and sensor response functions. The response functions have been used to develop synthetic IBEX data products.

The vast majority of ISOC tools are complete. These tools allow the processing of raw data, ephemeris products, housekeeping products, Level 1 time-series, Level 2 energy and spin-phase distributions, and Level 3 maps. A suite of web-based processing tools are now available to the IBEX team for analysis.

As our understanding of the interstellar interaction improves, the ISOC tools for building global maps of the heliosphere will be modified and improved. Flexibility in data processing is a fundamental feature of our design. As opposed to processing data from lower to higher level products, and rigidly building upon previously set parameters, the ISOC software continually reprocesses *all* raw data into higher level products that are tagged according to the software revision. Therefore,we will iteratively modify and improve the data processing pipeline as we learn more about the necessary parameters that are critical for segregation of heliospheric sources, potential background contributions, sensor response functions, and other factors.

It is remarkable that the era of IBEX development covered a time when *both* Voyager spacecraft passed the termination shock, revealing how little we know about the remarkable interactions at the edge of our solar system. IBEX is poised to open an entirely new chapter in our understanding of the global heliosphere and its interaction with the local interstellar medium. The ISOC greatly looks forward to the coming years as software development gives way to supporting the IBEX team and the scientific community in building the first global maps of our solar wind's interaction with the local interstellar medium.

Acknowledgements This work was supported by the IBEX project.

References

M. Bzowski, Astron. Astrophys. **488**, 1057 (2008)
M. Bzowski et al., in *The Interstellar Boundary Explorer Mission*, ed. by D. M. et al. (Springer, Berlin, 2009)
H. Funsten et al., in *The Interstellar Boundary Explorer Mission*, ed. by D. M. et al. (Springer, Berlin, 2009)
S. Fuselier et al., in *The Interstellar Boundary Explorer Mission*, ed. by D. M. et al. (Springer, Berlin, 2009)
K.M. Górski et al., in *Evolution of Large Scale Structure: From Recombination to Charging*, ed. by A.J. Banday, R.K. Sheth, L.N. da Costa, 1999, p. 37
K.M. Górski, E. Hivon, A.J. Banday, B.D. Wandelt, F.K. Hansen, M. Reinecke, M. Bartelmann, Astrophys. J. **622**, 759 (2005)
M. Gruntman, E.C. Roelof, D.G. Mitchell, H.J. Fahr, H.O. Funsten, D.J. McComas, J. Geophys. Res. **106**, 15767 (2001)
B.G. Lindsay, R.F. Stebbings, J. Geophys. Res., Space Phys. **110**, 12213 (2005)
C. Prested, N. Schwadron, J. Passuite, B. Randol, B. Stuart, G. Crew, J. Heerikhuisen, N. Pogorelov, G. Zank, M. Opher, F. Allegrini, D.J. McComas, M. Reno, E. Roelof, S. Fuselier, H. Funsten, E. Moebius, L. Saul, J. Geophys. Res., Space Phys. **113**, 6102 (2008)
J.D. Richardson, J.C. Kasper, C. Wang, J.W. Belcher, A.J. Lazarus, Nature **454**, 63 (2008)
D. Ruciński, A.C. Cummings, G. Gloeckler, A.J. Lazarus, E. Möbius, M. Witte, Space Sci. Rev. **78**, 73 (1996)
V.M. Vasyliunas, G.L. Siscoe, J. Geophys. Res. **81**, 1247 (1976)
P. Wu, P. Gary, D. Winske, N. Schwadron, J. Geophys. Res. (2009, submitted)

Space Sci Rev (2009) 146: 235–273
DOI 10.1007/s11214-009-9502-0

The Galactic Environment of the Sun: Interstellar Material Inside and Outside of the Heliosphere

P.C. Frisch · M. Bzowski · E. Grün · V. Izmodenov · H. Krüger · J.L. Linsky · D.J. McComas · E. Möbius · S. Redfield · N. Schwadron · R. Shelton · J.D. Slavin · B.E. Wood

Received: 30 July 2008 / Accepted: 24 March 2009 / Published online: 14 May 2009

Abstract Interstellar material (ISMa) is observed both inside and outside of the heliosphere. Relating these diverse sets of ISMa data provides a richer understanding of both the interstellar medium and the heliosphere. The galactic environment of the Sun is dominated by warm, low-density, partially ionized interstellar material consisting of atoms and dust grains. The properties of the heliosphere are dependent on the pressure, composition, radiation field, ionization, and magnetic field of ambient ISMa. The very low-density interior of the Local Bubble, combined with an expanding superbubble shell associated with star formation in the Scorpius-Centaurus Association, dominate the properties of the local interstellar medium (LISM). Once the heliosphere boundaries and interaction mechanisms are understood, interstellar gas, dust, pickup ions, and anomalous cosmic rays inside of the heliosphere can be directly compared to ISMa outside of the heliosphere. Our understanding of ISMa at the Sun is further enriched when the circumheliospheric interstellar material is compared to observations of other nearby ISMa and the overall context of our galactic environment. The IBEX mission will map the interaction region between the heliosphere and ISMa, and improve the accuracy of comparisons between ISMa inside and outside the heliosphere.

P.C. Frisch (✉)
University of Chicago, Chicago, IL, USA
e-mail: frisch@oddjob.uchicago.edu

M. Bzowski
Space Research Centre PAS, Warsaw, Poland
e-mail: bzowski@cbk.waw.pl

E. Grün
Max-Planck-Institut fuer Kernphysik, Heidelberg, Germany
e-mail: Eberhard.Gruen@mpi-hd.mpg.de

V. Izmodenov
Moscow State University and Space Research Institute RAS, Moscow, Russia
e-mail: izmod@ipmnet.ru

H. Krüger
Max-Planck-Institut fuer Sonnensystemforschung, Katlenburg-Lindau, Germany
e-mail: krueger@mps.mpg.de

Keywords Interstellar material · Heliosphere · Local bubble

1 Introduction

The heliosphere is flooded with low-density interstellar neutral gas and dust that have a flow velocity of $\sim$95,000 km h^{-1} and an upstream direction towards the Scorpius-Centaurus Association. Both raw interstellar particles and byproducts of the interaction of interstellar neutrals with the solar wind plasma are found within 1 AU of the Sun. These particles provide an *in situ* sample of the cosmos. Voyager 1 (V1) crossed the solar wind termination shock at ecliptic latitude $\beta = +35°$ and distance 94 AU (Stone et al. 2005, e.g.), while Voyager 2 (V2) crossed the termination shock at $\beta = -31°$ and distance 84 AU (e.g. Stone et al. 2008; Richardson et al. 2008; Gurnett and Kurth 2008; Burlaga et al. 2008, Fig. 1). The two Voyager spacecraft provide *in situ* data on both solar wind and interstellar particle populations in the inner heliosheath regions. The IBEX mission (McComas et al. 2004a, 2005, 2006, 2009, this issue), will add to these data by mapping the interaction between interstellar neutrals and the solar wind using observations of energetic neutral atoms (ENAs) with energies of 10 eV–6 keV, formed from charge exchange between $H°$ and H^+. The ENA data will constrain the filtration of interstellar neutrals entering the heliosphere. Building a self-consistent picture of the interaction between the heliosphere and interstellar material (ISMa) from these diverse data requires knowledge of the interstellar cloud surrounding the heliosphere. The surrounding cloud is part of a dynamical flow of interstellar gas and

J.L. Linsky
University of Colorado and NIST, Boulder, CO, USA
e-mail: jlinsky@jila.colorado.edu

D.J. McComas
Southwest Research Institute, San Antonio, TX, USA
e-mail: DMcComas@swri.edu

E. Möbius
University of New Hampshire, Durham, NH, USA
e-mail: eberhard.moebius@unh.edu

S. Redfield
Wesleyan University, Middletown, CT, USA
e-mail: sredfield@wesleyan.edu

N. Schwadron
Boston University, Boston, MA, USA
e-mail: nathanas@bu.edu

R. Shelton
University of Georgia, Athens, Georgia
e-mail: rls@physast.uga.edu

J.D. Slavin
SAO-Harvard, Cambridge, MA, USA
e-mail: jslavin@cfa.harvard.edu

B.E. Wood
Naval Research Lab, Washington, DC, USA
e-mail: brian.wood@nrl.navy.mil

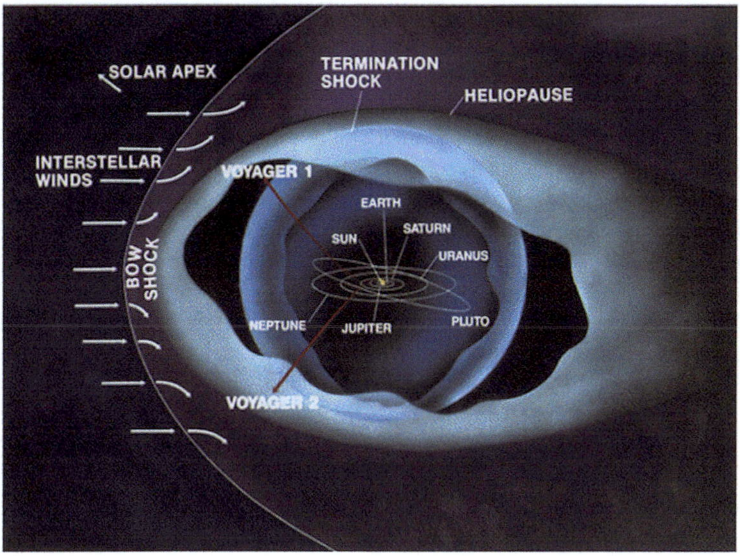

Fig. 1 Cartoon of the heliosphere showing the V1 and V2 trajectories. Interstellar gas flows into the heliosphere from the left. The solar wind termination shock, the heliopause, which is the contact discontinuity between solar wind and interstellar plasmas, and the bow shock are indicated. The ram pressures of the interstellar material, including the interstellar magnetic field, and the solar wind dominate the heliosphere configuration

dust through space. Once we know the physical properties of this flow, we will know the past and future galactic environment of the Sun. Extreme variations in the properties of the interstellar medium surrounding the Sun will have an extreme effect on the heliosphere, the flux of galactic cosmic rays (GCRs) onto the Earth, and possibly the terrestrial climate (Müller et al. 2006; Fahr et al. 2006; Florinski and Zank 2006; Frisch 2006). IBEX will provide key observations of the heliosphere response to our galactic environment.

The space age is old enough that we now have observations of ISMa inside of the heliosphere spanning almost four decades, and these data suggest that the cloud containing the circumheliospheric interstellar medium (CHISM) is homogeneous over spatial scales of at least ~115 AU in the downwind direction of the interstellar flow. The CHISM is defined as the parent cloud of the interstellar gas and dust flowing into the heliosphere. The CHISM appears to be at the edge of the Local Interstellar Cloud (LIC), which belongs to a flow of local interstellar medium (LISM, distance ≲40–70 pc) through space (Sect. 5.1). Interstellar gas was first discovered inside of the heliosphere through broad-band observations of the fluorescence of solar Lyα and 584 Å emissions from interplanetary H° and He^{o}, respectively (Thomas and Krassa 1971; Bertaux and Blamont 1971; Weller and Meier 1974, 1981; Adams and Frisch 1977; Ajello et al. 1987), and the importance of these data was quickly recognized (e.g. Holzer 1972; Fahr 1974; Wallis 1975; Thomas 1978). *Copernicus* made the first spectrum of the Lyα backscattered radiation and showed that interstellar gas inside of the solar system has a velocity similar to the velocity of interstellar clouds in the solar vicinity (Adams and Frisch 1977). The interplanetary Lyα glow from interstellar H° inside of the heliosphere provided the first glimpse that the CHISM is warm and tenuous, and quite different from the galactic cold dense H° that dominates the H° 21-cm hyperfine emission sky. The difference between the observed inflow velocities of H° and He^{o} was later shown to be due partly to secondary H° production from charge exchange between interstellar H°

and interstellar protons near the heliosphere boundaries, as well as solar radiation pressure (Sect. 2.1.2). Slavin and Frisch (2008, SF08) have compared the spectra of the H° Lyα interplanetary glow observed by *Copernicus* during 1975 (Adams and Frisch 1977) with spectra collected by the Hubble Space Telescope (HST) during the 1990s (Clarke et al. 1998), and concluded that the velocity of the interstellar cloud feeding H° into the heliosphere has not changed substantially over the intervening two decades, and that the circumheliospheric interstellar cloud extends at least $\sim$ 115 AU in the interstellar downwind direction. Ulysses observations of interstellar dust inside of the heliosphere during 1992–2002 can be fit with a constant dust density in the CHISM, also providing support for a homogeneous CHISM over scale sizes of $\sim$ 55 AU (Landgraf 2000).

The Sun resides in a region of space with very low average interstellar densities, the Local Bubble (LB), formed by ISMa associated with the Gould's Belt ring of young stars. Gould's Belt appears to belong to the 'Orion spur', protruding $\sim$ 1 kpc from the leading (convex) edge of the Sagittarius spiral arm into the adjacent interarm region (Frisch 2008a). The galactic setting of the Sun is dominated by the LB void in ISMa, and the LIC of which the CHISM is the nearest part. The LIC belongs to a low-density flow of interstellar material away from the Scorpius-Ophiuchus Association (Frisch 1981). Figure 2 shows the solar location inside of the LB void; the molecular clouds that border this void are associated with regions of star formation (for an alternate LB representation see Sfeir et al. 1999).

This chapter presents the data that help us determine the properties of the interstellar cloud that now surrounds the heliosphere. We start with the discussion of ISMa deep inside of the heliosphere (Sect. 2), including both primary and processed interstellar neutrals (Sect. 2.1). This includes He° (Sect. 2.1.1), H° (Sect. 2.1.2), pickup ions (PUIs, Sect. 2.2) consisting of ionized interstellar neutrals, and anomalous cosmic rays (ACRs, Sect. 3.1) consisting of accelerated PUIs. Larger interstellar dust grains (ISDGs) also flow through the heliosphere, penetrating to the Earth and offering a new glimpse of stardust (Sect. 2.3). We then discuss the data that trace the boundary regions of the heliosphere (Sect. 3), including the increased velocity dispersion of interstellar H° in the hydrogen wall (Sect. 3.2) and deficit of small interstellar dust grains inside the heliosphere (Sect. 3.3), both caused by charged ISMa pushed against the magnetized plasma in the outer heliosheath. The small dust grains in this region also appear to trace the direction of the interstellar magnetic field (ISMF) at the Sun, giving us clues about the origin of the LIC in a superbubble shell (Sect. 3.4). We then show that the properties of our immediate galactic environment and the heliosphere boundary conditions are affected by the Sun's location inside of the Local Bubble (Sect. 4), which controls the interstellar radiation field (ISRF) and ionizes nearby ISMa (Sect. 4.1). An important part of understanding the ISRF at the Sun is to first understand foreground contamination of the LB X-ray emission by charge-exchange between interstellar neutrals and the solar wind (Sect. 4.2). Once the interstellar radiation field is understood, the detailed boundary conditions of the heliosphere can be reconstructed using radiative transfer models of the LIC and observations of ISMa inside and outside of the heliosphere (Sect. 4.3). We then look at the family of nearby interstellar clouds, and find a best fit velocity that indicates a flow of ISMa from an upwind direction towards the Scorpius-Centaurus Association (Sect. 5). The overall properties of this dynamical flow of gas within 30 pc have been known for some time, but more detailed information on the velocity structure and physical properties are now available (5.1); we discuss the origin (5.2) and physical properties (Sect. 5.3) of these clouds.

The acronyms used here are summarized at the end of the chapter. Note that the terms LIC and CHISM are not interchangable. The CHISM is the source of the interstellar gas and dust flowing into the heliosphere, whereas the LIC is the interstellar 'cloud' reconstructed

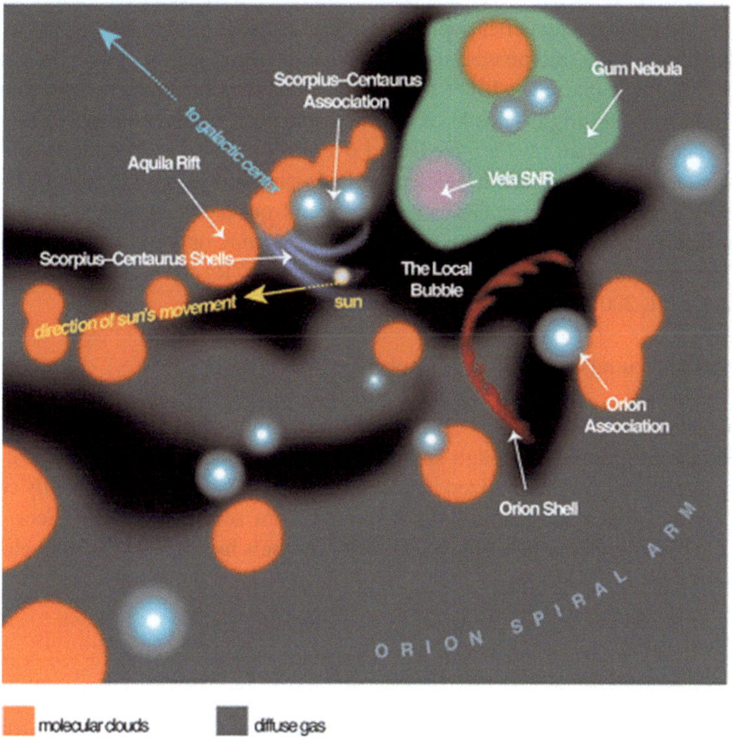

Fig. 2 The galactic setting of the Sun is shown out to distances of 400–500 pc, as viewed from a high latitude sightline in the second galactic quadrant. The galactic center is to the *upper left* (*blue arrow*), and the anticenter is below Orion (*lower-middle, right*). The superbubble shells expanding away from the Scorpius-Centaurus Association (*upper middle*) are shown as *blue arcs*; the CHISM is part of the S1 subshell of the Loop I superbubble (Sect. 3.4). The solar apex motion, and our location inside of the rim of the S1 shell, is shown in *yellow*. (The solar apex motion is defined as the velocity of the Sun with respect to the local standard of rest (LSR), e.g. with respect to the mean kinetic motion of a set of nearby cool, old, stars moving in tandem around the center of the galaxy.) The *large orange circles* show the locations of molecular clouds (CO data) bordering the Local Bubble. The *blue dots* show the locations of nearby star-forming associations (based on Hipparcos data). The S1 and S2 magnetic shells (Wolleben 2007) merge where Loop I (*gray arcs*) encounters the very dusty Aquila Rift. The Orion spur is often denoted the "Orion spiral arm", as is done in this figure. "Spurs" form from instabilities on magnetized, self-gravitating spiral arms (Kim and Ostriker 2002). The overall relation between the Orion Spur and the Sagittarius spiral arm can be seen in the reddening data in Fig. 10 of Lucke (1978). The Sagittarius spiral arm is approximately 1 kpc from the Sun in the direction of the galactic center (e.g. Russeil 2003). Figure copyrighted by American Scientist (Frisch 2000)

from the kinematics of an arbitrary set of velocity components towards nearby stars. The original definition of the LIC was intended to identify the interstellar cloud of which the CHISM is a part, but that definition is no longer universally accepted (Sects. 5.1, 5.1.2).

2 Interstellar Matter inside of the Heliosphere

Interstellar material inside of the heliosphere consists of neutral interstellar gas and large weakly charged dust grains. These atoms and grains interact with the solar wind, solar gravity, and the solar radiation field, and provide an *in situ* sample of the CHISM. These interstellar neutrals also form the parent population of PUIs and ACRs observed inside of the heliosphere. Additional interstellar populations observed at 1 AU, but not discussed in detail here, include interstellar micrometeorites and galactic cosmic rays (GCRs). The interstellar micrometeorites have masses $\sim 10^{-9.7}$ kg and velocities larger than the solar system escape velocity, indicating an origin outside of the heliosphere (see Baggaley 2000; Landgraf et al. 2000). Galactic cosmic rays are measured over 18 orders of magnitude in energy, and provide an *in situ* sample of stellar nucleosynthesis products as well as shock-accelerated interstellar material. Low energy GCRs are excluded from the heliosphere, while more energetic ions arrive at the surface of the Earth. Galastic cosmic rays are not discussed further, but excellent overviews of the origin and abundances of GCRs can be found elsewhere (e.g. Wiedenbeck et al. 2007).

2.1 Interstellar Neutrals inside the Heliosphere

Neutral interstellar atoms trace the CHISM neutrality at the point of entry to the heliosphere. These neutrals are observed through solar backscattered emission (H° and He^o), as PUIs formed by interstellar neutrals ionized inside of the heliosphere and subsequently convected outwards by the solar wind, and as ACRs formed from accelerated PUIs (Sect. 3.1). Interstellar He^o inside of the heliosphere provides the benchmark data on the velocity and temperature of the CHISM.

2.1.1 Helium

Neutral interstellar helium flows freely through the outer heliosphere, with $\leq 2\%$ filtration through charge exchange with H^+, and is ionized by photoionization and electron ionization inside of the Earth's orbit (e.g. see Cummings et al. 2002; Müller et al. 2004, for He^o filtration factors). Our knowledge of interstellar He^o inside the heliosphere is summarized from the present-day viewpoint by Möbius et al. (2004), and the classical viewpoint by Fahr (1974). Early observations of H° and He^o inside of the heliosphere found ratios $H^\circ/He^o \sim$ 6–7 (Ajello et al. 1987; Chassefiere et al. 1986) in contrast to the cosmic value $H/He = 10$. The relatively high heliospheric abundance of He is now known to be due partly to the high H° loss rate from charge exchange in the outer heliosphere, counteracted by the hard local interstellar radiation field that is more efficient at ionizing He^o than H° in the low-density surrounding cloud (Sect. 4.3). Photoionization and electron impact ionization dominate He^o loss mechanisms in the inner heliosphere. The trajectory of interstellar He^o inside of the heliosphere is unaffected by radiation pressure and charge exchange. Neutral interstellar He collects in a gravitational focusing cone extending ~ 8 AU downwind of the Sun (Sect. 2.2), which the Earth traverses early December each year (Fig. 3). Cone properties are derived self-consistently from He pickup ions and He^o backscattered 584 Å emission, for some combination of photoionization and electron impact ionization rates (Möbius et al. 2004, Sect. 2.2).

Interstellar helium has been observed in the heliosphere in four different ways, through fluorescence of solar 584 Å emissions (Weller and Meier 1974; Vallerga et al. 2004), direct particle counting by the GAS detector on Ulysses (Witte 2004), as PUIs formed from ionized interstellar He convected radially outwards by the solar wind (Sect. 2.2, Möbius et al. 1985; Geiss et al. 1995; Gloeckler et al. 2000), and as ACRs generated from accelerated PUIs (Sect. 2.2, e.g. Cummings et al. 2002). Interstellar helium is primarily neutral at the Earth, becoming ionized by photoionization and electron impact ionization within ~ 0.5 AU of the Sun. Möbius et al. (2004) have compared the He results from pickup ions, extreme ultraviolet (EUV) fluorescence, and direct particle detections, to obtain weighted mean values for

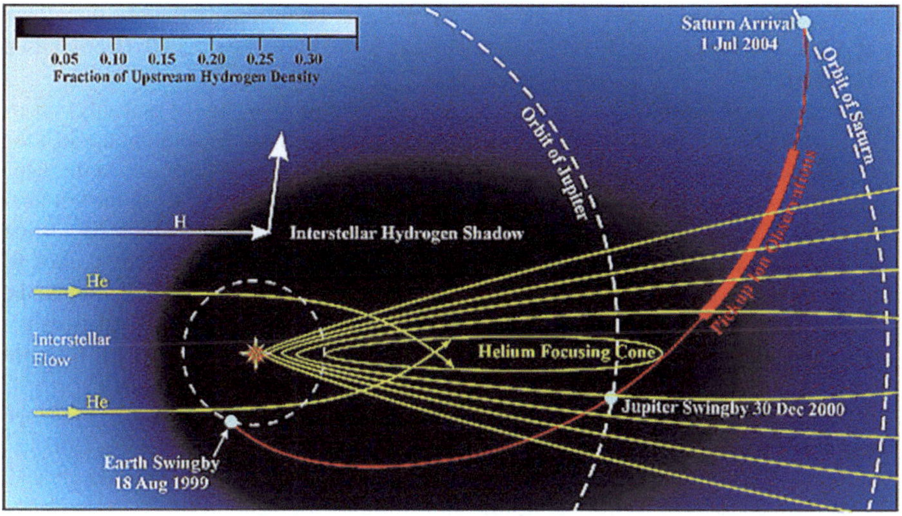

Fig. 3 Schematic diagram of the He focusing cone and the region of the Cassini trajectory between 6 and 8 AU where observations of PUIs were made. The interstellar hydrogen shadow, caused by radiation pressure on H° and photoionization, is also shown. (The figure is from McComas et al. 2004b)

interstellar He° at the termination shock. Since He° passes almost unimpeded through the outer heliosphere, these data give a CHISM velocity of -26.3 ± 0.5 km s^{-1}, a temperature of 6306 ± 390 K, and He° density $n(\mathrm{He}^\circ) = 0.0148 \pm 0.0020$ cm^{-3} (Möbius et al. 2004). The downwind direction for interstellar He° determined from Ulysses data and given in B1950 coordinates by Witte (2004) must then be corrected to J2000 coordinates for most purposes, yielding the direction $\lambda = 75.4^\circ \pm 0.5^\circ$ and $\beta = -5.1^\circ \pm 0.2^\circ$ (J2000, Witte, private communication, IBEX memo of August 6 2007). The early observations of the flow of interstellar H° and He° through the heliosphere found that the upwind directions of interstellar H° and He° directions are offset by $\sim 15^\circ$ (Weller and Meier 1974). The more precise He° and H° (Lallement et al. 2005) data now available indicate an offset angle of $\sim 4.9^\circ \pm 1.0^\circ$ (Frisch 2007, 2008a).

The CHISM parameters traced by He° define important details of the heliosphere interaction with the ISMa. For example, the sonic velocity for a perfect gas with solar composition at ∼6300 K is ∼8.2 km s^{-1}, while the Alfven velocity for a medium with $n(\mathrm{e}) = 0.06$ cm^{-3} and a magnetic field strength of 2.7 μG is 24.0 km s^{-1} (Sects. 2.1.2, 4.3, 3.4, Spitzer 1978). The relative motion between the ISMa and the Sun as indicated by the He° data (26.3 km s^{-1}) indicates that the heliosphere bow shock would be supersonic with Mach number $M_\mathrm{C} \sim 3$, if only acoustic velocities are considered. If Alfvenic wave propagation is also considered, the interaction would be barely super-magnetosonic with $M_\mathrm{A} \sim 1$ if the ISMF direction is approximately perpendicular to the shock normal (e.g. a perpendicular shock).

2.1.2 Hydrogen

Contrary to helium, neutral interstellar hydrogen is strongly coupled to protons in the interstellar gas by resonant charge exchange and thus is modified within the heliospheric interface. The basic understanding of the related heliosphere–LIC interaction has been summarized in early reviews (Axford 1972; Fahr 1974; Holzer 1989; Thomas 1978), as well as contemporary discussions (Fahr 1991; Baranov and Malama 1993; Ruciński et al. 1993; Zank et al. 1996; Izmodenov et al. 1999a, 2004; Fahr et al. 2000; Müller et al. 2000; Heerikhuisen et al. 2006).

On meeting the obstacle created by the expanding solar wind, the (most probably) supersonic interstellar plasma adapts to the changed flow conditions by decelerating through the heliospheric bow shock to a subsonic speed, and flowing along the heliopause. At the same time, the expanding solar wind plasma becomes compressed and heated up at the termination shock (TS), where $\sim$75% of the energy transfer goes into the pickup ion population (Richardson et al. 2008). Since the charge-exchange coupling length between the protons and neutral atoms in the CHISM is up to a few hundred AU, depending on proximity to the heliopause, i.e. on the order of the size of the entire heliosphere, the neutral gas does not adapt immediately to the changed flow conditions, and $\geq$50% of the neutral gas becomes kinematically decoupled from the charged component of interstellar gas.

The charge-exchange interaction does not stop, however, and since the charge exchange does not involve significant momentum exchange between the individual colliding particles, two new populations of particles between the bow shock and the heliopause are created. One of them consists of the former neutral atoms which lost their electrons due to charge exchange and became protons. Since they are subjected to electromagnetic interactions with the locally-thermalized ambient plasma, they quickly become incorporated and thermalized. At the same time, they are abundant enough to significantly affect the local thermodynamic parameters of the plasma, which becomes denser, faster, and cooler than in the absence of these new particles. The net pressure exerted on the plasma due to the charge exchange pushes the heliopause closer towards the Sun. Other important factors influencing the heliosphere configuration include the ISMF and the solar wind (e.g. Pogorelov et al. 2007; Opher et al. 2008).

The other charge-exchange products are the former protons from the outer heliosheath that acquired electrons from the flow of interstellar H° atoms and became decoupled from the local electromagnetic environment, while maintaining the velocities they had at the moment of interaction. Thus a new population of neutral atoms is created, the so-called secondary population. Since these neutral atoms inherit the local kinematic parameters of the plasma, they are much warmer than the primary interstellar atoms, but their bulk velocity is much lower than the velocity of the primary population. Since they are not tied to the ambient magnetic field, they can travel freely at large distances from their birthplaces without a change in their kinematic parameters. As a result, the local temperature and bulk velocity of the neutral component are different from the local temperature and bulk velocity of the ambient plasma. In particular, the secondary population becomes overdense, heated up, and slowed down as compared with the primary population. Thus, it forms a structure referred to as the "hydrogen wall", or alternatively as the "Baranov wall" (Baranov et al. 1991). The signature of this feature on interstellar Lyα absorption features is seen towards many stars (Sect. 3.2).

The original population of interstellar atoms is attenuated during the passage through the outer heliosheath, and since the charge-exchange process is more effective for the slow wing of this population, the primary population is accelerated by a few km s^{-1} and cooled by a few hundred K. Thus the velocity of the primary population at the nose of the TS will be greater than the bulk velocity of interstellar helium, and the temperature will be a little lower than the helium temperature. Details depend on the parameters of the LIC.

Contrary to interstellar protons, neutral atoms from both the primary interstellar population and the secondary population created by charge exchange in the outer heliosheath penetrate freely inside the heliopause, to the inner heliosheath region, where the solar wind is heated up and decelerated to a subsonic speed by transition through the TS. The solar wind density decreases as $\sim R^{-2}$ inside of the TS, where R is the distance to the Sun, so that the plasma density in this region is on the order of 10^{-4} cm^{-3} and lower by at least 3 orders of magnitude than the plasma density in the hydrogen wall. Thus the efficiency of charge exchange is reduced dramatically as compared with the efficiency in front of the heliopause, and the two populations flow through the inner heliosheath practically without a change. The few charge-exchange events occurring in this region result in the creation of another population of heliospheric neutral atoms, whose energy is equal to the typical energy of protons in the outer heliosheath, of about 0.5 keV. The distribution function of this population is strongly non-Maxwellian (Baranov et al. 1998; Izmodenov 2001; Izmodenov et al. 2001). This population, however tenuous it is, is important for the reasons of diagnostics of both heliospheric heliosheath and of the inner sheaths of astrospheres, because it can be observed spectroscopically (Linsky and Wood 1996; Wood et al. 2007). It is also the main target of observations by IBEX (McComas et al. 2004a, 2005, 2006). The atoms created by charge exchange inherit the local parameters of the plasma; owing to their relatively large speed they are able to penetrate close to the Sun with few losses, carrying the information on the conditions in their birth places.

The two thermal populations of neutral interstellar hydrogen at the nose of the TS, the primary and secondary, can be approximated (admittedly, less than perfect) by two Maxwellian functions shifted in velocity space (e.g. Baranov et al. 1998; Izmodenov 2000; Quémerais and Izmodenov 2002). Recent state-of-the-art models of the heliosphere tend to agree (Müller et al. 2008) that the combined density of the primary and secondary populations at this point is approximately two-fold lower than the neutral gas density in the CHISM (i.e., that the filtration factor of the hydrogen gas is equal to about 50%). The breakdown between the primary and secondary populations and exact values of their densities, bulk velocities, and temperatures depend on the conditions in the unperturbed CHISM. For example, in a recent report on the determination of the density of neutral interstellar hydrogen at the TS, Bzowski et al. (2008) adopted the following parameters in the CHISM: the upwind direction $\lambda_B = 254.68°$, $\phi_B = 5.31°$ in the B1950.0 ecliptic coordinates (Witte 2004; Möbius et al. 2004); bulk velocity $V_B = 26.3$ km s^{-1}; temperature $T_B = 6400$ K; density of neutral He equal to 0.015 cm^{-3}; and based on the He ionization degree in the LIC inferred by Wolff et al. (1999) on the level of ~30–40%, the density of He^+ equal to 0.008 cm^{-3}, in agreement with He ionization levels predicted by radiative transfer models of the CHISM (Sect. 4.3). With these parameters, they additionally modeled two cases of proton and H atom densities in the CHISM, and obtained the following modifications of parameters of the primary and secondary populations at the nose of the TS:

1. Input CHISM values: proton density $n_p = 0.06$ cm^{-3}; neutral gas density $n_H = 0.18$ cm^{-3}.
   - TS Primary:
     $n_{TS,pri} = 0.19\, n_{H,pri} = 0.035$ cm^{-3}; $V_{TS,pri} = 1.08\, V_B = 28.5$ km s^{-1}; $T_{TS,pri} = 6020$ K
   - TS Secondary:
     $n_{TS,sec} = 0.33\, n_H = 0.060$ cm^{-3}; $V_{TS,sec} = 0.71\, V_B = 18.7$ km s^{-1}; $T_{TS,sec} = 16\,300$ K
   - Hence the resulting net density of interstellar H at the TS:
     $n_{H,TS} = 0.53\, n_H = 0.095$ cm^{-3}.
2. Input CHISM values: proton density $n_p = 0.032$ cm^{-3}; neutral gas density $n_H = 0.2$ cm^{-3}.
   - TS Primary:
     $n_{TS,pri} = 0.29\, n_H = 0.059$ cm^{-3}; $V_{TS,pri} = 1.07\, V_B = 28.2$ km s^{-1}; $T_{TS,pri} = 6\,100$ K

- TS Secondary:
  $n_{\mathrm{TS,sec}} = 0.29\,n_{\mathrm{H}} = 0.059\ \mathrm{cm}^{-3}$, $V_{\mathrm{TS,sec}} = 0.70\,V_{\mathrm{B}} = 18.5\ \mathrm{km\,s}^{-1}$, $T_{\mathrm{TS,sec}} = 16500$ K
- Hence the resulting net density of interstellar H at the TS:
  $n_{\mathrm{H,TS}} = 0.59\,n_{\mathrm{H}} = 0.117\ \mathrm{cm}^{-3}$.

The CHISM input parameters for Case 1 above give the best match to the CHISM properties predicted by radiative transfer models (Sect. 4.3).

Inside the TS, the atoms can be treated as collisionless. They propagate under the influence of the joint action of solar gravity and solar radiation pressure, and they suffer losses due to ionization by solar wind and solar EUV radiation. Ionization results in creation of pickup ions (Sect. 2.2) in the solar wind and of another population of neutral atoms, the so-called Neutral Solar Wind, which is composed of solar wind protons that acquired electrons from the incoming interstellar gas and propagate away from the Sun while maintaining the original kinematic parameters of the parent protons.

The radiation pressure generally (over)compensates solar gravity. Its strength depends on the net intensity in the solar Lyα line and—since the line features a non-flat, self-reversed shape—on the instantaneous radial velocity of the atom. Since the solar EUV output varies with time on timescales from weeks (solar rotation period) to decades (the solar cycle period), the effective compensation of solar gravity varies from a factor of 0.9 at solar minimum to a factor of 1.6 at solar maximum (Bzowski 2001), the trajectories of the atoms are not Keplerian, and both density and bulk velocity of the atoms change with time (Ruciński and Bzowski 1995; Bzowski et al. 1997). Another factor modifying the local density and bulk velocity of the atomic ensemble is ionization, which differentiates the members of the ensemble by energy, yielding an apparent acceleration of the flow by several $\mathrm{km\,s}^{-1}$ within a few AU from the Sun, as well as a deflection of the local flow direction relative to the original interstellar flow vector. Ionization also varies on time scales from days to decades and additionally features a latitudinal anisotropy that depends on solar cycle phase and is especially pronounced during solar minimum period. The attenuation of density at the upwind axis within a few AU from the Sun is approximately exponential.

The photons of the solar Lyα radiation are scattered off the neutral H atoms and produce the heliospheric glow. The maximum of the signal when observed from 1 AU originates at a few AU (typically, when looking upwind, from 2–5 AU, depending on the phase of solar cycle), i.e. from the region where the distribution function of the gas is quite complex.

The net density of the gas at a point R on the inflow axis is composed of the densities of the two collisionless components, the primary and secondary atoms, that have different bulk velocities and temperatures, and which additionally vary with solar cycle phase and distance from the Sun. Their thermal spread along the inflow axis decreases between $\sim$ 10 AU and 1 AU from the original values at the TS, determined by the temperatures; the decrease is 3 $\mathrm{km\,s}^{-1}$ at solar minimum or 2 $\mathrm{km\,s}^{-1}$ at solar maximum. Simultaneously, the bulk velocity of the two components increases by 4 to 6 $\mathrm{km\,s}^{-1}$ except for the primary population during solar maximum, which is slowed down by 2 $\mathrm{km\,s}^{-1}$. This is illustrated in Fig. 4. Both populations show appreciable radial velocity gradients at the region of space where most of the spectral line of the heliospheric Lyα glow is created, which also varies during the solar cycle. In addition, the densities of these populations also feature a strong gradient, illustrated in Fig. 5. The Lyα backscattered line created by the primary and secondary H° populations during solar minimum conditions is shown in Quémerais and Izmodenov (2002), and the expected profile as a function of solar cycle phase is given in Quémerais et al. (2006). The difference in bulk velocities of the two populations in the region where the

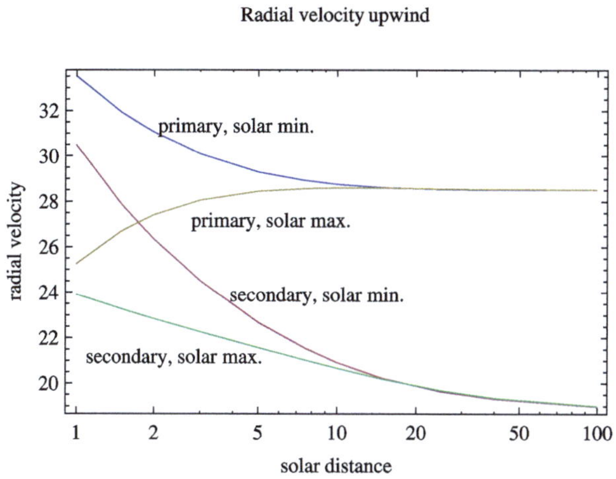

Fig. 4 Bulk velocity of the primary and secondary populations of neutral interstellar H for solar minimum and maximum conditions along the inflow axis, starting with initial conditions at the termination shock (figure right). Units of the horizontal axis are AU, the units of the vertical axis are $\mathrm{km\,s^{-1}}$. The profiles were calculated using the Warsaw time-dependent kinetic model of heliospheric hydrogen (Bzowski et al. 1997) with time-dependent radiation pressure and ionization rate, and assuming the boundary conditions at the termination shock according to case 2, which seem to be close to actual parameters in this region based on analysis of Ulysses pickup ion measurements by Bzowski et al. (2008). The radiation pressure for 1996.0, adopted as the solar minimum case, was equal to $\mu = 0.9$; for 1999.5, adopted as the solar maximum case, it was equal to $\mu = 1.4$

Fig. 5 Density of the primary and secondary populations of neutral interstellar H for solar minimum and maximum conditions along the inflow axis. The horizontal axis is scaled in AU, the vertical axis in $\mathrm{cm^{-3}}$. The profiles were calculated using identical assumptions as in the case of velocities shown in Fig. 4

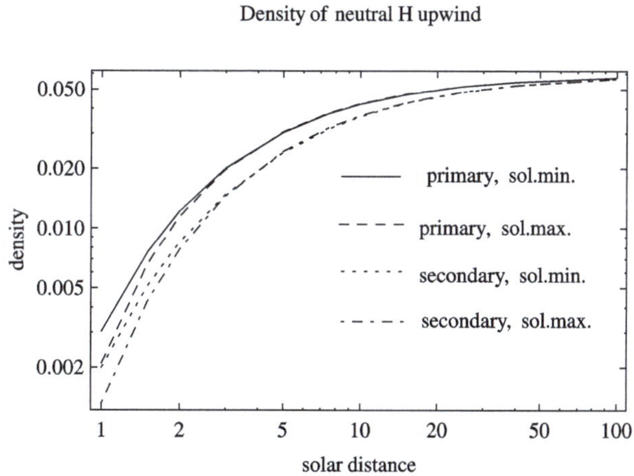

backscattered Lyα line is formed is approximately equal to half of the thermal spread. Thus the two populations overlap in the velocity space and create a spectral line with a single peak.

2.2 Pickup Ions and Anomalous Cosmic Rays

Cosmic ray data from the early 1970s reported particles at unexpected energies of $\sim$10 MeV/nucleon, showing anomalous compositions with enhanced He, H, O, N, Ne, and Ar abundances (e.g. Gloeckler and Wenzel 2001). Fisk et al. (1974) recognized that this population consisted of elements with high first ionization potentials (FIP) that were expected to be neutral in the ISMa. Neutral interstellar gas flows into the heliosphere, and is ionized through charge exchange with the solar wind (for interstellar H, N, O, Ne, and Ar), photoionization (for interstellar He, H, and Ne), or electron impact ionization (for interstellar He, N, and Ar, Ruciński et al. 1996; Cummings et al. 2002). The ionized interstellar atoms are picked up by the solar wind and convected outwards to the outer heliosphere, forming a suprathermal ion population known as pickup ions (PUIs). The PUIs then become accelerated to form the anomalous cosmic rays (ACRs, Sect. 3.1). Pickup He was discovered by Möbius et al. (1985). Pickup ions of H, ^{4}He and ^{3}He, N, O, Ar, and Ne have now been measured in the inner 5 AU by several different spacecraft, including Ulysses data obtained at all ecliptic latitudes (Möbius et al. 1985; Gloeckler and Fisk 2007; Gloeckler et al. 2004; Gloeckler and Geiss 2004). The observed PUI population consists of ions with an interstellar origin, that are mixed with an inner source of PUIs that peaks near the Sun and consists of low-FIP elements (Geiss et al. 1995; Schwadron and Gloeckler 2007). The inner source ions appear to originate from the recycling of solar wind on dust grains in the inner heliosphere, and thereby trace the solar composition. The unique signature of PUIs with an interstellar origin is the energy cutoff at twice the solar wind velocity, caused as the PUI energies are isotropized in the solar wind rest frame (e.g. Gloeckler and Fisk 2007).

The low charge exchange and similar photoionization cross sections of He and Ne (Ruciński et al. 1996) indicate that both interstellar elements reach the inner heliosphere as neutrals, and are gravitationally focused downwind of the Sun into the He focusing cone (Fig. 3, Möbius et al. 2004; McComas et al. 2004b). The He focusing cone is seen in 584 Å backscattered emission, as well as enhanced densities of He PUIs (Möbius et al. 2004; Gloeckler et al. 2004). Since the He focusing cone is formed by interstellar He$^{\circ}$ atoms arriving from all ecliptic latitudes, and ionization rates vary with ecliptic latitude β, solar activity proxies are required to reconstruct the He$^{\circ}$ ionization rate from the EUV radiation field over the full solar surface. The phase space density of the PUIs is proportional to the product of the local neutral density and the ionization rate, and can be used to estimate the maximum density of He$^{\circ}$ in the focusing cone. The He$^{\circ}$ density in the focusing cone varies over the solar cycle, from $\sim$0.01 cm^{-3} during solar maximum to $\sim$0.045 cm^{-3} during solar minimum, compared to the interstellar value of 0.015 cm^{-3} (Sect. 2.1.1). The higher focusing cone density during solar minimum conditions follows because the ionization rate varies approximately by a factor of 3 over the solar cycle, with higher rates during solar maximum because of increased solar EUV emission. During solar minimum, the interstellar upwind He density at 1 AU is 80% to 85% of the CHISM density ($\sim$0.015 cm^{-3}), while during solar maximum it is $\approx e^{-1}$ of the CHISM He density ($\sim$0.005 cm^{-3}). The result is that the product of density and ionization rate at 1 AU is approximately constant over the solar cycle to within 20–25% in the upwind direction. In the peak of the cone the He PUI phase space density is between a factor 2 (solar maximum) and a factor of almost 4 (solar minimum) higher than in the upwind direction defined by the interstellar flow.

In situ data from Cassini found H, He, and O PUIs out to distances of 6 to 8 AU downwind of the Sun (McComas et al. 2004b). The He data showed that the focusing cone extends beyond 8 AU downwind. The higher He PUI density in the focusing cone was associated with a decrease in H PUIs in the downwind hydrogen shadow caused by radiation pressure on the inflowing interstellar H. These *in situ* observations of the hydrogen shadow confirm the asymmetric distribution of H$^{\circ}$ in the inner heliosphere inferred from remote observations of the interplanetary Lyα glow (Sect. 2.1.2). The Cassini trajectory and a simulation of the He focusing cone are shown in Fig. 3.

The H, He, O, N, Ne, and Ar PUI data provide valuable constraints on the neutrality as well as the chemical composition of the CHISM (Sect. 4.3). At low energies, ≤ 20 MeV/nucleon, the isotopic composition of PUIs and ACRs show abundance ratios of $^{15}N/^{14}N$, $^{18}O/^{16}O$, and $^{22}Ne/^{20}Ne$, that are near the solar values of ~ 0.004, 0.002, and 0.07, respectively (Gloeckler and Wenzel 2001; Cummings and Stone 2007). Helium isotopes show $^4He/^3He \sim 1.7 \times 10^{-4}$, which is similar to meteoritic and HII region values (Salerno et al. 2003; Gloeckler and Fisk 2007). These data suggest the CHISM composition is solar, independent of uncertainties in actual solar abundances.

2.3 Large Interstellar Dust Grains in the Heliosphere

In the early 1990s, after its Jupiter flyby, the Ulysses spacecraft positively identified interstellar dust in the solar system penetrating deep into the solar system. The Ulysses spacecraft detected impacts predominantly from a direction that was opposite to the expected impact direction of interplanetary dust grains. It was found that, on average, the impact velocities exceeded the local solar system escape velocity (Grün et al. 1994). The motion of the interstellar grains through the solar system was parallel to the flow of neutral interstellar hydrogen and helium gas, both traveling at a speed of 26 $\mathrm{km\,s^{-1}}$ (Baguhl et al. 1995). The interstellar dust flow persisted at higher latitudes above the ecliptic plane, even over the poles of the Sun, whereas interplanetary dust is strongly depleted away from the ecliptic plane (Grün et al. 1997).

Ulysses has measured the interstellar dust stream at high ecliptic latitudes between 3 and 5 AU. Interstellar grains have also been observed with the in-situ dust detectors on board Cassini, Galileo and Helios (Altobelli et al. 2003, 2005, 2006), covering a heliocentric distance range between 0.3 and 3 AU in the ecliptic plane. The masses of clearly identified interstellar grains range from 10^{-18} kg to about 10^{-13} kg with a maximum at about 10^{-15} kg (Landgraf et al. 2000). In Fig. 6 the total interstellar dust mass distribution of 896 ISDGs observed by Ulysses, to date, is plotted, and compared to the power law size distribution of grain radii, $n(a) \sim a^{-3.5}$, that provides a nominal match to extinction data given abundance constraints (Mathis et al. 1977, e.g. MRN distribution). Ulysses has recorded significant differences in the particle sizes at different heliocentric distances. The measurements reveal a

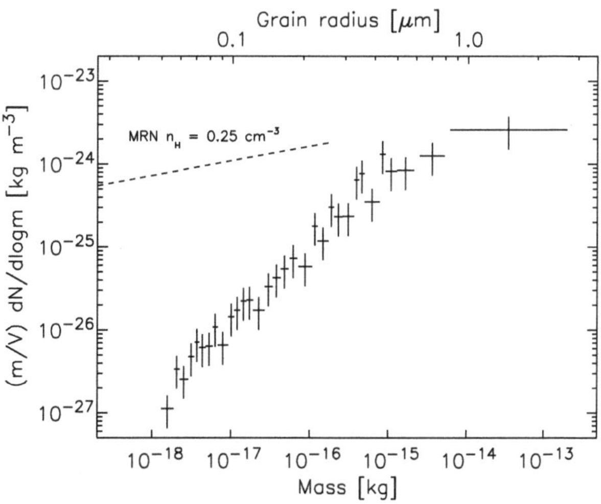

Fig. 6 The mass distribution of 896 interstellar dust grains detected by the Ulysses spacecraft inside the heliosphere (*crosses*) is shown as a function of grain mass. The corresponding grain radius is shown for the assumption of spherical silicate grains of density 2.5 $\mathrm{g\,cm^{-3}}$ (*top axis*). For comparison, an MRN interstellar dust grain distribution (Mathis et al. 1977) for an assumed total interstellar mass density of 0.25 $\mathrm{cm^{-3}}$ hydrogen atoms in the CHISM is plotted (*dashed line*)

lack of small (<3 × 10^{-16} kg) interstellar grains inside 3 AU heliocentric distance (Landgraf et al. 2000). Measurements by Cassini and Galileo in the distance range between 0.7 and 3 AU showed that interstellar particles are bigger than 10^{-15} kg, with increasing masses closer to the Sun (Altobelli et al. 2003, 2005). The trend of increasing particle masses close to the Sun was also seen by Helios, which recorded particles of ca. 10^{-14} kg mass down to 0.3 AU (Altobelli et al. 2006). These data are consistent with an interstellar dust stream filtered by solar radiation pressure. Interstellar particles with optical properties of astronomical silicates or organic refractory materials are consistent with the observed radiation pressure effects (Landgraf et al. 1999), however the overabundance of carbon in the CHISM (Sect. 4.3) appears to rule out a significant proportion of organics. The heliospheric trajectories of the largest ISDGs with radii ≥7 μm are determined by solar gravity. These large grains are predicted to form a "focusing cone" downwind of the Sun (Grogan et al. 1996; Landgraf 2000).

Ulysses has monitored the interstellar dust flow through the solar system for more than 16 years. This time period covers more than two and a half revolutions of the spacecraft about the Sun through more than 2/3 of a complete 22-year solar cycle. Thus, Ulysses measured interstellar dust during solar minimum and solar maximum conditions of the interplanetary magnetic field (IMF). In mid 1996, a decrease by a factor 3 of the interstellar dust flux density from initial values of 1.5×10^{-4} m^{-2} s^{-1} was observed. This decrease was attributed to the increased filtering of small grains by the solar wind magnetic field during solar minimum conditions (Landgraf 2000). Since early 2000, Ulysses has detected interstellar dust flux levels above 10^{-4} m^{-2} s^{-1} again (Landgraf et al. 2003). Until early 2005 the upwind direction of the interstellar grains was within a few tens of degrees in agreement with the interstellar helium flow direction (Landgraf and Grün 1998; Frisch et al. 1999; Krüger et al. 2006). During 2000 and 2006 Ulysses was traveling through almost the same spatial regions and with almost identical detection geometries for interstellar grains; however the upwind direction of the grains was somewhat wider and shifted by about 30 degrees away from the helium flow direction towards southern ecliptic latitudes during the latter set of observations (Krüger et al. 2007).

The flux variations and the deficiency of small grain masses (< 10^{-16} kg) indicate a time dependent depletion of small interstellar grains inside the heliosphere. Several groups have studied the alteration of the interstellar dust flux inside the heliosphere by solar radiation pressure force, gravitational focusing and the interactions with the time varying interplanetary magnetic field (Grogan et al. 1996; Landgraf et al. 1999; Landgraf 2000, 2003; Mann and Kimura 2000; Czechowski and Mann 2003b; Linde and Gombosi 2000).

3 The Edges of the Heliosphere: Interstellar Material and the Interstellar Magnetic Field

The outermost regions of the heliosphere that are closest to the ISMa are the least explored, but show visible evidence of the interaction between the ISMa and heliosphere. The diverse evidence for ISMa in the outer heliosphere regions include anomalous cosmic rays, decelerated interstellar hydrogen in the hydrogen wall, and possibly small excluded interstellar dust grains signaled by their absence inside of the heliosphere and polarization of light from nearby stars.

3.1 Anomalous Cosmic Rays and Heliospheric Asymmetries

Observations show that V2 crossed the TS much closer to the Sun (∼84 AU, Richardson et al. 2008) than V1 (∼94 AU, Stone et al. 2005). While the heliospheric boundaries are

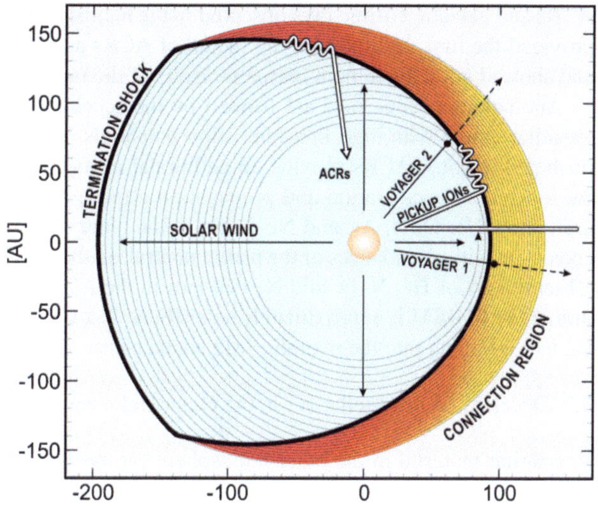

Fig. 7 Schematic diagram of an equatorial cut through the termination shock taken from McComas and Schwadron (2006). The Archimedean spiral flux tubes first cross the termination shock near the nose and their connection points move back along the flanks of the shock as they continue to propagate out through the heliosheath. The long connection times for the flanks and tail can allow diffusive shock acceleration to energize particles up to observed ACR energies

not time stationary, this large a difference almost certainly reflects at least some consistent north-south asymmetry of the TS. The ISMF at the heliosphere (Sect. 3.4) likely produces significant asymmetrical distortions of the TS and heliopause (HP, Linde et al. 1998; Ratkiewicz et al. 1998; Pogorelov and Matsuda 1998; Pogorelov et al. 2004) and is the most likely cause of this observed asymmetry.

Magnetohydrodynamic (MHD) models of the heliosphere that include an ISMF help quantify its effects on the global heliosphere. Opher et al. (2007) showed that a field tilted 60° from the galactic plane matches both the flow data and the radio emission data. This tilt creates a large north-south and some east-west asymmetries. This model also predicts that the distance of the TS in the direction of V2 is 8–11 AU closer than in the direction of V1, consistent with actual crossing distances. The same model further suggests that the HP distance may be 13–24 AU closer in the direction of V2. Pogorelov et al. (2008b) argue that the strength of the interstellar field can be increased to strengths of $\sim$4 μG to reproduce the heliospheric asymmetry seen by the V1 and V2 (also see Pogorelov et al. 2007). However, once the effects of neutrals are included, these asymmetries become much smaller and no longer match the large difference seen in TS crossing distances between V1 and V2 (Pogorelov et al. 2006; Heerikhuisen et al. 2008). An additional complication in the interpretation of the heliosphere asymmetry seen by V1 and V2 is that the solar wind dynamic pressure decreased significantly between the two crossings.

MHD models also reveal strong nose-to-tail asymmetries in the TS geometry (Zank 1999), as shown in Fig. 7. McComas and Schwadron (2006) showed that solar wind magnetic field lines will first make contact at the nose of the TS, because the location of the nose is much closer to the Sun than anywhere else along the shock. Diffusive acceleration of ions in the compressed magnetic field beyond the TS requires large amounts of time, typically on the order of a year, to achieve Anomalous Cosmic Ray (ACR) energies. Over this large timescale, the connection point between solar wind field lines and the TS moves from the nose to the flanks and tail. ACR acceleration initiates in this connection region. Therefore, the ACR energies should reach their highest intensities back on the flanks and tail of the TS. Thus, the absence of enhanced of ACR fluxes noted by V1 and V2 in the TS regions can be explained naturally by the large nose-to-tail asymmetry of the TS. Schwadron and McComas (2007) showed that this asymmetry also likely explains the strong correlation between

ACRs and galactic cosmic rays observed in the heliosheath. Finally, Schwadron et al. (2008) provided the first detailed analytic model of ACRs accelerated along a non-symmetric TS, and showed good agreement between results of the model and ACR observations by V1.

Anomalous cosmic rays are formed by the acceleration of PUIs, which in turn are interstellar atoms with high first ionization potentials, that survived passage through the heliosheath regions. ACRs provide an additional measurement of the ISMa, after correcting for ionization, acceleration and propagation effects. Measurements by the Voyager spacecraft show He, O, N, Ar, and Ne ACRs in the outer heliosphere, with abundances that are consistent with abundances of the parent neutrals in the CHISM (e.g. Cummings et al. 2002). Observations of He, N, O, and Ne isotopes in the ACR population show that the composition of the CHISM is approximately solar (Sect. 2.2, Cummings and Stone 2007), providing valuable insights into the overall ISMa abundances.

3.2 Decelerated Interstellar H° in the Outer Heliosheath Region

Interstellar material in the outer heliosphere can be studied directly using ultraviolet (UV) Lyα spectra of nearby stars taken by the HST. As it enters the heliosphere, neutral H from the CHISM is heated and decelerated through charge exchange processes in the hydrogen wall (Sect. 2.1.2). This heliospheric hydrogen produces a Lyα absorption signature that has been detected in some HST spectra (Linsky and Wood 1996; Wood et al. 2005). The heliospheric contribution to the hydrogen wall can be completely obscured by the interstellar absorption if the ISMa column density is too high and the ISMa absorption therefore too broad.

There are actually two distinct populations of interstellar hydrogen that are detectable. The dominant population consists of neutrals formed by charge exchange in the hydrogen wall between the heliopause and bow shock (Sect. 2.1.2). Most detections of heliospheric absorption sample this population, and originate near the upwind direction of the CHISM flow, because this is where the heating, compression, and deceleration of the ISMa neutrals is strongest. Deceleration of the H° flow in the radial direction is required to separate the heliospheric absorption from that of the ISMa (Wood et al. 2005). Several studies have demonstrated that hydrodynamic models of the heliosphere can reproduce the observed kinematic characteristics of the hydrogen wall absorption, depending on exactly what is assumed about the properties of the CHISM (Gayley et al. 1997; Wood et al. 2000a, 2007; Izmodenov et al. 2002).

Besides the dominant hydrogen wall component, a second population of heliospheric hydrogen is a heliosheath component formed by charge exchange in the inner heliosheath between the heliopause and termination shock (Sect. 2.1.2). The very broad, weak Lyα absorption that this population produces is much harder to detect than the stronger hydrogen wall signature. In upwind and sidewind directions, the inner heliosheath is too narrow for there to be detectable heliosheath absorption; in downwind directions, line-of-sight distances through this region are longer and the absorption is detectable through its kinematic properties (Izmodenov et al. 1999b; Wood et al. 2007).

Although harder to detect and study, it is the heliosheath population of neutrals that is most relevant to IBEX, because most of the ENAs that IBEX will study are from this population rather than the hydrogen wall population. While IBEX will make detailed all-sky maps of local ENA fluxes, and detailed measurements of their energy distributions from $\sim$10 eV to 6 keV, the heliosheath absorption measurements provide information about line-of-sight integrated energy distributions, particularly in the 0.01–0.2 keV energy range. Wood et al. (2007) have used constraints from the Lyα data to quote a very conservative flux upper limit of $F_{0.01-0.2\,\text{keV}} < 5000\ \text{cm}^{-2}\,\text{s}^{-1}\,\text{ster}^{-1}$ for the ENA flux that IBEX can expect to see from 0.01–0.2 keV emissions.

More precise constraints on ENA fluxes from the Lyα data (and future direct comparisons of IBEX and HST measurements) will require the assistance of kinetic heliospheric models that not only properly model the neutral velocity distributions throughout the heliosphere (Baranov and Malama 1993; Baranov and Malama 1995; Zank 1999; Izmodenov et al. 2005; Heerikhuisen et al. 2007; Pogorelov et al. 2007), but also extend at least 3000 AU downwind in order to be able to capture all the absorption in the downwind directions where the heliosheath absorption is detectable (Izmodenov and Alexashov 2003; Alexashov et al. 2004; Wood et al. 2008). These models should ideally use a sophisticated multicomponent treatment of the plasma in the heliosphere instead of the usual simple single-fluid treatment, as multicomponent models have been found to result in significantly different heliosheath neutral populations, and better reproduction of heliosheath absorption (Malama et al. 2006; Wood et al. 2007). In any case, once IBEX data are available it will be very interesting to see if models are capable of simultaneously reproducing both the local ENA spectra seen by IBEX and the Lyα absorption data.

3.3 Small Excluded Interstellar Dust Grains

The observed mass distribution of interstellar dust grains inside of the heliosphere is controlled by the original interstellar charge-to-mass ratios, Q/m, of incident ISDGs, grain-charging in heliosheath regions, and the relative strengths of gravitational focusing, radiation pressure, and Lorentz forces inside of the heliosphere (Landgraf 2000). Interstellar grains with large Q/m ratios are excluded from the heliosphere, as shown by the deficit of small grains in the Ulysses and Galileo data (Sect. 2.3, Fig. 6, Frisch et al. 1999) with radii $a \lesssim 0.2$ µm in comparison to the MRN power-law distribution of $n(a) \sim a^{-3.5}$ (Mathis et al. 1977). The very smallest carbonaceous grains, either graphites or polycyclic-aromatic-hydrocarbons (PAH) with radii <0.01 µm, appear to be missing from the CHISM entirely because all carbon is in the gas phase (Sect. 4.3, Slavin and Frisch 2008). The exclusion of small interstellar dust grains from the heliosphere has been discussed in many papers (Frisch et al. 1999; Czechowski and Mann 2003a; Mann and Kimura 2000; Grün et al. 2005).

The mass distribution of the excluded grains in principle can be obtained by comparisons between the observed grain mass distributions, and the 'true' interstellar mass distribution. Unfortunately, the ISDG mass distribution for very low-column-density ISMa such as the LIC has not been determined. LIC grains differ from grains in high-opacity clouds, in that they show high abundances of refractories in the gas due to grain destruction in shocks (Sect. 4.3, Frisch et al. 1999).

In contrast to the interstellar gas and small ISDGs that encounter the heliosphere, large ISDGs are not directly affected by the MHD boundaries. This is because the coupling of the grains to the gas is mediated primarily by the magnetic field, with only the relatively weak coupling of collisional and plasma drag acting directly on the grains. In the frame of reference of the heliosphere, the grains flow in with the interstellar gas and magnetic field at the relative speed of $\sim$26 km s^{-1}. Once in the outer heliosheath, the most important force on the grains is the Lorentz force as the gas and (frozen-in) field is slowed and deflected at the nose of the heliopause.

The primary determinant of the dynamics of small grains in the outer heliosphere is Lorentz coupling to the magnetic field, which is mediated by the charge-to-mass ratio of grains, Q/m. The charging rates for grains are very high, so we can assume that they are in equilibrium generally speaking. The primary sources of grain charging are the far UV radiation field, which charges the grains positively through the ejection of photoelectrons, and

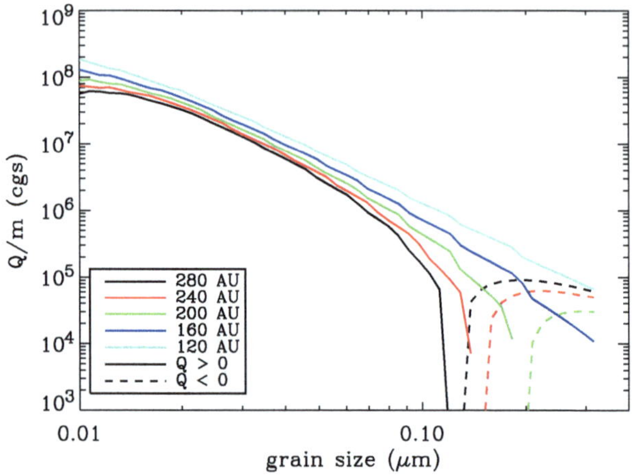

Fig. 8 Charge-to-mass ratio for silicate grains interacting with the heliosphere. As grains get closer to the Sun, solar UV radiation increases the charge of small grains through photoejection of electrons. The higher work function of larger grains impedes the photoejection of electrons, and larger collision rates of electrons drive the grain charge negative. Grains with positive and negative charges are denoted by *solid* and *dashed lines*, respectively. The distance of the grain from the Sun is denoted by the *line color*, with the most distant small grains having the least charge

the hitting and sticking of electrons on the grains. With these rates in balance, detailed calculations of the grain charging using rates and codes from Weingartner and Draine (2001), we find that the charge on a grain ranges from $Z = 14$ to 137 for grain radii ranging from $a_{\mathrm{gr}} = 0.01$–0.3 μm (Avinash et al. 2008). Grain charge increases as grains are exposed to higher fluxes of solar radiation (Fig. 8). The corresponding gyroradii depend on assumptions for the magnetic field and will vary as a function of position relative to the nose and within the heliosheath. For a 5 μG field just inside the bow shock and a relative velocity of 14 km s^{-1} between the inflowing gas and dust, we find gyroradii of 0.7–2000 AU for this range in grain sizes. Thus small grains with their relatively small gyroradii will couple tightly to the field and be diverted along with the field and gas, while the largest grains will pass through the heliopause relatively unaffected (Sect. 2.3).

3.4 Interstellar Magnetic Field at the Heliosphere

The Faraday rotation of local and extragalactic radio point sources show that the global ISMF in the interarm region surrounding the Orion spur (Fig. 2) is directed clockwise around the galactic center when viewed from the north galactic pole (Han 2006). A major distortion of the ordered global magnetic field occurs in the solar vicinity. This distortion is from expanding shock fronts generated by supernova and massive star winds, associated with star formation in the Scorpius-Ophiuchus Association over the past $\sim$ 15 Myrs. These shock fronts have displaced the global local field that is parallel to the galactic plane into a giant superbubble, and created the distinct Loop I feature that dominates the magnetic field in the northern hemisphere and at the Sun (Berkhuijsen 1971). Frisch (1981) concluded that the Loop I supernova remnant has expanded to the Sun, after comparing the kinetics of ISMa inside of the heliosphere and towards nearby stars, and evaluating refractory abundances indicative of dust grain destruction in shocks. Wolleben (2007) used polarizations traced by 1.4 GHz radio continuum data to define the magnetic field direction in the 'S1' subshell of Loop I, and concluded that the Sun is located in the rim of S1.

Four phenomena indicate the direction of the interstellar magnetic field (ISMF) at the heliosphere: (1) The offset angle between the upwind directions of H° and He° flowing into the heliosphere (Lallement et al. 2005). (2) The polarization of optical light from stars located 6–35 pc from the Sun (Tinbergen 1982; Frisch 2007). (3) The magnetic S1 subshell of the Loop I supernova remnant (Wolleben 2007; Frisch 2008b). (4) The 3 kHz emissions detected by the V1 and V2 spacecraft from beyond the heliopause (Gurnett et al. 2006). The first three of these phenomena can be characterized by a position angle. Position angles (PA) are a useful tool for describing a direction in the sky. By definition a PA is defined in the plane of the sky. The position angle PA_E in ecliptic coordinates is the angle between a vector and a great-circle meridian passing through the ecliptic north pole, and increasing to the east.

In this paragraph these four phenomena are summarized: (1) It is necessary to first convert the He° upwind direction to J2000 coordinates (this correction was inadvertently neglected in Lallement et al. 2005), in order to characterize the PA of the offset between the inflow directions of He° and H°. With this correction, the offset angle between the upwind directions of He° and H° is $4.9° \pm 1.0°$, and the PA defined by these two directions in the ecliptic coordinate system is $-38° \pm 12°$ (Frisch 2007). (2) The magnetic field direction in the plane of the sky, and with respect to the ecliptic plane, is given by $90°-PA_E$ for magnetically aligned ISDGs. The polarization of nearby stars towards the heliosphere nose yields $PA_E = -25° \pm 9°$. These two ISMF tracers then suggest that the ISMF is inclined to the ecliptic plane by $\sim 59° \pm 5°$, but do not provide any information on the angle between the cloud velocity and field direction. Regarding method (3) above, the ISMF direction near the Sun found from Faraday rotation data has large uncertainties because of the low densities. If the shell is assumed to be purely spherical, then the position angle of the magnetic field in the S1 shell at the Sun is $PA_E \sim -17°$ at the heliosphere nose (Frisch 2008b). This provides evidence that the ISMF in the solar vicinity does not have a purely spherical configuration.[1]

Method (4) above has interesting consequences that are not yet understood. The nature and extent of the TS's distortion depends on both the orientation and strength of the external ISMF. The sources of the heliospheric radio emissions are consistent either with a line roughly parallel to the galactic plane (Kurth and Gurnett 2003) with an average galactic latitude of 14.7°, or with an alternate line parallel to an ecliptic meridian with $\beta \sim 181°$ (Frisch 2007). The magnetic field orientation traced by the 3 kHz emissions has been interpreted in different ways. Recently, Gurnett et al. (2006) suggested that the radio emission should occur in regions where magnetic field lines are tangential to the surface of a shock, so that global merged interaction regions instigate particle beams that excite Langmuir waves in the surrounding ISMa plasma (Mitchell et al. 2005; Pogorelov et al. 2008a). If the primary location of the 3 kHz emissions are the correct locations for the origin of these events, this would indicate the ISMF is roughly perpendicular to the galactic plane. A meridinal distribution may instead be due to heliospheric asymmetries.

The direction of the ISMF at the heliosphere has been explored using the faint polarization of nearby stars, which peak in a direction offset along ecliptic longitude by $\Delta\lambda \sim +40°$ from the heliosphere nose (Frisch 2007), and the superbubble shell defined by the S1 radio continuum shell (Frisch 2008b; Wolleben 2007). An unconfirmed ENA peak detected in IMAGE data at ~1 keV (Collier et al. 2004; Wurz et al. 2004) is in the same direction of the optical polarization peak. This is shown in Fig. 9. The magnetic field directions traced by the polarizations and the S1 shell are consistent with the direction of angular offset between

[1]The sensitivity of the position angle of the He°-H° offset to the coordinate epoch mistake is illustrated by the value obtained, $PA_E = -32°$, by incorrectly comparing H° J2000 with He° B1950 coordinates.

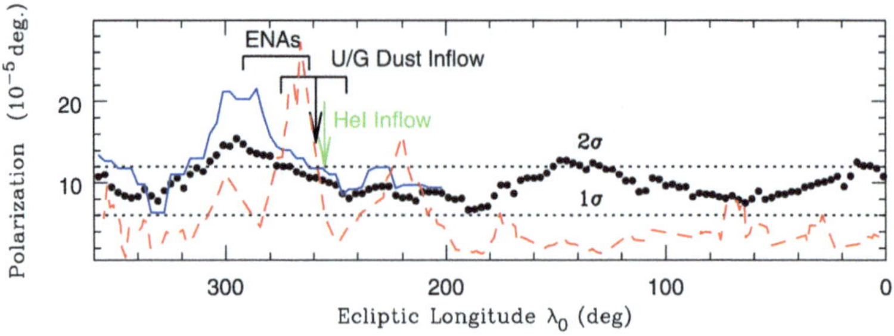

Fig. 9 Polarization properties as a function of ecliptic longitude: The *blue line* shows the averaged polarizations for stars with $|\beta| < 20°$, and the *dots* indicate averaged values for stars with $|\beta| < 50°$. The 1σ and 2σ polarization uncertainties are shown (from Tinbergen 1982). The polarization data for nearby stars are averaged over $\pm 20°$ in ecliptic longitude, λ. The direction of peak polarization is shifted by $\sim 40°$ from the He^{0} upwind direction. The *black arrow* shows the upstream direction, and 1σ uncertainties, for interstellar dust flowing through the heliosphere based on Ulysses and Galileo data (Frisch et al. 1999). The *green arrow* shows the upstream direction from observations of interstellar He^{0} in the inner heliosphere. A tentative upwind direction of an ENA flux from the outer heliosphere is shown (Collier et al. 2004; Wurz et al. 2004), although more recent data do not confirm these earlier conclusions. The *red dashed line* shows the polarizations, normalized by a factor of 20 and averaged over an interval of $\pm 5°$ around the central ecliptic longitude, of 184 stars 40–100 pc from the Sun with $|\beta| < 20°$

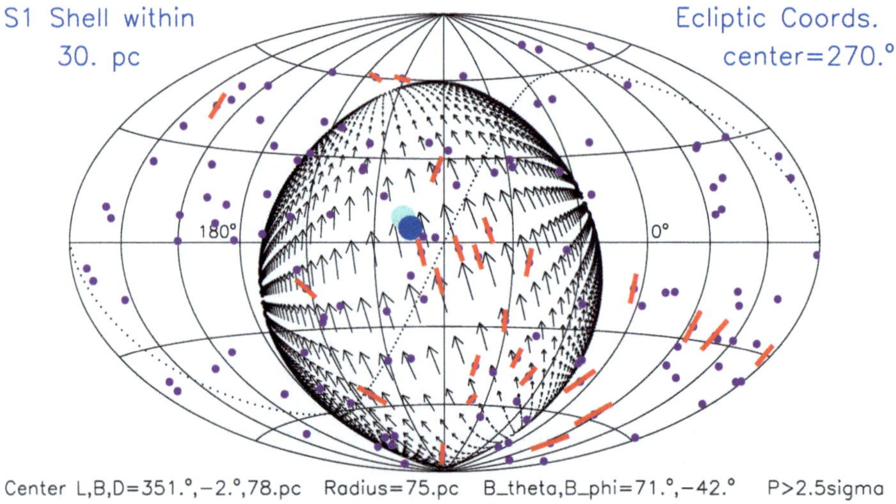

Fig. 10 The magnetic field associated with the parts of the S1 shell (Wolleben 2007) within 30 pc is plotted in ecliptic coordinates for an aitoff projection. The ecliptic plot is centered at $\lambda = 270°$. The S1 shell parameters have been varied within the range of allowed uncertainties to yield the best match to the polarization data of nearby stars (Tinbergen 1982; Frisch 2007). *Dark* and *light blue dots* show the inflow directions of interstellar H$^{\circ}$ and He$^{\circ}$. The *dots* show stars within 50 pc with polarization data, and the *red bars* show polarization vectors for stars where $P > 2.5\sigma$ (Tinbergen 1982; Piirola 1977; Frisch 2007)

the inflowing H$^{\circ}$ and He$^{\circ}$ velocities, given the uncertainties (Fig. 10, Frisch 2008b). The upwind direction of the bulk flow of local ISMa past the Sun is within $\sim 10°$ of the center of the S1 shell, supporting the conclusion that the S1 shell dominates the ISMa at the Sun.

There are no direct data on the polarity of the local ISMF, so we can only make an educated guess. The Loop I bubble is a distortion of the global local field, which is assumed to be the interarm field shown by, e.g., Han (2006). Therefore the near side of this feature, corresponding to the local B-field, is expected to be directed from southern to northern ecliptic latitudes at the solar location if possible magnetic turbulence is ignored.

Frisch (2007, 2008b) found the surprising result that the great circle meridian formed by the plane separating the hot and cold hemispheres of the cosmic microwave background dipole moment, which is generally thought to be dominated by the Doppler motion of the Sun through the cosmic microwave background, is an approximate predictor of the magnetic field direction at the heliosphere nose. This plane is $\sim 5°$ from the upwind direction and has a position angle within $\sim 20°$ of the magnetic field direction traced by the polarization and $H°$-He^{o} offset data. This combination is unlikely to be random.

4 The Boundary Conditions of the Heliosphere

The earliest comparisons between ISMa inside of the heliosphere and towards nearby stars showed that the CHISM is warm, low density, and partially ionized (Rogerson et al. 1973; McClintock et al. 1976; Adams and Frisch 1977). The CHISM properties are regulated by the low average interstellar densities inside of the Local Bubble, and the intensity and spectrum of the interstellar radiation field that follow from our location in a clump of low-opacity ISMa with $N(H^{o}) \lesssim 10^{18.5}$ cm^{-2}. Frisch and Slavin (2006b) concluded that our location in a tiny clump of ISMa indicates that the heliosphere boundary conditions will vary over timescales of less than $\sim 10^4$ yrs, as the Sun crosses the LIC density and ionization gradients at ~ 26 pc Myrs^{-1}. The boundary conditions of the heliosphere can not be measured directly, but can be modeled by combining ISMa data for a sightline through the LIC with data on interstellar neutrals from inside the heliosphere, combined with estimates of the filtration of interstellar neutrals. Interstellar He has minimal filtration ($\leq 2\%$, Sect. 2.1.1) and *in situ* data on interstellar He^{o} provide a valuable constraint on the interstellar densities. These diverse data on local ISMa gain a self-consistent interpretation through the radiative transfer models of the LIC, and together predict the boundary conditions of the heliosphere (Slavin and Frisch 2008).

The heliosphere boundary conditions are sensitive to the spectrum and intensity of the ambient interstellar radiation field, so below we first summarize the significant aspects of this radiation field, including the contamination of the observed soft X-ray background by emission generated by charge exchange between the solar wind and interstellar neutrals. The radiative transfer models of the LIC (Sect. 4.3) and heliospheric $H°$ data (2.1.2), combined with models of the attenuation of interstellar neutrals in the heliosheath regions (or 'filtration'), yield consistent results for the heliosphere boundary conditions.

4.1 Interstellar Radiation Field and the Heliosphere

Several spectral intervals of the interstellar radiation field are key to understanding the heliosphere boundary conditions. The first is the H ionizing radiation field, $E = 13.6$–24.6 eV. The second is the He^{o} ionizing field, $E = 24.6$–54.4 eV. Ulysses measurements of He^{o} in the heliosphere provide a key constraint on the density of the surrounding ISMa, so that the radiation field capable of ionizing He^{o} is important to understanding the heliosphere boundary conditions. Harder radiation, $E > 54.4$ eV up to ~ 100 eV, including the soft X-ray background radiation field, also may play an important role depending on the strength of the lower-energy ionizing radiation. The soft X-ray background is believed to originate in absorbed distant X-rays (e.g. from the galactic bulge and halo), absorbed X-rays with a local origin (e.g. from the Loop I superbubble), and essentially unabsorbed X-ray emission from the 10^6 K tenuous, $\sim$0.005 cm^{-3}, LB plasma (e.g. Snowden et al. 2000; Henley et al. 2007), with some foreground contamination from the heliosphere (Sect. 4.2). Models of the total soft X-ray spectrum at the Sun require modeling the temperature and spectrum of all sources.

The far-UV radiation field is also important for understanding the ionization of the LIC, in particular for low first ionization potential elements. Many A, B and late O stars contribute to the field at the Sun, with the stars in the third and fourth galactic quadrants, $\ell = 180°–360°$ dominating the spatial distribution (Gondhalekar et al. 1980). In contrast the stellar contribution to the extreme ultraviolet (EUV, $E = 13.6–100$ eV) radiation field is dominated by the B2 II star ε CMa (e.g. Vallerga 1996, 1998). New observations by the SPEAR instrument indicate that the diffuse, i.e. dust scattered, portion of the far ultraviolet (FUV) may be considerably larger than previously estimated (Kwang-Il Seon, private communication). If that is the case, the current photoionization models will need to be revised with possible ramifications for the electron density in the LIC and the abundance of C in the LIC.

The lower limit on the EUV radiation field is set by radiation from ε CMa, β CMa and nearby white dwarf stars, combined with the low energy tail of the soft X-ray background. The dominant source of EUV emission, however, may originate in the interface between the hot $T \sim 10^6$ K LB plasma and the warm LIC. Models of evaporation of the LIC into the LB caused by thermal conduction predict that there is strong EUV emission in a thin layer (Slavin 1989). The detailed results for the dynamical variables such as temperature, density and velocity of the gas as well as the emission in the interface region depend on the orientation and strength of the magnetic field (Slavin 1989; Slavin and Frisch 2008). The interface emission has not yet been directly observed, most of the emission coming in the low end of the EUV where the instruments that have been flown (EUVE, CHIPS) lack sufficient sensitivity for detection. The total radiation field at the Sun, based on a self-consistent radiative transfer model that includes all of these radiation sources, is shown in Fig. 11.

4.2 Heliospheric Contamination of the Soft X-Ray Background

Since the LB affects the LIC via photoionization and direct contact, understanding its spectrum and physical conditions is necessary. In the past, the LB's spectrum was found from shadowing observations, where interstellar clouds located tens to hundreds of pc away block distant X-rays, and the assumption that the LB was the sole source of locally produced diffuse X-rays. Recent observations of solar system X-rays, however, reveal that our solar system is a source of diffuse soft X-rays (Cravens 2000) that contaminate the hot LB's spectrum. These X-rays are emitted from solar wind ions after they have undergone charge exchange with inflowing interstellar neutrals in the solar system. The interaction mechanism is called solar wind charge exchange (SWCX). Here, we provide estimates for the strength and spectrum of the largest SWCX X-ray component, that are due to solar wind ions (excluding those from coronal mass ejections) interacting with neutral interstellar atoms. We also discuss whether SWCX affects our understanding of the photoionization of the LIC. There are two possibilities. First, can these photons directly ionize the CHISM? Second, did ignorance of the SWCX radiation field lead to an overestimation of the flux of the H-ionizing photons originating in the hot LB plasma?

Most of the SWCX emission is formed within 10 AU of the Sun, so that the brightness of heliospheric SWCX X-rays that reach the Earth's location is significant. Robertson and Cravens (2003) estimate the intensity of 0.1 to 1 keV SWCX X-rays found 1 AU from the

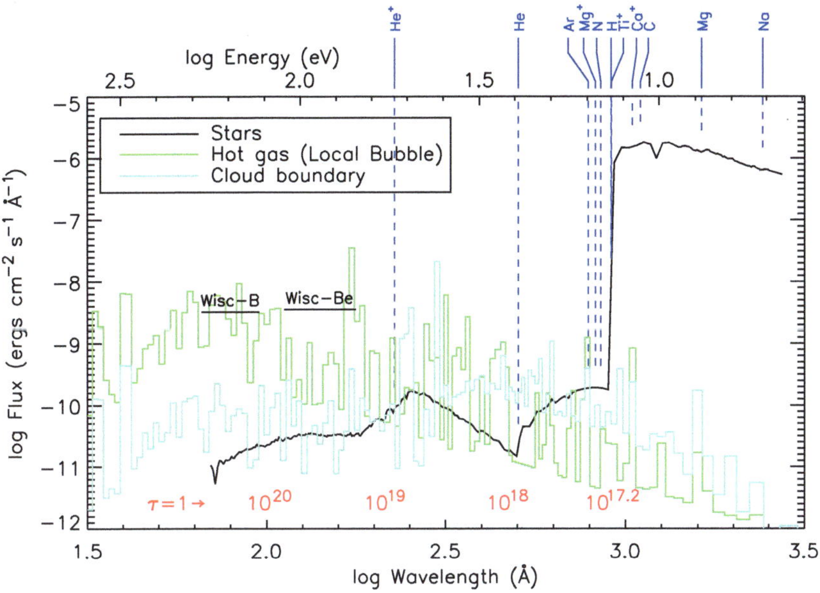

Fig. 11 The interstellar radiation field at the Sun, based on Model 26 in Slavin and Frisch (2008). The lower and top axes show, respectively, wavelength and energy. The *green* (*black*) histogram shows the modeled hot gas of the Local Bubble, ignoring possible foreground contamination from charge exchange in the solar wind. Figure 12 suggests that this contamination is negligible. The *cyan-blue* (*gray*) histogram is the cloud boundary contribution. The *remaining line* shows the stellar EUV/FUV background. The list of elements at the *top* of the plot identifies the ionization potentials for neutrals and ions of interest. The energies/wavelengths at which an optical depth of 1 is reached for several different column densities of H° are shown along the *bottom* of the plot. Additional information is given in Slavin and Frisch (2008)

Sun to be 3.5 to 6 keV $cm^{-2}\,s^{-1}\,sr^{-1}$, depending on look-direction and phase of the solar cycle. They find that the SWCX X-ray flux accounts for $\gtrsim 1/2$ of the X-rays previously attributed to the hot LB plasma. A second set of calculations was performed by Koutroumpa et al. (2008), who concentrated on emission lines in the 3/4 keV regime where the LB spectrum is relatively weak, but X-ray emission from SWCX is strong. They found that SWCX may account for all of the O VII and O VIII emission line intensity previously attributed to the hot LB plasma.

At the lower energies important to the LIC ionization, however, SWCX appears to be much less significant. A model spectrum has been provided by Koutroumpa et al. (2008). Fig. 12, top panel, shows the modeled spectrum for an observation of an ecliptic sight line during the maximum phase of the solar cycle. The solar wind is in its slow state in this case. For comparison, Fig. 12, bottom, shows the model spectrum of a thermal gas in collisional ionization equilibrium at a temperature of 10^6 K, which is the canonical model for the hot LB emission. We have scaled Fig. 12b so that it has the same intensity of O VII triplet (~570 eV) and O VIII Lyα (650 eV) photons as the SWCX spectrum. While it would be premature to calculate an ionization rate from the SWCX spectrum in Fig. 12, we can note that this SWCX spectrum appears to be harder than the canonical hot LB spectrum, and thus dimmer in the EUV and ~13.6 eV photons most important for ionizing H° in the LIC.

By their nature, SWCX X-ray estimates are uncertain and variable. The chief uncertainties are associated with the abundances of heavy elements in the solar wind, and the cross sections for charge exchange between highly charged ions and neutral material. As a re-

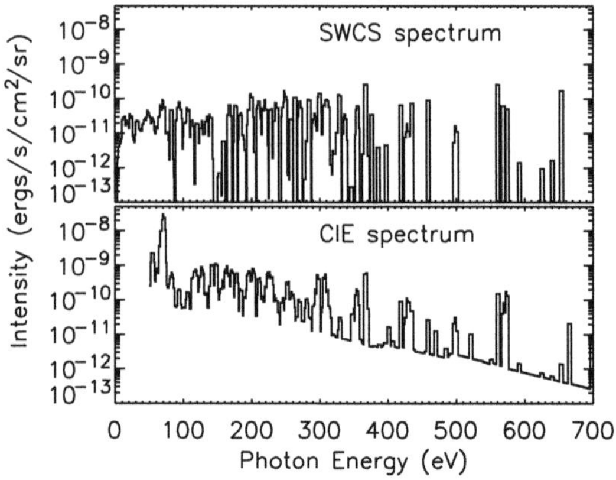

Fig. 12 *Top*: Solar wind charge exchange X-ray spectrum, courtesy of Koutroumpa et al. (2008). *Bottom*: Standard model of the Local Bubble spectrum: that of a thermal plasma in collisional ionization equilibrium at a temperature of 10^6 K. The lower spectrum has been scaled to yield the same intensity of O VII triplet (~570 eV) and O VIII Lyα (650 eV) photons as the SWCX spectrum

sult, Robertson and Cravens (2003) estimate that their calculation results are uncertain by roughly a factor of 2. In addition, the ionization level, density, and speed of the solar wind vary with the phase of the solar cycle and the location on the Sun (e.g. ecliptic latitude) from which the wind originated. As a result, the SWCX spectrum and intensity vary significantly with look-direction and time.

SWCX photons also are not effective direct ionizers of the LIC for two reasons: First, the SWCX spectrum appears to be harder than that from the hot plasma in the LB, and thus dimmer in EUV photons that could photoionize the LIC. Second, the intensity of SWCX photons affecting the LIC beyond the heliosphere is vastly diluted compared with the intensity of SWCX photons near the Earth and discussed above.

The discovery of SWCX X-rays also affects our understanding of the EUV radiation field capable of ionizing LIC He°, both through the properties of the conductive interface and the EUV radiation field external to the LIC (Sect. 4.1). If SWCX creates $\sim 1/2$ of low galactic latitude 0.25 keV X-ray emission that previously had been attributed to the hot LB plasma, then the LB radiation field may be weaker than previously thought. The theoretical properties of the conductive interface on the LIC have not yet been explored for cases where the LB plasma temperature differs significantly from 10^6 K, although as of yet there is no evidence for such a difference. One radiation transfer model of the LIC is successful without including interface emission, but most of the models that successfully match the observational constraints on the CHISM do include EUV emission from a conductive interface (Slavin and Frisch 2008). The reduction in the LB's X-ray intensity mandated by the discovery of SWCX X-rays implies that the thermal pressure of the hot gas in the LB is $\sim 1/\sqrt{2}$ less than previously reported. However, this decrease brings the LB pressure more in line with the LIC pressure used in the radiative transfer models, which assume a LB plasma temperature of $10^{5.9}$–$10^{6.1}$ K and LB thermal pressure of $P/k_B \sim 2000$–6600 cm^{-3} K.

4.3 Circumheliospheric Interstellar Material and the Local Interstellar Cloud

Slavin and Frisch (2008) have evaluated the physical properties of CHISM using a series of radiative transfer models that simultaneously model the properties of the integrated LIC absorption components towards ε CMa, and the CHISM neutrals traced by *in situ* heliospheric data. Note that the velocity of the CHISM observed inside of the heliosphere agrees with the LIC velocity projected towards ε CMa. The models are based on a cloud in photoionization equilibrium, using a self-consistent interstellar radiation field which includes, for most models, a conductive interface between the LIC and LB plasma. For those models, the conductive interface is the dominant source of most EUV emission that ionizes He° (Sect. 4.1). There presently is no observational proof of the LIC conductive interface, which is predicted to have column densities for O VI on the order of log $N(O^{+5}) = 12.3$–$12.8\ cm^{-2}$. These values are at the low end of detected O VI column densities towards nearby white dwarf stars that sample the nearest interstellar gas, but there is some debate as to whether the observed features sample interstellar instead of stellar features (Oegerle et al. 2005; Savage and Lehner 2006; Welsh 2008). At least one model is successful without interface emission, and in that case the EUV emission arises in the LB plasma. The LIC opacity is significant near energies of 13.6 eV, so the radiation field is propagated self-consistently in the LIC to match radiation field data and interface models. The most useful heliospheric constraints on the CHISM are the velocity, temperature, and He° density inside of the heliosphere (Sect. 2.1.1).

The Slavin and Frisch (2008) radiative transfer models yield the following heliosphere boundary conditions: $n(H^\circ) = 0.19$–$0.20\ cm^{-3}$, $n(e) = 0.07 \pm 0.01\ cm^{-3}$, $T = 6340$ K, and fractional ionizations for H and He of $\chi(H) = H^+/(H^\circ + H^+) = 0.19$–$0.26$ and $\chi(He) = He^+/(He^\circ + He^+) = 0.36$–$0.43$. These results are in excellent agreement with independent estimates of $n(H^\circ)$ and $n(H^+)$ based on observations of H pickup ions, combined with heliosphere models that predict H filtration (Sect. 2.1.2, Bzowski et al. 2008). The interstellar electron density in the CHISM is expected to be somewhat lower than the sightline averaged value, because of radiative transfer effects, and higher values are found towards stars sampling other parts of the CHISM (Sect. 5.3.3).

The key constraints on the total LIC electron density are the interstellar ratios of Mg°/Mg^+ and C^{+*}/C^+. The key constraints on the CHISM neutrality, and therefore ionization, are data on interstellar neutrals in the heliosphere, particularly He°, and also H, O, N, Ne, and Ar pickup ions. Better values for the filtration of O, N, Ne, and Ar are desirable to improve the comparisons between models and *in situ* data. The data constraining the models, and the model predictions, are listed in Table 1. The success of the radiative transfer models in reproducing LIC data supports the assumption that the surrounding cloud is in photoionization and thermal equilibrium. The modeling procedure is sensitive to the He° density inside the heliosphere.

The radiative transfer models also provide deeper insight into the heating and cooling mechanisms, as well as the chemical composition and dust mass, of the LIC. Heating of the LIC is dominated by the photoionization of H° (~66%) and He° (~25%). Cooling is dominated by the emission from the collisionally excited fine-structure lines of C^+ (~43%) and other ions (~35%). In the LIC, carbon appears to be overabundant by a factor of 2–3, possibly indicating destruction of carbonaceous grains in interstellar shocks (Slavin and Frisch 2006, 2008). The LIC abundances of refractory elements found from these radiative transfer models are typical abundances for interstellar clouds where dust grains have been destroyed by processing through interstellar shock fronts, i.e. the Routly-Spitzer effect (Routly and Spitzer 1952; Jones et al. 1994), and in the case of the LIC indicate grain processing by a shock with velocity ~90–140 km s^{-1} (Frisch et al. 1999). A problem with this interpretation is that models (e.g., Jones et al. 1994) predict that silicate dust would be more destroyed than carbonaceous dust. Dust destruction in shocks remains an active area of research that these results on the LIC can help to guide.

These models can also be used to predict the gas-to-dust mass ratio at the heliosphere boundary, and those predictions can be compared with values determined from *in situ* observations of dust in the heliosphere (Sect. 2.3). The prediction of $R_{g/d}$ from the radiative

Table 1 Neutral densities at heliosphere boundary from *in situ data and models*

Element	TS * cm^{-3}	Interstellar† cm^{-3}	Filtration‡ Ratios
H°	0.11 ± 0.02	0.194 ± 0.008	~ 0.56
He°	0.015 ± 0.003	0.015 ± 0.001	[1]◇
N°	$5.47 \pm 1.37 \times 10^{-6}$	$8.90^{+2.80}_{-1.99} \times 10^{-6}$	0.68–0.95
O°	$4.82 \pm 0.53 \times 10^{-5}$	$7.16^{+2.14}_{-1.45} \times 10^{-5}$	0.64–0.99
Ne°	$5.82 \pm 1.16 \times 10^{-6}$	$5.20^{+0.79}_{-0.67} \times 10^{-6}$	0.84–0.95
Ar°	$1.63 \pm 0.73 \times 10^{-7}$	$1.88^{+0.16}_{-0.25} \times 10^{-7}$	0.53–0.95

* Densities at the termination shock are based on pickup ion and He° data (Gloeckler and Fisk 2007; Bzowski et al. 2008; Möbius et al. 2004)

† Interstellar densities are based on radiative transfer models 14, 26–42, in Table 4 of Slavin and Frisch (2008)

‡ The filtration ratio is the percentage of the neutral that survives traversal of the heliosheath regions
The source for these filtration ratios is listed in Slavin and Frisch (2008)

◇ He° was assumed to have no filtration (although up to ~2% filtration is possible, Müller et al. 2004)

transfer models requires some assumption about the underlying abundances in the CHISM. The ACR isotopic data (Sect. 3.1) suggest that the surrounding interstellar cloud has solar composition. If solar abundances from Lodders (2003) are used, then we find $R_{g/d} \sim$ 160–240 for the CHISM, versus values $R_{g/d} < 115$–125 that are determined using *in situ* observations of the interstellar dust inside of the heliosphere (Slavin and Frisch 2008). The origin of the difference between these two values is not yet understood, but may indicate decoupling of interstellar gas and dust over very small spatial scales.

5 Complex of Local Interstellar Clouds

The Sun is embedded in a flow of ISMa that moves through the local standard of rest (LSR) with a bulk motion of ~17 km s^{-1} (Frisch and Slavin 2006a). The kinematics of this flow are determined by diffuse clouds in front of stars within 30 pc, and is typical of low-density interstellar clouds. In contrast, low velocities of ~0–2 km s^{-1} are found for more distant, cold, dense molecular clouds. Fast, low-density ISMa has long been known to show enhanced abundances of Ca^{+} and other refractory elements, caused by the destruction of dust grains in interstellar shocks (Routly and Spitzer 1952; Jones et al. 1994). Based on the kinematics and enhanced refractory abundances measured for ISMa within 14 pc, Frisch (1981) originally proposed that the CLIC is part of the Loop I superbubble that has expanded to the solar location. More recently Wolleben (2007) used 1.4 GHz, radio-continuum-polarization data and found that the Sun is located in the rim of the S1 subshell of Loop I (see Figs. 10 and 13).

5.1 Dynamics of the Complex of Local Interstellar Clouds

Two approaches have been used to study the dynamics of local ISMa: (1) The bulk motions of the ensemble of nearby clouds have been derived from some combination of optical and UV data (Crutcher 1982; Frisch and York 1986; Bzowski 1988; Vallerga et al. 1993; Frisch et al. 2002). (2) Individual dynamical clumps in the flow of ISMa past the Sun have been identified based on pure kinematics (Frisch 1981; Lallement et al. 1986; Frisch et al. 2002), or on both kinematical and temperature properties (Redfield and Linsky 2008). Many papers report cloud velocities in the solar system barycenter. However, for proper comparison with structural features in the global ISMa, velocities must be converted into the LSR.

Ultimately, all identifications of nearby interstellar clouds rely on absorption lines interpreted with the assumption that the absorbing gas has a thermal distribution of velocities around a central velocity, combined with non-thermal microturbulence. Absorption line spectroscopy is a powerful technique that, due to the Doppler shift of interstellar structures, can measure several distinct absorption components along a single line of sight. This technique has been used extensively, particularly at high spectral resolution (e.g. Crawford 2001; Welty et al. 1999; Gry and Jenkins 2001; Crawford 2001; Redfield and Linsky 2002, 2004a; Welty et al. 2003; Frisch et al. 2002). One basic property of LISM dynamics is that most sightlines show multiple distinct absorption profiles, particularly near the upwind-downwind axis where the radial components of the Doppler motions dominate. This argues for structured, coherently moving collections of gas in the LISM. For example, the median difference in radial velocity for LISM clouds within 100 pc and along a single line of sight is $\sim$7.5 km s^{-1}, whereas the typical line width (including thermal and microturbulent broadening mechanisms) is less than half of that at $\sim$2.8 km s^{-1} (Redfield and Linsky 2004a). One fact not yet understood about interstellar absorption lines is that the number of identified components increases as the spectral resolution of the observations increases. Welty and collaborators (Welty et al. 1996a; Welty and Hobbs 2001) argue that $\sim$40–48% of absorption components in distant sightlines,where the average number of absorbers is 5.0–9.5 per sight line, remain undetected because they are severely blended and observed at too low of a spectral resolution. For the LISM, on average, there are only 1.7 absorbers per sight line (Redfield and Linsky 2004a), yet absorption components are generally blended in velocity.

The recent studies of the kinematical structure of the LISM, described below, present the same general picture but do not agree in the detailed descriptions of cloudlets. Future studies of this topic need to include high-resolution observations in the UV, that can simultaneously provide information on the ionization, neutral densities, and kinematical structure of each sightline.

5.1.1 Bulk Dynamics of Local Interstellar Gas

Various studies of the CLIC properties and dynamics have concluded that the LISM within $\sim$30 pc flows away from the central regions of the Loop I supernova remnant (Frisch 1981; Crutcher 1982; Frisch and York 1986; Bzowski 1988; Vallerga et al. 1993; Frisch et al. 2002). Recent studies sampling ISMa within $\sim$30 pc show that the bulk flow of ISMa past the Sun has an LSR vector motion of -17 km s^{-1} from the upwind direction of $(\ell, b) \sim (358°, -5°)$. This LSR upwind direction of the CLIC is near the center of the S1 subshell of the Loop I supernova remnant, at $\ell = 346° \pm 5°$, $b = 3° \pm 5°$, identified by Wolleben (2007) based on 1.4 GHz polarization data. This LSR upwind direction, combined with the solar location inside the shell rim, suggest the bulk flow velocity vector represents the mean motion of the S1 shell expanding past the Sun. Fig. 13 shows the bulk motion of the CLIC through the LSR, together with the solar apex motion and the S1 shell location. The S1 center distance (78 $\pm$ 10 pc) and rim thickness ($\sim$20 pc) place the Sun inside of the shell rim. Comparisons of CLIC velocities for stars within $\sim$ 10 pc show that the bulk flow of ISMa past close to the Sun is decelerating (Frisch and York 1986; Redfield and Linsky

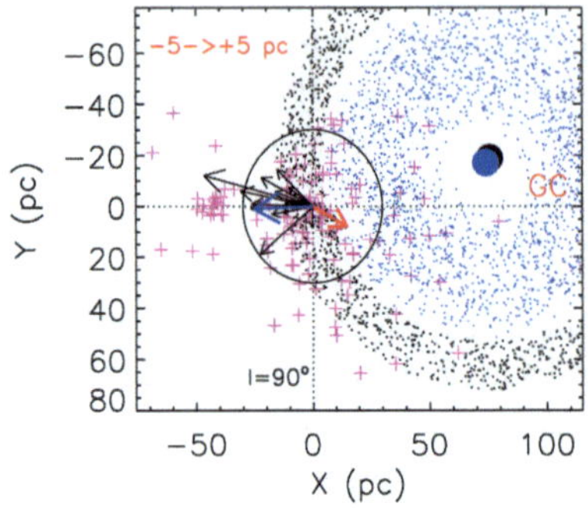

Fig. 13 Slices of the S1 shell (*black dots, large ring*) and S2 shell (*blue dots, small ring*) within 5 pc of the galactic plane ($|Z| < 5$ pc). The shell properties are displayed in Cartesian coordinates with the galactic center at the right and the direction of LSR rotation at the bottom. See Wolleben (2007) and Frisch (2008b) for additional information on the S1 and S2 shells. The *red* and *blue arrows* show the LSR velocities of the Sun and CLIC from Frisch and Slavin (2006a), and the *black arrows* show the 15 cloudlets of Redfield and Linsky (2008, RL8). The *black circle* shows the 30 pc radius region represented by the CLIC bulk flow vector. The *pink crosses* show the locations of the stars used to derive the velocity model with 15 cloudlets. The dot density scales with column density through the S1 and S2 slices. The coldest portions of the S1 shell are also seen as arcs protruding from the Scorpius-Centaurus Association in Fig. 2. This cold gas is observed, for instance, in Na° absorption towards δ Cyg (e.g. Hobbs 1969)

2001; Frisch et al. 2002). This effect vanishes when more distant stars, within ~100 pc, are included.

5.1.2 Structure and Cloudlets in the Complex of Local Interstellar Clouds

The inhomogeneous LISM was first apparent in individual 'clouds' identified 45 years ago towards stars within 20 pc (Munch and Unsold 1962; Hobbs 1969). The first spectrum of interstellar gas inside of the heliosphere was a *Copernicus* observation of the H° Lyα backscattered emission obtained during 1975, which determined an H° velocity of ~ -24.7 km s^{-1} (neglecting heliospheric acceleration and after correcting for the recent Ulysses measurement of the He° upstream direction, Sect. 2.1.1, Adams and Frisch 1977; Frisch 2008a). The nearest star for which optical data were then available, α Oph at 14 pc, showed interstellar Ca$^+$ velocities of -24 km s^{-1} and -26 km s^{-1} (Marschall and Hobbs 1972; Crawford 2001). In contrast, the heliospheric H° velocity projected in the α Oph direction was ~ -21.1 km s^{-1}. Although some of that difference is now recognized to be due to the deceleration of H° at the hydrogen wall (Sect. 3.2), this mismatch became the first evidence of velocity inhomogeneities in the LISM.

The earliest studies of the LISM structure were based on the kinematics of absorption components identified towards individual nearby stars. Frisch (1994) assumed that the LIC velocity vector is parallel to a cloud surface normal and used ISMa data towards α CMa at 2.7 pc to estimate that the Sun has entered the LIC within the past several thousand years. Using the original definition of the LIC velocity, Lallement et al. (1995) compared the ISMa velocity inside the heliosphere with ISMa velocities towards upwind stars, including α Cen at 1.3 pc, and concluded that the Sun will emerge from the LIC in the next 10,000 years. Again using the original definition of the LIC velocity, the absence of a LIC component towards the upwind stars (Wood et al. 2000b) and α Cen (the nearest star, Linsky and Wood 1996) set an upper limit for the distance to the LIC edge in the upwind direction of ~0.1 pc. These studies then suggested that the Sun is inside of a tiny distinct kinematic structure, and will exit this kinematical structure in $\leq$3,800 yrs. The most recent definition of the LIC velocity by Redfield and Linsky (2008, see below), using absorption components towards stars up to 100 pc away, instead places the Sun between two cloudlets. In this later case, the CHISM itself would be an independent kinematic structure, sampled only by the heliosphere. At this point the discussion enters the philosophical territory as to "what is an interstellar cloud" (a question posed by Eugene Parker). We note that 30 years ago 'clouds' such as the LIC and CHISM were referred to as 'intercloud material'.

A kinematical study of the whole sky requires both high-resolution optical Ca^+ data and medium-resolution UV absorption lines, because the UV resonance lines are far more opaque than Ca^+ and thus allow measurements of warm ISMa in the galactic anticenter and north galactic pole directions where Ca^+ column densities are very low. Individual cloudlets have been identified in the bulk flow of ISMa past the Sun based on the fitting of clouds with thermal Maxwellian velocity distributions combined with non-thermal microturbulence. Based on kinematics, Lallement et al. (1986) identified five cloudlets in the CLIC. The subsequent inclusion of UV absorption lines showed that the gas flow in the galactic anticenter hemisphere can be fit with a single velocity vector (originally called the AG Cloud and now called the LIC flow vector), but the galactic center hemisphere required a slightly different velocity vector (called the G Cloud flow vector). A more extended set of data, focused on stars that traced ISMa within 30 pc, allowed identification of seven distinct kinematical structures (Frisch et al. 2002). One difficulty in assigning absorption components to cloudlets in the CLIC is that many features have similar upwind directions, e.g. the LIC and G-cloud, yielding ambiguous assignments for stars located $\sim$90° from the bulk flow direction.

A more detailed cloudlet structure based on kinematical arguments has been developed by Redfield and Linsky (2008), who assembled a database with 270 radial-velocity measurements for 157 sightlines towards stars located within 100 pc, including the data from Lallement et al. (1986) and Frisch et al. (2002). The locations of these stars are plotted in Fig. 13, and include sightlines that sample both the S1 and S2 shells of Wolleben (2007). About 55% of this database consists of velocities measured in UV absorption lines of D°, C^+, O°, Si^+, Mg^+, Fe^+, measured by the GHRS and STIS instruments on HST, and 45% of the data are from ground-based Ca^+ spectra. Nearly half of these radial velocities, primarily in the galactic anticenter hemisphere, are consistent with a single flow vector to within $\pm 1\ \mathrm{km\,s^{-1}}$, which has also been named the LIC, and is denoted here the $V_{\mathrm{LIC,RL}}$. The $V_{\mathrm{LIC,RL}}$ flow vector in the LSR has a velocity amplitude of $23.8 \pm 0.9\ \mathrm{km\,s^{-1}}$, and is directed towards galactic coordinates $l = 187.0° \pm 3.4°$ and $b = -13.5° \pm 3.3°$. Alternate definitions of the LIC velocity vector are based on the velocity of interstellar He° inside of the heliosphere (Sect. 2.1.1) and interstellar absorption components in nearby stars in the downwind direction (Lallement and Bertin 1992). The temperatures of the components assigned to the $V_{\mathrm{LIC,RL}}$ component vary from 5,200 K (κ^1 Cet) to 12,050 K (V368 Cep), with a weighted mean of 7500 ± 1300 K, based on 19 independent temperature measurements. Temperatures are determined by assuming mass-dependent thermal and mass-independent turbulent broadening of absorption lines, with the low mass measurements generally based on the D° Lyα absorption line, which is superimposed on the blue wing of the stronger H° Lyα feature.

The morphology of the LIC, according to Redfield and Linsky (2008), is shown in Fig. 14. Using components with radial velocities that are inconsistent with the $V_{\mathrm{LIC,RL}}$ vector, Redfield and Linsky (2008) then searched for patches of galactic coordinates that contain radial velocities consistent with different velocity vectors to within measurement errors. The largest such patch is the G cloud based on 21 sightlines, but there are 13 other clouds each based on 4–15 sightlines. All of these clouds lie within 15 pc of the Sun as determined by

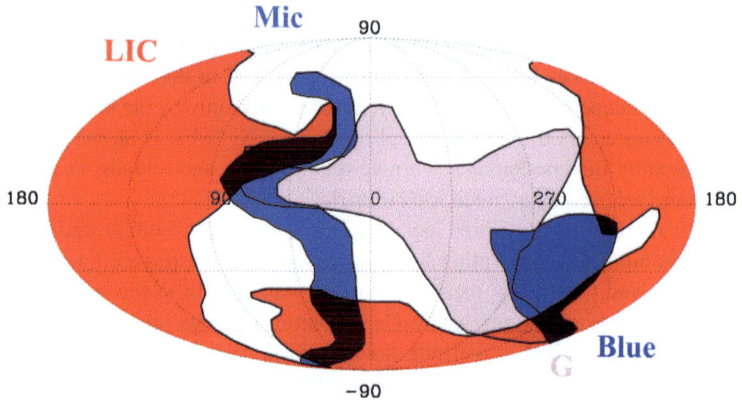

Fig. 14 The location in galactic coordinates of four interstellar clouds that are closest to the heliosphere. The Mic Cloud, which is likely located between the LIC and G Clouds with a similar shape, may be compressed and heated by the collision of the LIC and G Clouds. The G Cloud is seen in front of α Cen (1.3 pc), and the Blue Cloud is seen in front of Sirius (2.6 pc). Eleven other warm clouds located within 15 pc of the Sun are not plotted, but together the 15 clouds cover the whole sky (Redfield and Linsky 2008)

the nearest star with an interstellar velocity component consistent with the cloud vector. By this criterion, six of the clouds are closer than 3.5 pc, and the G cloud lies within 1.3 pc, the distance to α Cen. Approximately 19% of the velocity components could not be assigned to a cloud, indicating that they are located in a poorly sampled region of the sky or that the absorbing gas cloud subtends too small an angle to determine a reliable velocity vector either because the cloud is intrinsically small or too far away. Half of these 15 cloudlets overlap kinematical entities identified in earlier studies (Frisch 1981; Lallement et al. 1986; Frisch et al. 2002), and the differences between the original and newly derived vectors show that the three-dimensional kinematics of a cloud are highly sensitive to the set of stars, and absorption components, that have been selected to identify the clouds.

The velocity of ISMa inside of the heliosphere is well established by interstellar He^{o} flowing into the heliosphere at -26.3 ± 0.5 km s^{-1} (Sect. 2.1.1), which is between the flow speed of the $V_{LIC,RL}$ (-23.8 ± 0.9 km s^{-1}) and the G Cloud (-29.6 ± 1.1 km s^{-1}). When the new velocity vector $V_{LIC,RL}$ is used, the direction of the inflowing He^{o} is intermediate between that of the G and RL8-LIC clouds, and the temperature of the inflowing He^{o} gas (6300 ± 390 K, Witte 2004; Möbius et al. 2004) is also intermediate between the RL8-LIC (7500 ± 1300 K) and the G Cloud (5500 ± 400 K). Redfield and Linsky (2008) therefore concluded that heliosphere is in the transition region between the LIC and G Clouds. In this case, however, the CHISM is seen nowhere except inside of the heliosphere.

The identification of distinct, rigidly moving, kinematically defined cloudlets in the CLIC flow is subject to several different uncertainties. As noted above, the similar upwind directions for most of the local cloudlets identified so far means that absorption components towards stars in the sidewind direction frequently have velocities consistent with more than one cloudlet. A second uncertainty follows from apparent deceleration of the nearest parts of the CLIC. In addition, many cloud identifications rest on blended components where uncertainties in component velocities are comparable to the velocity differences between the cloudlets. An example of this later case is the sightline towards α Oph. High-resolution observations of the most positive absorption component towards α Oph, at 14 pc, find heliocentric velocities that differ by several km s^{-1} for components forming $V_{LIC,RL}$, with values of -22.2 km s^{-1}(Welty et al. 1996b), -23.6 km s^{-1} (Crawford and Dunkin 1995), and

$-24.0\,\mathrm{km\,s^{-1}}$ (Crawford 2001). The uncertainty range of this component velocity is comparable to the difference of $\sim 2.4\,\mathrm{km\,s^{-1}}$ between the interstellar He^{o} velocity and $V_{\mathrm{LIC,RL}}$, projected in the α Oph direction. These various uncertainties can only be overcome by building a comprehensive high-resolution database of ISMa within 30 pc, inclusive of UV spectral data, so that cloud physical properties can also be investigated.

5.2 Origin of Local Interstellar Medium

The similarity in the upwind directions of the velocity vectors of LISM clouds argues for a collective history or driver in order to create the common motion of LISM clouds. Using information on the dynamics and abundances of nearby warm gas, Frisch (1981) originally found that the origin of local ISMa is directly related to the history and structure of the asymmetrically expanding Loop I bubble formed by the supernovae and winds from stars in the Scorpius-Centaurus Association. The averaged LSR upwind direction of LISM cloudlets is found to be $l \sim 357.8°$, $b \sim -5.2°$ using stars within 30 pc (Frisch and Slavin 2006a) or $l \sim 10.7°$, $b \sim 16.6°$ using stars within 100 pc (Redfield and Linsky 2008). These directions are coincident with the location of one of the nearest OB associations, the Scorpius-Centaurus Association, and within 20 degrees of the Upper Scorpius subgroup. The agreement is even stronger for the location of the associations when they were formed, from 5–10 Myr ago (de Geus et al. 1989; Maíz-Apellániz 2001), when the upwind direction is within 10 degrees of both the Upper Scorpius and Upper Centaurus Lupus subgroups. A related interpretation of the data is that the LISM is part of the S1 shell (Fig. 13, and Frisch 2008b).

The connection between Sco-Cen and the Local Bubble, as well as the warm LISM around the Sun, has been discussed by various authors (e.g. Frisch 1981, 1995; Frisch and York 1986; Bochkarev 1987; Maíz-Apellániz 2001; Fuchs et al. 2006; Wolleben 2007; Redfield and Linsky 2008). Shells swept up by the winds of young stars in the Scorpius-Centaurus Association may be coincident with the location of the Sun today, and may be responsible for the very local warm ISMa clouds observed in our immediate vicinity. Frisch (1995, 1996) modeled the Loop I shell at the Sun as formed 4–5 Myrs ago by stellar evolution in the Scorpius-Centaurus Association. Breitschwerdt and de Avillez (2006) proposed an alternative, but related, model where supernovae from a Pleiades subgroup led to the formation of the LB. The interaction of two such structures has led to theories of the formation of warm LISM clouds from Rayleigh-Taylor instabilities near the interaction point, which subsequently get driven by the prevailing flows from the Scorpius-Centaurus Association (Breitschwerdt et al. 2000; Egger and Aschenbach 1995). Finally, straightening magnetic flux tubes, pulling dense material from the LB boundary, has been discussed as a possible formation mechanism for the warm LISM clouds by Cox and Helenius (2003). Detailed morphological and physical models of both the warm LISM clouds and the hot gas in the LB need to be assembled in order to develop a coherent model for the origin and evolution of our most local interstellar environments.

5.3 Physical Properties of the Local Interstellar Medium

5.3.1 Temperature and Turbulence

High-resolution UV spectra obtained with the GHRS and STIS instruments on HST have provided the important data for inferring the physical properties of the clouds. The widths of interstellar absorption lines of ions with atomic masses differing by over an order of magnitude, e.g. D° and Fe^{+}, can in principle be used to separate purely thermal from non-thermal (turbulent) broadening components. This deconvolution of line widths is based on the assumption that the line full-width-half at half-maximum is given by FWHM $\sim 1.7 * \sqrt{V_{\text{thermal}}^2 + V_{\text{turbulent}}^2}$, where V_{thermal} represents the mass-dependent Doppler broadening of the line by thermal motions, and $V_{\text{turbulent}}$ is presumably mass-independent line broadening due to the turbulent motions. We note that this later assumption may break down for ions subject to magnetically-induced gyromotion. Redfield and Linsky (2004b) found a weighted mean temperature along fifty LISM sightlines of 6680 ± 1490 K, with a range between 1000 K and 12,500 K. The weighted mean nonthermal broadening is 2.24 ± 1.03 km s^{-1}, which is subsonic and provides less pressure support than the thermal component. The mean turbulence may be somewhat overestimated due to unresolved velocity components. Redfield and Linsky (2004b) also found a moderately significant negative correlation of T with nonthermal broadening, which, if real, could result from pressure equilibrium. A puzzling aspect of the negative correlation is found for stars within 10 pc, where the negative correlation between temperature and turbulence is accompanied by positive correlations between $N(\text{D}^\circ)$ and temperature (Frisch 2008a). In low-density, partially ionized ISMa such as the LIC, D° and Fe^+ will have different spatial distributions, so that radiative transfer models of the sightline are required to accurately compare abundances of these species with different ionization levels.

5.3.2 *Chemical Composition*

In principle, elemental abundances can be used to estimate depletions onto dust grains, based on some assumed reference abundance for the cloud. Isotopic abundances suggest that solar abundances are appropriate for the CHISM (Sect. 2.2). Abundances in partially ionized gas must be compared to $N(\text{H}^\circ + \text{H}^+)$, as has been done implicitly for the CHISM through the use of radiative transfer models (Sect. 4.3). However cloud ionizations are not known for most of the CLIC sightlines. In the CHISM, gas-phase, refractory-element abundances of Fe, Mg, and Si are enhanced over cold cloud values by factors of $\sim$12, $\sim$3, and $\sim$4, which is attributed to the destruction of ISDGs in shocks (Jones et al. 1994; Frisch et al. 1999; Slavin and Frisch 2008). Redfield and Linsky (2008) found that small depletions of Fe and Mg are correlated with high turbulence values. A further consideration is that since the Sun resides in the rim of the S1 shell, and the CLIC samples the S1 shell (Fig. 13), it is likely that depletions are similar throughout the shell. In this case, apparent local depletion variations would either be due to neglected ionization corrections, or the above-mentioned problems with the conventional line-broadening algorithm.

5.3.3 *Ionization*

Ionization has been determined for the LISM from collisionally excited C^{+*} lines and the Mg° line, which has a strength enhanced in warm gas (>6000 K) by dielectric recombination (York and Kinahan 1979). The range of interstellar electron densities found for gas within $\sim$100 pc, using either the $\text{Mg}^\circ/\text{Mg}^+$ or C^{+*}/C^+ ratios, is $n(\text{e}) = 0.02$–0.50 cm^{-3} (Frisch et al. 1990; Lallement and Ferlet 1997; Lehner et al. 2003; Redfield and Falcon 2008). The CHISM value $n(\text{e}) = 0.07 \pm 0.01$ cm^{-3} determined from radiative transfer models (Sect. 4.3) is thus consistent with the general values in the LISM. The nearest star with a high observed level of ionization is the white dwarf HD 149499B (37 pc), where $N(\text{N}^+)/N(\text{N}^\circ) \sim 2$ (Lehner et al. 2003). The ionizations levels of N and H are coupled by charge exchange in diffuse clouds.

5.3.4 *When Local Clouds Interact*

We now know that even over sightlines as short as a few parsecs there are on average 1.7 clouds. Redfield and Linsky (2008, RL08) estimated that warm clouds occupy 5.5–19% of the volume of space within 15 pc. The model with 15 LISM velocity vectors (Sect. 5.1.2) allowed RL08 to compute the relative velocity differences between clouds located along the same sightline. These velocity differences are often supersonic (>8 $\mathrm{km\,s^{-1}}$) and in some cases as large as 45 $\mathrm{km\,s^{-1}}$, although we do not know the locations of most clouds along lines of sight. These pieces of information make it likely that some clouds interact with each other and that the collisions can produce shocks. Linsky et al. (2008) showed that the three quasars with well-studied, large-amplitude intraday and annual scintillation variability are all located along lines of sight that pass through the edges of warm clouds where cloud-cloud interactions are likely. Since scintillation is produced by a turbulent ionized medium, this study reinforces the possibility that locally cloud-cloud interactions produce the turbulence. A related possibility is that the scintillations occur where the magnetic fields of the S1 and S2 shells interact.

6 Conclusions

The IBEX mission is now providing detailed *in situ* measurements of ENAs formed in the interface between the CHISM and the solar wind. The ENA data will constrain the region where the filtration of interstellar neutrals occurs. The advent of observations of ISMa inside of the heliosphere brought an absolutely unique perspective to understanding interstellar clouds. Observations of ISMa inside and outside of the heliosphere can be combined with extraordinary accuracy to constrain the questions concerning the boundary of the heliosphere. We find that the CHISM properties consistent with available heliospheric and astronomical data on the LIC indicate a warm, low-density, partially ionized gas, $n(\mathrm{H}^\circ) \sim 0.195\ \mathrm{cm}^{-3}$, $n(\mathrm{e}) \sim 0.06\ \mathrm{cm}^{-3}$, $T \sim 6300$ K, $\chi(\mathrm{H}) \sim 0.24$, $\chi(\mathrm{He}) \sim 0.40$. If thermal and magnetic pressures are equal, the magnetic field is weak, with $B \sim 2.7\ \mu\mathrm{G}$.

This chapter in the IBEX volume summarizes data on interstellar material inside and outside of the heliosphere, including H°, He°, and pickup ions and anomalous cosmic rays. Accurate corrections for filtration are the key to comparing data on interstellar neutrals inside of the heliosphere with data on interstellar neutrals in the CHISM. The energetic neutral atom data measured by IBEX will provide important data for constraining the heliosphere boundary regions and the filtration corrections for interstellar neutrals.

Consistent values for the orientation of the interstellar magnetic field at the heliosphere are given by several different indicators, and suggest a field inclined by ~59° with respect to the ecliptic plane; if the ISMF is part of the extended Loop I magnetic field, the field polarity may be directed from the southern to the northern hemisphere. Additional data on the polarizations of nearby stars would strengthen our understanding of the nearby interstellar magnetic field.

This chapter also provides the context for understanding the origin of the CHISM, and the family of cloudlets that surround the CHISM on our journey through space. Clouds such as the CHISM, e.g. warm, low-density, partially ionized gas, were once termed "intercloud medium". However the dynamical motion of the CHISM and other nearby clouds clearly indicate a common origin and history for a parcel of gas that moves through the LSR at a mean velocity of ~17 $\mathrm{km\,s^{-1}}$. Recent studies of distinct kinematical components among the clump of nearby cloudlets offers the opportunity to understand cloud-cloud interactions, as well as the paleoheliosphere. Identifying and modeling CLIC kinematics requires grappling with the fundamental concept of 'what is an interstellar cloud', in the context of a turbulent flow, or superbubble shell, containing very low-density gas. The dust grains in the CHISM appear typical for rapidly moving clouds with a history of grain destruction in shocks, and at the same time *in situ* dust data provide a means of direct measurement of the interstellar gas-to-dust mass ratio. The small dust grains missing from the inner heliosphere are clues to the interstellar and solar wind magnetic fields in the heliosheath regions, and the way the interstellar dust distribution is modified by the radiation pressure and the solar activity cycle. The origin of the LIC as part of an expanding superbubble shell driven by star formation in the Scorpius-Centaurus Association is consistent with data on the LISM.

Summary of acronyms used in this chapter The many terms and acronyms used in this chapter include: Interstellar Boundary Explorer mission (IBEX), Local Bubble (LB), Voyager 1 (V1), Voyager 2 (V2), anomalous cosmic rays (ACR), circumheliospheric interstellar medium (CHISM), complex of local interstellar clouds (CLIC), energetic neutral atom (ENA), extreme ultraviolet (EUV), far ultraviolet (FUV), heliopause (HP), Hubble Space Telescope (HST), interplanetary magnetic field (IMF), interstellar dust grain (ISDG), interstellar magnetic field (ISMF), interstellar material (ISMa), interstellar radiation field (ISRF), Local Interstellar Cloud (LIC), local interstellar medium (LISM), local standard of rest (LSR), magnetohydrodynamic (MHD), pickup ion (PUI), polycyclic aromatic hydrocarbons (PAH), position angle (PA), solar wind charge exchange (SWCX), solar wind termination shock (TS), and ultraviolet (UV).

Acknowledgements P. Frisch, D. McComas, E. Möbius and N. Schwadron gratefully acknowledge support from NASA through the IBEX Explorer mission. P. Frisch thanks NASA for support through grants NNG06GE33G and NNX08AJ33G. We would like to acknowledge Dimitra Koutroumpa for kindly sharing her SWCX spectrum with us in advance of the publication of her article. S. Redfield would like to acknowledge support provided by NASA through Hubble Fellowship grant HST-HF-01190.01 awarded by the Space Telescope Science Institute, which is operated by the Association of Universities for Research in Astronomy, Inc., for NASA, under contract NAS 5-26555.

References

T.F. Adams, P.C. Frisch, Astrophys. J. **212**, 300–308 (1977)
J.M. Ajello, A.I. Stewart, G.E. Thomas, A. Graps, Astrophys. J. **317**, 964–986 (1987)
D.B. Alexashov, S.V. Chalov, A.V. Myasnikov, V.V. Izmodenov, R. Kallenbach, Astron. Astrophys. **420**, 729–736 (2004)
N. Altobelli, S. Kempf, M. Landgraf, R. Srama, V. Dikarev, H. Krüger, G. Moragas-Klostermeyer, E. Grün, J. Geophys. Res. (Space Phys.) **108**, 7–1 (2003)
N. Altobelli, S. Kempf, H. Krüger, M. Landgraf, M. Roy, E. Grün, J. Geophys. Res. (Space Phys.) **110**, 7102 (2005)
N. Altobelli, E. Grün, M. Landgraf, Astron. Astrophys. **448**, 243–252 (2006)
K. Avinash, J. Slavin, G.P. Zank, P. Frisch, AGU Fall Meeting Abstracts pages B1602+ (2008)
W.I. Axford, in *The Solar Wind*, ed. by J.M.W.C.P. Sonnet, P.J. Coleman. NASA Spec. Publ., vol. 308 (NASA, Washington, 1972), p. 609
W.J. Baggaley, J. Geophys. Res. **105**, 10,353–10,362 (2000)
M. Baguhl, E. Grün, D.P. Hamilton, G. Linkert, R. Riemann, P. Staubach, H.A. Zook, Space Sci. Rev. **72**, 471–476 (1995)
V.B. Baranov, Y.G. Malama, J. Geophys. Res. **98**, 15,157–15,163 (1993)
V.B. Baranov, Y.G. Malama, J. Geophys. Res. **100**, 14,755–14,761 (1995)
V.B. Baranov, M.G. Lebedev, Y.G. Malama, Astrophys. J. **375**, 347–351 (1991)
V.B. Baranov, V.V. Izmodenov, Y.G. Malama, J. Geophys. Res. **103**, 9575–9585 (1998)
E.M. Berkhuijsen, Astron. Astrophys. **14**, 359–386 (1971)

J.L. Bertaux, J.E. Blamont, Astron. Astrophys. **11**, 200 (1971)
N.G. Bochkarev, Astrophys. Space Sci. **138**, 229–302 (1987)
D. Breitschwerdt, M.A. de Avillez, Astron. Astrophys. **452**, L1–L5 (2006)
D. Breitschwerdt, M.J. Freyberg, R. Egger, Astron. Astrophys. **361**, 303–320 (2000)
L.F. Burlaga, N.F. Ness, M.H. Acuña, R.P. Lepping, J.E.P. Connerney, J.D. Richardson, Nature **454**, 75–77 (2008)
M. Bzowski, Acta Astron. **38**, 443–453 (1988)
M. Bzowski, in *The Outer Heliosphere: The Next Frontiers*, ed. by K. Scherer, H. Fichtner, H.J. Fahr, E. Marsch. COSPAR Colloquia Series, vol. 11 (Elsevier/Pergamon, Amsterdam, 2001), pp. 69–72
M. Bzowski, H.J. Fahr, D. Ruciński, H. Scherer, Astron. Astrophys. **326**, 396–411 (1997)
M. Bzowski, E. Möbius, S. Tarnopolski, V. Izmodenov, G. Gloeckler, Astron. Astrophys. **491**, 7–19 (2008)
E. Chassefiere, J.L. Bertaux, R. Lallemant, V.G. Kurt, Astron. Astrophys. **160**, 229–242 (1986)
J.T. Clarke, R. Lallement, J. Bertaux, H. Fahr, E. Quemerais, H. Scherer, Astrophys. J. **499**, 482–488 (1998)
M.R. Collier, T.E. Moore, D. Simpson, A. Roberts, A. Szabo, S.A. Fuselier, P. Wurz, M.A. Lee, B.T. Tsurutani, Adv. Space Res. **34**, 166–171 (2004)
D.P. Cox, L. Helenius, Astrophys. J. **583**, 205–228 (2003)
T.E. Cravens, Astrophys. J. Lett. **532**, L153–L156 (2000)
I.A. Crawford, Mon. Not. R. Astron. Soc. **327**, 841–848 (2001)
I.A. Crawford, S.K. Dunkin, Mon. Not. R. Astron. Soc. **273**, 219–224 (1995)
R.M. Crutcher, Astrophys. J. **254**, 82–87 (1982)
A.C. Cummings, E.C. Stone, Space Sci. Rev. **130**, 389–399 (2007)
A.C. Cummings, E.C. Stone, C.D. Steenberg, Astrophys. J. **578**, 194–210 (2002)
A. Czechowski, I. Mann, Astron. Astrophys. **410**, 165–173 (2003a)
A. Czechowski, I. Mann, J. Geophys. Res. **108**, 13 (2003b)
E.J. de Geus, P.T. de Zeeuw, J. Lub, Astron. Astrophys. **216**, 44–61 (1989)
R.J. Egger, B. Aschenbach, Astron. Astrophys. **294**, L25–L28 (1995)
H.J. Fahr, Space Sci. Rev. **15**, 483–540 (1974)
H.J. Fahr, Astron. Astrophys. **241**, 251–259 (1991)
H.J. Fahr, T. Kausch, H. Scherer, Astron. Astrophys. **357**, 268–282 (2000)
H.J. Fahr, H. Fichtner, K. Scherer, O. Stawicki, in *Solar Journey: The Significance of Our Galactic Environment for the Heliosphere and Earth*, ed. by P. Frisch (Springer, Berlin, 2006), pp. 259–279
L.A. Fisk, B. Kozlovsky, R. Ramaty, Astrophys. J. **190**, L35–L38 (1974)
V. Florinski, G.P. Zank, in *Solar Journey: The Significance of Our Galactic Environment for the Heliosphere and Earth*, ed. by P. Frisch (Springer, Berlin, 2006), pp. 281–316
P.C. Frisch, Nature **293**, 377–379 (1981)
P.C. Frisch, Science **265**, 1423 (1994)
P.C. Frisch, Space Sci. Rev. **72**, 499–592 (1995)
P.C. Frisch, Space Sci. Rev. **78**, 213–222 (1996)
P.C. Frisch, Am. Sci. **88**, 52–59 (2000)
P.C. Frisch, Introduction: The Paleoheliosphere versus PaleoLISM, in *Solar Journey: The Significance of Our Galactic Environment for the Heliosphere and Earth*, ed. by P.C. Frisch (Springer, Berlin, 2006), pp. 1–22
P.C. Frisch, ArXiv e-prints: arXiv:0707.2970 (2007)
P.C. Frisch, Space Sci. Rev. 100 (2008a)
P.C. Frisch, Astrophys. J. (2008b, submitted). arXiv:0804.1901
P. Frisch, D.G. York, in *The Galaxy and the Solar System* (University of Arizona Press, Tucson, 1986), pp. 83–100
P.C. Frisch, J.D. Slavin, Short term variations in the galactic environment of the Sun, in *Solar Journey: The Significance of Our Galactic Environment for the Heliosphere and Earth*, ed. by P.C. Frisch (Springer, Berlin, 2006a), pp. 133–193
P.C. Frisch, J.D. Slavin, Astrophys. Space Sci. Trans. **2**, 53–61 (2006b)
P.C. Frisch, D.E. Welty, D.G. York, J.R. Fowler, Astrophys. J. **357**, 514–523 (1990)
P.C. Frisch, J.M. Dorschner, J. Geiss, J.M. Greenberg, E. Grün, M. Landgraf, P. Hoppe, A.P. Jones, W. Krätschmer, T.J. Linde, G.E. Morfill, W. Reach, J.D. Slavin, J. Svestka, A.N. Witt, G.P. Zank, Astrophys. J. **525**, 492–516 (1999)
P.C. Frisch, L. Grodnicki, D.E. Welty, Astrophys. J. **574**, 834–846 (2002)
B. Fuchs, D. Breitschwerdt, M.A. de Avillez, C. Dettbarn, C. Flynn, Mon. Not. R. Astron. Soc. **373**, 993–1003 (2006)
K.G. Gayley, G.P. Zank, H.L. Pauls, P.C. Frisch, D.E. Welty, Astrophys. J. **487**, 259–270 (1997)
J. Geiss, G. Gloeckler, L.A. Fisk, R. von Steiger, J. Geophys. Res. **100**, 23373–23378 (1995)
G. Gloeckler, J. Geiss, Adv. Space Res. **34**, 53–60 (2004)

G. Gloeckler, L. Fisk, *Composition of Matter*. Space Sciences Series of ISSI, vol. 27 (Springer, Berlin, 2007), pp. 489–513

G. Gloeckler, K.-P. Wenzel, *Acceleration Processes of Heliospheric Particle Populations*. Century of Space Science, vol. II (Kluwer Academic, Dordrecht, 2001), pp. 963–1006

G. Gloeckler, L.A. Fisk, J. Geiss, N.A. Schwadron, T.H. Zurbuchen, J. Geophys. Res. **105**, 7459–7464 (2000)

G. Gloeckler, E. Möbius, J. Geiss, M. Bzowski, S. Chalov, H. Fahr, D.R. McMullin, H. Noda, M. Oka, D. Ruciński, R. Skoug, T. Terasawa, R. von Steiger, A. Yamazaki, T. Zurbuchen, Astron. Astrophys. **426**, 845–854 (2004)

P.M. Gondhalekar, A.P. Phillips, R. Wilson, Astron. Astrophys. **85**, 272 (1980)

K. Grogan, S.F. Dermott, B.A.S. Gustafson, S. Jayaraman, Y.L. Xu, D. Hamilton, in *IAU Colloq. 150: Physics, Chemistry, and Dynamics of Interplanetary Dust*, ed. by B.A.S. Gustafson, M.S. Hanner. Astronomical Society of the Pacific Conference Series, vol. 104 (Astronomical Society of the Pacific, San Francisco, 1996), p. 325

E. Grün, B. Gustafson, I. Mann, M. Baguhl, G.E. Morfill, P. Staubach, A. Taylor, H.A. Zook, Astron. Astrophys. **286**, 915–924 (1994)

E. Grün, P. Staubach, M. Baguhl, D.P. Hamilton, H.A. Zook, S. Dermott, B.A. Gustafson, H. Fechtig, J. Kissel, D. Linkert, G. Linkert, R. Srama, M.S. Hanner, C. Polanskey, M. Horanyi, B.A. Lindblad, I. Mann, J.A.M. McDonnell, G.E. Morfill, G. Schwehm, Icarus **129**, 270–288 (1997)

E. Grün, R. Srama, H. Krüger, S. Kempf, V. Dikarev, S. Helfert, G. Moragas-Klostermeyer, Icarus **174**, 1–14 (2005)

C. Gry, E.B. Jenkins, Astron. Astrophys. **367**, 617–628 (2001)

D.A. Gurnett, W.S. Kurth, Nature **454**, 78–80 (2008)

D.A. Gurnett, W.S. Kurth, I.H. Cairns, J. Mitchell, in *Physics of the Inner Heliosheath*, ed. by J. Heerikhuisen, V. Florinski, G.P. Zank, N.V. Pogorelov. American Institute of Physics Conference Series, vol. 858 (Springer, New York, 2006), pp. 129–134

J.L. Han, Chin. J. Astron. Astrophys. Suppl. **6**, 211–217 (2006)

J. Heerikhuisen, V. Florinski, G.P. Zank, J. Geophys. Res. **111**(A10), 6110 (2006)

J. Heerikhuisen, N.V. Pogorelov, G.P. Zank, V. Florinski, Astrophys. J. Lett. **655**, L53–L56 (2007)

J. Heerikhuisen, N.V. Pogorelov, V. Florinski, G.P. Zank, J.A. le Roux, Astrophys. J. **682** , 679–689 (2008)

D.B. Henley, R.L. Shelton, K.D. Kuntz, Astrophys. J. **661**, 304–319 (2007)

L.M. Hobbs, Astrophys. J. **157**, 135 (1969)

T.E. Holzer, J. Geophys. Res. **77**, 5407–5431 (1972)

T.E. Holzer, Annu. Rev. Astron. Astrophys. **27**, 199–234 (1989)

V. Izmodenov, B.E. Wood, R. Lallement, J. Geophys. Res. (Space Phys.) **107**, 1308 (2002)

V. Izmodenov, Y. Malama, G. Gloeckler, J. Geiss, Astron. Astrophys. **414**, L29–L32 (2004)

V. Izmodenov, D. Alexashov, A. Myasnikov, Astron. Astrophys. **437**, L35–L38 (2005)

V.V. Izmodenov, Astrophys. Space Sci. **274**, 55–69 (2000)

V.V. Izmodenov, Space Sci. Rev. **97**, 385–388 (2001)

V.V. Izmodenov, D.B. Alexashov, Astron. Lett. **29**, 58–63 (2003)

V.V. Izmodenov, J. Geiss, R. Lallement, G. Gloeckler, V.B. Baranov, Y.G. Malama, J. Geophys. Res. **104**, 4731–4741 (1999a)

V.V. Izmodenov, J. Geiss, R. Lallement, G. Gloeckler, V.B. Baranov, Y.G. Malama, J. Geophys. Res. **104**, 4731–4742 (1999b)

V.V. Izmodenov, M.A. Gruntman, Y.G. Malama, J. Geophys. Res. **106**, 10,681–10,689 (2001)

A.P. Jones, A.G.G.M. Tielens, D.J. Hollenbach, C.F. McKee, Astrophys. J. **433**, 797–810 (1994)

W.-T. Kim, E.C. Ostriker, Astrophys. J. **570**, 132–151 (2002)

D. Koutroumpa, R. Lallement, V. Kharchenko, A. Dalgarno, ArXiv e-prints 805 (2008)

H. Krüger, N. Altobelli, B. Anweiler, S.F. Dermott, V. Dikarev, A.L. Graps, E. Grün, B.A. Gustafson, D.P. Hamilton, M.S. Hanner, M. Horányi, J. Kissel, M. Landgraf, B.A. Lindblad, D. Linkert, G. Linkert, I. Mann, J.A.M. McDonnell, G.E. Morfill, C. Polanskey, G. Schwehm, R. Srama, H.A. Zook, Planet. Space Sci. **54**, 932–956 (2006)

H. Krüger, M. Landgraf, N. Altobelli, E. Grün, Space Sci. Rev. **130**, 401–408 (2007)

W.S. Kurth, D.A. Gurnett, J. Geophys. Res. **108**, 2–16 (2003)

R. Lallement, P. Bertin, Astron. Astrophys. **266**, 479–485 (1992)

R. Lallement, R. Ferlet, Astron. Astrophys. **324**, 1105–1114 (1997)

R. Lallement, A. Vidal-Madjar, R. Ferlet, Astron. Astrophys. **168**, 225–236 (1986)

R. Lallement, R. Ferlet, A.M. Lagrange, M. Lemoine, A. Vidal-Madjar, Astron. Astrophys. **304**, 461–474 (1995)

R. Lallement, E. Quémerais, J.L. Bertaux, S. Ferron, D. Koutroumpa, R. Pellinen, Science **307**, 1447–1449 (2005)

M. Landgraf, J. Geophys. Res. **105**, 10,303–10,316 (2000)

M. Landgraf, E. Grün, in *IAU Colloq. 166: The Local Bubble and Beyond*, ed. by D. Breitschwerdt, M.J. Freyberg, J. Truemper. Lecture Notes in Physics, vol. 506 (Springer, Berlin, 1998), pp. 381–384
M. Landgraf, K. Augustsson, E. Grün, B.Å.S. Gustafson, Science **286**, 2319–2322 (1999)
M. Landgraf, W.J. Baggaley, E. Grün, H. Krüger, G. Linkert, J. Geophys. Res. **105**, 10343–10352 (2000)
M. Landgraf, H. Krüger, N. Altobelli, E. Grün, J. Geophys. Res. (Space Phys.) **108**, 5–1 (2003)
N. Lehner, E. Jenkins, C. Gry, H. Moos, P. Chayer, S. Lacour, Astrophys. J. **595**, 858–879 (2003)
T.J. Linde, T.I. Gombosi, J. Geophys. Res. **105**, 10,411–10,418 (2000)
T.J. Linde, T.I. Gombosi, P.L. Roe, K.G. Powell, D.L. DeZeeuw, J. Geophys. Res. **103**, 1889–1904 (1998)
J.L. Linsky, B.E. Wood, Astrophys. J. **463**, 254–270 (1996)
J.L. Linsky, B.J. Rickett, S. Redfield, Astrophys. J. **675**, 413–419 (2008)
K. Lodders, Astrophys. J. **591**, 1220–1247 (2003)
P.B. Lucke, Astron. Astrophys. **64**, 367–377 (1978)
E. Möbius, M. Bzowski, S. Chalov, H.-J. Fahr, G. Gloeckler, V. Izmodenov, R. Kallenbach, R. Lallement, D. McMullin, H. Noda, M. Oka, A. Pauluhn, J. Raymond, D. Ruciński, R. Skoug, T. Terasawa, W. Thompson, J. Vallerga, R. von Steiger, M. Witte, Astron. Astrophys. **426**, 897–907 (2004)
H.-R. Müller, G.P. Zank, J. Geophys. Res. 7104–7116 (2004)
J. Maíz-Apellániz, Astrophys. J. Lett. **560**, L83–L86 (2001)
Y.G. Malama, V.V. Izmodenov, S.V. Chalov, Astron. Astrophys. **445**, 693–701 (2006)
I. Mann, H. Kimura, J. Geophys. Res. **105**, 10,317–10,328 (2000)
L.A. Marschall, L.M. Hobbs, Astrophys. J. **173**, 43–62 (1972)
J.S. Mathis, W. Rumpl, K.H. Nordsieck, Astrophys. J. **217**, 425–433 (1977)
W. McClintock, R.C. Henry, H.W. Moos, J.L. Linsky, Astrophys. J. Lett. **204**, L103–L106 (1976)
D.J. McComas, N.A. Schwadron, Geophys. Res. Lett. **33**, 4102 (2006)
D. McComas, F. Allegrini, P. Bochsler, M. Bzowski, M. Collier, H. Fahr, H. Fichtner, P. Frisch, H. Funsten, S. Fuselier, G. Gloeckler, M. Gruntman, V. Izmodenov, P. Knappenberger, M. Lee, S. Livi, D. Mitchell, E. Möbius, T. Moore, D. Reisenfeld, E. Roelof, N. Schwadron, M. Wieser, M. Witte, P. Wurz, G. Zank, in *AIP Conf. Proc. 719: Physics of the Outer Heliosphere*, pp. 162–181 (2004a)
D.J. McComas, N.A. Schwadron, F.J. Crary, H.A. Elliott, D.T. Young, J.T. Gosling, M.F. Thomsen, E. Sittler, J.-J. Berthelier, K. Szego, A.J. Coates, J. Geophys. Res. (Space Phys.) **109**, 2104 (2004b)
D. McComas, F. Allegrini, L. Bartolone, P. Bochsler, M. Bzowski, M. Collier, H. Fahr, H. Fichtner, P. Frisch, H. Funsten, S. Fuselier, G. Gloeckler, M. Gruntman, V. Izmodenov, P. Knappenberger, M. Lee, S. Livi, D. Mitchell, E. Möbius, T. Moore, S. Pope, D. Reisenfeld, E. Roelof, H. Runge, J. Scherrer, N. Schwadron, R. Tyler, M. Wieser, M. Witte, P. Wurz, G. Zank, in *Solar Wind 11/SOHO 16, Connecting Sun and Heliosphere*. ESA Special Publication, vol. 592, pp. 689–692 (2005)
D.J. McComas, F. Allegrini, L. Bartolone, P. Bochsler, M. Bzowski, M. Collier, H. Fahr, H. Fichtner, P. Frisch, H. Funsten, S. Fuselier, G. Gloeckler, M. Gruntman, V. Izmodenov, P. Knappenberger, M. Lee, S. Livi, D. Mitchell, E. Möbius, T. Moore, S. Pope, D. Reisenfeld, E. Roelof, H. Runge, J. Scherrer, N. Schwadron, R. Tyler, M. Wieser, M. Witte, P. Wurz, G. Zank, in *Physics of the Inner Heliosheath*, ed. by J. Heerikhuisen, V. Florinski, G.P. Zank, N.V. Pogorelov. American Institute of Physics Conference Series, vol. 858 (Springer, Berlin, 2006), pp. 241–250
D.J. McComas et al., Space Sci. Rev. (2009, this issue)
J.J. Mitchell, I.H. Cairns, H.-R. Müller, G.P. Zank, Geophys. Res. Lett. **32**, 7101 (2005)
E. Möbius, D. Hovestadt, B. Klecker, M. Scholer, G. Gloeckler, Nature **318**, 426–429 (1985)
H.-R. Müller, G.P. Zank, A.S. Lipatov, J. Geophys. Res. **105**, 27,419–27,438 (2000)
H.-R. Müller, P.C. Frisch, V. Florinski, G.P. Zank, Astrophys. J. **647**, 1491–1505 (2006)
H.-R. Müller, V. Florinski, J. Heerikhuisen, V.V. Izmodenov, K. Scherer, D. Alexashov, H. Fahr, Astron. Astrophys. **491**, 43–51 (2008)
G. Munch, A. Unsold, Astrophys. J. **135**, 711–715 (1962)
W.R. Oegerle, E.B. Jenkins, R.L. Shelton, D.V. Bowen, P. Chayer, Astrophys. J. **622**, 377–389 (2005)
M. Opher, E.C. Stone, T.I. Gombosi, Science **316**, 875 (2007)
M. Opher, E.C. Stone, G. Toth, V. Izmodenov, V. Alexashov, T.I. Gombosi, AGU Fall Meeting Abstracts page A7 (2008)
V. Piirola, Astron. Astrophys. Suppl. Ser. **30**, 213 (1977)
N. Pogorelov, T. Matsuda, J. Geophys. Res. **103**, 237 (1998)
N.V. Pogorelov, G.P. Zank, T. Ogino, Astrophys. J. **614**, 1007–1021 (2004)
N.V. Pogorelov, G.P. Zank, T. Ogino, Astrophys. J. **644**, 1299–1316 (2006)
N.V. Pogorelov, E.C. Stone, V. Florinski, G.P. Zank, Astrophys. J. **668**, 611–624 (2007)
N.V. Pogorelov, J. Heerikhuisen, G.P. Zank, Astrophys. J. Lett. **675**, L41–L44 (2008a)
N.V. Pogorelov, G.P. Zank, T. Ogino, Adv. Space Res. **41**, 306–317 (2008b)
E. Quémerais, V. Izmodenov, Astron. Astrophys. **396**, 269–281 (2002)

E. Quémerais, R. Lallement, S. Ferron, D. Koutroumpa, J.-L. Bertaux, E. Kyrölä, W. Schmidt, J. Geophys. Res. **111**, A10/9114 (2006)
R. Ratkiewicz, A. Barnes, G.A. Molvik, J.R. Spreiter, S.S. Stahara, M. Vinokur, S. Venkateswaran, Astron. Astrophys. **335**, 363–369 (1998)
S. Redfield, R.E. Falcon, Astrophys. J. **683**, 207–225 (2008)
S. Redfield, J.L. Linsky, Astrophys. J. **551**, 413–428 (2001)
S. Redfield, J.L. Linsky, Astrophys. J. Suppl. Ser. **139**, 439–465 (2002)
S. Redfield, J.L. Linsky, Astrophys. J. **602**, 776–802 (2004a)
S. Redfield, J.L. Linsky, Astrophys. J. **613**, 1004–1022 (2004b)
S. Redfield, J.L. Linsky, Astrophys. J. **673**, 283–314 (2008)
J.D. Richardson, Y. Liu, C. Wang, Adv. Space Res. **41**, 237–244 (2008)
I.P. Robertson, T.E. Cravens, J. Geophys. Res. **108**, 6–1 (2003)
J.B. Rogerson, D.G. York, J.F. Drake, E.B. Jenkins, D.C. Morton, L. Spitzer, Astrophys. J. Lett. **181**, L110–L114 (1973)
P. Routly, J.L. Spitzer, Astrophys. J. **115**, 227 (1952)
D. Ruciński, M. Bzowski, Astron. Astrophys. **296**, 248–263 (1995)
D. Ruciński, H.J. Fahr, S. Grzędzielski, Planet. Space Sci. **41**, 773–783 (1993)
D. Ruciński, A.C. Cummings, G. Gloeckler, A.J. Lazarus, E. Mobius, M. Witte, Space Sci. Rev. **78**, 73–84 (1996)
D. Russeil, Astron. Astrophys. **397**, 133–146 (2003)
E. Salerno, F. Bühler, P. Bochsler, H. Busemann, M.L. Bassi, G.N. Zastenker, Y.N. Agafonov, N.A. Eismont, Astrophys. J. **585**, 840–849 (2003)
B.D. Savage, N. Lehner, Astrophys. J. Suppl. Ser. **162**, 134–160 (2006)
N.A. Schwadron, G. Gloeckler, Space Sci. Rev. **130**, 283–291 (2007)
N.A. Schwadron, D.J. McComas, Geophys. Res. Lett. **34**, 14,105 (2007)
N.A. Schwadron, M.A. Lee, D.J. McComas, Astrophys. J. **675**, 1584–1600 (2008)
D.M. Sfeir, R. Lallement, F. Crifo, B.Y. Welsh, Astron. Astrophys. **346**, 785–797 (1999)
J.D. Slavin, Astrophys. J. **346**, 718–727 (1989)
J.D. Slavin, P.C. Frisch, Astrophys. J. Lett. **651**, 37 (2006)
J.D. Slavin, P.C. Frisch, Astron. Astrophys. **491**, 53–68 (2008)
S.L. Snowden, M.J. Freyberg, K.D. Kuntz, W.T. Sanders, Astrophys. J. Suppl. Ser. **128**, 171–212 (2000)
L. Spitzer, *Physical Processes in the Interstellar Medium* (Wiley, New York, 1978)
E.C. Stone, A.C. Cummings, F.B. McDonald, B.C. Heikkila, N. Lal, W.R. Webber, Science **309**, 2017–2020 (2005)
E.C. Stone, A.C. Cummings, F.B. McDonald, B.C. Heikkila, N. Lal, W.R. Webber, Nature **454**, 71–74 (2008)
G.E. Thomas, Ann. Rev. Earth Planet. Sci. **6**, 173–204 (1978)
G.E. Thomas, R.F. Krassa, Astron. Astrophys. **11**, 218–233 (1971)
J. Tinbergen, Astron. Astrophys. **105**, 53–64 (1982)
J. Vallerga, Space Sci. Rev. **78**, 277–288 (1996)
J. Vallerga, Astrophys. J. **497**, 921–927 (1998)
J. Vallerga, R. Lallement, M. Lemoine, F. Dalaudier, D. McMullin, Astron. Astrophys. **426**, 855–865 (2004)
J.V. Vallerga, P.W. Vedder, N. Craig, B.Y. Welsh, Astrophys. J. **411**, 729–749 (1993)
M.K. Wallis, Nature **254**, 202 (1975)
J.C. Weingartner, B.T. Draine, Astrophys. J. Suppl. Ser. **134**, 263–281 (2001)
C.S. Weller, R.R. Meier, Astrophys. J. **193**, 471–476 (1974)
C.S. Weller, R.R. Meier, Astrophys. J. **246**, 386–393 (1981)
B.Y. Welsh, Space Sci. Rev. 57 (2008)
D.E. Welty, L.M. Hobbs, Astrophys. J. Suppl. Ser. **133**, 345–393 (2001)
D.E. Welty, L.M. Hobbs, J.T. Lauroesch, D.C. Morton, L. Spitzer, D.G. York, Astrophys. J. Suppl. Ser. **124**, 465–501 (1999)
D.E. Welty, L.M. Hobbs, D.C. Morton, Astrophys. J. Suppl. Ser. **147**, 61–96 (2003)
D.E. Welty, D.C. Morton, L.M. Hobbs, Astrophys. J. Suppl. Ser. **106**, 533 (1996a)
D.E. Welty, D.C. Morton, L.M. Hobbs, Astrophys. J. Suppl. Ser. **106**, 533–562 (1996b)
M.E. Wiedenbeck, W.R. Binns, A.C. Cummings, A.J. Davis, G.A. de Nolfo, M.H. Israel, R.A. Leske, R.A. Mewaldt, E.C. Stone, T.T. von Rosenvinge, Space Sci. Rev. **130**, 415–429 (2007)
M. Witte, Astron. Astrophys. **426**, 835–844 (2004)
B. Wolff, D. Koester, R. Lallement, Astron. Astrophys. **346**, 969–978 (1999)
M. Wolleben, Astrophys. J. **664**, 349–356 (2007)
B.E. Wood, C.W. Ambruster, A. Brown, J.L. Linsky, Astrophys. J. **542**, 411–420 (2000a)
B.E. Wood, J.L. Linsky, G.P. Zank, Astrophys. J. **537**, 304–311 (2000b)
B.E. Wood, S. Redfield, J.L. Linsky, H.-R. Müller, G.P. Zank, Astrophys. J. Suppl. Ser. **159**, 118–140 (2005)

B.E. Wood, V.V. Izmodenov, J.L. Linsky, Y.G. Malama, Astrophys. J. **657**, 609–617 (2007)
B.E. Wood, V.V. Izmodenov, Y.G. Malama, Space Sci. Rev. 75 (2008)
P. Wurz, M.R. Collier, T.E. Moore, D. Simpson, S. Fuselier, W. Lennartson, in *Physics of the Outer Heliosphere*, ed. by V. Florinski, N.V. Pogorelov, G.P. Zank. American Institute of Physics Conference Series, vol. 719 (Springer, Berlin, 2004), pp. 195–200
D.G. York, B.F. Kinahan, Astrophys. J. **228**, 127–146 (1979)
G.P. Zank, Space Sci. Rev. **89**, 413–688 (1999)
G.P. Zank, H.L. Pauls, L.L. Williams, D. Hall, J. Geophys. Res. **101**, 21,636–21,655 (1996)

Space Sci Rev (2009) 146: 275–294
DOI 10.1007/s11214-009-9522-9

Physical Processes in the Outer Heliosphere

**M.A. Lee · H.J. Fahr · H. Kucharek · E. Moebius ·
C. Prested · N.A. Schwadron · P. Wu**

Received: 3 October 2008 / Accepted: 4 May 2009 / Published online: 9 June 2009

Abstract This chapter covers the theory of physical processes in the outer heliosphere that are particularly important for the IBEX Mission, excluding global magnetohydrodynamic/Boltzmann modeling of the entire heliosphere. Topics addressed include the structure and parameters of the solar wind termination shock, the transmission of ions through the termination shock including possible reflections at the shock electrostatic potential, the acceleration and transport of suprathermal ions and anomalous cosmic rays at the termination shock and in the heliosheath, charge-exchange interactions in the outer heliosphere including mass and momentum loading of the solar wind, the transport of interstellar pickup ions, and the production and anticipated intensities of energetic neutral atoms (ENAs) in the heliosphere.

Keywords Outer heliosphere · Solar wind termination shock · Energetic neutral atoms

1 Introduction

The Interstellar Boundary Explorer (IBEX) measures the flux of energetic neutral atoms (ENAs) in the energy range 0.01–6 keV (McComas et al. 2004, 2009, this issue). The diffuse "glow" of atoms in this energy range arises primarily from the heliosheath protons, which are predominantly interstellar pickup protons that are accelerated and transmitted by the solar wind termination shock (TS). These heliosheath protons may in turn undergo a charge-exchange interaction with interstellar hydrogen and produce ENAs with trajectories pointed toward the inner heliosphere.

M.A. Lee (✉) · H. Kucharek · E. Moebius
University of New Hampshire, Institute for the Study of Earth, Oceans and Space, Morse Hall, 8 College Road, Durham, NH 03824, USA
e-mail: marty.lee@unh.edu

H.J. Fahr
Argelander Institut für Astronomie, Universität Bonn, Auf dem Hügel 71, 53121 Bonn, Germany

C. Prested · N.A. Schwadron · P. Wu
Department of Astronomy, Boston University, 725 Commonwealth Avenue, Boston, MA 02215, USA

The diffuse signal observed by IBEX depends primarily on the proton distribution function downstream of the TS, the evolution of the proton distribution function within the heliosheath, and the global structure of the heliosheath including its plasma flow velocity. The global structure of the heliosphere including the heliosheath is the subject of Izmodenov et al. (2009, this issue) and Zank et al. (2009, this issue), and will not be discussed in detail in this chapter. The heliosheath flow is sub-fast-magnetosonic downstream of the TS. It is generally thought to remain so throughout the heliosheath. As a result pressure variations are minimized. We do expect that the heliosheath gas cools due to charge exchange with the interstellar neutrals. However, the cooling should be compensated by compression to insure pressure balance. Thus we do not expect substantial evolution of the heliosheath proton distribution. The major focus of this chapter is therefore on the expected distribution function of protons downstream of the TS. Although there exists information on this proton distribution, which we present in this chapter, a major goal of IBEX is the determination of the downstream proton distribution in the energy range accessible to IBEX.

In Sect. 2 we review the average properties of the solar wind just inside the TS based on the recent observations of Voyagers 1 and 2 and the inferred parameters related to the local interstellar gas and its interaction with the solar wind. The properties of the solar wind certainly exhibit large variations, in particular with the 11-year solar activity cycle. With the exception of the solar cycle variation and the dependence of the solar wind ram pressure on heliographic latitude, the variations should not influence the ENA flux observed by IBEX since the flux involves an approximately radial integration through the heliosheath.

In Sect. 3 we utilize the Rankine-Hugoniot relations for quasi-perpendicular shocks to estimate the downstream single-fluid mass density, velocity, pressure and temperature. In Sect. 4 we discuss previous work on the downstream proton distribution function based on assumptions about its functional form. Finally, in Sect. 5 we outline the formation of the downstream proton distribution based on a superposition of the allowed transmitted ion trajectories including the effect of the shock potential.

The energetic particles associated with the TS at energies below ~3 MeV as observed by Voyagers 1 and 2 (the so-called TSPs) and the anomalous cosmic rays (ACRs) are at energies substantially above those observable by IBEX. Nevertheless, they originate from and extend the distribution of shock-processed ions. Their origin, spatial distribution, and energy spectrum is discussed in Sect. 6.

There are a number of published numerical simulations of the TS structure. Most are hybrid simulations, which do not correctly describe the electron dynamics. However, the width of the shock potential depends sensitively on electron dynamics and controls the number of multiply reflected ions at the shock ramp. To the extent that multiple reflections of ions at the shock ramp are important in forming the downstream proton distribution at IBEX energies, particle-in-cell simulations of the TS are necessary to describe the shock potential accurately. A review of the insights into TS structure obtained with simulations is presented in Sect. 7.

Based on a reasonable downstream proton distribution function and a simplified model of the global heliosphere, Sect. 8 presents a very simple calculation of the heliospheric ENA flux that illustrates the important processes and parameters that govern the ENA intensity as a function of look direction.

The overarching theme of the interaction of the solar wind with the local interstellar medium is the subject of excellent reviews by Zank (1999) and Fahr et al. (2007).

2 The Solar Wind Upstream of the Termination Shock

The material in this section, based on Lee (2006), concerns typical conditions in the solar wind upstream of the TS in the heliocentric radial range of the Voyager encounters with the TS: 83 AU (V2) to 94 AU (V1). The solar wind plasma cannot be measured by Voyager 1. Voyager 2 observed an average nucleon density $n_p \cong 1 \times 10^{-3}$ cm^{-3} and solar wind speed $V_{sw} \cong 300$ km/s upstream of the TS (Richardson et al. 2008a). Voyager 2 also finds somewhat surprisingly that the proton temperature is roughly constant, $T_p \cong 1 \times 10^4$ K, for $r \geq 10$ AU (Williams et al. 1995; Chashei et al. 2003; Smith et al. 2006; Richardson et al. 2008a). This value is far larger than expected from adiabatic expansion of the solar wind. The apparent reason for the high temperature is heating by solar wind turbulence driven in part by the pickup and isotropization of interstellar pickup hydrogen in the solar wind (Williams et al. 1995; Matthaeus et al. 1999; Smith et al. 2001, 2006; Isenberg et al. 2003). The corresponding mean thermal speed is $\cong$15.7 km/s, and the corresponding contribution of the solar wind core to the sound speed is $C_{s,core} \cong 11.7$ km/s.

The magnetic field strength B_0 observed by Voyager 1 during the year prior to the shock traversal varied between 0.01 and 0.1 nT (Burlaga et al. 2005). The average during the few days before the traversal was 0.03 nT. Although the field strength observed by Voyager 2 prior to the crossing of the TS was somewhat higher, we take the Voyager 1 result which is closer to that expected based on field strength measurements at 1 AU. These values imply that the Alfvén speed $V_A \cong 20.7$ km/s.

The final major constituent of the solar wind is the population of interstellar pickup ions, which are created in the solar wind by the ionization of interstellar neutrals. The temperature (and pressure) of these ions is substantial and is not included in the temperature of the core protons quoted above. The pickup ion pressure in the outer heliosphere is readily calculated from the three stationary equations for solar wind mass, momentum and energy flux, neglecting solar gravity and the solar wind magnetic field, and including mass-loading due to photoionization, and momentum and energy loss due to charge exchange (Lee 1997):

$$\frac{1}{r^2}\frac{d}{dr}(r^2\rho V) = m_p\left(\frac{r_0}{r}\right)^2 (\nu_{\rm H}^0 n_{\rm H,\infty} + 4\nu_{\rm He}^0 n_{\rm He,\infty}) \tag{1}$$

$$\frac{1}{r^2}\frac{d}{dr}(r^2\rho V^2) = -\frac{dP}{dr} - \sigma n_{\rm H,\infty}\rho V^2 \tag{2}$$

$$\frac{1}{r^2}\frac{d}{dr}\left[r^2 V\left(\frac{1}{2}\rho V^2 + \frac{\gamma}{\gamma-1}P\right)\right] = -\sigma n_{\rm H,\infty}\frac{1}{2}\rho V^3 \tag{3}$$

See Isenberg (1986) for a more complete set of equations. Here we neglect electron impact ionization, which is generally much less important than photoionization beyond 3 AU (Rucinski et al. 1996) and the momentum and energy of the interstellar atoms, since their speeds of ~20–30 km/s are small compared with the solar wind speed. We consider only interstellar hydrogen and helium. We use the notation: ρ is the solar wind mass density, V is the radial solar wind speed, P is the solar wind pressure (which is completely dominated by pickup ions), γ is the standard ratio of specific heats, r is heliocentric radial distance, m_p is the proton mass, $r_0 = 1$ AU, $\nu_{\rm H}^0$ and $\nu_{\rm He}^0$ are the photoionization rates of hydrogen and helium at 1 AU, and σ is the charge-exchange cross-section for hydrogen. In the outer heliosphere the neutral gas densities are approximately constant and equal to the interstellar value for helium, $n_{\rm He,\infty}$, and the "filtered" interstellar value for hydrogen, $n_{\rm H,\infty}$. The filtering occurs primarily beyond the heliopause.

If $P \ll \rho V^2$, then a solution of (1)–(3) is readily obtained to first order in $P/(\rho V^2)$ (Lee 1997):

$$P = \frac{1}{2}\frac{\gamma - 1}{2\gamma - 1}\frac{r_0^2}{r}\rho_0 V_0 \alpha \tag{4}$$

$$\alpha \equiv \sigma n_{\mathrm{H},\infty} V_0 + m_p \rho_0^{-1}(\nu_{\mathrm{H}}^0 n_{\mathrm{H},\infty} + 4\nu_{\mathrm{He}}^0 n_{\mathrm{He},\infty}) \tag{5}$$

$$C_S^2 = \frac{1}{2}\gamma\frac{\gamma - 1}{2\gamma - 1}V_0 \alpha r \tag{6}$$

$$V = V_0 - \left[1 - \frac{1}{2}\frac{\gamma - 1}{2\gamma - 1}\right]\alpha r \tag{7}$$

where V_0 and ρ_0 are the solar wind speed and mass density at 1 AU, P is the pressure due to pickup ions, and C_S^2 refers to the sound speed including only the pickup ion pressure in the numerator. The solar wind core pressure is additive and obtains as a solution of the homogeneous equations (1)–(3) omitting the pickup ion source terms. Since the pickup ion pressure dominates the solar wind core pressure in the outer heliosphere we neglect the latter. Although the expression for P in (4) appears to diverge for small r, the expression is not valid within $r \sim 4$ AU where n_{H} decreases precipitously from its value $n_{\mathrm{H},\infty}$ in the outer heliosphere. The deceleration of the solar wind due to "loading" by pickup ions is described by the first term in square brackets in (7); the second term describes the smaller acceleration due to the pickup ion pressure gradient. Finally, in terms of the ionization rates, the number density of pickup protons in the outer heliosphere is given by

$$N_p = r_0^2 (V_0 r)^{-1} n_{\mathrm{H},\infty}(\nu_{\mathrm{H}}^0 + \sigma V_0 \rho_0 m_p^{-1}) \tag{8}$$

Note that $N_p/n_p \propto r$ in the outer heliosphere, so that the importance of pickup ions grows with increasing r. The pickup ion pressure and number density given by (4) and (8), respectively, reproduces the pressure of the pickup proton distribution function calculated by Vasyliunas and Siscoe (1976). Reasonable values for the interstellar gas parameters are $n_{\mathrm{H},\infty} = 0.1\ \mathrm{cm}^{-3}$, $n_{\mathrm{He},\infty} = 0.015\ \mathrm{cm}^{-3}$, $\sigma = 2 \times 10^{-15}\ \mathrm{cm}^2$ and $\nu_{\mathrm{H}}^0 \cong \nu_{\mathrm{He}}^0 = 10^{-7}\ \mathrm{s}^{-1}$ (Rucinski et al. 1996; Gloeckler and Geiss 2001; Möbius et al. 2004). Using these values, $V_0 = 400$ km/s as a nominal solar wind speed at 1 AU, $r = 90$ AU as a typical heliocentric radial distance to the TS near its nose, and those parameters previously quoted, we find in the solar wind just upstream of the termination shock $P = 1.29 \times 10^{-13}\ \mathrm{dyn\,cm}^{-2}$, $N_p = 3.1 \times 10^{-4}\ \mathrm{cm}^{-3}$, $C_{S,PUI} = 113$ km/s, $\Delta V = V_0 - V = 116$ km/s, and $\rho V^2 = 13.5 \times 10^{-13}\ \mathrm{dyn\,cm}^{-2}$, where $C_{S,PUI}$ is the sound speed due to pickup ion pressure only. We note that the parameter values listed here and in the next section, quoted to three-figure accuracy, are representative rather than precise.

Based on these values it is clear that the perturbation solution of (1)–(3) in powers of $P/(\rho V_0^2)$ is valid, the deceleration of the solar wind is substantial, the internal pressure of the solar wind is dominated by pickup ions, and N_p/n_p ($\cong 0.3$) is large. These values imply that the plasma-$\beta = C_S^2/V_A^2 \gg 1$, so that the termination shock is essentially a hydrodynamic shock.

The estimated slowdown of the solar wind due to pickup ion mass loading and charge-exchange momentum and energy loss is larger than that inferred by Richardson et al. (2008b) from a comparison of the solar wind speed measured by Ulysses and that measured by Voyager 2 at ~79 AU. They find a slowdown of ~67 km/s between 5 AU and 78 AU. Over this radial range, (7) with our choice of parameters yields ~94 AU, a somewhat high value.

Richardson et al. (2008b) compare the measured result with a 1-D magnetohydrodynamic (MHD) model (Wang et al. 2000) that includes the slowdown due to charge exchange. For $n_{H,\infty} = 0.09\ cm^{-3}$ they find good agreement. However, (5) with our choice of parameters indicates that photoionization of hydrogen and helium increases the slowdown by 25%. With photoionization included and $n_{H,\infty} = 0.1\ cm^{-3}$, their model would yield a ~35% larger result of ~90.5 km/s, essentially identical with our result. The models appear to agree. What remains a puzzle is that the observed slowdown appears to be somewhat small. A perhaps related puzzle is the observed slowdown of the solar wind by ~100 km/s during the 150 days prior to the TS encounter by Voyager 2 (Richardson et al. 2008a).

The pickup ion distribution function in the frame of the solar wind arises primarily from a balance between the production rate of newborn pickup ions and adiabatic deceleration in the expanding solar wind. As a result, upstream of the TS the phase-space density varies as $\sim v^{-3/2}$ within $V/9 \leq v \leq V$; outside this range the density is small or vanishes (Vasyliunas and Siscoe 1976). The domain $v \leq V/9$ corresponds to ion pickup within $r \sim 4$ AU where ionization has depleted the neutral gas density. However, the phase-space density likely has a suprathermal "tail" due to acceleration mechanisms in the solar wind such as stochastic acceleration (Bogdan et al. 1991), transit-time damping (Fisk et al. 2000), and shock acceleration (Giacalone 2000). Suprathermal tails are observed in the inner heliosphere (Gloeckler et al. 1994, 2000).

Thus, the ion distribution function just upstream of the termination shock consists of a cold solar wind core with a characteristic thermal speed of ~20 km/s for protons, pickup ions extending from ~40 km/s to ~400 km/s in the solar wind frame, and a tail for $v \geq$ 400 km/s presumably dominated by pickup ions. Of course, there is also a tenuous halo of high-energy galactic and anomalous cosmic rays (ACRs). The ACRs originate at the termination shock and will be discussed later.

3 Rankine-Hugoniot Relations at the Termination Shock

From Sect. 2 we have $C_S^2 = C_{S,PUI}^2 + C_{S,Core}^2$, where we neglect the contribution of the suprathermal solar wind to the sound speed. With the values quoted we have $C_S =$ 113.6 km/s, which yields a plasma-$\beta = C_S^2/V_A^2 = 30$ just upstream of the termination shock. The solar wind in the outer heliosphere is thus a very high-β plasma. With a nominal solar wind speed of 400 km/s and $\Delta V = 116$ km/s, we find $V_{SW} = 284$ km/s, which is consistent with the Voyager 2 measurements upstream of the TS. On average the TS is stationary in the frame of the Sun. The characteristic timescale for a parcel of plasma to traverse the heliosheath near the nose of the heliosphere is ~40 AU/100 km/s ≈ 2 years. Since the ENA measurements by IBEX average over the line-of-sight through the heliosheath and the observed hydrogen atoms (0.01–6 keV) have speeds greater than ~50 km/s, a finite TS speed and variations in heliosheath parameters are only important for variations with timescales greater than the traversal timescale. Thus, the ENA signal near the nose of the heliosphere is expected to vary through the solar activity cycle with the variation of V_{SW}, particularly at higher latitudes. We note that the parameters describing the TS as observed by Voyagers 1 and 2 will differ from the average parameters due to the motion of the TS during the encounters.

The Rankine-Hugoniot relations follow from the conservation of mass, momentum, energy and magnetic flux across a stationary planar shock. They determine the downstream mass density ρ, single-fluid isotropic pressure P, bulk plasma velocity $\mathbf{V}$, and magnetic field $\mathbf{B}$ as functions of the upstream values. The appropriate ratio of specific heats for the solar wind plasma is $\gamma = 5/3$, where the downstream conditions are specified sufficiently far downstream of the shock ramp that the particle distribution functions are nearly isotropic in the bulk plasma frame.

Without loss of generality, we assume that the shock normal is in the x-direction and that the upstream value of $\mathbf{V} \times \mathbf{B}$ is in the y-direction. Continuity of the electric field $\mathbf{E}$, which follows from the stationary induction equation and the requirement that there be no net accumulation of charge at the shock layer, requires that the downstream value of $\mathbf{V} \times \mathbf{B}$ is also in the y-direction. The (x, z)-plane is the "coplanarity" plane. We then choose $\mathbf{V} \times \mathbf{B} = 0$ both upstream and downstream of the shock by shifting to the deHoffman-Teller frame. In that frame the conservation laws are

$$\rho_1 V_{x1} = \rho_2 V_{x2} \tag{9}$$

$$\rho_1 V_{x1}^2 + P_1 + \frac{B_{x1}^2}{8\pi}\frac{V_{z1}^2}{V_{x1}^2} = \rho_2 V_{x2}^2 + P_2 + \frac{B_{x2}^2}{8\pi}\frac{V_{z2}^2}{V_{x2}^2} \tag{10}$$

$$\rho_1 V_{x1} V_{z1} - \frac{B_{x1}^2}{4\pi}\frac{V_{z1}}{V_{x1}} = \rho_2 V_{x2} V_{z2} - \frac{B_{x2}^2}{4\pi}\frac{V_{z2}}{V_{x2}} \tag{11}$$

$$\rho_1 V_{x1}\left(\frac{1}{2}V_1^2 - \frac{\gamma}{\gamma - 1}\frac{P_1}{\rho_1}\right) = \rho_2 V_{x2}\left(\frac{1}{2}V_2^2 + \frac{\gamma}{\gamma - 1}\frac{P_2}{\rho_2}\right) \tag{12}$$

$$B_{x1} = B_{x2} \tag{13}$$

where subscripts 1 and 2 refer to upstream and downstream quantities respectively. In terms of the parameters $\cos\theta = B_{x1}/B_1$, $V_{A1}^2 = B_1^2/(4\pi\rho_1)$, $C_1^2 = \gamma P_1/\rho_1$, and $\beta_1 = C_1^2/V_{A1}^2$, we obtain

$$\frac{\rho_2}{\rho_1} = X \tag{14}$$

$$\frac{V_{x2}}{V_{x1}} = \frac{1}{X} \tag{15}$$

$$\frac{V_{z2}}{V_{z1}} = \chi = \frac{V_{x1}^2 - V_{A1}^2\cos^2\theta}{V_{x1}^2 - X V_{A1}^2\cos^2\theta} \tag{16}$$

$$\frac{P_2}{P_1} = \gamma\frac{V_{x1}^2}{C_1^2}\left(1 - \frac{1}{X}\right) + 1 + \gamma\frac{V_{A1}^2\sin^2\theta}{2C_1^2} - \gamma\frac{V_{A1}^2\sin^2\theta}{2C_1^2}\chi^2 X^2 \tag{17}$$

$$\frac{P_2}{P_1} = X + \frac{X(\gamma - 1)V_{x1}^2}{2C_1^2}\left[\left(1 - \frac{1}{X^2}\right) + (1 - \chi^2)\tan^2\theta\right] \tag{18}$$

where X is the plasma shock compression ratio. Equating expressions (17) and (18), and eliminating the factor $(X - 1)$, we obtain a cubic equation for X.

The Archimedes spiral magnetic field in the solar wind is tightly wrapped in the outer heliosphere. The TS is approximately spherical but not centered on the Sun. The heliocentric radial distance of the TS in the direction of the heliospheric tail is about twice that in the nose direction. The deviation of the TS from spherical symmetry about the Sun is important for the acceleration of the ACR and we shall return to this point in Sect. 6. As a result the TS deviates from being a perpendicular shock along its flanks with $\theta \geq 60°$. Thus, a reasonable approximation is that the TS is nearly perpendicular on average ($B_x \cong 0$). For a perpendicular shock the term in the cubic equation proportional to X^3 vanishes; the resulting quadratic equation yields

$$X_\perp = [2(2-\gamma)]^{-1} \times \{-2\beta_1 - \gamma - (\gamma-1)M_{A1}^2 + [(2\beta_1 + \gamma + (\gamma-1)M_{A1}^2)^2 + 4(2-\gamma)(\gamma+1)M_{A1}^2]^{1/2}\} \quad (19)$$

where $M_{A1}^2 = V_{x1}^2/V_{A1}^2$ is the Alfven Mach number. Using the values we have taken, we obtain $X_\perp = 2.67$. Including the corrections of order $\cos^2\theta$ to the cubic equation, we obtain $X = X_\perp(1 + X_1\cos^2\theta)$, where $X_1 = O(M_{A1}^{-2}, \beta_1^{-1})$ and $X_1 > 0$. Clearly X increases as θ decreases since the magnetic field contributes less to the downstream pressure, and the required increase in downstream plasma pressure is insured by larger plasma compression. As expected for the nearly hydrodynamic TS, the correction due to finite $\cos^2\theta$ is small.

As $V_A^2 \to 0$, (19) reduces to the hydrodynamic compression ratio $X_h = (\gamma+1)M_1^2/[2 + (\gamma-1)M_1^2]$, where $M_1^2 = (V_{x1}/C_1)^2$ is the sonic Mach number. With $V_{x1} = V_{SW} = 284$ km/s and $C_1 = 113.6$ km/s, we obtain 2.70, which is virtually identical to the magnetohydrodynamic result. This value for the compression ratio is very close to 2.6, the value inferred by Stone et al. (1996) based on the ACR energy spectra observed by Voyagers 1 and 2, and by Isenberg (1997a) based on (4)–(7) neglecting photoionization. Neglecting the magnetic field we obtain $P_2 = P_1 + \rho_1 V_{x1}^2(1 - X^{-1}) \cong 9.80 \times 10^{-13}$ dyn cm $^{-2}$. The corresponding temperature is $T_2 \cong 2.63 \times 10^6$ K, where we have neglected the electron contribution to the pressure. The average temperature T_2 measured by Voyager 2 downstream of the TS is 1.4×10^5 K (Richardson et al. 2008a), implying that the plasma instrument measures a solar wind "core" distribution and cannot detect the pickup ion "halo."

4 Models for the Proton Distribution Function Downstream of the Termination Shock

Since interstellar pickup protons dominate the (internal) pressure of the solar wind upstream of the TS, the implication of Sect. 3 was that protons, rather than electrons, dominate the downstream pressure. This feature of the TS actually follows from the magnitude of our derived upstream Mach number, $M = 2.46$ (expressed in terms of the MHD "fast" speed), which implies that the shock is "supercritical" since $M > M_1^*$, the first critical Mach number (Kennel et al. 1985). M_1^* is defined by the equality of the component of the downstream flow velocity normal to the shock and the downstream sound speed. For $M < M_1^*$, ions play a minor role in shock dissipative processes; joule heating by electrons carrying the current in the shock ramp provides the dissipation. However, for $M > M_1^*$, ions play a dominant role in shock dissipation. The oft-quoted value $M_1^* = 2.76$ for a perpendicular shock applies only for plasma-$\beta = 0$. In our case plasma-$\beta = C_S^2/V_A^2 \sim 30$, for which $M_1^* \le 1.1$ (Kennel et al. 1985). Thus, the termination shock is strongly supercritical and dominated by ion heating.

The form of the downstream proton distribution function poses a challenging question. Based on the downstream mass density, pressure and temperature, various assumptions can be made. In their illustrative prediction of the ENA flux from the heliosheath at 1 AU, Gruntman et al. (2001) assumed various schematic Maxwellian proton distributions and a uniformly populated sphere in velocity space centered on the heliosheath flow to model the heliosheath pickup proton distribution. In order to account for the accelerated pickup proton halo, Prested et al. (2008) assume a proton κ-distribution, which exhibits a power-law suprathermal tail. Choosing the parameters of the κ-distribution to be consistent with global MHD simulations of the heliosphere, Prested et al. (2008) have constructed maps of the ENA flux. As expected, they predict substantially higher fluxes at the higher energies detected by IBEX in comparison with the results of Gruntman et al. (2001).

The proton distribution just downstream of the TS, of course, need not be isotropic. Fahr and Siewert (2006) and Siewert and Fahr (2007) have computed the evolution of protons, conserving their first and second adiabatic invariants across a laminar TS, and find that the downstream distribution is anisotropic. Although the adiabatic assumption is presumably limited to weak shocks, it is clear that anisotropic downstream distributions are obtained in general. Similar behavior is observed at Earth's bow shock (Sckopke et al. 1990). However, these distributions are unstable (Liu et al. 2005; Fahr and Siewert 2007) and relax to nearly isotropic distributions further downstream of the TS. For the ENA intensities observed by IBEX, which averages over the heliosheath, we need to consider the broader shock transition between states in which the proton distributions are nearly isotropic.

5 Proton Reflection and Transmission at the Termination Shock

At supercritical quasi-perpendicular shocks the ion heating occurs by ion reflection in the shock "ramp" due to a combination of electric and magnetic fields (Leroy et al. 1982; Gosling and Robson 1985; Liu et al. 2005). A fraction of the incoming upstream ions are reflected back into the upstream plasma where they gyrate back to the shock ramp while they gain energy in the motional ($\mathbf{E} = -c^{-1}\mathbf{V} \times \mathbf{B}_0$) electric field. Those with larger normal components of their original incident velocity are able to penetrate the ramp potential during the second encounter; they gyrate around those downstream core protons that are directly transmitted by the shock potential, and form a "heated" distribution. Those incident ions with smaller normal components of their incident velocity may skip (or surf) along the ramp potential many times before they penetrate the potential; they gain substantially more energy in the motional electric field (Zank et al. 1996; Lee et al. 1996). The ramp potential adjusts its magnitude to reflect enough ions to provide the required heating of the downstream plasma. Since at the TS about half the pickup protons (this half comprising $\sim$12% of all protons) are reflected with increasing potential before any of the core protons, and some of these acquire substantial energy by multiple reflections at the shock, the pickup protons may provide all or most of the heating. Additional heating is provided by the reflected minor pickup ions and all the solar wind core minor ions, which pass through the ramp potential and gyrate about the core protons with much smaller speed than the reflected ions. Ion reflection has been observed by Cluster at Earth's bow shock (Möbius et al. 2001; Kucharek et al. 2004).

Figure 1 is a schematic diagram of a perpendicular shock in the frame of the shock, where x measures distance in the direction of the shock normal from upstream to downstream. The vectors $\mathbf{V}_u$ and $\mathbf{V}_d$ represent the bulk velocities of the upstream and downstream plasma, respectively, in the "normal-incidence" frame in which both vectors are parallel to the shock normal. The lower profile shows the shock potential $\Phi(x)$, while the upper profile shows the magnetic field magnitude $|\mathbf{B}|$. Both show a smaller increase in the "foot" region followed by a steep large increase in the "ramp" region, an "overshoot," and downstream structures on the scale of an advected proton gyroperiod. The "ramp" scale is intermediate between the proton and electron inertial lengths. The upper two curves show schematic trajectories of a transmitted core proton, and a reflected pickup proton, which has a sufficiently large normal component of velocity at its second encounter with the shock ramp to be transmitted and gyrate about the downstream core protons.

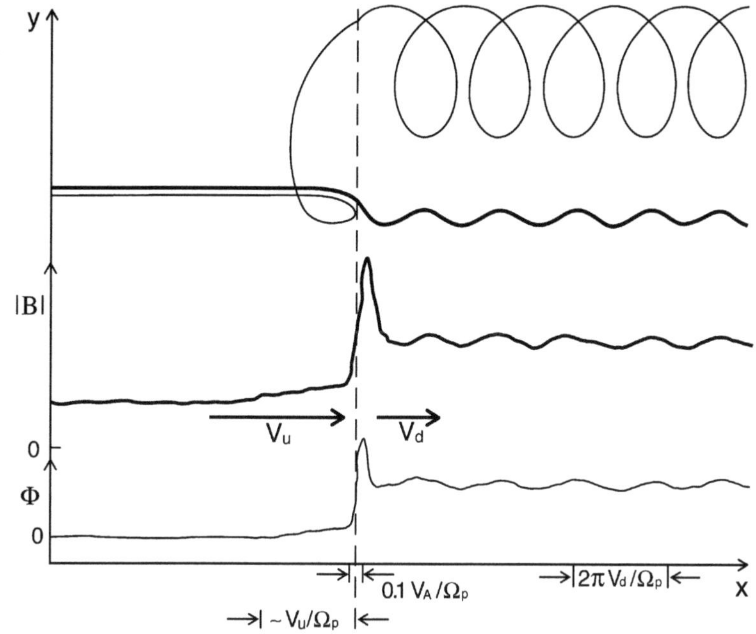

Fig. 1 Schematic diagram of a perpendicular shock with upstream and downstream flow vectors and, from *bottom* to *top*, the shock potential (*thin curve*), the magnetic field strength (*thick curve*), the trajectory of transmitted protons (*thick curve*) and a sample trajectory of an initially reflected proton (*thin curve*). The scale lengths at the *bottom*, from *left* to *right*, refer to the shock "foot," the shock "ramp," and the downstream quasi-periodic variations

We now outline the construction of the downstream pickup proton distribution function. The transport of the transmitted protons close to the shock ramp where the magnetic force can be neglected is governed by the Vlasov equation expressed in the shock frame

$$v_x \frac{\partial f}{\partial x} + \frac{eE_x}{m}\frac{\partial f}{\partial v_x} = 0 \tag{20}$$

where E_x is the electric field in the shock ramp. The solution is $f = g(W, v_y, v_z)$, where g is an arbitrary function of the three independent variables, and $W = 1/2mv_x^2 + e\Phi(x)$. The electric field $E_x = -d\Phi(x)/dx$, where Φ increases from zero to ϕ as x increases through the ramp and the overshoot ($\Phi = \Phi_m$). The distribution function of the pickup protons just across the shock ramp and overshoot is $F[(v_x^2 + 2e\phi/m)^{1/2}, v_y, v_z]$, where $F(v_x, v_y, v_z)$ is the distribution function of the upstream pickup protons neglecting the small fraction of reflected pickup protons upstream. The downstream distribution is only nonzero for $v_x > [2e(\Phi_m - \phi)/m]^{1/2}$. Otherwise the protons are reflected by the potential overshoot. The two free parameters, Φ_m and ϕ, are necessary to control the number of reflected protons and the speed of the heliosheath flow to be consistent with the Rankine-Hugoniot relations.

The Vlasov equation appropriate to the region downstream of the shock is

$$(V_d + \delta v_x)\frac{\partial f}{\partial x} + \Omega\frac{\partial f}{\partial \psi} = 0 \tag{21}$$

where V_d is the downstream flow speed, $\delta v_x = v_x - V_d$, Ω is the downstream cyclotron frequency, and ψ is the proton velocity phase angle about the magnetic field in the frame of the downstream flow. Subject to the distribution function just downstream of the ramp and overshoot, (21) yields the downstream transmitted distribution averaged over the downstream spatial periodicity of length $L^{-1} = \Omega/(2\pi V_d)$ (Liu et al. 2005), appropriate for the calculation of the ENA intensity. A complication for this treatment of the transmitted protons is that those with velocities (perpendicular to the magnetic field) substantially greater than V_d in the downstream fluid frame re-encounter the shock ramp during their first gyration. They accelerate in the negative x-direction in the ramp electric field, gyrate upstream parallel to the motional electric field, and again pass through the shock into the downstream domain. In principle these protons must be included in $F(v_x, v_y, v_z)$.

Those upstream ions described by $F(v_x, v_y, v_z)$ that satisfy $v_x > 0$ and $mv_x^2/2 < e\Phi_m$ are reflected back into the upstream plasma. The reflected portion of $F(v_x, v_y, v_z)$ divides itself into domains in which the ions are transmitted during the next shock encounter, or experience a total of one, two, three or more additional reflections. The number of reflections also depends on the width of the shock potential ramp. For finite ramp thickness the x-component of the magnetic force can eventually exceed the electrostatic force in the ramp and accelerate the ion downstream (Lee et al. 1996). It is therefore clear that electron dynamics need to be included for a complete description of the higher order reflections since the electrons control the ramp thickness. Either after the first or subsequent reflections an ion may satisfy $|v_{x0}|(|v_{x0}| + V)v_{y0}^{-2} \ll 1$, where v_{x0} and v_{y0} are the velocity components immediately following the last reflection. At this point the ion motion is adiabatic in the sense that v_y changes very little during each "bounce." The description of these skipping or "surfing" ions that attain the highest energies downstream is simplified (Lee et al. 1996; Zank et al. 1996). However, the ions that charge-exchange with interstellar hydrogen and produce ENAs in the IBEX energy range are generally those that experience only one or two reflections upstream of the shock. These protons and their downstream distribution function must be treated numerically.

Thus it appears that the dominant "heating" downstream of the termination shock is of the reflected pickup protons. The core solar wind protons may not be reflected at all, or a small fraction of them may be reflected to provide the required downstream heating. Interestingly, if the shock is modified by the pressure of the ACR component, resulting in a weaker subshock, then the solar wind core protons are less likely to participate in the shock dissipation. The temperature of the core protons does increase as they pass through the ramp potential gradient. As mentioned before, the solar wind minor ions, with larger mass per charge, experience less deceleration in the ramp and gyrate with a substantial fraction of the shock velocity decrease about the core protons. The minor pickup ions have a very large mass per charge; relatively few of them are reflected by the shock potential and heated by a large amount.

This qualitative structure of the termination shock, however, appears to be at odds with recent simulations, which indicate that the entropy increase at the shock (that is, heating beyond adiabatic compression) is due to both solar wind core proton reflection, and enhanced acceleration of pickup protons that traverse the shock magnetic field "overshoot" with an optimal gyrophase (Wu et al. 2009). General simulation results are discussed in Sect. 7. In addition, as discussed in the next section, the reduced heating of the pickup minor ions would seem to imply a reduced injection rate into the process of diffusive shock acceleration and an underabundance of the minor ions in the ACR. In fact the ACR minor ions appear to be overabundant.

6 The Termination Shock Energetic Particles and the Anomalous Cosmic Ray Component

The distribution functions of the reflected and transmitted ions discussed in Sect. 5 constitute only the lower-energy portion of the distributions. This portion dominates the production of ENAs in the IBEX energy range. Nevertheless, the ion distributions extend to much higher energies. The dominant energization process is almost certainly shock acceleration in which particles in the lower-energy portion of the distribution are sufficiently mobile to sample the shock compression while crisscrossing the shock by scattering or following stochastic field lines. Another possible acceleration process is stochastic acceleration, in which the ions are accelerated by the turbulence in the heliosheath (Fisk and Gloeckler 2006; Fisk et al. 2006; Ferreira et al. 2007). However, the fact that the heliosheath plasma is high-β would appear to limit the stochastic acceleration rate and the energy available in the hydromagnetic heliosheath turbulence.

During their traversals of the termination shock, Voyagers 1 and 2 observed energy spectral features that revealed two populations of energetic particles. The so-called Termination Shock Particles (TSPs) have a power-law spectrum from the lowest energies observed up to about 3 MeV/nucleon with a rollover beyond. At higher energies is the ACR component, which surprisingly is modulated at energies below ~20 MeV/nucleon. The most likely explanation for the two components is that the TSPs are accelerated locally by the shock, whereas the ACRs are accelerated at a more distant shock location favorable for the acceleration of a larger intensity of particles to higher energies. The power-law index of the TSP differential intensity at Voyager 1 after shock traversal was ~ -1.41 corresponding to the prediction of diffusive shock acceleration for $X = 2.6$. This compression ratio is consistent with that inferred from the observed jump in magnetic field strength and our simple estimate in Sect. 3. The power-law index measured by Voyager 2 after its traversal of the termination shock was ~ -1.1, corresponding to $X = 3.5$. This large compression ratio is inconsistent with that observed for each of three shock traversals by Voyager 2.

A reasonable explanation for the two components is as follows: a reasonable injection threshold for diffusive shock acceleration is $v_{th} = V \sec\theta$ (Jokipii et al. 2004), where θ is the angle between the local shock normal and the upstream magnetic field. The Archimedes spiral magnetic field is nearly perpendicular to the average shock normal near the nose of the heliosphere where the Voyager spacecraft encountered the shock. However, fluctuations of the large-scale magnetic field and warps in the shock surface lead to $v_{th} \geq \alpha V$, where α may be in the range $1 < \alpha < 10$. Only the multiply reflected pickup ions are able to exceed this threshold, which leads to a lower intensity of accelerating particles. The rollover may occur as the accelerating ions are advected toward the flanks of the TS, or escape along the upstream Archimedes spiral magnetic field, as the heliocentric distance to the shock increases with increasing longitude measured from the nose. Further toward the flanks of the shock θ and v_{th} decrease, and the larger density of once and twice reflected and of some transmitted pickup ions exceed the threshold and are accelerated. These ions can only reach the nose portion of the shock sampled by the Voyagers by diffusing along and across the heliosheath magnetic field, a process that suppresses the lower energy ions (McComas and Schwadron 2006; Schwadron et al. 2008). This transport leads naturally to a positive heliosheath radial gradient of ACRs at the location of Voyager 1, as observed. The idea that the injection rate of pickup ions is larger on the flanks of the TS was advocated by Chalov (1993).

Two other origins of the two components have been suggested. Florinski and Zank (2006) suggested that the two-component spectrum observed at Voyager 1 is the result of a solar wind merged interaction region disturbing the expected stationary configuration of energetic particles at the time of its shock crossing. This suggestion does not account for the similar energetic particle structure observed by Voyager 2 when it crossed the shock and no interaction region was present in the solar wind. Fisk and Gloeckler (2006), Fisk et al. (2006) and Ferreira et al. (2007) have suggested that both components are accelerated by stochastic acceleration both upstream of the shock and in the heliosheath. However, for the reasons given near the beginning of this section, this does not appear to be likely.

The local and distant shock interpretation advocated by McComas and Schwadron (2006) and Schwadron et al. (2008) may have difficulty accounting for the TSP spectral index ($= -1.1$) at Voyager 2. However, Voyager 2 measured a decrease in the solar wind speed during $\sim$150 days prior to shock passage from $\sim$400 km/s to $\sim$300 km/s, before it decreased to $\sim$125 km/s downstream of the shock (Richardson et al. 2008a). If the gradual decrease is a shock precursor that the accelerating TSPs sample, then a higher shock compression ratio ($\sim$3.5) is appropriate and consistent with the observed spectral index.

The pickup ion composition of both the TSP and ACR particles supports the idea described in Sect. 5 that the interstellar pickup ions dominate the dissipation at the shock; the heated pickup ions are more readily injected into the process of shock acceleration than the cold solar wind ions. One outstanding puzzle is that the ACR composition favors the pickup minor ions over hydrogen (Cummings and Stone 1996). The reflection process at the shock, however, favors hydrogen in the higher energy tail, which in turn is favored in the injection process. These two facts may be reconciled if the origin of the ACR is on the flanks of the termination shock (McComas and Schwadron 2006; Schwadron et al. 2008). Then the ACRs observed at longitudes near the nose and far from their origin are dominated by the minor ions, which are more mobile at the same energy per nucleon due to their higher rigidity.

It should be emphasized that the TSP distributions do not satisfy the condition of diffusive shock acceleration that anisotropies be small. The most obvious violation is the large anisotropy of the upstream ions. In view of the complex energetic particle structure at quasi-perpendicular shocks observed in Earth orbit, such complications could have been anticipated at the nearly perpendicular termination shock near its nose. It is quite possible that the crisscrossing of the shock by the particles is due primarily to the random walk of the magnetic field lines. In this case the "scattering" of the ions across the shock does not result in nearly isotropic distributions, and the energy gain may be restricted to the field-aligned velocity component (Jokipii and Giacalone 2007) or to the perpendicular components if the acceleration is primarily by gradient drift during each shock encounter along the stochastic field line. The crossing of the shock many times by each magnetic field line may be enhanced by the rippling of the shock front due to irregularities in the solar wind (Giacalone and Jokipii 2007).

7 Particle Simulations of Termination Shock Structure

Magnetohydrodynamic simulations are useful to investigate the global structure of the heliosphere and the large-scale structures within it. Global MHD models of the heliosphere are the subject of Izmodenov et al. (2009, this issue) and Zank et al. (2009, this issue). Kinetic simulations such as hybrid or even full particle simulations are ideal tools to investigate physical processes on length-scales of the order of the ion or electron gyroradius. Wave-particle interactions including minor ions such as interstellar pickup ions can be treated self-consistently with kinetic simulations upstream, downstream, and in the shock layer. In addition, the mechanism of ion injection into the process of diffusive shock acceleration can be studied. Kinetic simulations are of two types: hybrid simulations and Particle-In-Cell (PIC) simulations. We review application of both types of simulations to the solar wind termination shock.

Multi-dimensional hybrid simulations, in which electrons are treated as a massless fluid and ions as macro particles, play an important role in investigating the ion dynamics. Currently multi-component hybrid simulation codes are available in 1D, 2D, and even 3D. In hybrid simulations the ion equations of motion are solved. Electrons are treated as a charge-neutralizing fluid. Different ion species, such as solar wind protons, solar wind alpha particles, and interstellar pickup ions, can be included self-consistently. The initial particle distributions are based on observations as much as possible. Solar wind ions may be approximated with a Maxwellian or a kappa distribution (Scholer et al. 2002). Pickup ions are often described by a distribution that fills a sphere in velocity space centered on the solar wind velocity with a radius equal to the solar wind speed. The spherical distribution may be given by the isotropic distribution derived by Vasyliunas and Siscoe (1976), or the anisotropic distribution of Isenberg (1997b) with different values in the sunward and antisunward hemispheres. For all of these initial upstream distributions the reflection of a fraction of the ions at the shock ramp can be simulated, as was successfully demonstrated for both solar wind and pickup ions by Kucharek and Scholer (1995).

Hybrid simulations have been used to address for example the important issues of injection and acceleration of particles at different shocks in the heliosphere. A number of mechanisms have been proposed that may inject pickup ions into an acceleration process at quasi-perpendicular shocks such as the termination shock. These range from stochastic pre-acceleration by the turbulence in the outer solar system (Chalov et al. 1997; Chalov and Fahr 2000) to shock surfing (Lee et al. 1996; Zank et al. 1996).

Liewer et al. (1993) performed 1D hybrid simulations of quasi-perpendicular shocks corresponding to the termination shock. They showed that a ratio of 10–20% of pickup to solar wind protons in the upstream region affects the shock structure of a high Mach number shock ($M_A > 5$). A foot forms ahead of the shock ramp with a width of approximately the gyroradius of the pickup ions. Furthermore, their simulations provide evidence of a temperature increase of the pickup ions across the shock that is approximately adiabatic. This is probably due to the initial high energy of the pickup ions; their temperature increase at the shock is not expected to be much more than adiabatic. The thermal solar wind ions seem to provide most of the dissipation at the shock, i.e. "heating" beyond that associated with adiabatic compression. With an increasing fraction of pickup ions the dissipation provided by the pickup ions increases. However, it was found that even with 20% pickup hydrogen, the solar wind ions provide most of the dissipation. This is at odds with the scenario presented in Sect. 5 and remains a puzzle. It was also found that for magnetic obliquity $\theta \leq 60°$ some reflected pickup ions escape upstream and gain considerable energy in what appears to be an effective injection mechanism.

Kucharek and Scholer (1995) performed 1D hybrid simulations in order to study the dependence of the reflection rate of pickup ions on the angle θ and on the imposed level of upstream turbulence. Furthermore, since the ratio of pickup ions to solar wind ions is proportional to the heliocentric distance of the termination shock, they also investigated the reflection rates of pickup ions for different locations of the termination shock. They proposed direct injection into diffusive shock acceleration when θ is smaller. By comparison of their injection rate with the observed intensities of the ACR they inferred that the termination shock is located between 80–120 AU.

Interplanetary shocks can also be a source of the seed particle population for subsequent acceleration at the TS. Giacalone et al. (1997) have proposed that, in a first step, pickup ions are accelerated at interplanetary traveling shocks in the inner heliosphere; as a second step they are then further accelerated at the termination shock. Based on the magnetic field topology, the termination shock may be quasi-perpendicular or even quasi-parallel (for example in the polar regions). Scholer and Kucharek (1999) performed 1D hybrid simulations of quasi-parallel shocks with pickup ions included self-consistently. These simulations have shown that pickup ions are efficiently reflected at quasi-parallel interplanetary shocks, planetary bow shocks, and the termination shock. Scholer et al. (2002) also showed in a numerical study that a large increase in the relative abundance ratio of He^+ to He^{2+} can be achieved downstream of a strong parallel shock.

Giacalone and Ellison (2000) performed 3D hybrid simulations to investigate injection of pickup protons at quasi-perpendicular shocks. They investigated whether there is enough cross-field scattering to accelerate a fraction of the thermal particles and pickup ions to high energies. Indeed, they found that a fraction of the downstream particles are accelerated to energies well above the "thermal" energy. This occurs only if the system contains fluctuations with wavelengths that are considerably larger than the gyroradii of the particles of interest (the high-energy ones). The acceleration is due to the fact that the meandering of the magnetic field lines on large scales enhances the diffusive transport of the particles normal to the shock and counteracts the downstream convection. If the system does not contain these long-wavelength waves, the scattering is not sufficient to accelerate thermal particles or pickup ions. They also used a test-particle approach to confirm the above result and interpretation.

The electron scale is not included properly in hybrid simulations because the electrons are treated as a massless fluid. This deficiency may be important because the shock ramp structure and thickness may be controlled by the electron scales (Lembege et al. 2004; Scholer et al. 2003) and this may have an impact on injection and acceleration of ions at the shock. Lipatov and Zank (1999) studied the dynamics of pickup ions at high Mach number low-beta perpendicular shocks using a hybrid kinetic electromagnetic code [for ions and protons a kinetic/particle description was used, while for the finite mass electrons the hydrodynamic equations (including the electron pressure equation) were used]. Anomalous resistivity and electron inertia terms are included in this code. They found significant pickup ion acceleration; a possible explanation is the acceleration experienced during multiple reflections at the shock potential.

Particle-In-Cell simulations with reasonable ion-to-electron mass ratios require massive computational resources. Lee et al. (2005) performed PIC simulations in which pickup protons were included self-consistently. Using a proton-to-electron mass ratio of 20, they investigated reformation of a perpendicular shock in the presence of pickup ions. They concluded that reflected pickup protons could act as a suprathermal seed population for subsequent acceleration at the shock to generate ACRs. They also found some evidence of further acceleration of pickup protons downstream of the shock ramp, presumably due to stochastic acceleration in the downstream turbulence.

Matsukiyo et al. (2007) performed 1D PIC simulations with a more realistic proton to electron mass ratio (400 and 1024) in order to study shock reformation and ion acceleration with the inclusion of pickup ions. They found that the shock reforms periodically as in the low mass ratio case with a somewhat smaller time constant. Specular reflection of pickup protons results in an increase of the shock potential some distance ahead of the shock foot and ramp. The minimum scale of the cross shock potential during reformation is about 7 electron inertial lengths c/ω_{pe}, where ω_{pe} is the electron plasma frequency. However, no pickup proton acceleration in the ramp or downstream of the shock was found beyond the energy that specularly reflected ions gain by the motional electric field of the solar wind during their upstream gyration. The lack of further acceleration by multiple reflections may either be due to the high realistic mass ratio, or the small unrealistic value of ω_{pe}/Ω_e, or the combination of both. Here Ω_e is the electron gyrofrequency. Using a small value of ω_{pe}/Ω_e suppresses electrostatic effects; strong electric fields possibly occurring on electron scales are reduced or absent. It should also be noted that PIC simulations in general, and these in particular, are constrained by limited system size and run time.

In summary, MHD simulations provide a global picture of the termination shock and the heliosphere. They also provide an understanding of the effect of the interstellar magnetic field on the heliosheath and the termination shock. However, kinetic processes cannot be addressed with MHD simulations. Hybrid simulations have been very successful in providing insights into physical processes on the ion lengthscale at Earth's bow shock and interplanetary traveling shocks. The limiting factor for these simulations is the size of the simulation domain. In the last decade the size of the simulation domain has grown with computing power. Currently, sections of the termination shock can be simulated to study local structure, including for example local curvature of the shock front, but the entire heliosphere is not yet feasible. The computational box size for a 1D hybrid simulation is currently limited to approximately 20,000 c/ω_{pi}, which corresponds to ~1 AU at the termination shock (e.g. Giacalone 2004). This is far too small for a global simulation of the heliosphere or a small fraction of it, particularly if the description requires at least two dimensions. We expect these limitations to be less restrictive in the future.

Generally important physical processes are not limited to one scale (MHD, hybrid, or full particle). Therefore, a combination of large-scale MHD simulations and kinetic simulations is desirable to cover all the scales. The observations of Voyager 1, Voyager 2, and IBEX should be used to constrain the parameters used in numerical simulations and to identify physical processes that require further study.

8 An Illustrative Calculation of the ENA Intensity Expected at IBEX

We now consider the production of the ENAs in the heliosheath. We take the x-axis to be a heliocentric radial axis that measures distance from the inner heliosphere, intercepts the solar wind termination shock at $x = 0$, and finally intercepts the heliopause at $x = L$. If $f(\mathbf{r}, \mathbf{v})$ is the proton distribution function in the heliosheath, $F(\mathbf{r}, \mathbf{v})$ is the neutral hydrogen distribution in the heliosheath, and we consider ENAs propagating toward the inner heliosphere, then the stationary neutral distribution satisfies

$$v_x \frac{\partial F}{\partial x} = f(v_x \mathbf{i}) \int d^3\mathbf{v}' F(\mathbf{v}') \sigma |v_x \mathbf{i} - \mathbf{v}'| - F(v_x \mathbf{i}) \int d^3\mathbf{v}' f(\mathbf{v}') \sigma |v_x \mathbf{i} - \mathbf{v}'| \tag{22}$$

where σ is the charge exchange cross-section, $\mathbf{i}$ is the unit vector in the x-direction, and velocities are expressed in the frame of the heliosphere. The integrals in (22) are not sensitive to the forms of F and f. Since F and f are, respectively, dominated by the cold interstellar hydrogen and the cold heliosheath protons that were originally solar wind core protons directly transmitted at the termination shock, we write

$$v_x \frac{\partial F}{\partial x} = f(v_x \mathbf{i}) \sigma n_{\mathrm{H}} |v_x \mathbf{i} - \mathbf{U}| - F(v_x \mathbf{i}) \sigma n_p |v_x \mathbf{i} - \mathbf{V}| \tag{23}$$

where $n_{\mathrm{H}} (n_p)$ and $\mathbf{U}(\mathbf{V})$ are the density and bulk velocity of the neutral hydrogen (protons).

We first consider $v_x < 0$, corresponding to ENAs observable by IBEX. Assuming that neutrals in this velocity range are negligible beyond the heliopause, we have $F(v_x < 0, x = L) = 0$ and the solution of (23) is

$$F(v_x < 0) = \int_x^L dx' f(v_x \mathbf{i}) \sigma |v_x \mathbf{i} - \mathbf{U}| n_{\mathrm{H}} |v_x|^{-1} \exp\left[- \int_x^{x'} dx'' \sigma n_p |v_x \mathbf{i} - \mathbf{V}| |v_x|^{-1} \right] \quad (24)$$

The magnitude of the argument of the exponential function is bounded by $\sim L\sigma n_p$ unless $|v_x| \ll V$. With $\sigma \sim 2 \times 10^{-15}$ cm^2 and the heliosheath proton density $n_p \sim 2 \times 10^{-3}$ cm^{-3}, if $L \ll 10^4$ AU, then the exponential function can be replaced by unity. This is a valid approximation even deep in the tail of the heliosheath. The exception is for protons with $|v_x| \ll V \cong 150$ km/s, or ENA energies $\ll 100$ eV.

The proton distribution $f(\mathbf{v}, \mathbf{r})$ will in general evolve in the heliosheath due to charge exchange with the neutrals. For illustrative simplicity we now assume that we can neglect the components of the gradients of F and f transverse to x, and any variation in **U** and **V**. Then the equations for F and f, with **v** general, are

$$v_x \frac{\partial F(\mathbf{v}, x)}{\partial x} = H(f, F) \quad (25)$$

$$v_x \frac{\partial f(\mathbf{v}, x)}{\partial x} = O(f) - H(f, F) \quad (26)$$

where $H(f, F)$ is the right hand side of (23) for arbitrary **v**. The operator $O(f)$ may be taken to be rapid diffusion in θ' and ϕ', the spherical angular coordinates in velocity space in the plasma frame, to model rapid gyration and scattering. As a result the proton distribution is constrained to be nearly isotropic in the plasma frame. Adding (25) and (26), and averaging over solid angle in the plasma frame, we obtain

$$\frac{\partial}{\partial x}\left[\int \frac{d\Omega'}{4\pi} v_x F(\mathbf{v}, x) + \int \frac{d\Omega'}{4\pi} v_x f(\mathbf{v}, x) \right] = 0 \quad (27)$$

where we note that the x-derivative commutes with the solid-angle integration. Since $f(\mathbf{v}, x)$ is nearly isotropic in the plasma frame, we obtain

$$\int \frac{d\Omega'}{4\pi} v_x F(\mathbf{v}, x) + V_x f_0(v', x) = A(v') \quad (28)$$

where $A(v')$ is a function of v' alone, v' is particle speed in the plasma frame, and $f_0(v', x)$ is the proton distribution in the plasma frame. Setting $x = 0$ we have

$$A(v') = V_x f_0(v', 0) - \int_{v_x < 0} \frac{d\Omega'}{4\pi} |v_x| F(\mathbf{v}, 0) \quad (29)$$

where $f_0(v', 0)$ is the plasma-frame proton distribution just downstream of the termination shock. With

$$v_x F(v_x > 0, L) = \int_0^L dx' f(\mathbf{v}, 0) \sigma |\mathbf{v} - \mathbf{U}| n_{\mathrm{H}} \quad (30)$$

$$v_x F(v_x < 0, 0) = - \int_0^L dx' f(\mathbf{v}, 0) \sigma |\mathbf{v} - \mathbf{U}| n_{\mathrm{H}} \quad (31)$$

(where we first neglect the evolution of $f(\mathbf{v})$ with x) and with (28) and (29), we obtain

$$f_0(v', L) - f_0(v', 0) = -\int \frac{d\Omega'}{4\pi} \int_0^L dx' f_0(v', 0)\sigma|\mathbf{v} - \mathbf{U}|n_H V_x^{-1} \tag{32}$$

Thus the assumption that $f(\mathbf{v})$ does not vary with x and may be replaced by its value of $f(\mathbf{v}, 0)$ requires that the right hand side of (32) is small compared with $f_0(v', 0)$. Equivalently $L\sigma n_H v'/V_x \ll 1$ or $L \ll V_x(\sigma n_H v')^{-1}$. With $n_H \cong 0.1\ \text{cm}^{-3}$ and $V_x/v' \cong 0.1$, we obtain $L \ll 30$ AU. Equation (32) describes the reduction of $f(v', x)$ with increasing distance into the heliosheath to compensate for the production of ENAs and their loss at the surfaces of the heliosheath. The reduction of $f(v', x)$ at ENA energies and the replacement of those protons with protons of speed U leads to the cooling of the heliosheath plasma. Since the heliosheath is estimated to be at least as thick as 30 AU, the variation of $f(\mathbf{v}, \mathbf{r})$ is important. However, the reduction in the proton pressure is presumably compensated by the overall compression of the heliosheath to maintain approximate pressure balance. Clearly a precise calculation of $f(\mathbf{v}, \mathbf{r})$ requires global numerical modeling and is beyond the scope of this analysis.

According to (31), the sunward ENA flux is determined primarily by L and $\mathbf{V}$ since n_H is approximately constant and σ is insensitive to v_x. $\mathbf{V}$ is important since it determines which values of v' can contribute to a specified v_x of interest. From (31), with $\mathbf{v} = v_x\mathbf{i}$, we have

$$F(v_x < 0, x = 0) \cong L\sigma n_H f[(v_x^2 + V^2 - 2v_x V_x)^{1/2}] \tag{33}$$

Since $V_x > 0$ and $v_x < 0$, v' increases with increasing V and V_x, and $f(v')$ generally decreases.

As described in Sect. 4, the downstream proton distribution function is likely to consist of a cold solar wind core and a pickup proton halo, which dominates the pressure. Using the characteristic downstream parameters derived in Sect. 3, $N_p = 8.37 \times 10^{-4}\ \text{cm}^{-3}$, $P_d = 9.8 \times 10^{-13}\ \text{dyn/cm}^2$, and $V = 105$ km/s, we choose the representative pickup proton distribution

$$f(v') = C[1 + (v'/v_0)^\Gamma]^{-1} \tag{34}$$

which exhibits a relatively flat plateau for $v' < v_0$, as expected for the pickup protons, and a power-law tail as observed at the lowest energies detected by Voyagers 1 and 2. The parameters C and v_0 are fixed by N_p and P_d. A convergent pressure requires $\Gamma > 5$ and we choose $\Gamma = 5.5$. The value of Γ is larger than that observed by the Voyagers at energies below ~3 MeV; beyond this energy the observed spectra exhibit rollovers. The number density and pressure integrals require the integral

$$\int_0^\infty dx \frac{x^{a-1}}{1+x} = \frac{\pi}{\sin \pi a} \tag{35}$$

where $0 < a < 1$. Using (35) we obtain $v_0 = 2.33$ V and $C = 7.89 \times 10^{-27}\ \text{cm}^{-6}\,\text{s}^3$. By good fortune the value for v_0 appears to be reasonable, implying a downstream pickup proton "shoulder" of 245 km/s. With $n_H = 0.1\ \text{cm}^{-3}$, $\sigma = 2 \times 10^{-15}\ \text{cm}^2$, and $L = 40$ AU (characteristic of the heliospheric nose), we then obtain from (33)

$$F(v_x < 0, x = 0) \cong 9.47 \times 10^{-28} \times \{1 + [E/E_0 + 0.184 + 0.86(V_x/V)(E/E_0)^{1/2}]^{\Gamma/2}\}^{-1}\ \text{cm}^{-6}\,\text{s}^3 \tag{36}$$

where $E = mv_x^2/2$ and $E_0 = mv_0^2/2$.

Transforming expression (36) to differential intensity $J(E, x = 0)$ we obtain

$$J(E, x = 0) \cong 541(E/E_0) \times \{1 + [(E/E_0) + 0.184 + 0.86(V_x/V)(E/E_0)^{1/2}]^{\Gamma/2}\}^{-1}\ (\mathrm{cm}^2\,\mathrm{s\,str\,keV})^{-1} \quad (37)$$

In order to estimate what IBEX would observe for this simple case we ignore any further production of ENAs by suprathermal solar wind or energetic particle populations. However, we must include the exponential factor in (24). Although it is negligible in the heliosheath it is important in the inner heliosphere where n_p is large. This factor describes the loss of the ENAs due to charge-exchange with solar wind protons; we neglect photoionization and electron-impact ionization which are smaller. Adapting the exponential factor to the spherical geometry of the solar wind and performing the radial integration from Earth orbit to the termination shock we obtain

$$\exp[-\sigma n_{p,0} r_0 (E^{1/2} + E_{SW}^{1/2}) E^{-1/2}] \quad (38)$$

where $n_{p,0}$ is the solar wind proton density at 1 AU, $r_0 = 1$ AU, and $E_{SW} = mV_{SW}^2/2$. Combining (37) and (38), we obtain the predicted differential intensity of ENAs at 1 AU. With $n_{p,0} = 8\ \mathrm{cm}^{-3}$ and $V_{SW} = 400$ km/s we obtain

$$\begin{aligned} J(E, 1\ \mathrm{AU}) \cong{}& 1.72 \times 10^3 E \\ & \times \{1 + [(E/0.314) + 0.184 + 0.86(V_x/V)(E/0.314)^{1/2}]^{2.75}\}^{-1} \\ & \times \exp[-0.24(E^{1/2} + 0.91)E^{-1/2}]\ (\mathrm{cm}^2\,\mathrm{s\,str\,keV})^{-1} \end{aligned} \quad (39)$$

where E is expressed in keV. Equation (39) exhibits clearly the decrease of J with increasing E due to the form of the power-law tail in the heliosheath, the decrease in J for small E due to charge-exchange in the inner heliosphere, and the decrease of J with increasing V_x/V due to the outward convection of the heliosheath protons relative to the spacecraft. The order of magnitude of $J(E, 1\ \mathrm{AU})$ appears to be consistent with intensities predicted by Gruntman et al. (2001) using a Maxwellian distribution for the heliosheath protons. A more detailed comparison of this simple analytical model with numerical predictions is beyond the scope of this paper.

9 Conclusions

This chapter has attempted to highlight some of the physical processes that are important in determining the structure of the outer heliosphere. Understanding these processes will be crucial to an informed interpretation of the interstellar neutral gas and the ENAs observed by IBEX. We have also attempted to introduce the theoretical constructs and methods, which are useful in some of these investigations. However, much remains to be done as the IBEX team prepares for the most extensive global measurements of the boundary regions of the heliosphere to date.

Acknowledgements The authors wish to acknowledge informative discussions at several IBEX Science Working Team meetings, which stimulated much of the science addressed in this chapter. M.L. is particularly grateful to Maciej Bzowski, Mike Gruntman, Dave McComas and Gary Zank for specific conversations. The work was supported, in part, by the IBEX Project.

References

T.J. Bogdan, M.A. Lee, P. Schneider, J. Geophys. Res. **96**, 161 (1991)
L.F. Burlaga, N.F. Ness, M.H. Acuna, R.P. Lepping, J.E.P. Connerney, E.C. Stone, F.B. McDonald, Science **309**, 2027 (2005)
S.V. Chalov, Planet. Space Sci. **41**(2), 133 (1993)
S.V. Chalov, H.-J. Fahr, Astron. Astrophys. **363**, L12 (2000)
S.V. Chalov, H.-J. Fahr, V. Izmodenov, Astron. Astrophys. **320**, 659 (1997)
I.V. Chashei, H.-J. Fahr, G. Lay, Ann. Geophys. **21**(7), 1405 (2003)
A.C. Cummings, E.C. Stone, Space Sci. Rev. **78**, 117 (1996)
H.-J. Fahr, M. Siewert, Astron. Astrophys. **458**(1), 13 (2006)
H.-J. Fahr, M. Siewert, Astrophys. Space Sci. Trans. **3**(1), 21 (2007)
H.-J. Fahr, H. Fichtner, K. Scherer, Rev. Geophys. **45**(4), CiteID RG4003 (2007). doi:10.1029/2006RG000214
S.E.S. Ferreira, M.S. Potgieter, K. Scherer, J. Geophys. Res. **112**(A11), CiteID A11101 (2007). doi:10.1029/2007JA012477
L.A. Fisk, G. Gloeckler, Astrophys. J. **640**(1), L79 (2006)
L.A. Fisk, G. Gloeckler, T.H. Zurbuchen, Astrophys. J. **644**, 631 (2006)
L.A. Fisk, G. Gloeckler, T.H. Zurbuchen, N.A. Schwadron, in *Acceleration and Transport of Energetic Particles Observed in the Heliosphere: ACE 2000 Symposium*, ed. by R.A. Mewaldt et al. AIP Conference Proceedings, vol. 528 (AIP, New York, 2000), pp. 229–233
V. Florinski, G.P. Zank, Geophys. Res. Lett. **33**(5), CiteID L15110 (2006). doi:10.1029/2006GL026371
J. Giacalone, in *Acceleration and Transport of Energetic Particles Observed in the Heliosphere: ACE 2000 Symposium*, ed. by R.A. Mewaldt et al. AIP Conference Proceedings, vol. 528 (AIP, New York, 2000), pp. 258–265
J. Giacalone, Astrophys. J. **609**, 452 (2004)
J. Giacalone, D.C. Ellison, J. Geophys. Res. **105**, 12541 (2000)
J. Giacalone, J.R. Jokipii, Astrophys. J. **663**(1), L41 (2007)
J. Giacalone, D. Burgess, S.J. Schwartz, D.C. Ellison, L. Bennett, J. Geophys. Res. **102**, 19789 (1997)
G. Gloeckler, J. Geiss, Space Sci. Rev. **97**, 169 (2001)
G. Gloeckler, J. Geiss, E.C. Roelof, L.A. Fisk, F.M. Ipavich, K.W. Ogilvie, L.J. Lanzerotti, R. von Steiger, B. Wilken, J. Geophys. Res. **99**, 17,637 (1994)
G. Gloeckler, L.A. Fisk, T.H. Zurbuchen, N.A. Schwadron, in *Acceleration and Transport of Energetic Particles Observed in the Heliosphere: ACE 2000 Symposium*, ed. by R.A. Mewaldt et al. AIP Conference Proceedings, vol. 528 (AIP, New York, 2000), pp. 221–228
J.T. Gosling, A.E. Robson, in *Collisionless Shocks in the Heliosphere: Reviews of Current Research*, ed. by B.T. Tsurutani, R.G. Stone. Geophysical Monograph, vol. 35 (AGU, Washington, 1985), pp. 141–152
M. Gruntman, E.C. Roelof, D.G. Mitchell, H.J. Fahr, H.O. Funsten, D.J. McComas, J. Geophys. Res. **106**, 15767 (2001)
P.A. Isenberg, J. Geophys. Res. **91**(A9), 9965 (1986)
P.A. Isenberg, Geophys. Res. Lett. **24**(6), 623 (1997a)
P.A. Isenberg, J. Geophys. Res. **102**, 4719 (1997b)
P.A. Isenberg, C.W. Smith, W.H. Matthaeus, Astrophys. J. **592**, 564 (2003)
V.V. Izmodenov, D.B. Alexashov, S.V. Chalov, O.A. Katushkina, Y.G. Malama, E.A. Provornikova (2009, this issue)
J.R. Jokipii, J. Giacalone, Astrophys. J. **660**(1), 336 (2007)
J.R. Jokipii, J. Giacalone, J. Kóta, Astrophys. J. **611**(2), L141–L144 (2004)
C.F. Kennel, J.P. Edmiston, T. Hada, in *Collisionless Shocks in the Heliosphere: A Tutorial Review*, ed. by R.G. Stone, B.T. Tsurutani. Geophysical Monograph, vol. 34 (AGU, Washington, 1985), pp. 1–36
H. Kucharek, M. Scholer, J. Geophys. Res. **100**, 1745 (1995)
H. Kucharek, E. Moebius, M. Scholer, C. Mouikis, L. Kistler, T. Horbury, A. Balogh, H. Réme, J. Bosqued, Ann. Geophys. **22**(7), 2301 (2004)
M.A. Lee, in *Cosmic Winds and the Heliosphere*, ed. by J.R. Jokipii, C. Sonett, M.S. Giampapa (University of Arizona Press, Tucson, 1997), pp. 857–886
M.A. Lee, AIP Conf. Proc. **858**, 291 (2006)
M.A. Lee, V.D. Shapiro, R.Z. Sagdeev, J. Geophys. Res. **101**, 4777 (1996)
R.E. Lee, S.C. Chapman, R.O. Dendy, Ann. Geophys. **23**, 643 (2005)
B. Lembege, J. Giacalone, M. Scholer, T. Hada, M. Hoshino, V. Krasnoselskikh, H. Kucharek, P. Savoini, T. Terasawa, Space. Sci. Rev. **110**, 116 (2004)
M.M. Leroy, D. Winske, C.C. Goodrich, C.S. Wu, K. Papadopoulos, J. Geophys. Res. **87**, 5081 (1982)
P.C. Liewer, B.E. Goldstein, N. Omidi, J. Geophys. Res. **98**, 15221 (1993)

A.S. Lipatov, G.P. Zank, Phys. Rev. Lett. **82**, 3609 (1999)
Y.C.-M. Liu, M.A. Lee, H. Kucharek, J. Geophys. Res. **110**, A09101 (2005). doi:10.1029/2005JA011096
W.H. Matthaeus, G.P. Zank, C.W. Smith, S. Oughton, Phys. Rev. Lett. **82**, 3444 (1999)
S. Matsukiyo, M. Scholer, D. Burgess, Ann. Geophys. **25**, 283 (2007)
D.J. McComas et al., AIP Conf. Proc. **719**, 162–181 (2004)
D.J. McComas, N.A. Schwadron, Geophys. Res. Lett. **33**(4), CiteID L04102 (2006). doi:10.1029/2005GL025437
D.J. McComas et al. (2009, this issue)
E. Möbius, H. Kucharek, C. Mouikis, E. Georgescu, L.M. Kistler, M.A. Popecki, M. Scholer, J.M. Bosqued, H. Réme, C.W. Carlson, B. Klecker, A. Korth, G.K. Parks, J.C. Sauvaud, H. Balsiger, M.-B. Bavassano-Cattaneo, I. Dandouras, A.M. Dilellis, L. Eliasson, V. Formisano, T. Horbury, W. Lennartsson, R. Lundin, M. McCarthy, J.P. McFadden, G. Paschmann, Ann. Geophys. **19**(1), 1411 (2001)
E. Möbius et al., Astron. Astrophys. **426**, 897 (2004)
C. Prested, N.A. Schwadron, J. Passuite, B. Randol, B. Stuart, G. Crew, J. Heerikhuisen, N. Pogorelov, G. Zank, M. Opher, F. Allegrini, D.J. McComas, M. Reno, E. Roelof, S. Fuselier, H. Funsten, E. Moebius, L. Saul, J. Geophys, Res. **113**(A6), CiteID A06102 (2008). doi:10.1029/2007JA012758
J.D. Richardson, J.C. Kasper, C. Wang, J.W. Belcher, A.J. Lazarus, Nature **454**(3), 63 (2008a)
J.D. Richardson, Y. Liu, C. Wang, D.J. McComas, Astron. Astrophys. **491**, 1 (2008b)
D. Rucinski, A.C. Cummings, G. Gloeckler, A.J. Lazarus, E. Möbius, M. Witte, Space Sci. Rev. **78**, 73 (1996)
M. Scholer, H. Kucharek, Geophys. Res. Lett. **26**, 29 (1999)
M. Scholer, H. Kucharek, C. Kato, Phys. Plasmas **9**, 4293 (2002)
M. Scholer, H. Kucharek, I. Shinohara, J. Geophys, Res. **108**, CiteID 1273 (2003). doi:10.1029/2002JA009820
N.A. Schwadron, M.A. Lee, D.J. McComas, Astrophys. J. **675**(2), 1584 (2008)
N. Sckopke, G. Paschmann, A.L. Brinca, C.W. Carlson, H. Lühr, J. Geophys. Res. **95**, 6337 (1990)
M. Siewert, H.-J. Fahr, Astron. Astrophys. **463**(3), 799 (2007)
C.W. Smith, W.H. Matthaeus, G.P. Zank, N.F. Ness, S. Oughton, J.D. Richardson, J. Geophys. Res. **106**, 8253 (2001)
C.W. Smith, P.A. Isenberg, W.H. Matthaeus, J.D. Richardson, Astrophys. J. **638**, 508 (2006)
E.C. Stone, A.C. Cummings, W.R. Webber, J. Geophys. Res. **101**(A5), 11017 (1996)
V.M. Vasyliunas, G.L. Siscoe, J. Geophys. Res. **81**, 1247 (1976)
C. Wang, J.D. Richardson, J.T. Gosling, Geophys. Res. Lett. **27**(16), 2429 (2000)
L.L. Williams, G.P. Zank, W.H. Matthaeus, J. Geophys. Res. **100**, 17059 (1995)
P. Wu, D. Winske, S.P. Gary, N.A. Schwadron, M.A. Lee, J. Geophys, Res. (2009, submitted)
G.P. Zank, Space Sci. Rev. **89**(3/4), 413 (1999)
G.P. Zank, H.L. Pauls, I.H. Cairns, G.M. Webb, J. Geophys. Res. **101**, 457 (1996)
G.P. Zank, N.V. Pogorelov, J. Heerikhuisen, H. Washimi, V. Florinski, S. Borovikov, I. Krykov, H.R. Müller (2009, this issue)

Space Sci Rev (2009) 146: 295–327
DOI 10.1007/s11214-009-9497-6

Physics of the Solar Wind–Local Interstellar Medium Interaction: Role of Magnetic Fields

G.P. Zank · N.V. Pogorelov · J. Heerikhuisen · H. Washimi · V. Florinski · S. Borovikov · I. Kryukov · H.R. Müller

Received: 3 November 2008 / Accepted: 19 February 2009 / Published online: 7 May 2009

Abstract The interaction of the solar wind with the local interstellar medium is characterized by the self-consistent coupling of solar wind plasma, both upstream and downstream of the heliospheric termination shock, the interstellar plasma, and the neutral atom component of interstellar and solar wind origin. The complex coupling results in the creation of new plasma components (pickup ions), turbulence, and anomalous cosmic rays, and new populations of neutral atoms and their coupling can lead to energetic neutral atoms that can be detected at 1 AU. In this review, we discuss the interaction and coupling of global sized structures (the heliospheric boundary regions) and kinetic physics (the distributions that are responsible for the creation of energetic neutral atoms) based on models that have been developed by the University of Alabama in Huntsville group.

Keywords Solar wind boundaries · Interstellar medium · Pickup ions · Energetic neutral atoms

1 Introduction

In the last two decades, great progress has been made in our understanding of the physical processes thought to describe the outer heliosphere. Fundamental to these advances has been the recognition that the interstellar medium is coupled intimately to the heliosphere through a variety of mechanisms and that much of outer heliospheric physics cannot be understood independently of the local interstellar medium (LISM). As a result, this field truly lies at the crossroads of space physics and astrophysics. For the first time, *in situ* measurements by the spacecraft Voyager 1 (V1), Voyager 2 (V2) and remote sensing observations by IBEX will allow us to directly probe the boundaries of the solar wind–LISM interaction region, and will provide data of unprecedented quality about the nature of the interstellar medium.

G.P. Zank (✉) · N.V. Pogorelov · J. Heerikhuisen · H. Washimi · V. Florinski · S. Borovikov · I. Kryukov · H.R. Müller
Center for Space and Aeronomic Research (CSPAR) and Department of Physics, University of Alabama, Huntsville, AL, USA
e-mail: zank@cspar.uah.edu

Voyager 1 crossed the termination shock (TS) of the supersonic flow of the solar wind on 16 December 2004 at a distance of 94 astronomical units (AUs) from the Sun, becoming the first spacecraft to enter the heliosheath (Stone et al. 2005; Burlaga et al. 2005; Decker et al. 2005; Gurnett and Kurth 2005). Voyager 2 crossed the heliospheric TS several times between 30 August and 1 September 2007 (Stone et al. 2008; Richardson et al. 2008; Burlaga et al. 2008; Decker et al. 2008; Gurnett and Kurth 2008), at a distance of 84 AU, providing the first plasma measurements of the TS and heliosheath region. With IBEX, we are poised at the threshold of an exciting decade of new observations of the heliosheath boundaries that will provide not only information about the boundaries and the plasma regimes in which energetic neutral atoms (ENAs) are created, but will be able to provide this information on a global scale. The IBEX mission will provide new and unexpected insights into the solar wind and the LISM, as well as into the stellar winds of neighboring stars and their LISM.

The heliosphere is the region of space filled by the expanding solar corona; a vast region indeed extending perhaps 150–180 AUs in the direction of the Sun's motion through the interstellar medium and several thousand AU in the opposite direction. A convenient, if slightly vague, definition of the outer heliosphere, and one that we adopt here, is that it is that region of the solar wind influenced dynamically by physical processes associated with the LISM. Thus, for example, neutral interstellar hydrogen (H) is the dominant (by mass) constituent of the solar wind beyond an ionization cavity of $\sim$6–10 AUs in the upstream direction (the direction antiparallel to the incident interstellar wind). The neutral H is coupled weakly to the solar wind plasma via resonant charge exchange—a coupling that leads to the creation of pickup ions that come to eventually dominate the internal energy of the solar wind. The solar wind changes then from a small plasma beta β_{SW} (the ratio of plasma pressure to magnetic field pressure) environment to one in which $\beta_{SW} \simeq 4$. As a by-product of the creation of pickup ions, low-frequency turbulence is generated, replenishing the dissipated magnetic fluctuations originating at and near the Sun (Williams and Zank 1994; Zank et al. 1996a). This has important implications for both solar wind heating and charged particle transport (Williams et al. 1995; Zank et al. 1998). Our growing understanding of the physical processes governing the outer heliosphere has led to major discoveries about both our own solar system (the discovery of a "hydrogen wall" bounding our heliosphere, for example (Linsky and Wood 1996; Gayley et al. 1997), the interstellar medium (the composition of the ISM (Gloeckler and Geiss 1998)), and the astrospheres of other stars (such as the discovery of winds blown from stars like our Sun (Wood et al. 2001)). Since the Sun orbits our galactic center, the solar system has experienced many different interstellar environments, and it is entirely possible that this may have, and may have had, an important impact on the Earth's environment (Zank and Frisch 1999e; Scherer et al. 2001; Florinski et al. 2003b; Florinski and Zank 2006; Zank et al. 2006; Müller et al. 2006; Scherer et al. 2008).

The physics underlying the interaction of the solar wind with the LISM describes a complex relationship between several basic elements: the solar wind plasma and magnetic field, the LISM plasma and magnetic field, the LISM neutrals, interstellar pickup ions, and anomalous and galactic cosmic rays. The solar wind terminates at a shock whose location is determined by the balance of dynamical solar wind pressure and the external LISM pressure (Davis 1955; Parker 1963; Axford 1972). Galactic cosmic rays, which can diffuse deep into the heliosphere, are of lesser dynamical importance. Interstellar neutral H is coupled to both the LISM and solar wind plasmas through resonant charge exchange, providing an effective volumetric force and hence affecting the dynamical nature of the solar wind–LISM interaction profoundly. The two basic ways in which neutrals can modify the heliospheric and LISM plasmas are: (i) Interstellar neutrals decelerate the solar wind indirectly; (ii) Secondary very hot neutrals produced in the shocked solar wind (downstream of the TS) can, through secondary charge exchange, heat the LISM, as can fast neutrals produced through charge exchange with the supersonic solar wind. Newly created ions in the supersonic solar wind—the pickup ions—are very energetic ($\sim$ 1 keV) compared to typical solar wind protons and dominate the internal energy of the solar wind in the outer heliosphere. The pickup process itself is expected to generate significant levels of low frequency magnetohydrodynamic (MHD) turbulence, which will isotropize the pick-up ion beam, dissipate and heat the solar wind and scatter cosmic rays (Williams et al. 1995; Matthaeus et al. 1999; Smith et al. 2000; Zank et al. 1996b; Isenberg et al. 2003; Zank et al. 1998). Some small fraction of the pickup ions will be further energized and, possibly at the TS (although this remains a mystery despite, or perhaps because of, the recent Voyager crossings of the TS), be accelerated up to MeV energies to form the anomalous cosmic-ray (ACR) component. The energy density of the anomalous cosmic rays in the vicinity of the TS was expected to modify the structure and properties of the shock itself. Present day models hope to capture the fully non-linear coupled physics of the solar wind–LISM interaction, and progress has been made at several levels of approximation. The most comprehensive models today are fully 3D and time-dependent, and include solar wind and LISM plasma and magnetic fields at an MHD level, and are coupled self-consistently with interstellar neutral H at either a kinetic or multi-fluid level. Many key parameters remain nonetheless somewhat poorly constrained, particularly properties of the LISM such as the magnetic field strength and orientation and the ionization fraction.

Besides helping explain the observations returned by the Voyager spacecraft, an understanding of the plasma downstream of the TS, the inner heliosheath plasma, is critical to interpreting observations that will be made by the IBEX mission. IBEX (McComas et al. 2006) will measure ENAs at 1 AU that are created by charge exchange between interstellar neutral atoms of low energy and inner heliosheath plasma. The working assumption is that the heliosheath plasma will be very hot (model temperatures suggest $\sim 10^6$ K). An observational (if not entirely theoretical) surprise was the discovery by V2 (Richardson et al. 2008) that only 20% of the incident solar wind flow energy went into heating the solar wind ions, making them a relatively cool component of the downstream shocked inner heliosheath plasma (to the extent that the downstream flow appears to be supersonic, at least as determined from solar wind ions!). As we shall discuss, the form of the heliosheath distribution function can make a profound difference in the measured ENA distribution. Heerikhuisen et al. (2008a) find significant differences in the ENA distribution resulting from assuming either a Maxwellian or a kappa (κ) distribution for the solar wind ions in the heliosheath. Undoubtedly, since the downstream or heliosheath plasma distribution is determined by the processing of the supersonic solar wind by the TS, a cold solar wind ion heliosheath component will have an important impact on interpreting the IBEX ENA measurements.

With the upcoming IBEX mission, it is appropriate to elucidate the basic physics underlying models that describe the self-consistent interaction of the solar wind with the LISM. IBEX will return all-sky maps and distributions integrated over lines-of-sight, and the deconvolution and interpretation of these observations will rely heavily on complementary modeling. IBEX will undoubtedly return data that taxes both our models and the assumptions underlying them and we will need to understand the theory thoroughly. This review discusses the assumptions underlying the most sophisticated models that we currently possess—a fully 3D time-dependent model that incorporates solar and LISM plasma and magnetic fields with interstellar neutral H self-consistently (both kinetically and using a multi-fluid description). This review is based on two much more detailed and lengthy discussions of the interaction of the solar wind with the LISM (Zank 1999; Zank et al. 2008).

2 Mathematical Formulation for Modeling the Solar Wind–LISM Interaction

Key to understanding the structure and properties of the heliosphere is the effect that interstellar neutral gas, primarily hydrogen, plays in the physics of the solar wind–LISM interaction. Interstellar neutral gas flows into the heliosphere relatively unimpeded and can penetrate to within several AU of the sun. Neutral atoms scatter solar radiation resonantly so that the distribution of H and Helium (He) in the heliosphere can be studied by observing sky background radiation in HI $\lambda 1216$ and HeI $\lambda 584$. The distribution of interstellar H only is addressed here in any detail since He and the other elements have a negligible dynamical influence on the solar wind–LISM interaction. Nonetheless, He in particular provides important and direct information about the conditions in the LISM (Möbius et al. 2004; Witte 2004), such as the interstellar flow speed, direction, and temperature.

The heliospheric-LISM plasma environment is composed of three thermodynamically distinct regions: (i) the supersonic solar wind, expanding more-or-less radially from the Sun at speeds of $\sim$ 400–800 km/s, with a density that decays as $\sim r^{-2}$ (r denoting radial heliocentric distance from the Sun), and a relatively low temperature ($\sim 10^4$ K in the vicinity of the HTS) which appears to decay adiabatically with an effective adiabatic index of $\gamma \simeq 1.1$ within about 20 AU (see Williams et al. 1995; Matthaeus et al. 1999); (ii) the shock heated subsonic solar wind, which possesses much higher "effective" temperatures ($\sim 10^6$ K), somewhat higher densities ($\sim 10^{-3}$ cm^{-3}) and lower flow speeds ($\sim$ 100 km/s); and finally (iii) the LISM, where the plasma flow speed is low ($\sim$ 26 km/s) and the temperature is several $\times 10^3$K. Here and henceforth, region 1 is taken to be the region beyond the heliopause (HP) i.e., the LISM, and H atoms whose origin lies in region 1 shall be referred to as component 1 neutrals. One can further subdivide the LISM region into a possibly shocked region (if a bow shock exists), region 1b, and an unshocked region, region 1a. Region 2 is that region occupied by the shocked solar wind, plasma and component 2 neutrals are those created in region 2. Finally, region 3 refers to the supersonic solar wind, and the component 3 neutrals are those born there (i.e., the "splash" component). Quite clearly, the neutral populations, components 1a, 1b, 2 and 3, possess distinct characteristics, and the complete local neutral H distribution will be highly anisotropic.

Several interactions between H atoms, protons and electrons are possible and some important points can be made (Zank 1999). (1) Photoionization is important within several AU, but is otherwise not (assuming that the LISM is in ionization equilibrium with the local stellar UV radiation field, as suggested (Frisch 1995); see also Slavin and Frisch (2002). (2) Two charge-exchange cross-sections are used in the literature (Maher and Tinsley 1977; Fite et al. 1962; Maher and Tinsley 1977; Fite et al. 1962), but that published by Fite et al. (1962) is a fit to experimental data. Nonetheless, at 1 eV, a 40% discrepancy exists between the cross-sections and this can effect the heliospheric H number density by a similar amount. However, the cross-sections provided by Lindsay and Stebbings (2005) are now regarded as the "accepted" values. (3) The charge-exchange cross-sections assume that no momentum transfer occurs during the interaction but, of course, non-charge-exchange H^+–H interactions can occur. The H^+–H cross-section computed by Dalgarno (1960) is the total momentum transfer cross-section and includes charge-exchange. Thus, an exclusively charge-exchange treatment of the H–H^+ interaction may underestimate the efficiency of the coupling. (4) H–H collisions have a cross-section comparable to that of charge-exchange. It is presently unclear how and where such collisions may be important, but two possibilities are in the heliotail and the region upstream of the HP where hot component 2 and cooler component 1 neutrals can equilibrate Williams et al. (1997). (5) Electron-H collisions, electron impact ionization, and recombination are unlikely to be dynamically important on 1000 AU scales. However, because electron impact ionization is sensitive to the electron distribution, it may occur in the solar wind (Isenberg and Feldman 1995). (6) Since region 1 H^+–H^+ and e–e mean free paths are ~ 0.15 AU, the LISM is Coulomb collisional and the electrons equilibrate to the proton temperature.

The distribution of neutral H, both LISM and that created in the various boundary regions of the heliosphere, drifting through the heliosphere may be calculated directly from the Boltzmann equation,

$$\frac{\partial f}{\partial t} + \mathbf{v} \cdot \nabla f + \left(\frac{\mathbf{F}}{m} \cdot \nabla_{\mathbf{v}} \right) f = P - L, \tag{1}$$

where $f(\mathbf{x}, \mathbf{v}, t)$ is the neutral H particle distribution function expressed in terms of position $\mathbf{x}$, velocity $\mathbf{v}$ and time t. $\mathbf{F}$ is the force acting on a particle of mass m, typically gravity and radiation pressure. The terms P and L describe the production and loss of particles at $(\mathbf{x}, \mathbf{v}, t)$, and both terms are functions of the assumed plasma and neutral distributions. In all cases of interest here, the loss term may be expressed as

$$L = f(\mathbf{x}, \mathbf{v}, t)\beta(\mathbf{x}, \mathbf{v}, t), \tag{2}$$

where β is the total loss rate in s^{-1}. The neutral H loss rate due to charge-exchange is obtained by integrating over the proton distribution function, thus

$$\beta_{ex}(\mathbf{x}, \mathbf{v}, t) = \int f_p(\mathbf{x}, \mathbf{v}_p, t) V_{rel,p} \sigma_{ex}(V_{rel,p}) \mathrm{d}^3\mathbf{v}_p, \tag{3}$$

where f_p and $\mathbf{v}_p$ refer to proton quantities, $V_{rel,p} \equiv |\mathbf{v} - \mathbf{v}_p|$ is the relative speed between an H atom and a proton, and σ_{ex} denotes the charge-exchange cross-section.

The charge-exchange neutral H production term is given by

$$P_{ex}(\mathbf{x}, \mathbf{v}, t) = f_p(\mathbf{x}, \mathbf{v}, t) \int f_H(\mathbf{x}, \mathbf{v}_H, t) V_{rel,H} \sigma_{ex} \left(V_{rel,H} \right) \mathrm{d}^3\mathbf{v}_H, \tag{4}$$

where $V_{rel,H} \equiv |\mathbf{v} - \mathbf{v}_H|$ and f_H is the neutral H distribution.

The initial or boundary data is assumed typically to be a Maxwellian distribution parameterized by the bulk LISM density, velocity and temperature, and the boundary condition is imposed at "infinity".

All models of the solar wind interaction with the LISM assume a simple model of the plasma which incorporates both a solar wind component and a suprathermal component. Although number densities are too low for the direct interaction of the solar wind plasma flow with the neutral flux, appreciable momentum and energy exchange, particularly in the supersonic solar wind, is possible nonetheless through charge exchange of solar wind protons and neutral H. In a plasma possessing a sufficiently strong magnetic field, a newly charged particle acquires a gyrospeed equal to its initial velocity relative to the plasma and convects with the plasma flow within a gyroperiod. Although the microscopic details of this process are complicated and depend on the plasma-magnetic field configuration (Williams and Zank 1994; Isenberg and Lee 1996; Zank 1999), the net result of ion pickup on hydrodynamic scales is qualitatively unique—there is a change in the density, momentum, and energy of the plasma flow for each act of charged particle production or destruction.

The dynamical or ram pressure (ρu^2) and thermal pressure p of the solar wind decrease with increasing heliocentric distance and must reach a value which eventually balances the pressure exerted by the LISM. The relaxation towards pressure equilibrium between the solar and interstellar plasmas is characterized by (i) a transition of the supersonic solar wind flow to a subsonic state, and (ii) a divergence of the interstellar flow about the heliospheric obstacle. The transition of the supersonic solar wind is accomplished by means of a shock transition, the TS. Voyager 1 has now crossed the TS (at 94 AU) (Stone et al. 2005; Burlaga et al. 2005) and V2 crossed at 84 AU (Stone et al. 2008; Richardson et al. 2008; Burlaga et al. 2008). The divergence of the LISM flow about the heliosphere may be accomplished either adiabatically (if the relative motion between the Sun and the LISM is subsonic) or by means of a bow shock in the case of supersonic relative motion. Although one can estimate the location of the TS and the HP, the discontinuity separating solar wind material from the interstellar plasma (a contact/tangential discontinuity in the case of gas/MHD), using simple pressure balance arguments, the problem of the interaction of the solar wind with the LISM is fundamentally multi-dimensional. Thus, the main advances in our understanding of global heliospheric structure since the pioneering work of Davis (1955), Parker (1958, 1961, 1963), Axford et al. (1963) and Baranov et al. (1971) have been more recent and based largely on computer simulations. The initial simulations were based on pure one-fluid gas dynamic models, but now the self-consistent inclusion of neutral interstellar hydrogen models is recognized as the key to understanding the structure and properties of the large-scale heliosphere. We discuss the basic MHD model, and the kinetic and multi-fluid approaches to including neutral H self-consistently in heliospheric models of the solar wind–LISM interaction.

Interplanetary and interstellar magnetic fields have only recently been fully incorporated into global heliospheric models. The most detailed models (fully 3D and time-dependent and including neutral H both kinetically and in the multi-fluid approximation—see below) are those described here and developed by the University of Alabama in Huntsville group. To model the interaction of the solar wind with a partially ionized LISM, the following 3D set of MHD equations must be solved,

$$\frac{\partial \rho}{\partial t} + \nabla \cdot \rho \mathbf{u} = Q_\rho; \tag{5}$$

$$\rho \frac{\partial \mathbf{u}}{\partial t} + \rho \mathbf{u} \cdot \nabla \mathbf{u} + (\gamma - 1)\nabla e + (\nabla \times \mathbf{B}) \times \mathbf{B} = \mathbf{Q}_m; \tag{6}$$

$$\frac{\partial}{\partial t}\left(\frac{1}{2}\rho u^2 + e + \frac{B^2}{8\pi}\right) + \nabla \cdot \left[\left(\frac{1}{2}\rho u^2 + \gamma e\right)\mathbf{u} + \frac{1}{4\pi}\mathbf{B} \times (\mathbf{u} \times \mathbf{B})\right] = Q_e; \tag{7}$$

$$\frac{\partial \mathbf{B}}{\partial t} - \nabla \times (\mathbf{u} \times \mathbf{B}) = 0; \tag{8}$$

$$\nabla \cdot \mathbf{B} = 0, \tag{9}$$

$\rho = m_p n$, together with the equation of state $e = \alpha n k_B T/(\gamma - 1) = p/(\gamma - 1)$. Here the choice of $\alpha = 2$ (or greater if incorporating the contribution of cosmic rays, dust indirectly) corresponds to a plasma population comprising electrons and protons. The remaining variables have their usual definitions, and the source terms Q_ρ, $\mathbf{Q}_m$, and Q_e serve to couple the neutral H and proton populations. Certain subtleties in the formulation of an MHD model in the presence of neutral H need to be recognized, and this was discussed by Florinski and Zank (2003d) in the context of non-ideal MHD. Subject even to the assumption of an isotropic solar wind, the formulation (5)–(9) is inherently 3D thanks to the solar magnetic field and the current sheet.

As discussed, pickup ions are created in the solar wind through charge-exchange of LISM neutrals with solar wind protons, but they do not thermalize with the background solar wind plasma (Isenberg 1986; Zank 1999) and are not therefore equilibrated with the solar wind. Pickup ions constitute a separate suprathermal particle population in the SW (Möbius et al. 1985; Gloeckler et al. 1993; Gloeckler 1996; Gloeckler and Geiss 1998) and contribute to the power-law tails observed almost universally in the solar wind plasma distribution (Mewaldt et al. 2001; Fisk and Gloeckler 2006). Heerikhuisen et al. (2008a) and Prested et al. (2008) suggested that a simple way to incorporate a power-law tail, and thereby model the proton, energetic particle, and pickup ion populations as a single distribution, is to assume a generalized Lorentzian, or κ-, distribution (Bame et al. 1967; Summers and Thorne 1991; Collier 1995) given by

$$f_p(\mathbf{v}) = \frac{n_p}{\pi^{3/2}\theta_p^3}\frac{1}{\kappa^{3/2}}\frac{\Gamma(\kappa+1)}{\Gamma(\kappa-1/2)}\left[1+\frac{1}{\kappa}\frac{(\mathbf{v}-\mathbf{u})^2}{\theta_p^2}\right]^{-(\kappa+1)}, \tag{10}$$

where θ_p is a typical speed related to the effective temperature of the distribution, and is evaluated using the pressure moment, discussed further below. The distribution function (10) possesses a Maxwellian core, a power-law tail that scales as $v^{-2(\kappa+1)}$, and reduces to a Maxwellian in the limit of large κ. Although the core and tail features agree qualitatively with observations, a limitation of the κ formalism is that it does not allow for the relative abundances of core and tail to be adjusted for a particular choice of κ. Obviously, the observed flat-topped pickup ion population is also absent in the κ approximation. In Fig. 1, we plot a κ-distribution for $\kappa = 1.63$, comparing it to a Maxwellian distribution (Heerikhuisen et al. 2008a).

As recognized by Heerikhuisen et al. (2008a), the MHD equations for the plasma do not change if a κ-distribution for the solar wind protons is assumed—this is because the basic fluid conservation laws do not assume a specific form of the distribution function (see for example Burgers 1969). Closure at the second moment is possible if the distribution is isotropic, since the heat flux and the off-diagonal components of the stress tensor are then identically zero. The only difference from conventional fluid dynamics is that the collision integrals do not vanish as they would for a Maxwellian distribution. However, collisional frequencies are so low for the SW that we may neglect these collisional terms and treat the distribution function (10) as "frozen" into the plasma. Even though the solar wind is effectively collisionless, an MHD approach is still warranted since the plasma has fluid properties perpendicular to the magnetic field, while various wave phenomena help isotropize the distribution (see for example Khabibrakhmanov et al. 1996; Kulsrud 1984). For these reasons, Heerikhuisen et al. (2008a) solve the regular MHD equations to determine the bulk plasma quantities, but in the inner heliosheath they interpret these as having come from (10). For simplicity, Heerikhuisen et al. assume $\kappa = 1.63$ for the solar wind plasma, which is a value consistent with the data analysis of Decker et al. (2005). We may anticipate that observations by the IBEX mission can be used to estimate or constrain κ in the heliosheath, and we follow the discussion by Heerikhuisen et al. (2008a) to indicate how this can be done.

The κ distribution is an approximation in the sense that it is relatively straightforward to include a tail on the plasma distribution, which emerges directly from a thermal core.

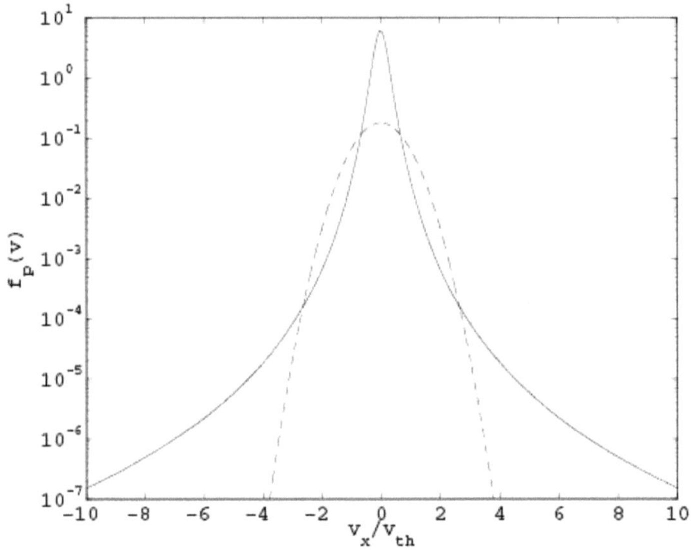

Fig. 1 A 1D cut of the velocity distribution function in the plasma frame for $\kappa = 1.63$ (*solid line*), together with a Maxwellian distribution (*dashed line*). Note that the core of the κ-distribution is narrower than the Maxwellian, but the zeroth and second moments are the same for both distributions. Here, we have defined $v_{th} = \theta_p \sqrt{\kappa/(\kappa - 3/2)}$—see (12). Heerikhuisen et al. (2008a)

In this, it mimics the pickup ion distribution. However, the core and tail of a κ distribution are coupled self-similarly, meaning that broadening of the core leads to a corresponding broadening of the tail, and this may not be a feature of separate pickup ion and thermal solar wind distributions. We note that the magnetic field orientation and solar wind speed are unlikely to have much effect on the overall pickup ion distribution. In principle, the subsequent scattering of the pickup ions shortly after birth onto (most probably) a bispherical distribution should wash out much of the initial magnetic field orientation characteristics, and the gradual cooling of the pickup ions, and thus creation of the filled shell distribution, is likely to weaken the dependence on solar wind flow speed. Thus, in summary, the κ distribution probably captures adequately the dynamic range of the pickup ion distribution and its underlying symmetry, but it introduces a coupling between the core distribution and the tail that may not accurately reflect the physics of a separate thermal solar wind core and pickup ion distribution.

The κ- and Maxwellian plasma distribution functions are linked through the choice of θ_p in equation for the isotropic plasma pressure. Thus, using

$$p = \frac{4\pi m_p}{3} \int_0^\infty v^4 f_p(v) dv = \frac{\rho\theta_p^2}{2} \frac{\kappa}{\kappa - 3/2}, \tag{11}$$

allows us to introduce an effective temperature for the κ-distribution as

$$T_{\text{eff}} = \frac{p}{nk_B}, \tag{12}$$

so that

$$p = \frac{2k_B T_{eff}}{m_p} \frac{1}{\theta_p^2} \frac{\kappa - 3/2}{\kappa}.$$

Note the singularity $\kappa \longrightarrow 3/2$, which corresponds to a v^{-5} tail (Fisk and Gloeckler 2006) and an unbounded pressure.

As discussed above, for a Maxwellian distribution (Pauls et al. 1995) the charge exchange loss terms depend on the assumed proton distribution. For a κ-distribution, Heerikhuisen et al. follow a derivation analogous to that of Pauls et al. (1995) and Zank (1999) to find that the charge exchange loss rate is now given by

$$\beta_{ex} = n\sigma_{ex}\sqrt{\frac{4\Gamma^2(\kappa+1)\,\theta_p^2}{\pi\kappa(\kappa-1)\,\Gamma^2(\kappa-1/2)} + (\mathbf{v}-\mathbf{u})^2}, \tag{13}$$

which, in the limit of large κ i.e., a Maxwellian, reduces to distribution,

$$\beta_{ex} = n\sigma_{ex}\sqrt{\frac{4}{\pi}\theta_p^2 + (\mathbf{v}-\mathbf{u})^2}. \tag{14}$$

It is of interest that the Heerikhuisen et al. (2008a) approach has the consequence that the locations of the TS, HP and bow shock change when pickup ions react self-consistently on the plasma, a result that is consistent with the 5-fluid model of Fahr et al. (2000) that introduces PUIs as an explicit plasma component. This distinguishes the self-consistent approach from the "post-processing" approach used by Prested et al. (2007), and is discussed further below.

The mean free path for charge-exchange collisions can be very large in the heliospheric boundary regions and in the supersonic solar wind. To illustrate this point, we plot in Fig. 2 the average number of charge-exchange events per particle as a function of spatial location for a steady-state kinetic based simulation (Heerikhuisen et al. 2006a, 2006b; 2008a). The Sun is located at the origin. Clearly, the number of charge exchange events varies throughout the heliosphere and the solar wind–LISM boundary regions, with no more than two charge exchanges being typical for the region extending from beyond the nose of the HP to 500 AU from the Sun in the heliotail. Other regions can experience more charge exchange events; the flanks for example can have as many as 5 events per particle on average. Thus, ideally one should calculate the interstellar neutral distribution at a kinetic level since the Knudsen number $Kn \simeq 1$ for neutral H throughout large regions of the heliosphere, or recognize that a single set of gas dynamical equations cannot adequately describe the different neutral H populations created through charge exchange. However, a multiple set of gas dynamics equations corresponding to various neutral particle populations that are created in different thermodynamical regions of the heliosphere and LISM can be a good approximation to the fully kinetic description, subject to certain caveats.

Neutral H gas is coupled to the plasma through appropriate source terms. When a fluid description for the neutrals is assumed, source terms are calculated using approximations to integrals over the distribution functions convolved with the charge-exchange cross-section. Since the form of the distribution functions originating from all possible source regions is known at all points in time, the coupling between the codes can be done at every time-step for each species in the code. This gives rise to very tight, accurate coupling, so that both steady-state problems such as the one considered here, as well as fully dynamic problems, can be solved (Zank and Müller 2003; Florinski et al. 2005a).

Coupling a Monte-Carlo code to a plasma code is more difficult than the corresponding coupling to a multi-fluid model (Baranov and Malama 1993; Heerikhuisen et al. 2006). To avoid making assumptions about the nature of the distribution function, source terms for the

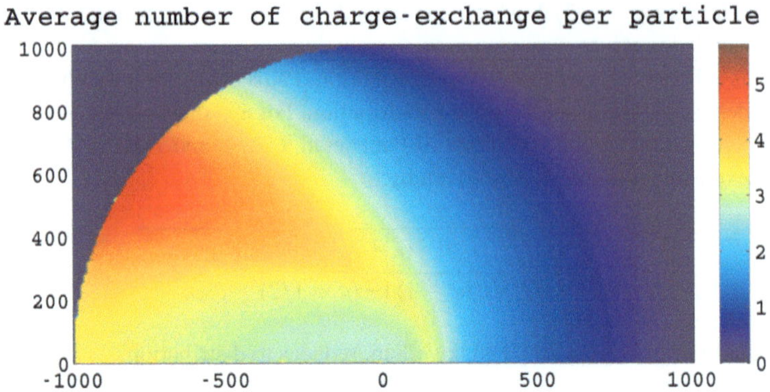

Fig. 2 The average number of charge-exchange events per particle as a function of spatial location for a steady-state model of the heliosphere interaction with the LISM. This simulation is based on a Monte-Carlo treatment (Heerikhuisen et al. 2006a, 2006b) of the kinetic neutral H, and is discussed further below. This figure illustrates the regions where the Knudsen number $Kn \simeq 1$. The distance along both axes is measured in AU, centered on the Sun (the origin)

plasma must be gathered in terms of individual charge-exchange events. Each event contributes to the plasma source at one location. It therefore requires a large number of charge-exchange events within each grid-cell in order to generate smooth and accurate sources for the plasma. When solving a steady-state problem, we may simply compute as many particle trajectories through the domain as is necessary for smooth sources. Time-dependent problems may be solved by collecting sources over a time interval which is shorter than the shortest time-scales we are trying to resolve. Under this constraint, we typically need a large number of particles to retain accuracy if the timescales present in the solution are small.

Extending the original hydrodynamic-like model of Baranov et al. (1981) and Baranov and Malama (1993, 1995) used a Monte-Carlo approach to solve the neutral H Boltzmann equation and coupled this self-consistently to a steady-state 2D hydrodynamic model of the solar wind and LISM plasma. The method of coupling is that of "global iterations". Here the plasma code is run to a steady-state iteratively, using the source term generated by running the neutrals on the preceding plasma state until successive states converge. The model has been used by Izmodenov et al. (1999) to try to infer the local interstellar electron density from the passage of interstellar H through the heliospheric boundaries.

However, as discussed below, a time-dependent model of the heliosphere develops physical (not numerical) instabilities similar to a Rayleigh-Taylor instability at the nose of the HP and analogous to a Kelvin-Helmholtz instability on the flanks (Zank et al. 1996d; Liewer et al. 1996; Wang and Belcher 1998; Zank 1999d; Florinski et al. 2005a, Borovikov et al. 2008a, 2008b), these being driven by the "frictional" coupling of neutrals to plasma in the vicinity of a contact or tangential discontinuity. We therefore need to treat the steady-state heliospheric structure problem as time-dependent, with a time-scale of the order of the instability growth time. To improve statistics, and since the physical problem is devoid of time-scales, we can run the Monte-Carlo code for longer times, while running the plasma code for a much shorter time (of the order of a few years) to suppress the instability.

The Boltzmann code developed independently by Heerikhuisen et al. (2006, 2006b, 2008b) traces a large number of macro-particles on their orbit through the heliosphere. The H atoms move ballistically for a distance of the order of their (local) mean free path, before charge-exchanging with a proton picked at random from the local plasma. At this point the process repeats with the new H atom inheriting the momentum of its parent proton. A splitting procedure is utilized to improve source term statistics in regions of large plasma gradients.

The loss term (2) may be simplified somewhat since both the Baranov & Malama and the original Heerikuisen et al. formulations assumed a Maxwellian distribution for the protons. In this case we can use the expressions developed in Ripken and Fahr (1983) and Pauls et al. (1995). Heerikhuisen et al. (2008a, 2008b) have extended their original formulation to draw particles from a κ distribution rather than a Maxwellian, and this is discussed further below.

During a charge-exchange event, a proton partner must be drawn from the plasma distribution. In the case of a Maxwellian plasma distribution, the probability distribution for the velocity of this partner will not simply be a Maxwellian, due to a selection effect caused by the distribution of individual particle velocities. Using a 3D analog of the 2D procedure described in Lipatov et al. (1998) yields the charge exchange frequency as a function of proton velocity of the form

$$\nu \propto |\mathbf{v}_H - \mathbf{v}_p|^2 \exp\left[-\frac{(\mathbf{v}_p - \mathbf{u}_p)^2}{v_{th,p}^2} \right], \tag{15}$$

where $\mathbf{v}_H$ and $\mathbf{v}_p$ are the atom and ion velocities and $\mathbf{u}_p$ is the bulk averaged plasma velocity (see also Malama 1991). Physically, this result confirms the intuitive notion that fast neutrals encounter charge-exchange partners more frequently.

As discussed above, the heliosphere–LISM environment can be described as either three or four thermodynamically distinct regions; the supersonic solar wind (region 3), the very hot subsonic solar wind (region 2), and the LISM itself (region 1a and 1b). Each region acts as a source of neutral H atoms whose distribution reflects that of the plasma distribution in the region. Accordingly, Zank et al. (1996d) identified neutral components 1, 2, and 3 with neutral atoms originating from regions 1, 2, and 3. Each of these three neutral components is represented by a distinct Maxwellian distribution function appropriate to the characteristics of the source distribution in the multi-fluid models. This observation allows the simplification of the production and loss terms (2) and (3) for each neutral component. The complete highly non-Maxwellian H distribution function is then the sum over the three components, i.e.,

$$f(\mathbf{x}, \mathbf{v}, t) = \sum_{i=1}^{3} f_i(\mathbf{x}, \mathbf{v}, t), \tag{16}$$

and for each component, the integral equation (1) must, in principle, be solved. Instead of solving (1), Zank et al. (1996d) use (16) in (1) to obtain three Boltzmann equations corresponding to each neutral component. This is an extension of the procedure developed in Pauls et al. (1995). For component 1, both losses and gains in the interstellar medium need to be included, but only losses are needed in the heliosheath and solar wind. This applies similarly to components 2 and 3. Thus, for each of the neutral H components i ($i = 1, 2$ or 3)

$$\frac{\partial f_i}{\partial t} + \mathbf{v} \cdot \nabla f_i = \begin{cases} P_1 + P_2 + P_3 - \beta_{ex} f_i & \text{region } i \\ -\beta_{ex} f_i & \text{otherwise} \end{cases}, \tag{17}$$

where $P_{1,2,3}$ means that P_{ex} is to be evaluated for the parameters of components 1, 2, or 3 respectively. Under the assumption that each of the neutral component distributions is approximated adequately by a Maxwellian, one obtains immediately from (17) an isotropic hydrodynamic description for each neutral component,

$$\frac{\partial \rho_i}{\partial t} + \nabla \cdot (\rho_i \mathbf{u}_i) = Q_{\rho i}; \tag{18}$$

$$\frac{\partial}{\partial t}(\rho_i \mathbf{u}_i) + \nabla \cdot \left[\rho_i \mathbf{u}_i \mathbf{u}_i + p_i \mathbf{I}\right] = \mathbf{Q}_{mi}; \tag{19}$$

$$\frac{\partial}{\partial t}\left(\frac{1}{2}\rho_i u_i^2 + \frac{p_i}{\gamma - 1}\right) + \nabla \cdot \left[\frac{1}{2}\rho_i u_i^2 \mathbf{u}_i + \frac{\gamma}{\gamma - 1}\mathbf{u}_i p_i\right] = Q_{ei}. \tag{20}$$

The source terms Q are appropriate moments of (2) and (4) and are listed in Pauls et al. (1995) and Zank et al. (1996d). The subscript i above refers to the neutral component of interest ($i = 1, 2, 3$), ρ_i, $\mathbf{u}_i$, and p_i denote the neutral component i density, velocity, and isotropic pressure respectively, $\mathbf{I}$ the unit tensor and γ ($= 5/3$) the adiabatic index.

Heerikhuisen et al. (2006) explore the similarities and differences between multi-fluid and Monte-Carlo models of the heliosphere and compare their Monte-Carlo code (similar in many respects to that developed by Malama 1991) with a four-neutral fluid code, which is a straightforward extension of the original 3-fluid neutral code of Zank et al. (1996d) that subdivides region 1 into regions 1a and 1b. They also compare their models to those of Alexashov and Izmodenov (2005), using the same parameters and boundary conditions for a variety of fluid and kinetic models. For this comparison, Heerikhuisen et al. (2006) used relatively simple axially symmetric models without magnetic fields and considered only steady-state solutions. Surprisingly good comparative results were found between kinetic and multi-fluid models, especially in global features, structure, and location, and between the UAH and Moscow groups models. While the axially symmetric models have now been superseded by more sophisticated 3D models, they nonetheless allow one to probe much of the basic physics of the solar wind-LISM interaction simply and efficiently, and in particular, allow for more accurate kinetic modeling since the neutral H statistics can be done very accurately. Pogorelov et al. (2008a, 2008) have extended the comparison of multi-fluid and kinetic approaches to 3D, and some of these results are discussed below.

3 Role of Magnetic Fields and Neutral H in Determining Heliospheric Structure

Both the interplanetary and interstellar magnetic field affect the shape and position of the HP relative to the Sun and interstellar plasma velocity vector, originally discussed by Fahr et al. (1988), Linde et al. (1998), Pogorelov and Semenov (1997), Pogorelov and Matsuda (2000), and Ratkiewicz et al. (1998). The interplanetary magnetic field (IMF), by virtue of the current sheet, introduces a corresponding asymmetry inside the inner heliosheath that also affects the shape and position of the HP (Washimi and Tanaka 1996, 2001; Linde et al. 1998; Zank 1999). The asymmetry in the HP position affects the shape of the heliospheric TS, and the different distances at which the V1 and V2 TS crossings occurred suggests that some asymmetry in the TS position exists (Stone et al. 2008). Prior to the V1 and V2 crossings of the TS, Lallement et al. (2005) suggested that an asymmetry in the plasma distribution on the interstellar side of the HP might be responsible for the deviation of the H-atom and He-atom flows observed in the inner heliosphere by the SOHO SWAN experiment.

In an effort to more carefully identify the role of the ISMF in introducing an asymmetric heliospheric structure, Pogorelov et al. (2004) reconsidered the ISMF coupling with the IMF at the HP using an ideal MHD plasma model in the absence of neutral H.

A range of different orientations of the ISMF $\mathbf{B}_\infty$ with respect to the interstellar flow vector $\mathbf{V}_\infty$ and the ecliptic plane was considered to determine the resulting TS and HP location and shape. Like other related studies (Ratkiewicz et al. 1998; Opher et al. 2006; Washimi et al. 2006), the ISMF can have a significant effect on the global structure and symmetry of the TS and HP. However, all these studies neglect the self-consistent inclusion of neutral H, which acts to symmetrize the heliosphere, therefore reducing the asymmetric effects of the ISMF (Pogorelov et al. 2006; Pogorelov and Zank 2006; Pogorelov et al. 2007). Global models of the heliosphere that properly include neutral H have relatively small hemispheric asymmetries compared to models that neglect neutrals, and if larger asymmetries are observed, then the likely cause is probably the result of a highly temporal solar wind (Zank and Müller 2003; Washimi et al. 2007b).

The SOHO SWAN experiment observations by Lallement et al. (2005) showing a $4^\circ \pm 1^\circ$ deflection of the H-atom flow from the interstellar He-atom flows in the inner heliosphere offers an interesting possible opportunity to extract some information about the strength and orientation of the ISMF. Izmodenov et al. (2005), estimated the direction of the H-atom flow in the inner heliosphere by analyzing statistically averaged trajectories of neutral H atoms for a model that neglected the IMF and assumed a spherically-symmetric solar wind. Not surprisingly, this ensured that the heliosphere was symmetric with respect to the plane formed by $\mathbf{B}_\infty$ and $\mathbf{V}_\infty$ (a BV-plane). Consequently, an average neutral H trajectory which starts at a LISM point in the BV-plane will remain in that plane. By contrast, those trajectories that start at two points lying symmetrically above and below the BV-plane will acquire out-of-plane velocity components oriented in the opposite directions, making the average out-of-plane deflection zero. A better and more careful analysis by Pogorelov et al. (2006) and Pogorelov and Zank (2006) showed that the flow of neutral H never preserves its original unperturbed LISM orientation, even for $\mathbf{B}_\infty \parallel \mathbf{V}_\infty$. Moreover, for models that include the IMF, the H deflection inevitably takes place both within and perpendicular to the BV-plane. Thus, the H deflection plane (HDP) and the BV-plane do not typically coincide.

It does not necessarily follow that the observed $4^\circ \pm 1^\circ$ deflection of H-atoms from its LISM vector implies a highly obliquely oriented ISMF. The deflection of H parallel and perpendicular to the BV-plane can be comparable if the angle between $\mathbf{B}_\infty$ and $\mathbf{V}_\infty$ is not large (Pogorelov and Zank 2006). However, to produce a $4^\circ \pm 1^\circ$ deflection requires a strong ISMF (perhaps greater than 5 μG) for $\mathbf{B}_\infty$ lying at small angles to $\mathbf{V}_\infty$. Although not commonly assumed as ISMF values, such strengths should not be disregarded. The Cox and Helenius (2003) theory for the origin of the Local Bubble suggests a strong ISMF lying nearly parallel to the LISM velocity vector. Obviously, since magnetic pressure acts perpendicularly to magnetic field lines, an increase in $\mathbf{B}_\infty$ will not unduly affect the TS and HP stand-off distances in the upstream LISM directions (Baranov et al. 1971; Florinski et al. 2004a). However, increasing $\mathbf{B}_\infty$ may well yield a LISM flow speed that is either sub-fast magnetosonic or even sub-Alfvénic. Even for this case, Florinski et al. (2004a) (2D) and Pogorelov et al. (2006) (3D) found global heliospheric solutions that had the HP at a finite distance from the Sun, provided interstellar neutral H was included self-consistently.

To include the self-consistent coupling of neutral H atoms and a magnetized plasma, Pogorelov et al. (2006) extended the two-fluid and four-fluid models of Pauls et al. (1995) and Zank et al. (1996b) to include magnetic fields in a full MHD description. As already discussed, this is a reasonable approximation whenever we are interested primarily in the role of the magnetic field. The 4-fluid interaction model has proved effective in modeling 3D (Pauls and Zank 1997b), magnetized but axi-symmetric (Florinski et al. 2003a), and non-stationary aspects of the solar wind-LISM interaction (Zank 1999a; Zank and Müller 2003). The review by Zank et al. (2008) compares purely MHD models and models that include neutral H self-consistently for a variety of ISMF $\mathbf{B}_\infty$ orientations relative to the interstellar flow vector $\mathbf{V}_\infty$. We may anticipate, on the basis of numerous gas dynamical calculations, that the influence of charge exchange on global heliospheric structure is likely to be stronger than the effect of a weak ISMF.

The numerical calculations are based on a Cartesian coordinate system with origin at the Sun. The x-axis is oriented along the Sun's rotation axis, perpendicular to the ecliptic plane (yz-plane), for simplicity. The z-axis belongs to the plane defined by the x-axis and $\mathbf{V}_\infty$, and is directed toward the LISM. The y-axis completes the coordinate system.

Here we consider two cases, the first being $\mathbf{B}_\infty \perp \mathbf{V}_\infty$, and the other with $\mathbf{B}_\infty$ in the HDP.

Consider the case of $\mathbf{B}_\infty \perp \mathbf{V}_\infty$ and $\mathbf{B}_\infty \perp 0x$ modeled using a two-fluid approach. Figure 3 shows the plasma density (Fig. 3a; logarithmic scale), magnetic field strength (Fig. 3b), plasma temperature (Fig. 3c; logarithmic scale), and neutral H density (Fig. 3d) distributions. The structure of the heliosphere when neutral H is included self-consistently is substantially different from that derived from a purely MHD model. The distance to the TS is decreased considerably and the bullet-shape, so characteristic of models (both gas dynamic and MHD) that exclude the neutral H, is absent. Both effects are of course due to the deceleration of supersonic flows by the pickup process.

From the plot of the magnetic field strength (Fig. 3b), it is seen that the heliospheric current sheet (HCS), assumed to lie in the ecliptic instead of being tilted, bends into the lower hemisphere. As a result, some IMF lines from the upper hemisphere are carried by the solar wind to the lower hemisphere. The IMF and ISMF vectors at the upwind section of the HP are oriented approximately in the same direction. Further downstream, the HCS width increases, it interacts with the (numerically diffusive) HP, and we observe ripples on its surface. Since the IMF and ISMF are oppositely directed in the lower hemisphere, the ripples most likely result from a combination of numerical magnetic reconnection between the IMF and ISMF lines (see Pogorelov et al. 2004) and possibly a Kelvin-Helmholtz-type instability of the HP, perhaps driven by neutral H (Borovikov et al. 2008b; Zank 1999d; Florinski et al. 2003b). In particular, the temperature of the LISM protons increases due to the energy released by reconnection. The HP is asymmetric with respect to the ecliptic plane, and it also appears that the maximum IMF magnitude on the inner side of the HP is about 10% greater in the upper hemisphere than in the lower, and the region of the enhanced IMF is more extended there.

The other $\mathbf{B}_\infty \perp \mathbf{V}_\infty$ example that we discus is that of $\mathbf{B}_\infty$ tilted 60° to the ecliptic plane. Figure 4 shows the distributions of the plasma density (Fig. 4a), the magnetic field strength (Fig. 4b), the plasma density (Fig. 4c), and the population 1 neutral H (Fig. 4d) in the meridional plane for the four-fluid model. Comparing two-fluid and four-fluid models (Zank et al. 1996b; Pogorelov et al. 2006) shows that the bow shock standoff distance from the HP is larger in the four-fluid approximation, while the maximum values of the plasma density and magnetic field are noticeably smaller. The width of the hydrogen wall is greater, while the increase in the neutral H density is smaller. The distribution of the IMF strength is qualitatively similar for both four- and two-fluid models. Worth noting is an essentially 3D effect showing an asymmetry in the population 2 hydrogen distribution with respect to the ecliptic plane. A more detailed analysis of the distribution of quantities in the upwind direction is presented in Figs. 9–11 of Pogorelov et al. (2006). They find that the bow shock, HP, and TS distances to the Sun obtained with the four-fluid model are somewhat larger than those calculated with the two-fluid model, with the differences consecutively decreasing (295, 137, and 95 AU and 329, 152, and 102 AU for the two- and four-fluid models, respectively). The TS distances to the Sun in the downwind direction are about 211 AU for both models. This result differs significantly from that presented in the paper of Alexashov and Izmodenov (2005), where a two-fluid model (a one-fluid neutral model in their terminology) gave

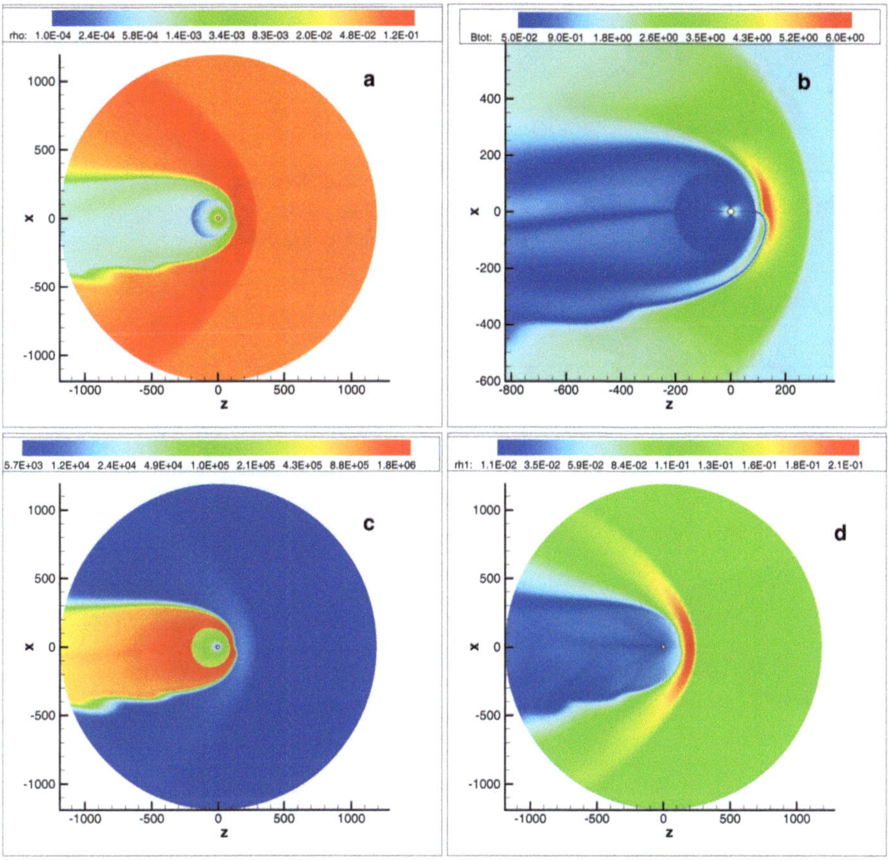

Fig. 3 Distributions in the meridional plane for the solar wind–LISM interaction the case of $\mathbf{B}_\infty \perp \mathbf{V}_\infty$ and $\mathbf{B}_\infty \perp 0x$ in the two-fluid model approximation: (**a**) plasma density (logarithmic scale) in particles per cm^{-3}, (**b**) magnetic field strength in μG, (**c**) plasma temperature (logarithmic scale) in K, and (**d**) neutral H density in particles per cm^{-3}. $B_\infty = 1.5$ μG. The distance along both axes is measured in AU, centered on the Sun (the origin). (Pogorelov et al. 2006)

a difference in the TS downwind locations greater than 50% of that in the kinetic model. A possible reason might be their somewhat different treatment of the secondary hydrogen atoms. Although both models take into account momentum loading of the solar wind plasma properly, Alexashov and Izmodenov (2005) neglect only the secondary hydrogen atoms that originate in the supersonic solar wind, whereas our model neglects them everywhere in the solar wind and heliosheath. This may be a possible reason why the one-, two-, and three-fluid hydrogen models of Alexashov and Izmodenov (2005) give results so different from those in the four-fluid model. The discrepancy is smaller if our approach is applied. On the other hand, discrepancies between different multi-fluid models may potentially increase with increasing number density of interstellar H atoms.

Figure 5 shows the distributions of the magnetic field magnitude B_{tot} (solid lines) and the Cartesian components B_x (dashed lines), B_y (dot-dashed lines), and B_z (double-dot-dashed lines) of the magnetic field vector in the meridional plane for lines of sight slightly above (left) and slightly below the ecliptic plane (right). Thicker lines correspond to a two-fluid

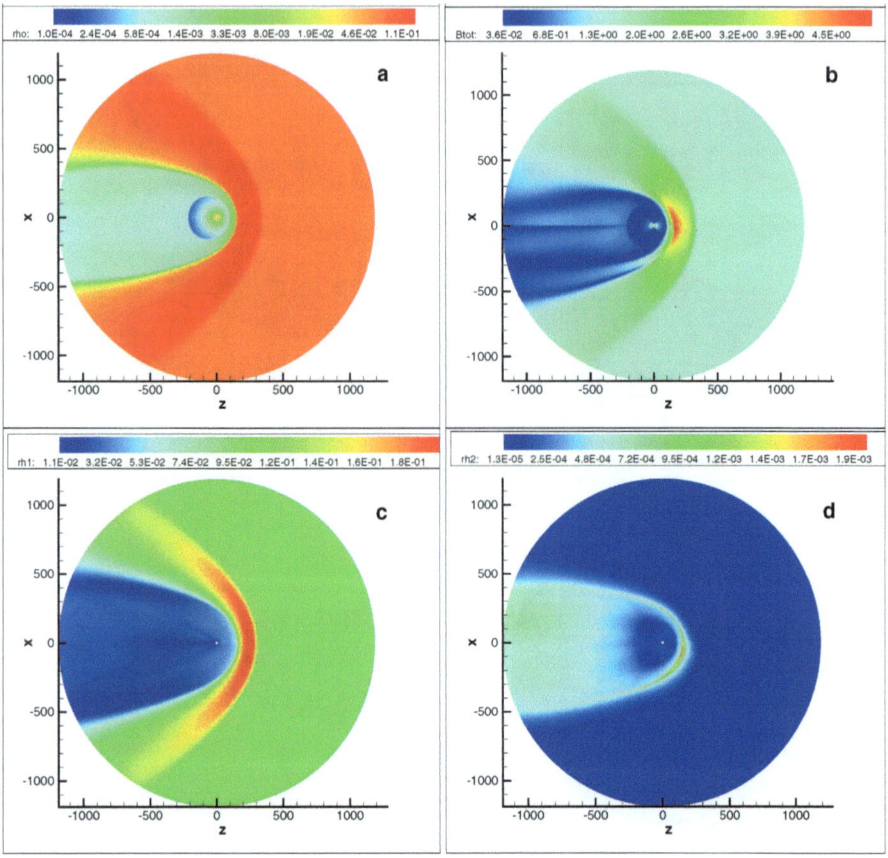

Fig. 4 Distributions in the meridional plane for the solar wind–LISM interaction for the case of $\mathbf{B}_\infty \perp \mathbf{V}_\infty$ and tilted 60° to the solar ecliptic plane in the four-fluid model approximation: (**a**) plasma density (logarithmic scale) in particles per cm^{-3}, (**b**) magnetic field strength in μG, (**c**) plasma temperature (logarithmic scale) in K, and (**d**) neutral H density in particles per cm^{-3}. $B_\infty = 1.5$ μG. The distance along both axes is measured in AU, centered on the Sun (the origin). (Pogorelov et al. 2006)

model. A decrease in the magnetic field strength occurs when the line of sight intersects the HCS. The distribution of the magnetic field is asymmetric with respect to the ecliptic plane, which may well lead to related asymmetries in the cosmic-ray modulation in opposite hemispheres.

On the basis of a kinematic magnetic field model, Zank (1999) suggested the possibility that the IMF may cross the TS multiple times. Jokipii and Giacalone (2004), Jokipii et al. (2004), and Stone et al. (2005) have suggested this as an explanation for the V1 observations of several month-long increases in energetic particle fluxes in late 2002–2003. Since the shape of the TS does not coincide with the shape of the Parker spiral magnetic field in the upwind region, it is therefore possible (Zank 1999; McDonald et al. 2003) that some IMF lines can reappear in the upstream solar wind region after crossing the TS from the downstream side. This possibility was analyzed carefully by Pogorelov et al. (2006) using their self-consistent 3D IMF distribution, finding that this is indeed the case. It was shown that the self-consistent inclusion of neutral H ensures that V2 could not be directly connected

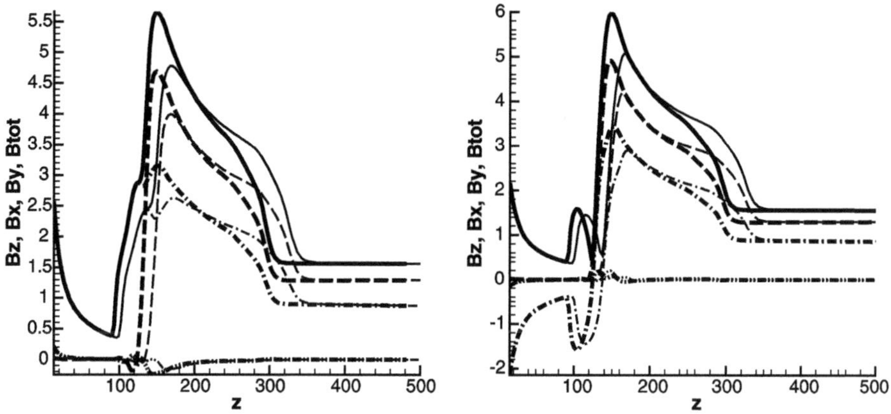

Fig. 5 (*Left*) Distributions of the magnetic field magnitude (*solid lines*), B_x (*dotted lines*), B_y (*dot-dashed lines*), and B_z (*double-dot-dashed lines*) components of the magnetic field vector in the meridional plane along a line of sight slightly above the ecliptic plane (in μG as a function of AU). (*Right*) Corresponding distributions in the meridional plane along a line of sight slightly below the ecliptic plane. The two-fluid results are shown by *thick lines*. (Pogorelov et al. 2006)

magnetically to the TS if the radial separation between V2 and the TS was greater than 3 AU, even for $\mathbf{B}_\infty$ as large as 3 μG (Pogorelov et al. 2007).

A critical observational result was the measurement of the deflection of H atoms from the interstellar He trajectory by Lallement et al. (2005). Pogorelov et al. (2008c), building on earlier studies by Heerikhuisen et al. (2006, 2007) and Pogorelov et al. (2008a), used a fully time-dependent 3D MHD-kinetic model to consider the effect of $\mathbf{B}_\infty$ lying in the HDP. This code incorporates both solar wind and LISM magnetic fields, and couples an MHD code self-consistently to a time-dependent 3D Monte-Carlo code that determines the kinetic distribution of neutral H. The LISM plasma velocity, temperature and density for their simulations are assumed to be $\mathbf{B}_\infty = 26.4\ \mathrm{km\,s^{-1}}$, $T_\infty = 6527$ K, and $n_\infty = 0.06\ \mathrm{cm^{-3}}$, respectively, and the density of neutral H is $n_{H\infty} = 0.15\ \mathrm{cm^{-3}}$. The solar wind is assumed to be spherically symmetric with the following parameters at 1 AU: $V_E = 450\ \mathrm{km\,s^{-1}}$, $T_E = 51,100$ K, and $n_E = 7.4\ \mathrm{cm^{-3}}$. The magnitude of the ISMF vector is $\mathbf{B}_\infty = 3$ μG. The radial component of the IMF at 1 AU is set to 37.5 μG, and we assume a Parker spiral at 1 AU. For convenience and to allow for easy visualization of the various planes and symmetries, we plot the coordinate system in Fig. 6. The direction of the LISM velocity is aligned with the vector $\mathbf{l}_{\mathrm{He}} = (-0.087156, 0, -0.9962)$. The HDP is defined by $\mathbf{l}_{\mathrm{He}}$ and the vector $\mathbf{l}_{\mathrm{H}} = (-0.1511, -0.04049, -0.9877)$ corresponding to the direction that LISM neutral H enters the inner heliosheath determined observationally by Lallement et al. (2005). We assume that $\mathbf{B}_\infty$ is aligned with the vector $\mathbf{l}_B = (-0.5, -0.2678, -0.82356)$. Thus, $\mathbf{B}_\infty$ belongs to the observed HDP and is directed into the southern hemisphere at an angle of $\sim 30°$ to the ecliptic plane. This direction gave one of the largest V1–V2 asymmetries of the TS in the two-fluid calculations of Pogorelov et al. (2007). In fact, $\mathbf{B}_\infty$ in the plane tilted at 60° to the ecliptic plane is very close to the observed HDP (Pogorelov et al. 2007, 2006).

Figure 7 shows the distribution of the plasma temperature in the V1–V2 plane, illustrating that, in agreement with the two-fluid calculations of Pogorelov et al. (2007), the asymmetry of the TS is minor. Both two-fluid and multi-fluid models show that the asymmetry of the TS is rather small, unlike the ideal MHD calculations of Opher et al. (2006) where the absence of neutral H exaggerates the asymmetry. In Fig. 8, we show the distributions of the proton

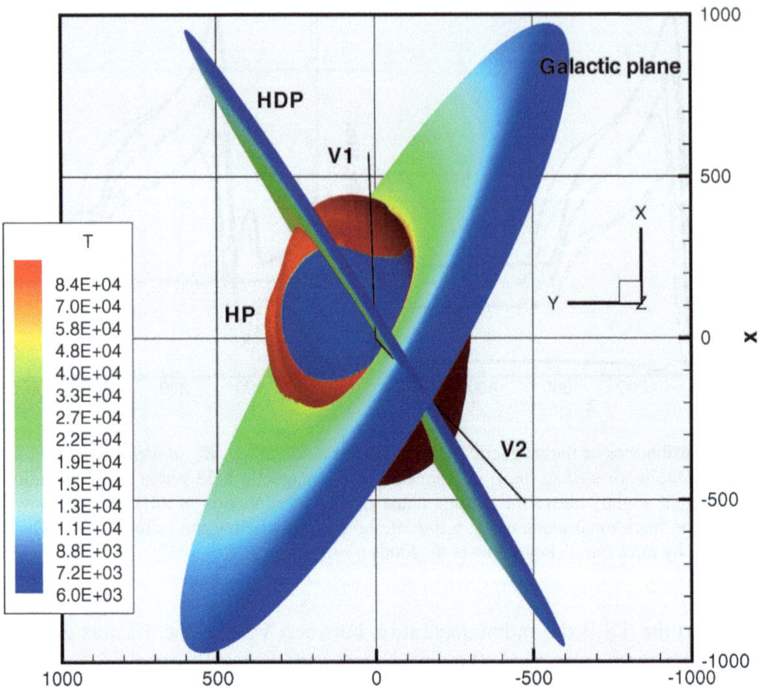

Fig. 6 Frontal view of the HP, HDP, Galactic plane, and V1 and V2 trajectories. *Blue* and *red colors* on the surface of the HP correspond to $B_R < 0$ and $B_R > 0$, respectively. The planes are colored according to the plasma temperature (K) distributions. The HP is clearly asymmetric with respect to the BV-plane. The axes are in AU, measured from the Sun (the origin). (Pogorelov et al. 2008b)

temperature along sight lines in the direction of V1 (solid black lines) and V2 (solid red lines). For the sake of comparison, Pogorelov et al. (2006) also use dashed lines to show the same distributions obtained with a five-fluid model.[1] In Fig. 9, we compare radial profiles of the plasma density and magnetic field magnitudes obtained using a five-fluid and MHD-kinetic model respectively (Pogorelov et al. 2008a, 2008). The colored lines correspond to the directions $\phi = 180°$, $\theta = 35°$ (black), $\phi = 0°$, $\theta = 0°$ (red), $\phi = 0°$, $\theta = 90°$ (green), and $\phi = 0°$, $\phi = 180°$ (purple). The angles ϕ and θ are measured from the x-axis in the xy-plane and from the z-axis, respectively. Clearly, the results obtained from these two different models are in a excellent qualitative and even quantitative agreement.

An interesting result to emerge from this simulation is that the kinetic treatment of charge exchange leads to the disappearance of the bow shock, which is essential in an equivalent ideal MHD model. The disappearance of the bow shock is a direct consequence of the secondary charge exchange of hot neutral H created in the inner heliosheath in the LISM. Hot neutral H created in the inner heliosheath has high thermal velocities, so these particles move easily into the LISM upstream of the bow shock. The subsequent secondary charge

[1]The five-fluid model extends the original multi-fluid model of Zank et al. (1996d) by using the ideal MHD equations to model the flow of protons and four (instead of three) coupled sets of Euler equations to simulate the flow of separate neutral H fluids. These consist of the parent LISM neutrals (population 1a) and those born in the outer heliosheath (population 1b), inner heliosheath (population 2), and supersonic solar wind (population 3).

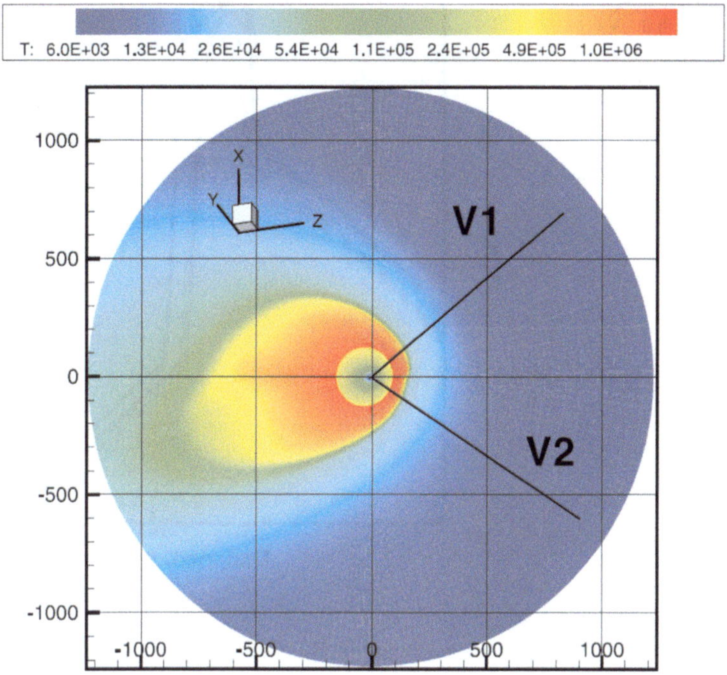

Fig. 7 Plasma temperature (K) distribution in the V1–V2 plane. The *straight lines* show the V1 and V2 trajectories and the axes are plotted in AU. (Pogorelov et al. 2008a)

exchange leads to anomalous heating of the LISM plasma ahead of the HP and the bow shock, which increases the sound speed and thus reduces the interstellar flow Mach number (to the point that the flow becomes subsonic!).

The symmetrizing effect of charge exchange in the presence of an ISMF is easily understood. For the chosen solar wind and LISM parameters, neglecting the neutral particles results in the HP rotating so that the nose is shifted to the south while the tail is shifted to the north (Pogorelov et al. 2007). As a result, the stagnation point of the LISM plasma on the HP moves above the ecliptic plane. This creates an asymmetric distribution of the LISM plasma in the outer heliosheath, which results in enhanced charge exchange in this region. Charge exchange created pickup ions exert additional pressure, which acts to decrease the asymmetry of the HP by counterbalancing the ISMF pressure. A similar mechanism works on the solar wind side of the HP. Pogorelov et al. (2008a) find that the TS is closer to the Sun by only about 3 AU in the V2 direction than the V1 direction in their MHD-kinetic simulation—see Fig. 8. The steady-state asymmetry is far too small to explain a V2 crossing of the TS at a distance to the Sun closer by 10 AU than V1. Instead, it is likely that temporal variations in the solar wind ram pressure modify the TS location significantly (Scherer and Fahr 2003; Zank and Müller 2003; Borrmann and Fichtner 2005; Pogorelov et al. 2007; Washimi et al. 2007a, 2007b), offsetting the effect of the ISMF pressure.

The MHD-neutral analysis of Pogorelov et al. (2007) shows that east-west asymmetry of the TS due to the ISMF in the observed HDP is also insufficiently large to allow V2 to be directly (by less than a full winding of the IMF spiral) connected to the TS at distances larger than about 3 AU ahead of the TS. Energetic protons with energies less than 7 MeV cannot stream along a field line from the TS to V2 directly.

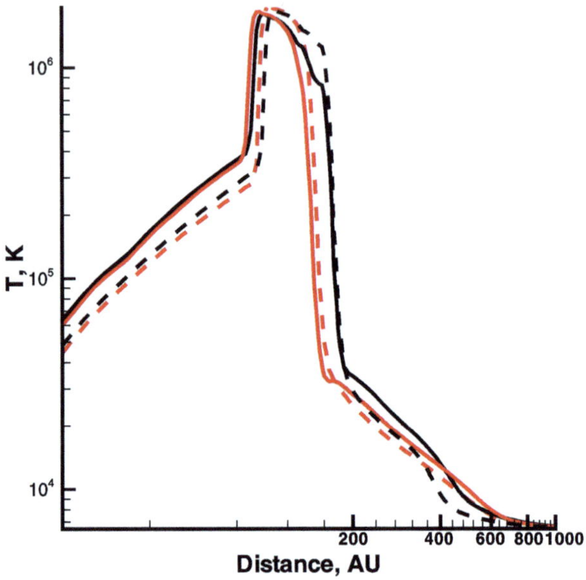

Fig. 8 Distribution of plasma temperature (K) in the V1 (*black lines*) and V2 (*red lines*) directions. The results shown with *solid* and *dashed lines* were obtained using an MHD-kinetic and a five-fluid model respectively. (Pogorelov et al. 2008a)

Experience shows that a larger component of $\mathbf{B}_\infty$ parallel to the ecliptic plane increases the TS asymmetry. This may result, however, in an H flow deflection greater than that observed in the SOHO SWAN experiment since the two effects are mutually related. To quantify the effect of the neutral H flow deflection, Pogorelov et al. (2008a) used their kinetic neutral-atom code to collect statistics on the H-atom velocity distribution in the solar wind. They recorded the deflection from $\mathbf{V}_\infty$ of all H-atoms within a 45° cone about $\mathbf{V}_\infty$ out to 80 AU, both in the BV-plane and perpendicular to it, thus creating a two dimensional distribution of deflections. In Fig. 10, we show the primary (population 1a) LISM H-atoms (left panel), secondary (population 1b) H-atoms (middle panel), and the total (weighted) distribution (right panel) in the plane perpendicular to $\mathbf{V}_\infty$. Although the primary LISM distribution is initially Maxwellian, its interaction with the heliosphere results in a distribution of deflections that is obviously not isotropic. This is because charge-exchange losses preferentially cull a particular part of the distribution due to asymmetric plasma flow and the dependence of the charge-exchange rate on the relative plasma flow speed. Secondary H-atoms, and thus the combined distribution, are clearly not isotropic, and the mean of the distribution does not coincide with its center, making it more difficult to quantify the overall deflection. We find that the average deflection of primary neutrals is about 1.8° in the BV-plane and $-0.18°$ perpendicular to this plane. For secondary neutrals, the corresponding values are 4.7° and 0.15°, and for the combined population these are 3.8° and 0.05°. The peaks of the distributions are not at these locations; and instead, the primary population shows a peak close to zero deflection, and the most common deflection of the secondary neutrals is around 7° in the BV-plane and 1° out of it. For this particular example, the average deflection takes place almost entirely in the BV-plane. Thus, the actual angle between the BV-plane and the HDP is determined by the accuracy in measuring the H-flow direction by the SOHO SWAN experiment, which is to within 1°. Although Pogorelov et al. (2008a) assume that the BV-plane is parallel to the average HDP, an additional deflection of the order of $\pm 1°$ perpendicular to the average HDP cannot be excluded. This gives us an estimate for the angle between the HDP and BV-plane as $\arctan(\tan 1°/\tan 4°) = 15°$ (Pogorelov et al. 2007).

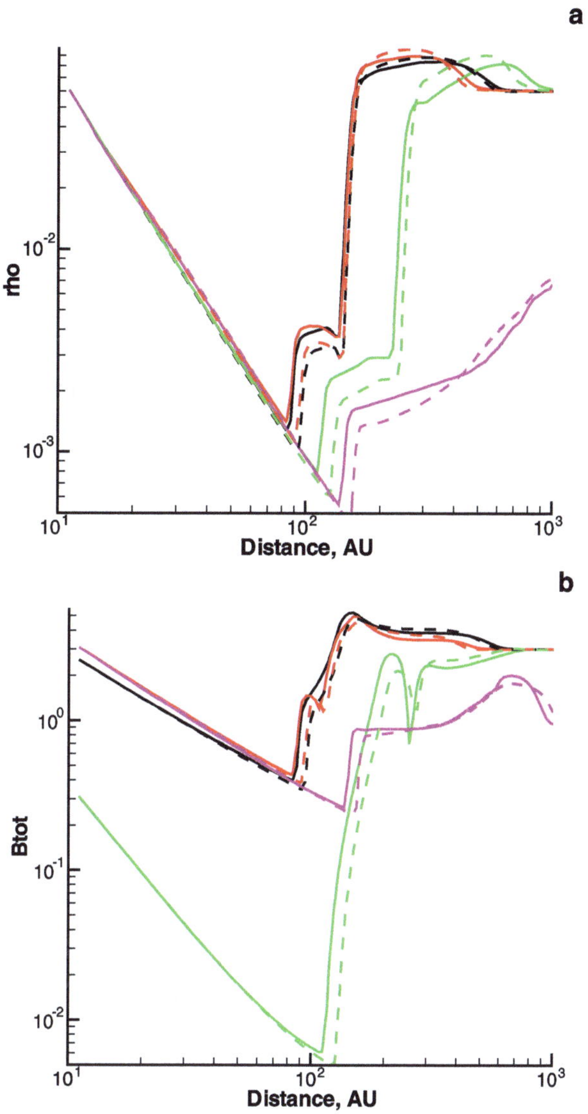

Fig. 9 Distribution of (**a**) plasma number density (in cm^{-3}) and (**b**) magnetic field magnitude (in μG) along the directions $\phi = 180°$, $\theta = 35°$ (*black lines*), $\phi = 0°$, $\theta = 0°$ (*red lines*), $\phi = 0°$, $\theta = 90°$ (*green lines*), and $\phi = 0°$, $\phi = 180°$ (*purple lines*). The results shown with *solid* and *dashed lines* correspond to an MHD-kinetic and a five-fluid model, respectively. (Pogorelov et al. 2008)

The effect of a stronger IMF was discussed by Pogorelov et al. (2008c), who found that $B_\infty \sim 7$ μG can increase the TS asymmetry to 8 AU.

In summary, MHD-only models that neglect the self-consistent inclusion of neutrals at either a multi-fluid or kinetic level cannot adequately describe global heliospheric structure, and the ideal MHD models of Pogorelov et al. (2004) and Opher et al. (2006) are inappropriate in this context. The Opher et al. model, to arrive at some quantitative results, uniformly scales the ideal MHD solution to the known distance of the TS crossing by V1, but as we have shown explicitly above for both gas dynamic and MHD models, the TS heliocentric distance can be 1.5 times larger when neutral particles are neglected, even for identical SW and LISM conditions (Zank 1999).

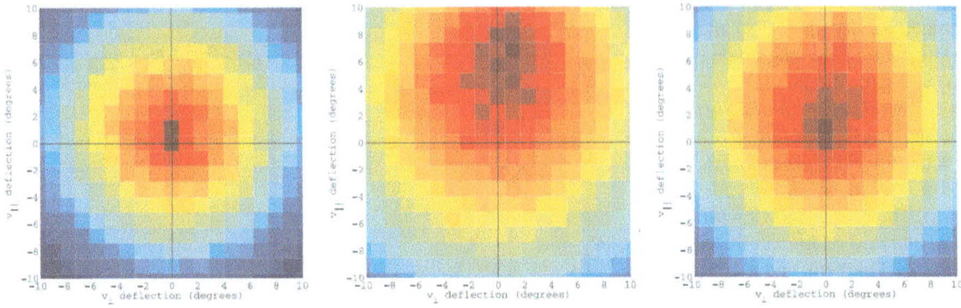

Fig. 10 Two-dimensional distribution of H-atom deflections (measured in degrees) from $\mathbf{V}_\infty$ in the plane perpendicular to the LISM BV-plane (the interstellar perspective), showing (*left*) primary interstellar H-atoms, (*middle*) secondary H atoms (i.e., the last charge-exchange occurred in the outer heliosheath), and (*right*) the combined distribution. The normal is determined by the vector product $\mathbf{r}_H \times \mathbf{r}_{He}$. The *darkest red color* corresponds to a particle count twice larger than that of the *darkest blue color*. (Pogorelov et al. 2008)

4 Termination Shock (TS) Response to Interplanetary Disturbances

Recent 3D MHD simulations by Washimi et al. (2007a, 2007b) confirmed that solar-wind ram-pressure variability contributed to the variability in R_{TS}, the distance of the TS from the Sun. The heliopause too responds to supersonic solar wind disturbances, although with a delay corresponding to the transit-time of transmitted disturbances in the heliosheath (Karmesin et al. 1995; Zank and Müller 2003). The variability of the solar wind therefore plays an important role in the global heliospheric dynamics. Washimi et al. (2007b) discuss the response of R_{TS} to solar wind ram-pressure changes. They initially consider a stationary heliosphere assuming fixed inner (solar wind) and outer boundary (ISM) conditions, and then analyze the heliosheath response by generating simplified solar-wind ram-pressure pulses in the supersonic solar wind. This simplified simulation reveals the considerable effect on R_{TS} by large amplitude incident, transmitted, and reflected heliosheath disturbances. A realistic and time-varying inner boundary using V2 plasma data is then incorporated in their 3D MHD simulation, which yields a temporal R_{TS}. By using the most current V2 data, Washimi et al. (2007b) somewhat boldly attempted to forecast the future TS position after making some reasonable assumptions about the overall solar wind ram-pressure for the next year. In part, a plausible forecast can be made because of the long transit time and response time of the global heliosphere to short timescale supersonic solar wind disturbances. The Washimi et al. (2007b) model did not include neutral H self-consistently but this is now done in a multi-fluid framework in their most recent (but unpublished) models.

Washimi et al. (2007a, 2007b) begin with a stationary global 3D MHD heliosphere. At the inner boundary of the initially stationary global 3D heliosphere, Washimi et al. (2007b) assign a set of standard values at 1 AU, i.e., density 5/cc, velocity 400 km/s (the ram-pressure of the standard set is $P_{ram,0} = 1.3 \times 10^9$ Pa at 1 AU), and toroidal magnetic field of 2.8 nT. The solar magnetic moment is assumed to be non-tilted, so that the toroidal magnetic field intensity on the inner boundary has a $\sin\theta$-dependence and its polarity reverses from northern to southern hemispheres, where θ is the colatitude measured from the solar-rotation axis z. The LISM flow speed relative to the Sun is 26.3 km/s along the x-axis, and proton density is 0.105/cc. As discussed already at length above, Washimi et al. assume that the ISMF is in the HDP (Lallement et al. 2005) has an intensity of 0.24 nT and components $\mathbf{B} = (0.163, -0.100, -0.145)$ nT. The LISM temperature is 10^4 K. Under these conditions, R_{TS} of the stationary heliosphere along the Sun–V1 line is $R_{TS,0} = 86$ AU. To obtain a reasonably realistic and time-varying heliosphere, Washimi et al. (2007a, 2007b) used

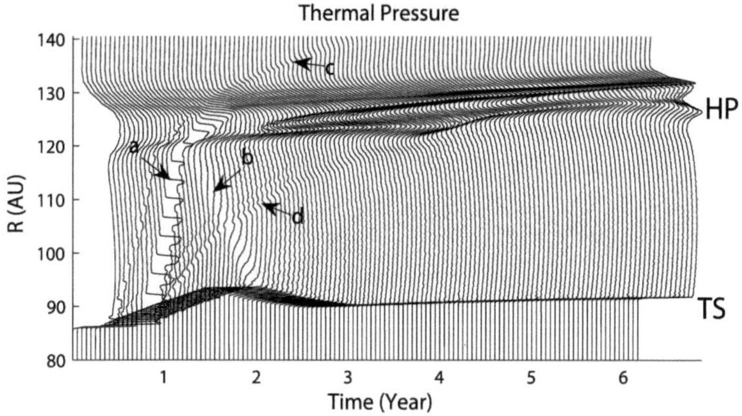

Fig. 11 Stacked space–time plots showing the propagation of the thermal pressure in the heliosheath and beyond when the solar wind ram-pressure steps increases abruptly from $P_{ram,0}$ to $1.5 \times P_{ram,0}$ at year 0. The *arrows* labeled (a), and (b), denote the forward and reflected thermal pressure pulses, and (c) and (d) the transmitted and convected pulses, respectively. The movement of the TS in response to both the incident and reflected pressure pulses is evident, as is the gradual relaxation of the TS to a new steady-state location because of the now higher (constant) solar wind ram pressure. (Washimi et al. 2007b)

V2 observed daily averaged solar wind velocity and density during the period 2001.9.10–2007.3.24 as their inner boundary condition. Under these conditions Washimi et al. (2007b) use an iterative method of simulation to determine a LISM density of 0.1107/cc that yields a V1-crossing at the observed time with an error of ≈ 10 days. Consequently, they adopt this parameter in their analysis. This of course does not correspond necessarily to a precise LISM number density but is a device to ensure that the TS location corresponds to the V1 data point, and is a method to mimic the presence of H atoms. Note that the magnetic field orientation is the same as used in Opher et al. (2006), whose plasma parameters result in a relatively large north–south (N–S) asymmetry. However, as already noted, the self-consistent inclusion of neutral H in the simulations strongly reduces the asymmetry, and Washimi et al. (2007b) are therefore extending their model to properly include interstellar atoms on the basis of a multi-fluid model (Zank et al. 1996d).

In this subsection, all analyzed quantities of the 3D Washimi et al. simulations are measured along the Sun–V1 or Sun–V2 line.

Starting from a stationary heliosphere, Washimi et al. (2007b) consider the response of the TS to discrete disturbances such as the solar wind ram-pressure changing step-wise from $P_{ram,0}$ to $1.5P_{ram,0}$. Figure 11 shows explicitly how and why the TS oscillates in location. A disturbance incident on shock must emit a series of waves and convected structures such that the shock satisfies the evolutionary conditions (e.g., McKenzie and Westphal 1968). For the case considered here, this is a fast mode wave and a convected entropy-vorticity wave. The emitted wave propagates to the HP at the magneto-acoustic speed $C_a \approx 524$ km/s, where it is partially reflected (the heliosheath is subsonic) and transmitted at the HP. The reflected wave propagates back to the TS at a speed $C_b \approx 210$ km/s. When the HP reflected pulse strikes the TS, the outward motion of the TS is reduced and R_{TS} begins to decrease. An increase in the magnetic pressure accompanies the thermal-pressure pulse. The reflected thermal-pressure pulse plays a surprisingly important role in determining the change of R_{TS}, being the reason that R_{TS} decreases continually for nearly a year even though the overall ram pressure of the solar wind increased. This effect is a consequence of both the propagation of

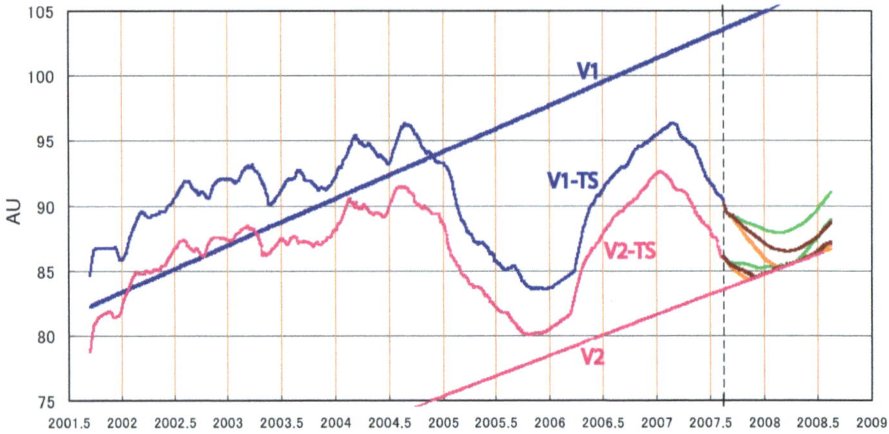

Fig. 12 TS position along the Sun–V1 (*blue*, labeled V1-TS) and Sun–V2 (*pink*, labeled V2-TS) directions based on V2 plasma data from 2001.9.10–2007.8.14 (*dashed vertical line*), together with forecasts for a year after the available V2 plasma data under the assumption that the solar wind ram-pressure is 1.0 (*orange*), 1.25 (*brown*), and 1.5 (*green*) $\times P_{ram,0}$. Note that the TS positions below the V2 line during the forecasting period are not shown. (Washimi et al. 2007b)

the pulse and the reflections at the HP occurring across the whole TS surface. The same is true for the initial increase in R_{TS} because of the solar wind ram-pressure pulse, which also continues for about a year.

Beyond the HP, the transmitted pulse, identified by (c) in Fig. 11, propagates in the interstellar medium at a speed of about 30 km/s. This pulse continues to propagate in the outer heliosheath and will eventually reach the bow shock (BS) (Zank and Müller 2003). In principle, the bow chock will partially reflect the incident wave which could travel back to the HP and further to the TS, affecting the TS position at several tens of years later. However, the damping of waves in the heliosheath makes this unlikely to be an important factor in determining TS and HP location. Another interesting result in Fig. 11 is the formation of a pulse structure at the HP, which splits into 2 humps after ≈ 2 years when the convection pulse arrives at the HP. Because the convection entropy-vorticity modes cannot be transmitted through the HP, these structures accumulate at the HP until, in principle, a steady-state is achieved with an adjusted thermal pressure/magnetic pressure jump across the HP.

A small but finite convection pulse, identified by (d) in Fig. 11, also propagates in the heliosheath. Because the shocked solar wind flow direction in the heliosheath is not radial but is bent poleward, disturbances driven at the TS in lower latitudes are convected to higher latitudes. Consequently, these disturbances cross the sun-V1 line. This speed v_d in Fig. 11 is 130 km/s near the TS and ≈ 14 km/s or less near the HP.

Figure 12 shows the simulated time-varying TS position along the Sun–V1 (blue) and Sun–V2 (pink) trajectories. R_{TS} changes from 84 AU to ≈ 100 AU during the simulation period along the Sun–V1 line. The rapid increases of R_{TS} are due to collisions of ram-pressure pulses with the TS (Washimi et al. 2007a, 2007b). The time-varying TS location along the Sun–V1 line is almost parallel to the V1 trajectory before the V1 crossing. At the end of a series of ram-pressure pulses in October 2004, R_{TS} began to decrease, and V1 crossed the TS in late 2004. R_{TS} continued to decrease rapidly, reaching 84 AU near the end of 2005. This small value of R_{TS} is due to the overshoot discussed above. A maximum in the incoming speed of the TS occurs around 2005.15, and is ≈ 300 km/s, which is almost the same as speed as the outgoing solar wind. This suggests that the reflected thermal-pressure pulses, which were driven by the series of the strong ram-pressure pulses prior to the V1 crossing, were anomalously strong. After October 2005, the TS started to move out again. At the beginning of this phase, the subsequent increase is rather gradual due to the gradual increase in ram-pressure of the solar wind: these times correspond to solar-minimum when the ram-pressure is relatively high. The increase in R_{TS} is punctuated by a rapid outward acceleration associated with the March 2006 event (Richardson et al. 2006), and R_{TS} reaches a maximum of about 96 AU by the middle of February 2007, then begins to decrease to 90 AU in response to solar wind conditions measured by the most recently available (14 August, 2007) V2 plasma data used in their analysis. Due to the N–S asymmetry of the heliosphere, R_{TS} along the Sun–V2 trajectory is generally less than that along the Sun–V1 line, but the asymmetry is relatively small ($\approx$ 3–6 AU most of the time) and variations in R_{TS} are dominated by temporal solar wind effects. R_{TS} along the Sun–V2 trajectory shows very similar changes when compared to those occurring along the Sun–V1 line. Because the simulation did not include the effects of the neutral interstellar gas, the N–S asymmetry in the Washimi et al. (2007b) simulation is probably overestimated. However, for this low ram-pressure case, R_{TS} along the Sun–V1 line is virtually the same as that along the Sun–V2 trajectory, as illustrated in Fig. 12.

Just before the end of the V2 plasma data, R_{TS} in Fig. 12 begins to decrease. In view of our remarks above concerning reflected heliospheric pulses, this decrease indicated that the TS was moving inward again because of the reflected thermal-pressure pulse associated with the March 2006 event. To quantify this effect, Washimi et al. (2007b) extended the simulation. Because the R_{TS} decrease will continue for at least a year, Washimi et al. forecast the TS location for a year, assuming that the solar wind ram-pressure can take 3 different values, viz., 1.0, 1.25, and 1.5 $\times P_{ram,0}$. As shown in Fig. 12, R_{TS} continued to decrease sharply for all assumed solar wind ram-pressure cases, and R_{TS} was a minimum from October 2007 to March 2008, depending on the assumed ram pressure. The Washimi et al. (2007b) simulations suggested that if the solar wind ram-pressure was 1.0 or 1.25 $\times$ $P_{ram,0}$, a V2-TS crossing would occur in October or November 2007. Rather remarkably, the crossing of the V2 occurred several times between 30 August and 1 September 2007 (Stone et al. 2008; Burlaga et al. 2008; Richardson et al. 2008), which was within a month of the predicted crossing by Washimi et al. (2007b).

A simple nose-cone-type of outer heliospheric model of the kind used by Washimi et al. (2007a, 2007b) will be asymmetric or modified by several possible effects. The ideal MHD simulation of Washimi et al. (2007b) introduces a N-S asymmetry by taking into account the possibility of an obliquely oriented ISMF that is consistent with the HDP observations (Lallement et al. 2005). The response of the TS to the returned pulse reflected at the HP was included properly in the simulations. However, there should certainly be additional effects such as the coupling of neutral H atoms to the plasma, possible anisotropy of the solar wind, and nonlinear magnetic field effects that will contribute to the asymmetry (or absence) of the heliosphere. Besides contributing to the basic symmetrization of the heliosphere, as discussed already, neutral H will introduce slightly different propagation characteristics into both the transmitted waves and shocks in the heliosheath as well as the HP reflected pulses. The solar wind ram pressure anisotropy during solar minimum observed by Ulysses (Phillips et al. 1995; McComas et al. 1999, 2000) will introduce a latitudinal asymmetry in the TS position (Pauls and Zank 1996; Tanaka and Washimi 1999). However, the latitudinal asymmetry is offset somewhat by the presence of neutral H since, as shown by Pauls and Zank (1997a, 1997b), the smaller filtration at higher latitudes results in a greater deceleration of the high latitude solar wind flow

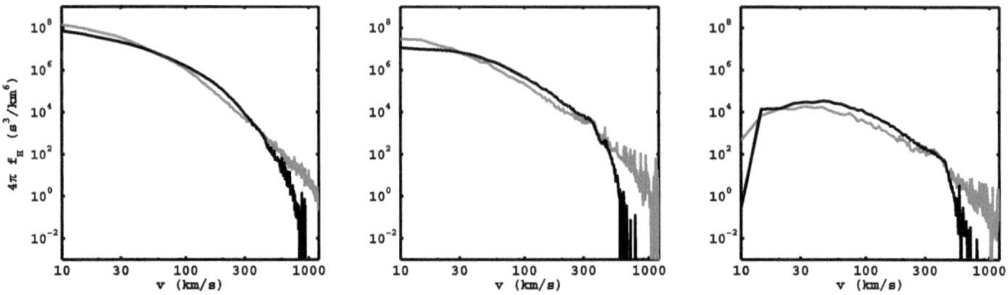

Fig. 13 Velocity distributions of ENA's at three locations along the axis defined by the LISM flow vector with the Sun at the origin: -400 AU in the heliotail (*left*), 180 AU upstream in the hydrogen wall (*middle*), 600 AU in the nearby LISM (*right*). The *black line* is for ENA's obtained from a Maxwellian distribution of heliosheath ions (the parent population of ENA's), and the *gray line* is from a $\kappa = 1.63$ distribution for heliosheath protons in the same steady-state configuration. For small κ fewer medium energy ENA's are present, but more result at low and high energies, which is consistent with the distributions shown in Fig. 1. (Heerikhuisen et al. 2008a)

and an overall isotropization of the solar wind ram pressure throughout the heliosphere. Tanaka and Washimi (1999) also suggested a "dipped" heliospheric structure around the equator which could result in an earlier crossing for V2 than V1, even within a purely MHD model. Thus, additional effects that might modify the predicted V2 crossing distance and time by Washimi et al. (2007b) include both the effect of a time-delay due to neutral particles modifying propagation speeds and the effects of an anisotropic solar wind ram pressure and/or nonlinear magnetic field effects. Nonetheless, the ideal MHD-only forecast of the TS crossing by V2 was remarkably accurate, with the predicted crossing time only in error from that observed by a few months.

A more elaborate multi-fluid model that includes neutral H self-consistently is under development Washimi et al. (2007b).

5 Energetic Neutral Atoms (ENAs)

The use of an isotropic κ-distribution (10) (Fig. 1) to describe the heliospheric plasma has important implications for both the global structure of the heliosphere and the properties and characteristics of the ENA distributions. Heerikhuisen et al. (2008a) discuss the effects of κ-distributed neutral atoms originating from the heliosheath on the global heliosphere-interstellar medium structure, and compute ENA spectra and skymaps. Figure 13 shows the velocity distribution of heliosheath hydrogen at various locations along the LISM flow vector. It is clear from this figure that for a $\kappa = 1.63$ distribution, significantly more H-atoms with energies above 1 keV result than for a Maxwellian ion population in the heliosheath. It is also important to note that ENAs in the heliotail (left plot) show a clear power-law tail ($\sim v^{-5}$), mirroring the plasma, when a κ-distribution is assumed for heliosheath protons. These tails persist even outside the heliosphere (middle and right plots) for energies above 1 keV.

Figure 14 compares plasma density and temperature along radial lines in the nose, polar and tail directions for the Maxwellian and equilibrated $\kappa = 1.63$ heliosheath cases. Secondary charge-exchange of neutrals created in the hot heliosheath was identified by Zank et al. (1996b) as a critical mechanism for the anomalous transport of energy from the shocked solar wind to the shocked and unshocked LISM. In particular, the upwind region abutting the HP experiences considerable heating as a result of secondary charge-exchange of hot

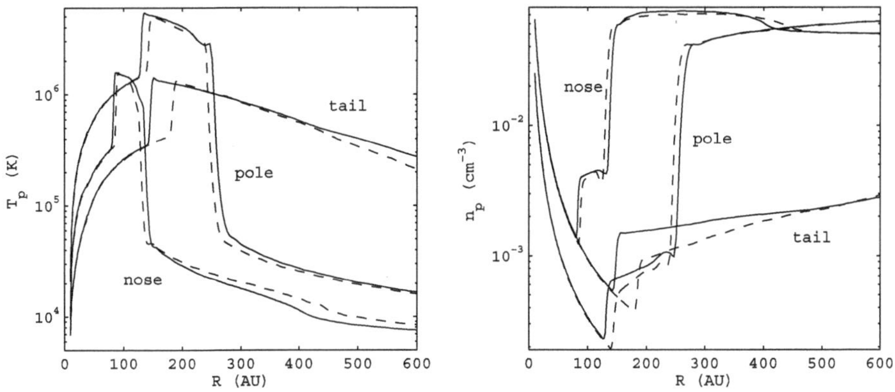

Fig. 14 Radial profiles of effective plasma temperature (*left*) and density (*right*) in the nose, polar (i.e. in the meridional plane), and tail directions. The *solid line* represents the values obtained by using a Maxwellian distribution function for the proton distribution. The *dashed line* is obtained by assuming that the proton distribution in the supersonic and subsonic solar wind can be described as an isotropic κ-distribution with $\kappa = 1.63$. (Heerikhuisen et al. 2008a)

($\sim 10^6$ K) neutrals with the cold LISM protons. The efficiency of this mechanism of anomalous heat transfer is increased with a κ-distribution in the inner heliosheath, resulting simultaneously in a shrinking of the inner heliosheath and an expansion of the outer heliosheath. The inner heliosheath plasma temperature (defined in terms of pressure) remains unchanged, because the Maxwellian and κ-distributions have the same second moment. Heerikhuisen et al. find that the TS moves out by about 4 AU in the nose direction, and the HP moves inward by about 9 AU. The bow shock stand-off distance increases by 25 AU, and the shock itself is weakened by the additional heating of the LISM plasma by fast neutrals from the solar wind. For the anticipated crossing of the HP by the V1 and V2 spacecraft, it is important to note that the inner heliosheath thickness shrinks from 56 AU (Maxwellian plasma distribution based description) to 44 AU (κ plasma distribution), a reduction of nearly 20%. This would reduce the expected V1 crossing time by as much as 3 years.

The filtration rate of hydrogen changes at the HP for a Maxwellian compared to a κ-distribution based model. For the Maxwellian case, the hydrogen density at the TS is about 63% of the interstellar value, whereas the density drops slightly to 60% for the κ-distribution model.

Figure 15 shows three energy spectra for ENA's originating from the nose, tail and polar directions. To obtain these spectra, Heerikhuisen et al. divide the flux measured at 1 AU by the survival probability for each energy band to undo the ionization losses. For the three directions considered, the energy spectrum tends toward the value of -κ above about 1 keV. This result shows that the IBEX data, in spite of being line-of-sight integrated, should be able to determine the spectral slope of the heliosheath protons in the 0.01–6 keV range. Figure 15 also shows that the spectra in the three directions considered have very similar properties. This will not necessarily be true for the real heliosphere, where the post-shock solar wind may develop different high energy tails in different directions. The dotted line (labeled "nose2") is for a spectrum in the nose direction obtained using 32 energy bins (compared to about 10 non-overlapping IBEX bins). The agreement between this curve and the green markers shows that, for $\kappa = 1.63$ at least, the number of IBEX bins is sufficient to reproduce the spectrum.

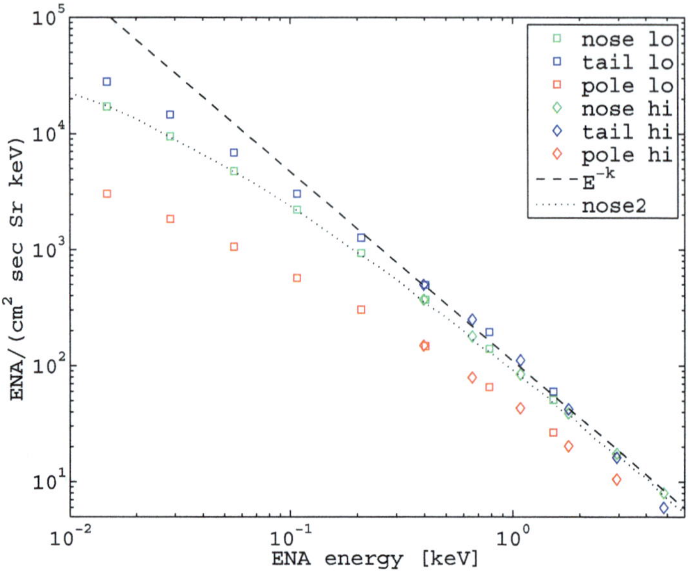

Fig. 15 ENA energy spectra as observed at 1 AU along various lines of sight. The *squares* and *diamonds* represent data using approximate IBEX energy bins obtained by dividing the IBEX-lo and IBEX-hi energy ranges (0.01–2.0 keV and 0.3–6 keV) into 8 and 6 equal bins on a logarithmic scale (see also Prested et al. 2007). The *dotted line* was obtained using narrower bins (32 total), and demonstrates that the IBEX bin widths are sufficiently narrow to maintain accuracy. The *dashed line* has a slope of -κ, which represents the plasma spectrum at a particular point, and appears reasonably well reproduced along the lines of sight considered. (Heerikhuisen et al. 2008a)

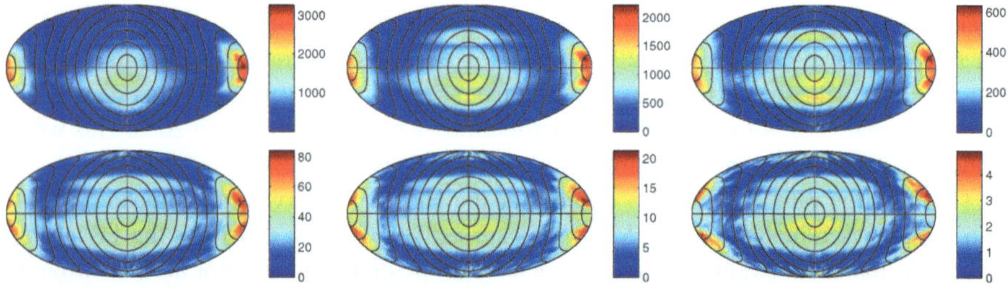

Fig. 16 All-sky maps of ENA flux at 1 AU, in units of $(\mathrm{cm}^2\,\mathrm{Sr\,keV\,sec})^{-1}$, generated in the inner heliosheath through charge-exchange between an interstellar neutral atom and a heliosheath proton drawn from a κ-distribution with $\kappa = 1.63$. The direction of the interstellar flow is at the center of the plot, with the poles top and bottom, and the heliotail on the far sides. *Contour lines* have been drawn at 15° intervals. Maps are generated by binning ENAs, which intersect the 1 AU sphere on radially inward trajectories. The maps from top left to bottom right correspond to the following energies and bin-widths (in eV): 10 ± 2, 50 ± 10, 200 ± 20, 1000 ± 100, 2400 ± 200, and 6000 ± 400. (Heerikhuisen et al. 2008a)

Heerikhuisen et al. (2007) used a steady-state heliosphere to compute all-sky ENA maps. They follow the same procedure for a κ-distribution based model, obtaining ENA's up to several keV in energy. Figure 16 shows all-sky ENA maps obtained from their steady-state solution with a κ distribution for heliosheath protons. The top right plot shows the ENA map for 200 eV, which can be compared with their previous work (Heerikhuisen et al. 2007), where they did not self-consistently couple the plasma and kinetic neutral atoms, and assumed a Maxwellian proton distribution. In the case of a κ-distribution, the ENA flux at 200 eV is two to three times smaller than for the Maxwellian case because more protons are found in the wings of the distribution for a κ-distribution. The decrease of medium energy (100s of eV) ENAs is compensated by an increased ENA flux above 1 keV, and Heerikhuisen et al. (2008a) find a count rate of about 3 atoms per (cm^2 sr s keV) at 6 keV.

Heerikhuisen et al. (2008a) assumed "solar minimum" conditions, with clearly defined high speed wind originating at the poles. The high speed wind creates hotter heliosheath plasma, which in turn increases the energy of ENAs generated in the subsonic polar solar wind. The all-sky maps of Fig. 16 show that at energies above about 1 keV, these streams of hot solar wind dominate the ENA flux, while at lower energies the central tail region is the major source of ENAs. Figure 16 compares skymaps at different energies, showing that the qualitative properties do not vary widely over the IBEX energy range. This contrasts sharply with the results for a Maxwellian heliosheath (Heerikhuisen et al. 2007), where more flux generally come from the tail than the nose at low energies, and the reverse at high energies. This is easily understood from the fewer particles that are found in the wings of the Maxwellian distribution, compared to the much broader κ-distribution. Thus, the relatively cool plasma in the distant heliotail can still be a significant source of high energy ENAs, if we assume it has a κ-distribution.

Acknowledgements The authors acknowledge the partial support of NASA grants NNG04GF83G, NNG05GH38G, NNG05GM62G, NNG05GH48G, NNG06GD43G, and NSF grants ATM0317509, and ATM0428880.

We acknowledge the use of the solar-wind data from the MIT plasma experiment on V2 during the period from day of year 253, 2001 to 226, 2007, for our simulations.

Supercomputer time allocations were provided by the NASA High-End Computing program, DOE's INCITE project PSS001, NCSA project MCA07S033, and by collaborative agreement with the Solar-Terrestrial Environment Laboratory of Nagoya University. Numerical computations were also carried out on the SX-8 machine at the National Institute of Computer Technology in Japan.

References

D. Alexashov, V. Izmodenov, Kinetic vs. multi-fluid models of H atoms in the heliospheric interface: a comparison. Astron. Astrophys. **439**, 1171–1181 (2005)

W.I. Axford, The interaction of the solar wind with the interstellar medium. Solar Wind, NASA Spec. Publ. **SP-308**, 609 (1972)

W.I. Axford, A.J. Dessler, B. Gottlieb, Astrophys. J. **137**, 1268 (1963)

S.J. Bame, J.R. Asbridge, H.E. Felthauser, E.W. Hones, I.B. Strong, J. Geophys. Res. **72**, 113 (1967)

V.B. Baranov, Y.G. Malama, J. Geophys. Res. **98**, 15157 (1993)

V.B. Baranov, Y.G. Malama, J. Geophys. Res. **100**, 14755 (1995)

V.B. Baranov, K. Krasnobaev, A. Kulikovsky, Sov. Phys. Dokl. **15**, 791 (1971)

V.B. Baranov, M.K. Ermakov, M.G. Lebedev, Three-component gasdynamic model of the interaction of the solar wind with the interstellar medium. Sov. Astron. Lett. **7**, 206 (1981)

T. Borrmann, H. Fichtner, Adv. Space Sci. **35**, 2091 (2005)

S.N. Borovikov, J. Heerikhuisen, N.V. Pogorelov, I.A. Kryukov, G.P. Zank, in *Numerical Modeling of Space Plasma Flows: ASTRONUM-2007*, ed. by N.V. Pogorelov, E. Audit, G.P. Zank (ASP, San Francisco, 2008a), pp. 197–203

S.N. Borovikov, N.V. Pogorelov, G.P. Zank, I.A. Kryukov, Astrophys. J. **682**, 1404 (2008b)

J.M. Burgers, *Flow Equations for Composite Gases* (Academic Press, New York, 1969)

L.F. Burlaga et al., Science **309**, 2027–2029 (2005)

L.F. Burlaga et al., Nature **454**, 75 (2008)

M.R. Collier, Geophys. Res. Lett. **22**, 2673 (1995)

D.P. Cox, L. Helenius, Flux-tube dynamics and a model for the origin of the Local Fluff. Astrophys. J. **583**, 205 (2003)

A. Dalgarno, Proc. Phys. Soc. **75**, 374 (1960)

L.E. Davis, Interplanetary magnetic fields and cosmic rays. Phys. Rev. **100**, 1440 (1955)
R.B. Decker et al., Science **309**, 2020 (2005)
R.B. Decker et al., Nature **454**, 67 (2008)
H.J. Fahr, S. Grzedzielski, R. Ratkiewicz, Ann. Geophys. **6**, 337 (1988)
H.J. Fahr, T. Kausch, H. Scherer, Astron. Astrophys. **357**, 268 (2000)
L.A. Fisk, G. Gloeckler, Astrophys. J. **640**, L79 (2006)
W.L. Fite, A.C.H. Smith, R.F. Stebbings, Charge transfer in collisions involving symmetric and asymmetric resonance. Proc. R. Soc. Lond. Ser. A **268**, 527 (1962)
V. Florinski, G.P. Zank, N.V. Pogorelov, J. Geophys. Res. **108**(A6), 1228 (2003a). doi:10.1029/2002JA009695
V. Florinski, G.P. Zank, Comment on "On nonideal MHD properties of the partially ionized interstellar gas" by V.B. Baranov and H.J. Fahr. J. Geophys. Res. **108**(A12), 1438 (2003d). doi:10.1029/2003JA009950
V. Florinski, G.P. Zank, The galactic cosmic ray intensity in the heliosphere in response to variable interstellar environments, in *Solar Journey: The Significance of Our Galactic Environment for the Heliosphere and Earth*, ed. by P.C. Frisch (Springer, Netherlands, 2006), p. 281
V. Florinski, G.P. Zank, W.I. Axford, The solar system in a dense interstellar cloud: implications for cosmic ray fluxes at Earth and 10Be records. Geophys. Res. Lett. **30**, 2206 (2003b). doi:10.1029/2003GL017566
V. Florinski, N.V. Pogorelov, G.P. Zank, B.E. Wood, D.P. Cox, Astrophys. J. Lett. **604**, 700 (2004a)
V. Florinski, G.P. Zank, N.V. Pogorelov, J. Geophys. Res. (Space), **110**(A7) (2005a). CiteID A07104
P.C. Frisch, Characteristics of nearby interstellar matter. Space Sci. Rev. **72**, 499 (1995)
G.P. Gayley, G.P. Zank, H.L. Pauls, P.C. Frisch, D.E. Welty, One-shock or two-shock heliosphere: Lyman-α observations and measurements. Astrophys. J. **487**, 259 (1997)
G. Gloeckler, Space Sci. Rev. **78**, 335 (1996)
G. Gloeckler, J. Geiss, Space Sci. Rev. **86**, 127 (1998)
G. Gloeckler, J. Geiss, H. Balsiger, L.A. Fisk, A.B. Galvin, F.M. Ipavich, K.W. Ogilvie, R. von Steiger, B. Wilken, Detection of interstellar pick-up hydrogen in the solar system. Science **261**, 70 (1993)
D.A. Gurnett, W.S. Kurth, Electron plasma oscillations upstream of the solar wind termination shock. Science **309**, 2025–2027 (2005)
D.A. Gurnett, W.S. Kurth, Nature **454**, 67 (2008)
J. Heerikhuisen, V. Florinski, G.P. Zank, Interaction between the solar wind and interstellar gas: a comparison between Monte-Carlo and fluid approaches, J. Geophys. Res. **111**(A6) (2006). CiteID A06110
J. Heerikhuisen et al., in *Numerical Modeling of Space Plasma Flows: ASTRONUM-2006*, ed. by N.V. Pogorelov. ASP Conf., vol. 359 (Astronomical Society of the Pacific, San Francisco, 2006b), p. 251
J. Heerikhuisen, N.V. Pogorelov, G.P. Zank, V. Florinski, Astrophys. J. **655**, L53 (2007)
J. Heerikhuisen, N.V. Pogorelov, V. Florinski, G.P. Zank, J. le Roux, Astrophys. J. **682**, 679 (2008a)
J. Heerikhuisen et al., in *Numerical Modeling of Space Plasma Flows: ASTRONUM-2008*, ed. by N.V. Pogorelov, E. Audit, G.P. Zank. ASP Conf. Ser. (Astronomical Society of the Pacific, San Francisco, 2008b)
P.A. Isenberg, Interaction of the solar wind with interstellar neutral hydrogen: Three-fluid model. J. Geophys. Res. **91**, 9965 (1986)
P.A. Isenberg, W.C. Feldman, Electron-impact ionization of interstellar hydrogen and helium at interplanetary shocks. Geophys. Res. Lett. **22**, 873 (1995)
P.A. Isenberg, M.A. Lee, A dispersive analysis of bispherical pickup ion distributions. J. Geophys. Res. **101**(11), 055 (1996)
P.A. Isenberg, C.W. Smith, W.H. Matthaeus, Turbulent heating of the distant solar wind by interstellar pickup protons. Astrophys. J. **592**, 564–573 (2003)
V.V. Izmodenov, J. Geiss, R. Lallement, G. Gloeckler, V.B. Baranov, Y.G. Malama, Filtration of interstellar hydrogen in the two-shock heliospheric interface: Inferences on the local interstellar cloud electron density. J. Geophys. Res. **104**, 4731 (1999)
V.V. Izmodenov, D. Alexashov, A. Myasnikov, Astron. Astrophys. **437**, L35–L38 (2005)
J.R. Jokipii, J. Giacalone, Astrophys. J. **605**, L145–L148 (2004)
J.R. Jokipii, J. Giacalone, J. Kota, **611**, L141–L144 (2004)
S.R. Karmesin, P.C. Liewer, J.U. Brackbill, Geophys. Res. Lett. **22**, 1153 (1995)
I.K. Khabibrakhmanov, D. Summers, G.P. Zank, H.L. Pauls, Solar wind flow with interstellar hydrogen pick-up. Astrophys. J. **469**, 921 (1996)
R.M. Kulsrud, in *Basic Plasma Physics*, ed. by A.A. Galeev, R.N. Sudan. vol. 1 (North Holland Publ., Amsterdam, 1984), p. 115
R. Lallement, E. Quemerais, J.L. Bertaux, S. Ferron, D. Koutroumpa, R. Pellinen, Science **307**, 1447 (2005)
P.C. Liewer, S.R. Karmesin, J.U. Brackbill, Hydrodynamic instability of the heliopause driven by plasma-neutral charge-exchange interactions. J. Geophys. Res. **101**, 17119 (1996)

T. Linde, T.I. Gombosi, P.L. Roe, K.G. Powell, D.L. DeZeeuw, The heliosphere in the magnetized local interstellar medium: results of a 3D MHD simulation. J. Geophys. Res. **103**, 1889 (1998)

B.J. Lindsay, R.F. Stebbings, J. Geophys. Res. **110**, A12213 (2005)

J.L. Linsky, B.E. Wood, The α Cen line of sight: D/H ratio, physical properties of local interstellar gas, and measurement of heated hydrogen near the heliopause. Astrophys. J. **463**, 254 (1996)

A.S. Lipatov, G.P. Zank, H.L. Pauls, The interaction of neutral interstellar H with the heliosphere: A2.5D particle-mesh Boltzmann simulation. J. Geophys. Res. **103**, 20,631 (1998)

L. Maher, B. Tinsley, Atomic hydrogen escape rate due to charge exchange with hot plasmaspheric ions. J. Geophys. Res. **82**, 689 (1977)

Yu.G. Malama, Monte-Carlo simulation of neutral atoms trajectories in the solar system. Astrophys. Space Sci. **176**, 21 (1991)

W.H. Matthaeus, G.P. Zank, C.W. Smith, S. Oughton, Turbulence, spatial transport, and heating of the solar wind. Phys. Rev. Lett. **82**, 3444 (1999)

D.J. McComas, H.O. Funsten, J.T. Gosling, W.R. Pryor, ULYSSES measurements of variations in the solar wind-interstellar hydrogen charge exchange rate. Geophys. Res. Lett. **26**(17), 2701 (1999)

D.J. McComas, B.L. Barraclough, H.O. Funsten, J.T. Gosling, E. Santiago-Muñoz, R.M. Skoug, B.E. Goldstein, M. Neugebauer, P. Riley, A. Balogh, Solar wind observations over Ulysses' first full polar orbit. J. Geophys. Res. **105**, 10419 (2000)

D.J. McComas, F. Allegrini, L. Bartolone, P. Bochsler, M. Bzowski, M. Collier, H. Fahr, H. Fichtner, P. Frisch, H. Funsten, S. Fuselier, G. Gloeckler, M. Gruntman, V. Izmodenov, P. Knappenberger, M. Lee, S. Livi, D. Mitchell, E. Möbius, T. Moore, S. Pope, D. Reisenfeld, E. Roelof, H. Runge, J. Scherrer, N. Schwadron, R. Tyler, M. Wieser, M. Witte, P. Wurz, G. Zank, The Interstellar Boundary Explorer (IBEX): Update at the end of phase B, in *The Physics of the Inner Heliosheath*, ed. by J. Heerikhuisen, V. Florinski, G.P. Zank, AIP Conference Proceedings, vol. 858 (2006), pp. 241–250

F.B. McDonald, E.C. Stone, A.C. Cummings, B. Heikkila, N. Lal, W.R. Webber, Nature **426**, 48 (2003)

J.F. McKenzie, K.O. Westphal, Phys. Fluids **11**, 2350 (1968)

R.A. Mewaldt et al., in Joint SOHO/ACE workshop "Solar and Galactic Composition, ed. by R.F. Wimmer-Schweingruber. American Institute of Physics Conference Series, vol. 598 (2001), p. 165

E. Möbius, D. Hovestadt, B. Klecker, M. Scholer, G. Gloeckler, Nature **318**, 426 (1985)

E. Möbius, M. Bzovski, S. Chalov, H.-J. Fahr, G. Gloeckler et al., Astron. Astrophys. **426**, 897 (2004)

H.-R. Müller, P.C. Frisch, G.P. Zank, Astrophys. J. **647**(2), 1491 (2006)

M. Opher, E.C. Stone, P.C. Liewer, Astrophys. J. **640**, L71 (2006)

E.N. Parker, Dynamics of interplanetary gas and magnetic fields. Astrophys. J. **123**, 644 (1958)

E.N. Parker, The stellar-wind regions. Astrophys. J. **134**, 20 (1961)

E.N. Parker, *Interplanetary Dynamical Processes* (Interscience, New York, 1963)

H.L. Pauls, G.P. Zank, Interaction of a nonuniform solar wind with the local interstellar medium. J. Geophys. Res. **101**, 17081 (1996)

H.L. Pauls, G.P. Zank, Interaction of a nonuniform solar wind with the local interstellar medium. 2. A two-fluid model. J. Geophys. Res. **102**, 19,779 (1997a)

H.L. Pauls, G.P. Zank, Modelling the solar wind/interstellar wind interaction, in *Proc. 25th Int. Cosmic Ray Conf.*, vol. 2 (1997b), p. 241

H.L. Pauls, G.P. Zank, L.L. Williams, Interaction of the solar wind with the local interstellar medium. J. Geophys. Res. **100**, 21,595 (1995)

Phillips et al., Science **268**, 1030 (1995)

N.V. Pogorelov, T. Matsuda, Nonevolutionary MHD shocks in the solar wind and interstellar medium interaction. Astron. Astrophys. **354**, 697 (2000)

N.V. Pogorelov, A.Y. Semenov, Solar wind interaction with the magnetized interstellar medium. Shock-capturing modeling. Astron. Astrophys. **321**, 330 (1997)

N.V. Pogorelov, G.P. Zank, in *Numerical Modeling of Space Plasma Flows*, ed. by N.V. Pogorelov, G.P. Zank. ASP Conf. Ser., vol. 359 (ASP, San Francisco, 2006), p. 184

N.V. Pogorelov, G.P. Zank, T. Ogino, Three-dimensional features of the outer heliosphere due to coupling between the interstellar and interplanetary magnetic fields I. Magnetohydrodynamic model: interstellar perspective. Astrophys. J. **614**, 1007 (2004)

N.V. Pogorelov, G.P. Zank, T. Ogino, Astrophys. J. **644**, 1299 (2006)

N.V. Pogorelov, E.C. Stone, V. Florinski, G.P. Zank, Astrophys. J. **668**, 611 (2007)

N.V. Pogorelov, J. Heerikhuisen, G.P. Zank, Astrophys. J. **675**, L41 (2008a)

N.V. Pogorelov, J. Heerikhuisen, G.P. Zank, S.N. Borovikov, Space Sci. Rev. (2008)

N.V. Pogorelov, G.P. Zank, S.N. Borovikov, V. Florinski, J. Heerikhuisen, I.A. Kryukov, in *Numerical Modeling of Space Plasma Flows: ASTRONUM-2007*, ed. by N.V. Pogorelov, E. Audit, G.P. Zank. ASP Conf. Ser., vol. 385 (Astronomical Society of the Pacific, San Francisco, 2008b), pp. 180–188

N.V. Pogorelov et al., in *Particle Acceleration in the Heliosphere and Beyond*, ed. by Gang Li, Qiang Hu, and G.P. Zank, AIP Conference Proceedings (2008c)
C. Prested et al., J. Geophys. Res. (2007, accepted)
C. Prested, N. Schwadron, J. Passuite, B. Randol, B. Stuart, G. Crew, J. Heerikhuisen, N. Pogorelov, G. Zank, M. Opher, F. Allegrini, D.J. McComas, M. Reno, E. Roelof, S. Fuselier, H. Funsten, E. Moebius, L. Saul, Implications of solar wind suprathermal tails for IBEX ENA images of the heliosheath. J. Geophys. Res. **113**(A6), (2008)
R. Ratkiewicz, A. Barnes, G.A. Molvik, J.R. Spreiter, S.S. Stahara, M. Vinokur, S. Venkateswaran, Effect of varying strength and orientation of local interstellar magnetic field on configuration of exterior heliosphere: 3D MHD simulations. Astron. Astrophys. **335**, 363 (1998)
J.D. Richardson, C. Wang, M. Zhang, in 5th *International IGPP Conf.* AIP Conf. Proc., vol. 85 (2006). p. 110
J.D. Richardson et al., Nature **454**, 63 (2008)
H.L. Ripken, H.J. Fahr, Astron. Astrophys. **122**, 181 (1983)
K. Scherer, H. Fichtner, O. Stawicki, The heliosphere, cosmic rays, climate, in *The Outer Heliosphere: The next Frontiers*, ed. by K. Scherer, H. Fichtner, H.-J. Fahr, E. Marsch (Pergamon, New York, 2001), p. 493
K. Scherer, H.J. Fahr, Geophys. Res. Lett. **30**(2), 1045 (2003)
K. Scherer, H. Fichtner, B. Heber, S.E.S. Ferreira, M.S. Potgieter, Adv. Space Res. **41**(8), 1171 (2008)
J.D. Slavin, P.C. Frisch, The ionization of nearby gas. Astrophys. J. **565**, 364 (2002)
C.W. Smith, W.H. Matthaeus, G.P. Zank, N.F. Ness, S. Oughton, J.D. Richardson, Heating of the low-latitude solar wind by dissipation of turbulent magnetic fluctuations. J. Geophys. Res. Space **106**, 8253–8272 (2000)
E.C. Stone et al., Science **309**, 2017 (2005)
E.C. Stone et al., Nature **454**, 71 (2008)
D. Summers, R.M. Thorne, Phys. Fluids B **3**, 1835 (1991)
T. Tanaka, H. Washimi, J. Geophys. Res. **104**, 12605 (1999)
C. Wang, J.W. Belcher, Numerical investigation of hydrodynamic instabilities of the heliopause. J. Geophys. Res. **103**, 247 (1998)
H. Washimi, T. Tanaka, 3-D magnetic field and current system in the heliosphere. Space Sci. Rev. **78**, 85 (1996)
H. Washimi, T. Tanaka, Adv. Space Res. **27**, 509 (2001)
H. Washimi, G.P. Zank, T. Tanaka, *5rd International IGPP Conf.* AIP Conf. Proc. vol. 858 (2006), p. 58
H. Washimi, G.P. Zank, Q. Hu, T. Tanaka, *6th International IGPP Conf.* AIP Proc. vol. 932 (2007a), p. 153
H. Washimi, G.P. Zank, Q. Hu, T. Tanaka, K. Munakata, Astrophys. J. Lett. **670**, L139 (2007b)
L.L. Williams, G.P. Zank, The effect of magnetic field geometry on the wave signature of the pickup of interstellar neutrals. J. Geophys. Res. **99**, 19,229 (1994)
L.L. Williams, G.P. Zank, W.H. Matthaeus, Dissipation of pick-up induced waves: A solar wind temperature increase in the outer heliosphere? J. Geophys. Res. **100**, 17,059 (1995)
L.L. Williams, D.T. Hall, H.L. Pauls, G.P. Zank, The heliospheric Hydrogen distribution: A multi-fluid model. Astrophys. J. **476**, 366 (1997)
M. Witte, Kinetic parameters of interstellar neutral helium. Review of results obtained during one solar cycle with the Ulysses/GAS-instrument. Astron. Astrophys. **426**, 835 (2004)
B.E. Wood, J.L. Linsky, H.-R. Müller, G.P. Zank, Observational estimates for the mass-loss rates of alpha Centauri and Proxima Centauri using HST Ly-alpha spectra. Astrophys. J. Lett. **547**, L49–L52 (2001)
G.P. Zank, Interaction of the solar wind with the local interstellar medium: A theoretical perspective. Space Sci. Rev. **89**, 413 (1999)
G.P. Zank, The dynamical heliosphere, in *Solar Wind Nine*, ed. by S.R. Habbal, R. Esser, J.V. Hollweg, P.A. Isenberg, AIP Conf. Proc., vol. 471 (1999d), pp. 783–786
G.P. Zank, P.C. Frisch, Consequences of a change in the galactic environment of the Sun. Astrophys. J. **518**, 965 (1999e)
G.P. Zank, H.-R. Müller, The dynamical heliosphere. J. Geophys. Res. **108**(A6), 1240 (2003). doi:10.1029/2002JA009689
G.P. Zank, W.H. Matthaeus, C.W. Smith, Evolution of turbulent magnetic fluctuation power with heliospheric distance. J. Geophys. Res. **101**, 17,081 (1996a)
G.P. Zank, H.L. Pauls, I.H. Cairns, G.M. Webb, Interstellar pick-up ions and perpendicular shocks: implications for the termination shock and interplanetary shocks. J. Geophys. Res. **101**, 457 (1996b)
G.P. Zank, H.L. Pauls, L.L. Williams, D.T. Hall, Interaction of the solar wind with the local interstellar medium: A multifluid approach. J. Geophys. Res. **101**(21), 639 (1996d)

G.P. Zank, W.H. Matthaeus, J.W. Bieber, H. Moraal, The radial and latitudinal dependence of the cosmic ray diffusion tensor in the heliosphere. J. Geophys. Res. **103**, 2085 (1998)
G.P. Zank, H.-R. Mueller, V. Florinski, P.C. Frisch, Heliospheric variation in response to changing interstellar environments, in *Solar Journey: Significance of our Galactic Environment for the Heliosphere and Earth*, ed. by G.P. Zank (Springer, Berlin, 2006), p. 23
G.P. Zank et al., Space Sci. Rev. (2008, in preparation)

Space Sci Rev (2009) 146: 329–351
DOI 10.1007/s11214-009-9528-3

Kinetic-Gasdynamic Modeling of the Heliospheric Interface

V.V. Izmodenov · Y.G. Malama · M.S. Ruderman · S.V. Chalov · D.B. Alexashov · O.A. Katushkina · E.A. Provornikova

Received: 2 September 2008 / Accepted: 4 May 2009 / Published online: 17 June 2009

Abstract Heliospheric energetic neutral atoms (ENAs) that will be measured by the Interstellar Boundary Explorer (IBEX) originate from the heliosheath. The heliosheath is formed as a result of the interaction of the solar wind (SW) with the circum-heliospheric interstellar medium (CHISM). The expected fluxes of ENAs are strongly dependent on the nature of this interaction. In turn, the interaction of the solar wind with the local interstellar cloud has a complex and multi-component nature. Detailed theoretical modeling of the interaction between the SW and the local interstellar medium is required to understand the physics of the heliosheath and to predict and explain the heliospheric ENAs. This paper summarizes current state-of-art kinetic-gasdynamic models of the SW/CHISM interaction. We shall restrict our discussion to the kinetic-gasdynamic and kinetic-magnetohydrodynamic (MHD) models developed by the Moscow group. This paper summarizes briefly the main results of the first self-consistent, two-component, kinetic-gasdynamic model by Baranov and Malama (J. Geophys. Res. 98:15157–15163, 1993), presents new results obtained from the 3D kinetic-MHD model by Izmodenov et al. (Astron. Astrophys. 437:L35–L38, 2005a), describes the basic formulation and results of the model by Malama et al. (Astron. Astrophys. 445:693–701, 2006) as well as reports current developments in the model. This self-consistent model considers pickup protons as a separate non-equilibrium component. Then we discuss a stochastic acceleration model for pickup protons in the supersonic solar wind and in the heliosheath. We also present the expected heliospheric ENA fluxes obtained in the framework of the models.

V.V. Izmodenov (✉) · Y.G. Malama · M.S. Ruderman · S.V. Chalov · D.B. Alexashov · O.A. Katushkina · E.A. Provornikova
Lomonosov Moscow State University, Moscow, Russia
e-mail: izmod@iki.rssi.ru

V.V. Izmodenov · Y.G. Malama · M.S. Ruderman · S.V. Chalov · D.B. Alexashov · O.A. Katushkina · E.A. Provornikova
Space Research Institute (IKI), Russian Academy of Sciences, Moscow, Russia

V.V. Izmodenov · Y.G. Malama · M.S. Ruderman · S.V. Chalov · D.B. Alexashov · O.A. Katushkina · E.A. Provornikova
Institute for Problems in Mechanics, Russian Academy of Sciences, Moscow, Russia

Keywords Inner heliosheath · Interstellar H atoms · Heliospheric ENA

1 Introduction

The first decade of the 21st century is characterized by great progress in exploration of the heliospheric boundary—the region of the solar wind (SW) interaction with the circum-heliospheric interstellar medium. Two spacecraft, Voyager 1 and Voyager 2, crossed the inner frontier of the heliospheric interface—the solar wind termination shock (TS)—at ~94 AU in December 2004 and ~84 AU in August 2007, respectively. The difference of 10 AU can be partly explained by the time-variation of the TS distance due to the variation of the solar wind ram pressure (e.g. Izmodenov et al. 2008). However, this variation can not completely account for the observed difference of 10 AU. As it has been argued in Izmodenov (2009), the most probable physical reason for this difference is the existence of a substantial interstellar magnetic field inclined to the direction of the interstellar gas flow with respect to the Sun. An indication of the inclined interstellar magnetic field was also found from analysis of backscattered Lyman-alpha radiation measured by the SWAN instrument on board of SOHO (Lallement et al. 2005). More precisely, the analysis of the SOHO/SWAN data has shown that the direction of the flow of interstellar hydrogen atoms inside the heliosphere is deflected by a few degrees from the flow direction of interstellar helium atoms that penetrate through the heliospheric interface freely due to their large mean free path. Contrary to interstellar helium atoms, interstellar hydrogen atoms penetrating in the heliosphere are perturbed. The perturbation of the interstellar H atom flow occurs in the heliospheric interface region due to coupling with the plasma component, mainly by charge exchange. Photoionization and electron impact ionization play important roles as well.

Since the parameters of H atoms are disturbed in the heliospheric interface before they penetrate inside the heliosphere (where their parameters can be measured directly or indirectly), these atoms can serve as a remote diagnostics of the heliospheric interface. In particular, the deflection of H atom flow detected by SOHO/SWAN is associated with the interstellar magnetic field (Lallement et al. 2005; Izmodenov et al. 2005a).

The heliospheric interface is divided into 4 regions (Fig. 1) by the heliopause (HP), which is a contact discontinuity separating heliospheric and interstellar plasma components, the termination shock that decelerates the heliospheric plasma, and the bow shock (BS) that decelerates the interstellar plasma. Plasma parameters are essentially different in these regions. As described in Sect. 3, the bow shock does not exist in the case of strong interstellar magnetic field. In this case there is no definite boundary between regions 3 and 4, but plasma properties in the vicinity of the heliopause are still different from the properties of the undisturbed interstellar medium.

Heliospheric energetic neutral atoms (ENAs) that will be measured by the Interstellar Boundary Explorer (IBEX; McComas et al. 2004, 2005, 2006, 2009, this issue) originate from the inner heliosheath—the region between the heliospheric termination shock (TS) and the heliopause (HP) (Fig. 1). The inner and outer heliosheaths are dynamically coupled. Therefore, IBEX measurements will provide constraints on the outer heliosheath and circum-heliospheric interstellar medium (CHISM). In addition, the IBEX-Lo instrument will be able to make, for the first time, direct measurements of interstellar oxygen atoms (Möbius et al. 2009).

In this paper we summarize current state-of-art, kinetic-gasdynamic models of the SW/CHISM interaction. We shall restrict our discussion to the kinetic-gasdynamic and kinetic-magnetohydrodynamic (MHD) models developed by the Moscow group. A summary of the

Table 1 Modern multi-component kinetic-gasdynamic and kinetic-MHD models of the heliospheric interface developed by the Moscow group

Component or effect	Reference
Interstellar H atoms (kinetic description)	Baranov and Malama (1993), Izmodenov et al. (2001a) in all models below in the table
Interstellar plasma: protons, electrons + helium ions	Baranov and Malama (1993) Izmodenov et al. (2003)
Interstellar magnetic field	Aleksashov et al. (2000), Izmodenov et al. (2005a), Izmodenov and Alexashov (2006)
Galactic Cosmic Rays	Myasnikov et al. (2000)
Anomalous Cosmic Rays	Alexashov et al. (2004)
Solar wind (protons, electrons) + alpha particles	Baranov and Malama (1993) Izmodenov et al. (2003)
Pickup ions (kinetic description)	Malama et al. (2006)
Solar Cycle Variations of the solar wind	Izmodenov et al. (2005b), Izmodenov et al. (2008)
Latitudinal variations of the solar wind	Alexashov & Izmodenov, in preparation

See also recent reviews by Baranov and Izmodenov (2006), Baranov (2009) and a book edited by Izmodenov and Kallenbach (2006). For comparison of kinetic and multi-fluid models see Alexashov and Izmodenov (2005)

relevant publications is given in Table 1. Models developed by other groups are described in other papers of this volume. The next section of this paper summarizes briefly the main results of the first self-consistent, two-component, kinetic-gasdynamic model by Baranov and Malama (1993). Section 3 presents new results obtained from the 3D kinetic-MHD model by Izmodenov et al. (2005a). Section 4 describes the basic formulation and results of the model by Malama et al. (2006) as well as reports current developments in the model. This self-consistent model considers pickup protons as a separate non-equilibrium component. Section 5 presents a stochastic acceleration model of pickup protons in the supersonic solar wind and in the heliosheath, and the expected heliospheric ENA fluxes obtained in the framework of the model.

2 Axisymmetric Kinetic-Gasdynamic Model of the Heliospheric Interface

2.1 Baranov–Malama Model

The first self-consistent model of the SW/CHISM interaction was developed by Baranov and Malama (1993). This is an axisymmetric and stationary two-component model. The plasma component is quasi-neutral and consists of electrons and protons. It is assumed that pickup

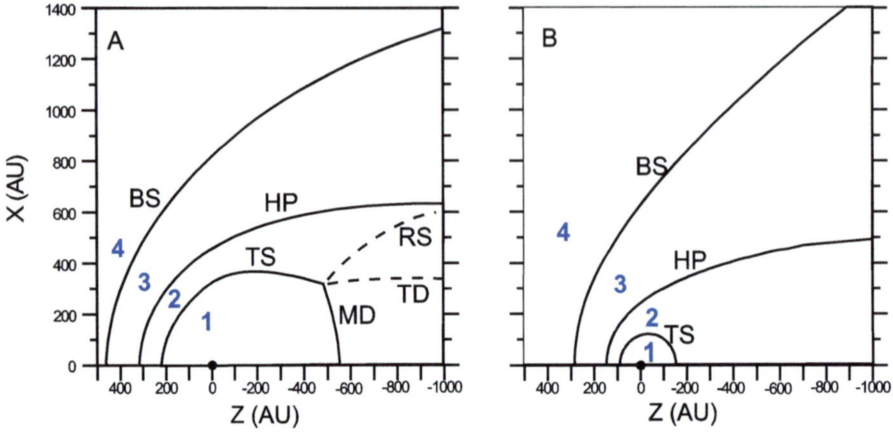

Fig. 1 Effect of the interstellar H atoms on the geometrical pattern of the interface. (**A**) The heliospheric interface pattern in the case of a fully ionized circum-heliospheric interstellar medium (CHISM), (**B**) the case of partly ionized CHISM. Here BS, HP, TS are the bow shock, the heliopause and the termination shock, respectively. MD and TD are the Mach disk and the tangential discontinuity; RS is the reflected shock that is formed in the case of fully ionized plasma. These results were obtained initially by Baranov and Malama (1993). Region 1 is the supersonic solar wind, 2 is the inner heliosheath between the TS and HP, 3 is the outer heliosheath between the HP and BS. (From Izmodenov and Alexashov 2003)

protons are assimilated into the plasma component immediately after ionization. The plasma component is described as a fluid, and the Euler equations are solved to get the spatial distribution of the plasma number density $n_{\mathrm{p}}(\mathbf{r})$, bulk velocity $\mathbf{V}_{\mathrm{p}}(\mathbf{r})$ and pressure $P_{\mathrm{p}}(\mathbf{r})$. The neutral component consists of hydrogen atoms and is described kinetically. The two components interact by charge exchange. Photoionization and electron impact ionization are taken into account in the model as well.

The main results of the model can be described as follows. The interstellar atoms strongly influence the heliospheric interface structure. The heliospheric interface is much closer to the Sun when the H atoms are taken into account in the model, as compared to the pure gas dynamical case (Fig. 1). The termination shock becomes more spherical and the flow in the region between HP and TS becomes subsonic (sonic lines disappear). The Mach disk and the complicated tail shock structure, consisting of the reflected shock (RS) and the tangential discontinuity (TD), disappear.

The supersonic flows upstream of the bow and termination shocks are disturbed due to the charge exchange with the neutral component. The supersonic solar wind flow (region 1 in Fig. 1) is disturbed by charge exchange with the interstellar neutrals. The new protons created by charge exchange are picked up by the solar wind magnetic field. The Baranov-Malama model assumes immediate assimilation of the pickup ions into the solar wind plasma. The solar wind protons and pickup protons are treated as one fluid, called the solar wind. The number density, velocity, temperature, and Mach number of the solar wind are shown in Fig. 2A. The effect of charge exchange on the solar wind is significant. By the time the solar wind flow reaches the termination shock, it is decelerated by 15–30%, strongly heated by a factor of 5–8, and loaded with the pickup proton component (approximately 20–50%).

The interstellar plasma flow is disturbed upstream of the bow shock by charge exchange of the interstellar protons with secondary H atoms. These secondary atoms originate in the solar wind. This leads to heating (40–70%) and deceleration (15–30%) of the interstellar

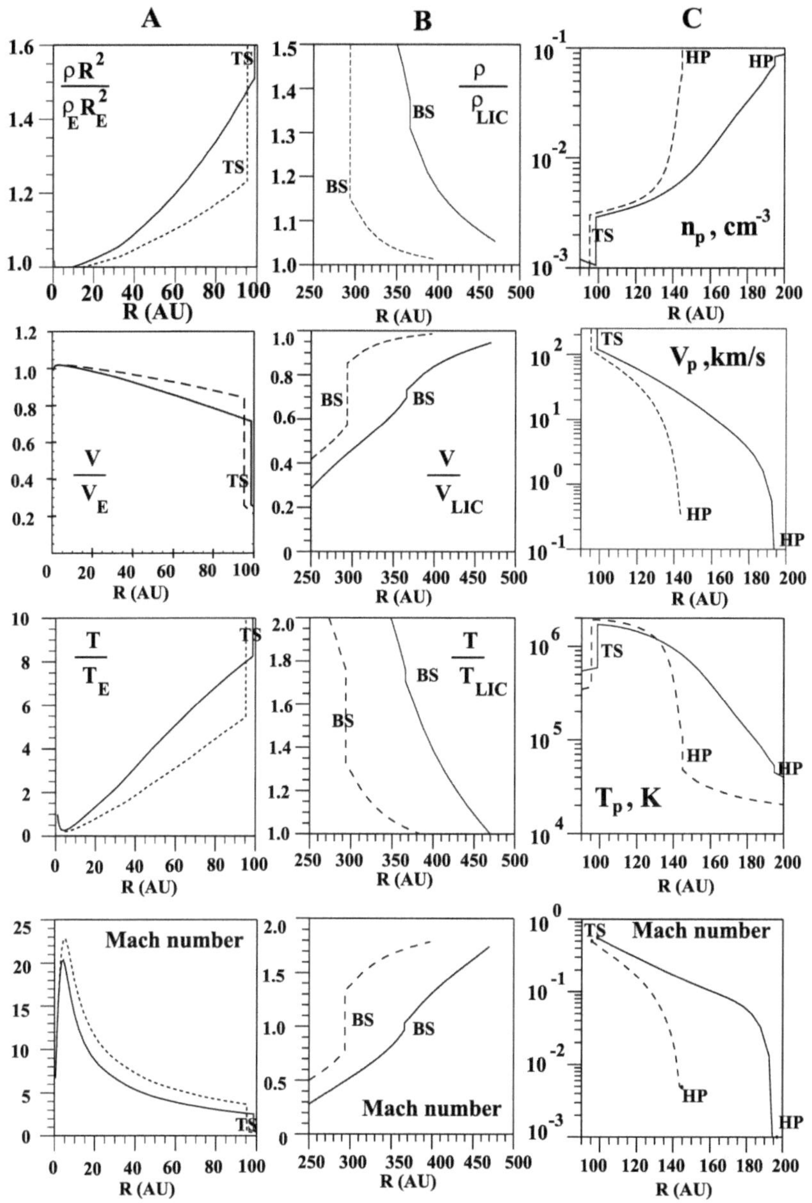

Fig. 2 Plasma density, velocity, temperature and Mach number upstream of the termination shock (**A**), upstream of the bow shock (**B**), and in the heliosheath (**C**). The distributions are shown for the upwind direction. The *solid curves* correspond to $n_{\mathrm{H,CHISM}} = 0.2\ \mathrm{cm}^{-3}$, $n_{\mathrm{p,CHISM}} = 0.04\ \mathrm{cm}^{-3}$. The *dashed curves* correspond to $n_{\mathrm{H,CHISM}} = 0.14\ \mathrm{cm}^{-3}$, $n_{\mathrm{p,CHISM}} = 0.10\ \mathrm{cm}^{-3}$. $V_{\mathrm{CHISM}} = 25.6$ km/s and $T_{\mathrm{CHISM}} = 7000$ K in both cases. (From Izmodenov 2000)

plasma before it reaches the bow shock. The Mach number decreases upstream of the BS and for a certain range of interstellar parameters ($n_{\mathrm{H,CHISM}} \gg n_{\mathrm{p,CHISM}}$) the bow shock may disappear. The solid curves in Fig. 2B correspond to a small ionization degree of CHISM ($n_p/(n_p+n_H)=1/6$); the bow shock almost disappears in this case.

The interstellar neutrals also modify the plasma structure in the inner heliosheath. In a pure gasdynamic case (without neutrals) the density and temperature of the post-shock plasma are nearly constant. However, the charge exchange process leads to a large increase in the plasma number density and decrease in its temperature (Fig. 2C). The electron impact ionization process may influence the heliosheath plasma flow by increasing the gradient of the plasma density from the termination shock to the heliopause (Baranov and Malama 1996). The influence of interstellar atoms on the heliosheath plasma flow is important, in particular, for the interpretation of kHz-radio emissions detected by Voyager and for analysis of the heliospheric ENA fluxes.

Charge exchange significantly alters the interstellar atom flow. Atoms newly created by charge exchange have the velocity of their ion counterparts in the charge exchange collisions. Therefore, the velocity distribution of these new atoms depends on the local plasma properties in the place of their origin. It is convenient to distinguish four different populations of atoms, depending on the region in the heliospheric interface where the atoms were formed. Population 1 are the atoms created in the supersonic solar wind up to the TS (region 1 in Fig. 1), population 2 are the atoms created in the inner heliosheath (region 2 in Fig. 1), and population 3 are the atoms created in the outer heliosheath (region 3 in Fig. 1). The atoms of population 3 are often called the secondary interstellar atom component. We will call the original (or primary) interstellar atoms population 4. The number densities and mean velocities of these populations are shown in Fig. 3-I as functions of the heliocentric distance. The distribution function of H atoms, $f_H(\mathbf{r},\mathbf{w}_H)$, can be represented as a sum of the distribution functions of these populations: $f_H = f_{H,1} + f_{H,2} + f_{H,3} + f_{H,4}$. The Monte Carlo method allows us to calculate these four distribution functions. These distributions were presented by Izmodenov (2001) and Izmodenov et al. (2001a) at 12 selected points in the heliospheric interface. As an example, the distribution functions at the termination shock in the upwind direction are shown in Fig. 3-II for the four introduced populations of H atoms. It is seen from this figure that the distribution functions of all H-atom populations are not Maxwellian inside the heliosphere, i.e. the fluid approach is not correct for describing the motion of neutral atoms. Comparisons of kinetic and different multi-fluid approaches show significant differences in the results (Alexashov and Izmodenov 2005; Muller et al. 2008).

Original (or primary) interstellar atoms (population 4) are significantly filtered (i.e. their number density is reduced) before reaching the termination shock (Fig. 3I-A). The outer heliosheath is the main “filter” for these atoms. Since slow atoms have a small mean free path (due to both larger charge exchange cross section and smaller velocities) in comparison with the fast atoms, they undergo larger losses. This kinetic effect, called *selection*, results in a deviation of the interstellar distribution function from Maxwellian (Fig. 3II-A). The selection also results in $\sim$10% increase in the primary atom mean velocity towards the termination shock (Fig. 3I-C).

The secondary interstellar atoms (population 3) are created in the disturbed interstellar medium by charge exchange of primary interstellar neutrals with protons decelerated in the vicinity of the heliopause. The secondary interstellar atoms collectively make up the *hydrogen wall*, a density increase at the heliopause. The *hydrogen wall* has been predicted by Baranov et al. (1991) and detected in the direction of α Cen (Linsky and Wood 1996) on the Hubble Space Telescope. At the termination shock, the number density of secondary neutrals is comparable to the number density of the primary interstellar atoms (Fig. 3I-A, dashed curve). The relative abundances of secondary and primary atoms entering the heliosphere vary with the degree of interstellar ionization. The bulk velocity of population 3 is

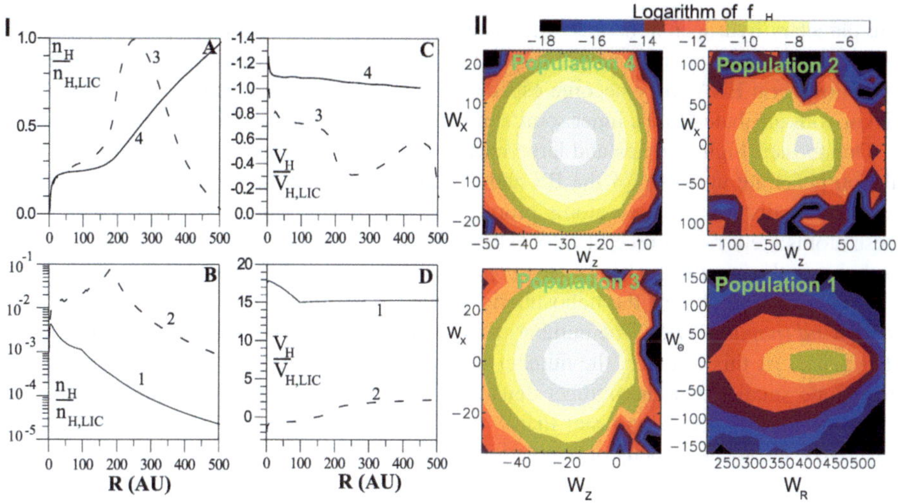

Fig. 3 **I.** Number densities and velocities of the four atom populations as functions of heliocentric distance in the upwind direction. *1* designates atoms created in the supersonic solar wind, *2* atoms created in the heliosheath, *3* atoms created in the disturbed interstellar plasma, and *4* original (or primary) interstellar atoms. Number densities are normalized to $n_{H,CHISM}$, and velocities are normalized to V_{CHISM}. It is assumed that $n_{H,CHISM} = 0.2$ cm^{-3}, $n_{p,CHISM} = 0.04$ cm^{-3}. **II.** Velocity distributions of the four atom populations—primary interstellar atoms (population 4), secondary interstellar atoms (population 3), atoms created in the inner heliosheath (population 2), and atoms created in the supersonic solar wind (population 1)—at the termination shock in the upwind direction; w_z is the projection of the velocity vector on the axis parallel to the LIC velocity vector. Negative values of w_z indicate approach to the Sun. w_x is the magnitude of the projection of the velocity vector on the plane perpendicular to the interstellar velocity vector; w_R, w_θ are radial and tangential velocity components. All velocities are in km/sec. (From Izmodenov et al. 2001a)

about $-18 \div -19$ km/s. The sign "–" means that the population approaches the Sun. One can see that the distribution function of this population is not Maxwellian (Fig. 3II-B). The reason for the abrupt behavior of the distribution function for $w_z > 0$ is that the particles with significant positive w_z velocities can reach the termination shock only from the downwind direction. The distribution functions of different H atom populations were calculated by Izmodenov et al. (2001a) for different directions from upwind. The fine structures of the distribution functions of the primary and secondary interstellar populations vary with direction. The directional variation of the velocity distribution reflects the geometrical pattern of the heliospheric interface. The distribution functions of the interstellar atoms can be a good diagnostic of the global structure of the heliospheric interface.

Another population (population 2) of the heliospheric hydrogen atoms are *the atoms created in the inner heliosheath* by charge exchange with hot and compressed solar wind and pickup protons. The number density of this population is by an order of magnitude smaller than the number densities of the primary and secondary interstellar atoms. This population has a minor importance for the interpretation of Lyman-α and pickup ion measurements inside the heliosphere. Some atoms of this population may be detectable by a Lyman-α hydrogen cell experiment due to their large Doppler shifts (Quemerais and Izmodenov 2002). Recently it was pointed out by Chalov and Fahr (2003) that charge exchange of these atoms with solar wind protons may produce tails in the distribution function of pickup ions that are measured at one or several AU during quiet time periods. Gruntman and Izmodenov (2004) have shown that this population of H-atoms is a major contributor to the density of interplanetary hydrogen at heliocentric distances <1 AU and could dominate in the downwind (interstellar wind) region under typical solar and interstellar conditions. Mass transport by heliospheric ENAs may become especially important for determining the origin of the pickup ions attributed to the inner source of neutral particles in the Sun's vicinity.

Due to their high energies and large mean free path, a portion of the atoms in this population penetrate upstream of the BS and disturb the pristine interstellar medium at large heliocentric distances. These atoms propagate freely in the supersonic solar wind. Thus, these atoms are a rich source of information on the plasma properties at the place of their birth, i.e. at the inner heliosheath. *This population of atoms will be measured by IBEX.*

The last population of the heliospheric atoms are *the atoms created in the supersonic solar wind (population 1)*. The number density of this atom population takes its maximum at ~ 5 AU. At this distance, the number density of the population is about two orders of magnitude smaller than the number density of interstellar atoms. Outside the termination shock the density decreases faster than r^{-2}, where r is the heliocentric distance (curve 1, Fig. 3I-B). The mean velocity of population 1 is about 450 km/sec, which corresponds to the bulk velocity of the supersonic solar wind. The distribution function of this population is also not Maxwellian (Fig. 3II-D). The extended "tail" in the distribution function is caused by the solar wind plasma deceleration upstream of the termination shock. This "supersonic" atom population penetrates the interface and charge exchanges with interstellar protons beyond the BS. The process of charge exchange leads to heating and deceleration upstream of the bow shock and, therefore, to a decrease of the Mach number ahead of the bow shock.

2.2 Role of the Solar Wind Alpha Particles and Interstellar Helium Ions

Since 2003 the standard Baranov-Malama model was slightly modified by adding solar wind alpha particles and interstellar helium ions to plasma components (Izmodenov et al. 2003). Previously, helium ions and alpha particles were not included in the alternative models of other groups.

To evaluate possible effects of both interstellar ions of helium and solar wind alpha particles, Izmodenov et al. (2003) performed parametric model calculations. It was shown that the heliopause, and the termination and bow shocks, are closer to the Sun when the influence of interstellar helium ions is taken into account. This effect is partially compensated by additional solar wind alpha particle pressure that we also took into account in our model. The net result is as follows: the heliopause, termination and bow shocks are closer to the Sun by $\sim$ 12 AU, $\sim$ 2 AU, $\sim$ 30 AU, respectively in the model taking into account both interstellar helium ions and solar wind alpha particles as compared to the model ignoring these ionized helium components. Despite the fact that the net effect of interstellar helium ions and solar alpha particles is rather small ($\sim$7% displacement for the heliopause, $\sim$10% the BS and $\sim$2% for the TS), it still can be important for the interpretation and prediction of experimental data related to the inner heliosheath region.

2.3 ENA Fluxes

The Baranov–Malama model was used to estimate heliospheric ENA fluxes in the paper by Gruntman et al. (2001). Here we present results of parametric studies made within the framework of the Baranov-Malama model by Izmodenov et al. (2001b). As described in Sect. 4 the ENA fluxes predicted in the framework of a more realistic model for pickup ions are different from those predicted in the framework of the Baranov-Malama model. However, this section will demonstrate how the ENA fluxes vary with the interstellar proton and hydrogen number densities that may be useful in the analysis of the IBEX data.

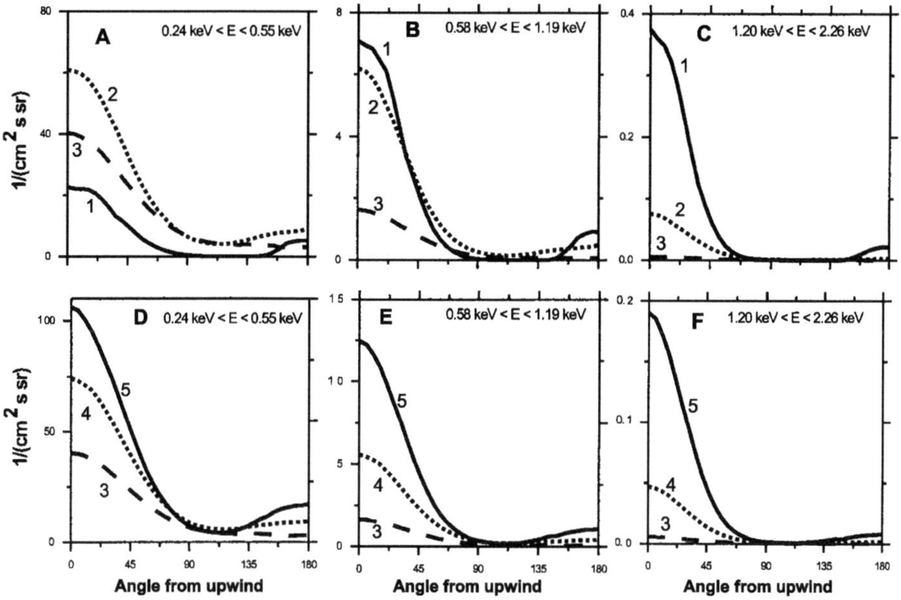

Fig. 4 Heliospheric ENA fluxes in three energy ranges: 0.24–0.55 keV, 0.58–1.19 keV and 1.20–2.26 keV. Different curves correspond to different values of $n_{\mathrm{H,CHISM}}$ and $n_{\mathrm{p,CHISM}}$ (see the text)

The directional differential ENA flux is the line of sight integral

$$j_{\mathrm{s,ENA}}(r_0, E) = \int_{R_{\mathrm{TS}}}^{R_{\mathrm{HP}}} j_{\mathrm{p}}(s, E)\sigma(E) n_{\mathrm{H}}(s) \exp[-D(s, E)] ds. \tag{1}$$

Here, $j_{\mathrm{p}}(s, E)$ is the directional differential flux of protons in the heliosheath with the energy $E, n_{\mathrm{H}}(r)$ is the local number density of interstellar H atoms and $\sigma(E)$ is the energy-dependent charge exchange cross section. The extinction of ENAs between points r and r_0 is determined by $D(s, E) = \int(\beta_{\mathrm{p}} + \beta_{\mathrm{e}} + \beta_{\mathrm{v}})dt$, where the loss processes include charge exchange with the solar wind protons (rate β_{p}), electron-impact ionization (rate β_{e}) and solar photoionization (rate β_{v}). The integration is performed along the atom trajectory with $dt = dl/(2E/m)^{1/2}$, dl being the element of trajectory.

In order to obtain the heliospheric ENA fluxes, we compute $j_{\mathrm{p}}(s, E)$ and $n_{\mathrm{H}}(s)$ in the heliosheath using the Baranov-Malama model of the heliospheric interface. The heliospheric interface structure is obtained for five sets of parameters. In all calculations, the same solar wind velocity, density and temperature, and the CHISM temperature and velocity relative to the Sun, were used (see Izmodenov et al. 1999). Interstellar proton and H atom number densities (n_{p} and n_{H}, respectively) were varied from 0.04 to 0.2 cm^{-3}.

In sets 1 through 3: $n_{\mathrm{p,CHISM}} + n_{\mathrm{H,CHISM}} = 0.24$ cm^{-3} and $n_{\mathrm{H,CHISM}} = 0.04$ cm^{-3} (set 1), 0.14 cm^{-3} (set 2), 0.2 cm^{-3} (set 3). This means that in sets 1 through 3 the interstellar ionization degree $\chi = n_{\mathrm{p}}/(n_{\mathrm{H}} + n_{\mathrm{p}})$ has been varied. In sets 4 and 5 $n_{\mathrm{H,CHISM}} = 0.2$ cm^{-3} and $n_{\mathrm{p,CHISM}} = 0.1$ cm^{-3} (set 4) and 0.2 cm^{-3} (set 5).

Figure 4 presents heliospheric ENA fluxes in three energy ranges: 0.24–0.55 keV, 0.58–1.19 keV, and 1.20–2.26 keV. In order to understand the result we note that there are three major factors influencing the ENA fluxes: 1) thickness of the inner heliosheath (i.e. distance between the TS and HP) and heliocentric distance, 2) number densities of H atoms and protons, and 3) temperature of protons in the heliosheath that affects the tails of the velocity distributions. These factors may interplay with each other, creating complex dependence of ENA fluxes.

Figure 4B and C demonstrate that ENA fluxes decrease with the increasing interstellar H atom number density. At first view this result is counter-intuitive since the more H atoms penetrate into the inner heliosheath, the more charge exchange events with shock-processed solar wind protons appear and, therefore, more ENAs are produced. However, other factors start to play their roles. In fact, set 1 corresponds to the high interstellar proton number density of $n_{\mathrm{p,CHISM}} = 0.20\ \mathrm{cm}^{-3}$. In this case the dynamic pressure of the interstellar plasma component is high, therefore the heliopause and the termination shock are closer to the Sun as compared to sets 2 and 3. Due to the small distance to the Sun the proton number density in the inner heliosheath is larger for set 1 than for sets 2 and 3, which partially compensates the decrease of the flux due to small number density of interstellar H atoms. There is also the second factor that leads to the increase the ENA fluxes in set 1 as compared to sets 2 and 3. An energetic atom with energy of $\sim$1 keV is produced by charge exchange of an interstellar atom and a proton from the flanks of the velocity distribution function. The number of flank protons depends on the temperature of the plasma component since the Baranov–Malama model assumes a one-fluid approach for the heliospheric plasma. If the influence of the neutral component on plasma is negligible or small (as it is for set 1) then the plasma temperature is nearly constant in the entire region between the TS and HP. The interaction with interstellar neutrals results in effective cooling of the heliosheath plasma (Fig. 2C). This leads to significant reduction of the flanks of the proton velocity distribution and the corresponding fluxes of ENAs. By this effect we explain the results demonstrated in Fig. 4B and C. The reduction of the number density of the protons with smaller energies due to the temperature effect will be smaller. Therefore, in the lower energy interval of 0.24–0.56 keV the effect of H number density increase becomes more important. As a result we see that the ENA fluxes for sets 2 and 3 are higher than for set 1 in Fig. 2A.

Figure 4 also demonstrate that there is a strong directional asymmetry of the ENA fluxes. The fluxes peak in the upwind direction. The minimum fluxes are expected at the angles $100° < \theta < 150°$ from the upwind. The relative asymmetry of the ENA fluxes becomes more pronounced with the increasing ENA energies. This dependence is valid for the range of the considered ionization states of the CHISM (sets 1–5). The asymmetry is the highest for the largest ionization degree, χ.

The presented results were obtained for not particularly realistic scenarios of the post-shocked solar wind plasma. The Baranov-Malama model considers the original solar wind protons and the pickup protons as a single fluid. As described later, multi-component consideration of the heliospheric plasma changes the expected ENA fluxes significantly (see Malama et al. 2006 and Sect. 4 of this paper).

3 2D and 3D Kinetic-MHD Models of the Heliospheric Interface: Role of the Interstellar Magnetic Field

Aleksashov et al. (2000) explored the problem of the solar wind interaction with the CHISM for the case when the magnetic field is parallel to the relative Sun/CHISM velocity vector. In this case, the model remains axisymmetric (2D). It was shown that the effect of the interstellar magnetic field on the positions of the TS, BS and HP is significantly smaller when H atoms are included (e.g., Baranov and Zaitsev 1995). The calculations were performed with various Alfvén Mach numbers in the undisturbed CHISM. It was found that the bow shock straightens out with decreasing Alfvén Mach number $M_A = V_{CHISM}\sqrt{4\pi\rho}/B_{CHISM}$ (increasing magnetic field strength in the CHISM). It gets closer to the Sun near the symmetry axis, but recedes from it on the flanks. By contrast, the nose of the heliopause recedes from the Sun due to the tension of magnetic field lines, while wings of the heliopause get closer to the Sun due to magnetic pressure. As a result, the region of the compressed interstellar medium at the heliopause (or "pileup region") decreases by almost 30%, as the magnetic field increases from zero to 3.5×10^{-6} Gauss. It was also shown that H atom filtration and heliospheric distributions of primary and secondary interstellar atoms are virtually unchanged over the entire considered range of the interstellar magnetic field (0–3.5×10^{-6} Gauss). The magnetic field has the strongest effect on the density distribution of population 2 of H atoms, which increases by a factor of almost 1.5 as the interstellar magnetic field increases from zero to 3.5×10^{-6} Gauss.

Izmodenov et al. (2005a) and Izmodenov and Alexashov (2005, 2006) have studied the problem assuming that the interstellar magnetic field (IsMF) is inclined to the direction of the interstellar flow. In this case the SW/CHISM interaction region becomes asymmetric and the flow pattern becomes essentially three-dimensional. Here we present new results obtained within the framework of this model for the interstellar magnetic field $B_{CHISM} =$ 4.4 μG and $\alpha = 15°$, where α is the angle between the CHISM flow velocity and the IsMF direction.

Figure 5 shows the results of the model calculations in the xz plane. The xz plane is determined by the Sun-CHISM relative velocity and the interstellar magnetic field vector. The direction of the z axis is chosen to be opposite to the interstellar gas velocity vector.

The plasma streamlines and isolines of plasma density are shown in plot A, the plasma temperature in plot B, and the total (plasma thermal plus magnetic field) pressure in plot D. Plot C shows the magnetic field lines and isocontours of the interstellar magnetic field. The heliopause and the termination shock become asymmetric with respect to the direction of the interstellar flow due to the asymmetric pressure of the IsMF (plot C). The interstellar magnetic field pressure pushes the heliopause and the termination shock towards the Sun as compared to the model without magnetic field. Quantitatively, the effect of the IsMF on the location of the TS in the directions of Voyager 1 and Voyager 2 is shown in Table 2. The TS distances in this table are obtained in the framework of the stationary 3D kinetic-MHD model that is under discussion, but corrected according to a kinetic-gasdynamic model of the interface by Izmodenov et al. (2008) that takes into account solar cycle effects (see, also, Izmodenov 2009). It is seen from the table that the models with $B_{CHISM} = 4.375$ μG and $\alpha = 15°$–$20°$ are in good agreement with the actual distances of 94 AU and 84 AU of the Voyager 1 and 2 crossings of the TS. This is the reason why this model is presented here.

The bow shock disappears for such a strong magnetic field because the Mach numbers calculated with respect to Alfven and fast magnetosonic waves are smaller than one.

Figure 5A presents also the streamlines of the plasma component. The stagnation point is located in the upper half of the zx plane shifted away from the z axis. It is important to note that the velocity vector of the plasma passing through the region of maximum plasma density has a noticeable V_x component. The secondary interstellar atoms that originate in the vicinity of the HP should have the properties of the plasma in this region. Figure 5E presents the number density of this interstellar atom component. The maximum density appears in the region between the HP and the BS. This is the so-called hydrogen wall as discussed above. It is seen in the figure that the maximum of the hydrogen wall is also slightly shifted to the upper half of the xz plane and reflects the behavior of the plasma distributions. The streamlines of the H atom component are also shown in Fig. 5E. The streamlines were plotted based on the mean velocity field distribution of the interstellar H atoms. The velocity

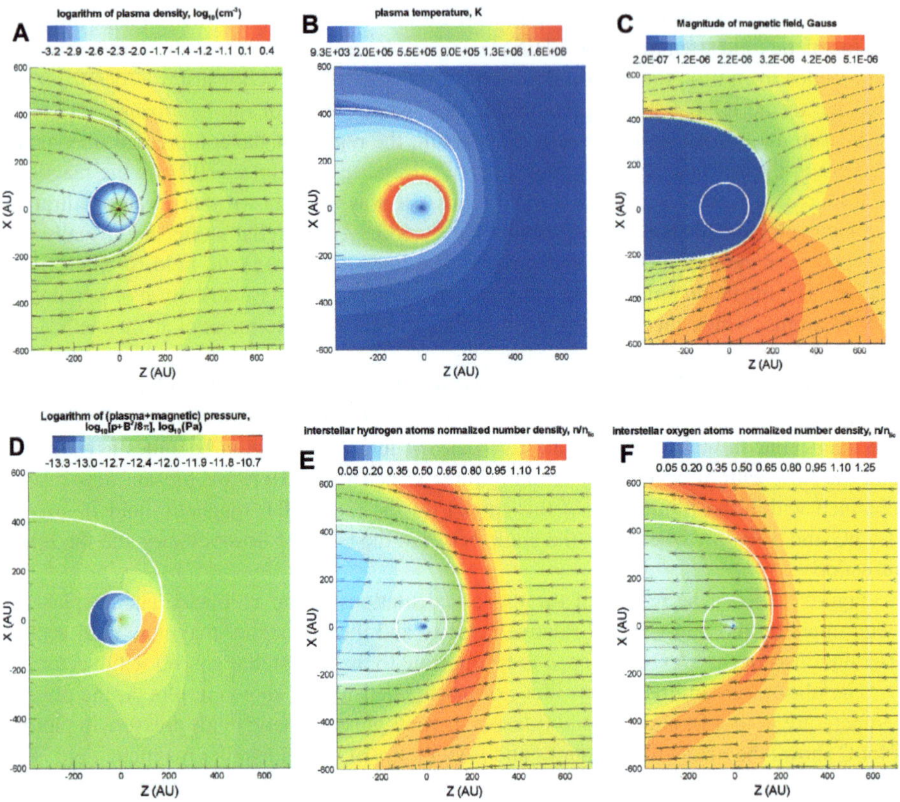

Fig. 5 Isocontours of the proton number density (plot **A**), plasma temperature (plot **B**), magnetic field (plot **C**), total thermal plasma and magnetic pressure (plot **D**), normalized hydrogen (plot **E**) and oxygen (plot **F**) number densities in the plane determined by the interstellar velocity and magnetic field vectors. The streamlines of the plasma component (plot **A**), the magnetic field lines (plot **C**) and "streamlines" of the hydrogen (plot **E**) and oxygen (plot **F**) atoms are shown. The termination shock and heliopause are shown as *white curves*

vector $\mathbf{V}_{\mathrm{H}}$ determines the direction of the H atom flow on average. It is seen from the figure that in the heliosphere the velocity vector $\mathbf{V}_{\mathrm{H}}$ has a noticeable V_x component even very close to the Sun. The effect of the shift of the direction of the interstellar hydrogen flow as compared with the direction of the interstellar helium was observed by the SOHO/SWAN H cell instrument (Lallement et al. 2005).

To draw conclusions on the magnitude and direction of the interstellar magnetic field one needs to take into account the magnitude of the deflection of the interstellar H atom flow as compared with the direction of the interstellar helium flow. Estimates performed by Izmodenov et al. (2005a) and Izmodenov and Alexashov (2006) have shown that the models with $B_{\mathrm{CHISM}} = 2.5\ \mu\mathrm{G}$, and $\alpha = 30$–$45°$ provide levels of deviation in agreement with the level obtained from the analysis of the SOHO/SWAN data. However, recent numerical results (Alexashov et al. 2008) have shown that the deflection angle is very sensitive to the charge exchange cross section. The results of calculations with the updated cross section from Lindsay and Stebbings (2005) show a larger deflection angle as compared with the previous calculations that employed the cross section from Maher and Tinsley (1977). More

Table 2 Positions of the TS in the directions of Voyager 1 and Voyager 2

B_{CHISM}	0	2.5	2.5	5.0	4.375	4.375	2.5	2.5	1.25	2.5	2.5
$\alpha(B_{\mathrm{CHISM}}, V_{\mathrm{CHISM}})$		0	15	15	15	20	30	45	45	60	90
cross section	MT	MT	MT	MT	Steb	Steb	MT	MT	MT	MT	MT
V1	98	99.9	97.4	92.2	96.0	93.3	91.5	85.8	92.1	84.5	83.1
V2	98	99.7	94.2	82.3	86.5	82.4	87.6	82.5	90.6	82.2	83.2

MT = charge-exchange cross-section by Maher and Tinsley (1977); Steb = charge-exchange cross-section by Lindsay and Stebbings (2005)

detailed parametric model calculations with the Lindsay and Stebbings (2005) cross section are required and the work is currently in progress.

3.1 Effect of Interstellar Magnetic Field on the Maps of Heliospheric ENAs

The kinetic-MHD model has been used to calculate the heliospheric ENA fluxes at 1 AU. The results are presented in Figs. 6 and 7. Figure 6 shows the fluxes as they would be measured in the plane determined by the directions of the interstellar wind and interstellar magnetic field. The fluxes are shown in three energy ranges as functions of the angle θ in the plane, where $\theta = 254.7°$ corresponds to the upwind direction. Plots A–C in Fig. 6 and Fig. 7 show the dependence of the fluxes on the magnitude of the interstellar magnetic field for $\alpha(\mathbf{B}_{\mathrm{CHISM}}, \mathbf{V}_{\mathrm{CHISM}}) = 15°$. It is seen from the figures that the larger the magnetic field is, the more the maxima of the fluxes are shifted from the upwind direction. Plots D–F in Fig. 6 show the ENA fluxes for different values of α between the directions of the interstellar flow and interstellar magnetic field for $B_{\mathrm{CHISM}} = 2.5$ μG. It is seen that the orientation of the IsMF influences the location of maximum fluxes as well as their absolute values. The latter fact is related to the width of the inner heliosheath, which depends on the IsMF orientation. Therefore, if the location of the maximum ENA fluxes is different from the upwind direction, this will be an immediate indication of the asymmetry of the heliosphere induced by the interstellar magnetic field.

Plots G–I in Fig. 6 show the heliospheric ENA fluxes calculated under the assumption that the velocity distribution function of protons (solar wind protons and pickup protons) can be represented as the kappa-function

$$f_{\mathrm{p}}(\mathbf{u}) = \frac{n_{\mathrm{p}}}{\pi^{3/2} c_{\mathrm{p}}^3} \frac{1}{k^{3/2}} \frac{\Gamma(k+1)}{\Gamma(k-1/2)} \left[1 + \frac{1}{k} \frac{(\mathbf{u} - \mathbf{u}_{\mathrm{p}})^2}{c_{\mathrm{p}}^2}\right]^{-(k+1)},$$

where c_{p} is the speed related to the effective temperature of the protons in the Maxwellian distribution, and the parameter k can be determined by comparing the theory and observations (Heerikhuisen et al. 2008). As will be seen in the next section of this paper, such an approach is very crude, and more detailed models should be applied to the distribution function of pickup protons. Nevertheless, this approach can give a crude estimate of how the heliospheric ENA fluxes vary with the parameter k that mimic the level of suprathermal pickup proton tails. The plots show that the change from the Maxwellian to kappa distribution leads to increased fluxes in all energy ranges. The parameter k does not change the asymmetry effects that are observed at the chosen direction of the IsMF $\alpha = 45°$ for the three energy bands (see Plots G–I in Fig. 6).

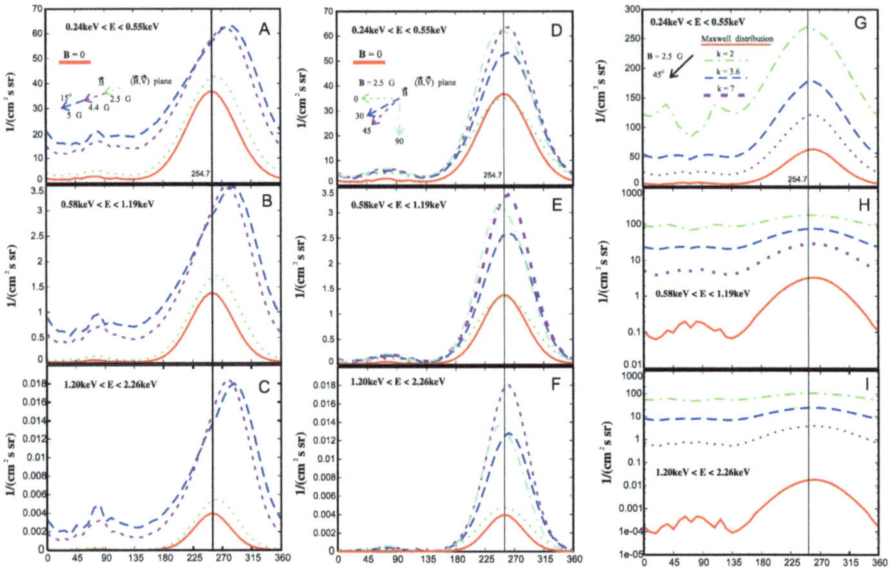

Fig. 6 Fluxes of heliospheric ENAs as functions of the angle θ in the plane determined by $\mathbf{B}_{\text{CHISM}}$ and $\mathbf{V}_{\text{CHISM}}$. θ=254.7° corresponds to the upwind direction. Plots **A–C** show the dependence of the fluxes on the magnitude of the interstellar magnetic field for $\alpha(\mathbf{B}_{\text{CHISM}}, \mathbf{V}_{\text{CHISM}}) = 15°$. Plots **D–F** show the dependence of the fluxes on the angle α for $B_{\text{CHISM}} = 2.5$ μG. Plots **G–I** show the dependence of the fluxes on the parameter k in the kappa-function

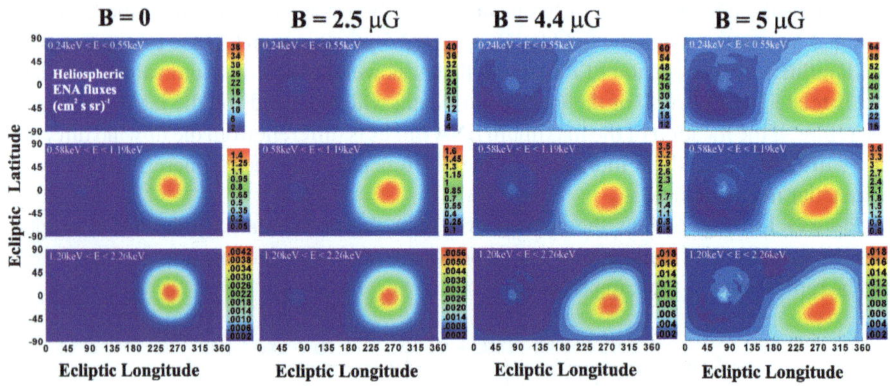

Fig. 7 Maps of heliospheric ENA fluxes that were calculated using the 3D kinetic-MHD model by Izmodenov et al. (2005a). Results are shown for $\alpha = 15°$ and different values of the interstellar magnetic field

Figure 7 presents sky maps of the ENAs fluxes as would be measured at 1 AU. The maps are presented for three energy bands. The left column corresponds to the axisymmetric case with no interstellar magnetic field. Other columns demonstrate results obtained in the frame of 3D MHD-kinetic model calculations. All models have the same angle between the direction of the IsMF and the direction of the interstellar flow of 15 degrees, and the magnitude of the IsMF increases from left column to right column in the figure. It is clearly seen that ENA maps become clearly asymmetric under the influence of the IsMF, and the asymmetry is growing with the magnitude of the IsMF. The asymmetry is clearly pronounced for all energy ranges. Therefore, from the figure we conclude that ENA maps that will be obtained by IBEX will be good diagnostics of the interstellar magnetic field.

Note, that the model employed in this section does not take into account ENAs from the outer heliosheath. As described in the next section, these ENAs may have significant contribution to the total heliospheric ENAs at 1 AU. Therefore, these ENAs should be considered in the analyses of the interstellar magnetic field effects in the heliospheric ENA maps. This is a task for future studies.

4 Modeling of the Multi-Component Nature of the Heliospheric Plasmas

The Baranov-Malama model and more recent models described above assume immediate assimilation of pickup protons into the solar wind plasma and consider the mixture of solar wind and pickup protons as a single component.

However, measurements of the pickup proton distribution function onboard the Ulysses and ACE spacecraft show that the pickup proton distribution function is non-Maxwellian, though close to isotropic. These data point to the absence of thermodynamic equilibrium between the pickup and solar-origin protons, although their mean velocities are equal. In the improved model of the solar wind/interstellar medium interaction developed by Malama et al. (2006), the pickup protons are considered as a separate component with thermodynamic parameters different from those of the solar wind. Since in the solar-wind reference frame the pickup proton distribution function is isotropic, the following angle-averaged distribution function can be introduced:

$$f^*_{\mathrm{pui}}(\mathbf{r}, w) = \frac{1}{4\pi} \int\int f_{\mathrm{pui}}(\mathbf{r}, \mathbf{v}) \sin\theta d\theta d\phi. \tag{2}$$

Here, $\mathbf{v}$ is the individual velocity of a pickup proton, $\mathbf{w} = \mathbf{v} - \mathbf{V}$ is the pickup proton velocity in the solar-wind reference frame ($\mathbf{V}$ is the solar wind velocity in the heliocentric coordinate system), w, θ, and ϕ are the coordinates of the vector $\mathbf{w}$ in the spherical coordinates, and $f_{\mathrm{pui}}(\mathbf{r}, \mathbf{v})$ is the pickup proton distribution function. The kinetic equation for $f^*_{\mathrm{pui}}(\mathbf{r}, w)$ can be written in the form:

$$\frac{\partial f^*_{\mathrm{pui}}}{\partial t} + \mathbf{V} \cdot \frac{\partial f^*_{\mathrm{pui}}}{\partial \vec{r}} = \frac{1}{w^2} \frac{\partial}{\partial w}\left(w^2 D \frac{\partial f^*_{\mathrm{pui}}}{\partial w} \right) + \frac{w}{3} \frac{\partial f^*_{\mathrm{pui}}}{\partial w} \mathrm{div}\mathbf{V} + S(\mathbf{r}, w). \tag{3}$$

Here, $D(\mathbf{r}, w)$ is the diffusivity in the velocity space, while the source of the pickup protons $S(\mathbf{r}, w)$ reflects their birth and loss due to charge exchange, photoionization, and the electron impact ionization of H atoms. In the Malama et al. (2006) model the kinetic equation (3) was solved together with the Euler equations written for the sum of all the charged components and the kinetic equation for the interstellar hydrogen atoms. The use of Euler equations for the sum of the charged components is justified by the fact that all the components (solar protons and electrons, as well as pickup protons) move at the same velocity, and for all the components the distribution functions are isotropic. In this case, the pressure p is equal to the sum of partial pressures, while the effective pressure of the pickup protons is determined in terms of the distribution function f^*_{pui}, so that

$$p = p_{\mathrm{e}} + p_{\mathrm{p}} + p_{\mathrm{pui}}, \quad p_{\mathrm{pui}} = \frac{4\pi}{3} \int m_{\mathrm{p}} w^4 f^*_{\mathrm{pui}}(\mathbf{r}, w) dw.$$

To treat the passage of pickup protons through the termination shock we assume the conservation of their magnetic moment. The reflections of pickup protons at the shock potential is not taken into account. A self-consistent solution of the problem just formulated was obtained by Malama et al. (2006) for the case $D = 0$ corresponding to a quiescent solar wind in which the magnetic field fluctuation level is low.

The thermodynamic non-equilibrium of the pickup and solar protons leads to thinning of the inner interface region (Fig. 8A), which is attributable to the decrease of the total pressure of the charged component. As compared with the calculations performed in accordance with the corresponding equilibrium model, the TS is 5 AU farther from the Sun in the direction of the oncoming flow, whereas the HP is 12 AU closer. In the tail region the heliocentric distance of the TS increases by 70 AU.

Recently we elaborated on the Malama et al. (2006) model by including the pickup proton component outside the heliopause (Malama, Izmodenov, Chalov, in preparation, 2009). Indeed, an energetic atom of hydrogen that originated in the supersonic solar wind and the inner heliosheath can penetrate through the heliopause and enter into the interstellar plasma. Outside the heliopause the atom can suffer charge exchange with an interstellar proton. As a result a new energetic interstellar pickup proton is created. Having energy of 1 keV and above, these pickup protons have large mean free path as compared with thermal interstellar protons and can be considered as collisionless with non-Maxwellian velocity distribution. The pickup protons in the outer heliosheath can charge exchange with interstellar H atoms. New ENAs from the outer heliosheath will be created in this secondary charge exchange. Similar to ENAs from the inner heliosheath, the ENAs of outer heliosheath origin can penetrate at small heliocentric distances and be measured there.

Figure 8B shows the calculated fluxes of H atoms at 1 AU obtained within the framework of the new non-equilibrium model (details of the calculations will be given in our future paper by Malama, Izmodenov, Chalov, 2009). The figure presents total fluxes of all heliospheric ENAs (solid curve), fluxes of ENAs originating in the inner heliosheath from solar wind protons (black dots), fluxes of ENAs originating in the inner heliosheath from pickup protons (blue triangles), and fluxes of ENAs originating in the outer heliosheath from pickup protons (blue dots). It is seen from the figure that the ENAs originating in the inner heliosheath from solar wind protons are dominated at energies of ~0.02–0.2 keV. For this energy range the fluxes of ENAs from the outer heliosheath (blue dots in figure) are smaller but still comparable. In the energy range of ~0.4–1.0 keV, the ENAs that originated in the inner heliosheath from pickup protons (blue triangles in figure) dominate, while for energies above 2 keV, fluxes of the ENAs from the outer heliosheath (blue dots) become higher. Therefore, the model suggests that the ENAs with energies above 1 keV may be of interstellar (outer heliosheath) origin. The ENAs of such energies have been measured by several spacecraft. Upper limits of the heliospheric ENA fluxes estimated from the measurements on boards of SOHO/CELIAS (Hilchenbach et al. 1998, 2000), Cassini/INCA (Krimigis et al. 2009) and Venus Express (Brandt et al. 2009) are shown in Fig. 8B as green dots. The figure shows that generally there is a good agreement between the model results and data.

Here we should also notice that the calculations were performed in the framework of the model that has no stochastic mechanisms of the pickup proton acceleration (see, Malama et al. 2006). As described in the next section the stochastic acceleration mechanisms would generally result in an increase of ENA fluxes at energies above 1 keV. In this case the model fluxes are expected to be higher than the upper limits established from observations. Note that other model restrictions, such as the absence of the interstellar magnetic field in this particular model or latitudinal variation of the solar wind parameters, may play a role.

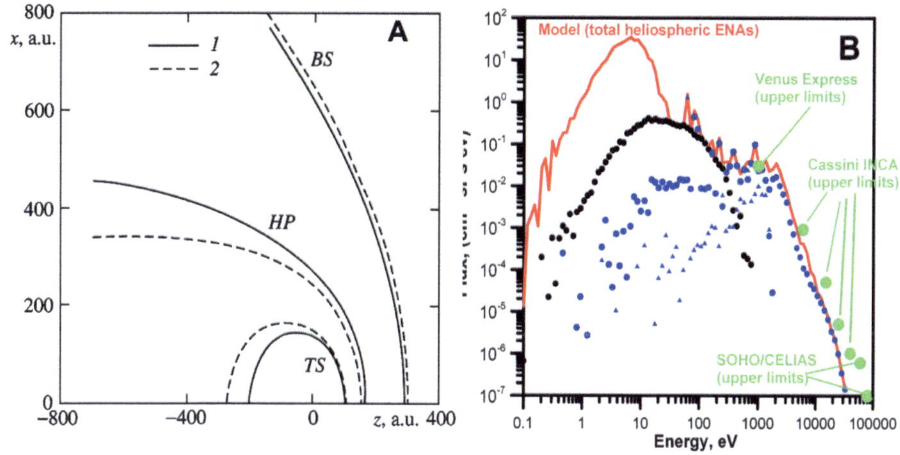

Fig. 8 **A** Strong discontinuity surfaces calculated taking into account the pickup proton non-equilibrium (*curves 1*), and in accordance with the equilibrium plasma model (*curves 2*). **B** Fluxes of H atoms at 1 AU calculated with pickup proton non-equilibrium model: total fluxes of all heliospheric ENAs (*solid curve*), ENAs originating in the inner heliosheath from solar wind protons (*black dots*), ENAs originating in the inner heliosheath from pickup protons (*blue triangles*), ENAs originating in the outer heliosheath from pickup protons (*blue dots*). Upper limits of heliospheric ENA fluxes measured by SOHO/CELIAS, Cassini/INCA and Venus Express are shown by the green dots

5 Energy Distributions of Pickup Protons in the Upwind Region of the Inner Heliosheath and ENA Fluxes

Downstream of the termination shock the physical conditions for the pickup process are different from the inner heliospheric conditions in many respects, e.g. the plasma flow is decelerated then becomes subsonic, its flow geometry is essentially non-radial, and the speed of plasma becomes very small when it approaches the heliopause (at least in the upwind region). In this section we consider the phase-space transport of pickup protons downstream of the shock along or close to the stagnation line up to the heliopause. We include the effects of energy diffusion and of charge exchange processes of these pickup protons with H-atoms from the local interstellar medium (LISM).

In the region of the inner heliosheath pickup protons can be considered as consisting of two populations: 1) pickup protons created in the supersonic solar wind then convected into the heliosheath and 2) pickup protons created in the heliosheath. The first population is accelerated by the solar wind turbulence in the whole region from the Sun up to the termination shock, therefore these pickup protons can already have a high velocity tail when they reach the termination shock.

The transport of the pickup protons in the solar wind is governed by the Fokker-Planck equation (3) for the isotropic velocity distribution function $f^*_{\mathrm{pui}}(\mathbf{r}, w)$. We assume in this section that $S = S_{\mathrm{inj}} + S_{\mathrm{ex}}$, where $S_{\mathrm{inj}}(\mathbf{r}, w)$ is the source of freshly ionized pickup protons and the term $S_{\mathrm{ex}}(\mathbf{r}, v)$ describes the charge exchange process between pickup protons and interstellar H-atoms. This latter process does not change the number density of pickup protons but results in a redistribution of the energy in the pickup proton spectra.

In the upwind region, downstream of the termination shock up to the heliopause, the behavior of the velocity distribution function of pickup protons can be approximately described by a spherically symmetric version of (3). At $V_{\mathrm{LISM}} = 25.6\ \mathrm{km\,s^{-1}}$, $n_{\mathrm{H,LISM}} = 0.2\ \mathrm{cm^{-3}}$,

$n_{\rm p,LISM} = 0.05$ cm^{-3} considered here and for typical parameters of the solar wind ($V_{\rm E} = 430$ km s^{-1}, $n_{\rm eE} = 6.5$ cm^{-3}) the termination shock and heliopause are located, according to the Baranov-Malama model, at 88 AU and 166 AU, respectively. To calculate the efficiency of the stochastic acceleration by means of wave-particle interactions in the inner heliosheath, for the sake of simplicity, we restrict our consideration to Alfvénic turbulence, though there are observational evidences that compressible turbulence is also prominent just downstream of the TS (Burlaga et al. 2006). The corresponding velocity diffusion coefficient D is given by Chalov et al. (2003). This coefficient depends on the mean-square amplitude of magnetic fluctuations $\langle \delta B_{\rm A}^2 \rangle$. Sacking specific information, we assume here that the quantity $\zeta_{\rm H} = \langle \delta B_{\rm A}^2 \rangle / B^2$ is a constant in the heliosheath.

The term $S_{\rm ex}(r, w)$ in (3), which describes the redistribution of the energy of pickup protons due to charge exchange with the interstellar H-atoms, can be written as

$$S_{\rm ex}(r, v) = -\sigma_{\rm ex}(u) n_{\rm H} u \left[f(r, v) - \frac{n_{\rm PUI}}{4\pi v^2} \delta(v - V_{\rm H}) \right], \tag{4}$$

where $\sigma_{\rm ex}$ is the charge-exchange cross section, $n_{\rm H}$ is the number density of atoms, $V_{\rm H} = V + V_{\rm LISM}$ is the relative velocity of a LISM H-atom with respect to the local wind frame, and $u(w)$ is the average relative velocity of pickup protons with velocity w with respect to the H-atoms. It is calculated to be equal to $V + w^2/3V$ for $V \geq w$, and to $w + V^2/3w$ for $V \leq w$. For simplicity, the thermal velocities of H-atoms are ignored here.

Since the parameters of solar wind turbulence are known only in a narrow region of the heliosheath just behind the termination shock we will consider $\zeta_{\rm H}$ as a free parameter. At first, let us consider the case of a low level of Alfvénic turbulence in the heliosheath: $\zeta_{\rm H} = 0.4 \times 10^{-3}$ (for more details see Chalov et al. 2003). Such a low level of turbulence is realized when no generation of waves in the region upstream of the termination shock takes place and when fluctuations are simply convected from the inner region of the heliosphere outwards.

Figure 9a shows downstream fluxes of pickup protons (in the solar wind rest frame) at different distances from the Sun when $\zeta_{\rm H} = 0.4 \times 10^{-3}$. The fluxes are steady-state solutions of (3). In order to obtain the solutions we transform this equation to a coupled system of stochastic ordinary differential equations, which describe the motion of the pickup protons in phase space. Since the pickup protons are comoving with a solar wind parcel, the characteristic time during which they experience stochastic acceleration is equal to the corresponding convective time. The spectrum given by curve 1 is just downstream of the shock. The peak at low energies is formed due to ionization of interstellar atoms in the downstream region (charge exchange with solar wind protons, photoionization and electron collision ionization) and due to charge exchange between the interstellar atoms and pickup protons, which results in the creation of energetic neutral atoms and cold pickup protons with velocities close to the local solar wind velocity (see (4)).

The relative contributions of pickup protons with different regions of origin to the resulting fluxes can be seen in Fig. 9b, which shows downstream fluxes of pickup protons originating in the supersonic solar wind (solid lines) and in the heliosheath (dashed lines) at 98 AU (lines 1) and 164 AU (lines 2). The subsequent spatial evolution of pickup proton spectra in the heliosheath given by curves 2, 3, and 4 in Fig. 9a have the following interesting features. Firstly, the high-energy tail shifts to higher energies. This shift is not connected with the stochastic acceleration, which is not efficient at such a low level of Alfvénic turbulence (calculations give identical spectra at much lower values of $\zeta_{\rm H}$), but is a consequence of the adiabatic heating due to the compression of decelerated plasma when it moves from

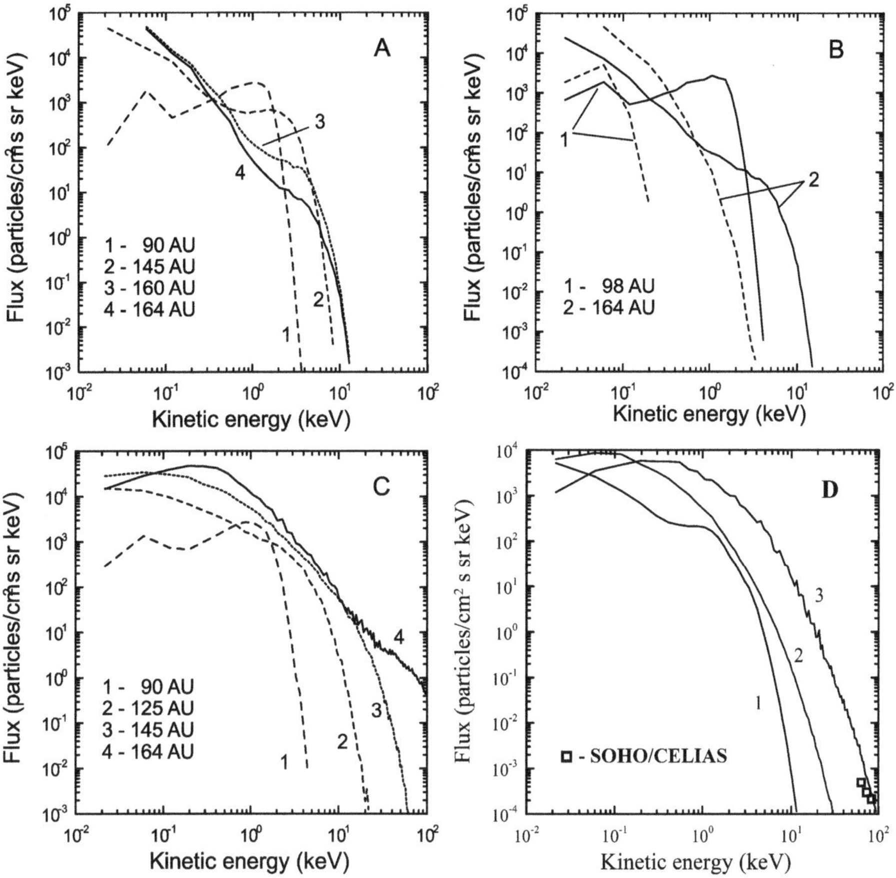

Fig. 9 **A** Spatial evolution of downstream pickup proton fluxes in the solar wind rest frame at $\zeta_H = 0.4 \times 10^{-3}$. **B** Downstream fluxes of pickup protons originating in the supersonic solar wind (*solid lines*) and in the heliosheath (*dashed lines*) at $\zeta_H = 0.4 \times 10^{-3}$. **C** The same as **A** but with an enhanced level of turbulence in the heliosheath: $\zeta_H = 0.7 \times 10^{-2}$. **D** Differential fluxes of ENAs from the inner heliosheath at 1 AU originating from pickup protons at low level of turbulence in the heliosheath with $\zeta_H = 0.4 \times 10^{-3}$ (*curve 1*), at medium level with $\zeta_H = 0.14 \times 10^{-2}$ (*curve 2*), and enhanced level with $\zeta_{AH} = 0.7 \times 10^{-2}$ (*curve 3*). The squares in the right bottom corner are fluxes of ENAs from the apex direction observed with SOHO/CELIAS at 1 AU (Hilchenbach et al. 2000). The termination shock and heliopause are located at 88 AU and 166 AU, respectively

the termination shock to the heliopause (see Chalov et al. 2007). Secondly, there is a spectral gap at medium energies (about several keV), which is more pronounced at large distances (close to the heliopause). This gap is formed due to charge exchange of pickup protons with the interstellar atoms, which results in the spectral energy redistribution: protons with low energies are created while medium-energy protons are removed. This process is one of the important causes for the increase of spectral fluxes at low energies, which is clearly seen in Fig. 9a. The other cause for this increase is the production of low-energy pickup protons due to charge exchange of the solar wind protons with the H-atoms in the heliosheath. The fact that the gap is pronounced at medium energies and absent at high energies is explained by the decrease of the charge exchange cross section with the energy increase.

Let us assume now that local generation of Alfvénic waves takes place upstream of and at the termination shock. The wave generation is possible due to the beam instability caused by reflection of ions from the shock front (e.g. Lembege et al. 2004). As a result, the level of turbulence in the heliosheath will be enhanced as compared with the case considered above. Figure 9c shows spectral fluxes of pickup protons in the heliosheath at different distances from the Sun but with a more than 10 times enhanced level of turbulence in the heliosheath ($\zeta_H = 0.7 \times 10^{-2}$). One can see in Fig. 9c that the high-energy tails of pickup protons, due to the effective stochastic acceleration, are now considerably more populated than those in Fig. 9a. Note that stochastic acceleration in the upwind portion of the heliosheath is much more efficient than in the inner heliosphere under identical parameters of the solar wind turbulence. This is due to the large exposure times during which the particles suffer the acceleration while being slowly convected towards the flanks of the heliosheath (Chalov et al. 2007). It is important also that the velocity divergence in this region is negative, so that adiabatic cooling can not be considered now as a competing mechanism. The divergence in the upwind part of the inner heliosheath is negative due to deceleration of the plasma flow when it approaches the heliopause (the velocity equals zero at the stagnation point). The deceleration is connected partly with the process of charge exchange between the interstellar atoms and protons and with the mass loading effect. The theoretical fluxes of pickup protons presented here were calculated in the test particle approximation. In other words, dumping of turbulence by virtue of the stochastic acceleration was ignored. In fact the back-reaction effect of pickup ions on the turbulence level can be very prominent (Chalov et al. 2004) and should be taken into account in advanced models.

The interstellar pickup protons in the inner heliosheath with the spectral distributions given in Figs. 9a,c suffer charge exchanges with the interstellar atoms. As a product of this charge exchange, H-atoms (ENAs) with velocities represented by the pickup distribution functions are created. These H atoms can then partly enter the inner heliospheric regions and can be detected there. The first detections of such ENAs in the energy range from 60 to 100 keV/nuc have been carried out with the SOHO/CELIAS instrument (Hilchenbach et al. 1998).

After their creation ENAs move on nearly straight ballistic trajectories suffering losses due to ionization, which is mainly important in the inner heliosphere. The expression for the directional differential flux of H-ENAs from the inner heliosheath in ($\mathrm{cm}^{-2}\,\mathrm{s}^{-1}\,\mathrm{sr}^{-1}\,\mathrm{keV}^{-1}$) is given by (1) (see Gruntman 1992, 1997; Hsieh and Gruntman 1993), with $j_{\mathrm{pui,H}}$ substituted for j_{p}. Here, sr is stereoradians (solid angle). We assume here that the ionization rate is inversely proportional to the square of the distance from the Sun and adopt the value $\beta_E = 6 \times 10^{-7}\,\mathrm{s}^{-1}$ at 1 AU in our calculations of ENA fluxes.

Figure 9d shows the differential fluxes of H-ENAs from the inner heliosheath at 1 AU originating from the pickup protons. The curves labeled 1, 2 and 3 show the fluxes when the level of turbulence in the heliosheath is low with $\zeta_H = 0.4 \times 10^{-3}$, medium with $\zeta_H = 0.14 \times 10^{-2}$, and enhanced with $\zeta_H = 0.7 \times 10^{-2}$, respectively. One can see in this figure that the fluxes of H-ENAs are very sensitive to the level of solar wind turbulence in the inner heliosheath over the whole range of energies. The squares in the right bottom corner in Fig. 9d are the fluxes of H-ENAs from the apex direction observed with SOHO/CELIAS at 1 AU (Hilchenbach et al. 2000). We emphasize that our predicted fluxes are only due to pickup protons. Low energy atoms, which are produced in the heliosheath due to charge exchange between the solar protons and H-atoms, are not shown.

The theoretical pickup proton and ENA fluxes presented in this section are calculated in the framework of the model, which does not take into account shock-drift acceleration of multiple reflected particles (see, e.g., Chalov 2005). In other words, energetic particles crossing the shock front are only adiabatically heated at the shock. Thus our ENAs are mainly due to stochastic acceleration of pickup protons in the heliosheath or due to their adiabatic heating in this region, which, as it was clearly demonstrated here, can be very efficient. What type of acceleration is more responsible for the ENA production remains an open question.

It can be clearly concluded from these results that the combined observations of ENA fluxes in the energy range from 1 keV up to several tens of keV can give strong limits to the level of solar wind turbulence near the termination shock and in the inner heliosheath. More generally, these results are very important for understanding the mechanism of energization of pickup ions at the termination shock and the physics of the multi-component plasma in the heliosheath.

6 Summary and Perspectives

Imaging of the heliospheric boundaries in heliospheric ENAs is a powerful tool for the remote sensing of this region. In this paper we have summarized basic relevant results of the kinetic-gasdynamic models of the SW/CHISM interaction that theoretically predict global 3D structure of the heliosphere and underline effects of different interstellar and heliospheric components such as the interstellar H atoms, plasma and magnetic field, and heliospheric pickup protons. We use these models to explore how different physical phenomena are evident in fluxes of ENAs at 1 AU. It was shown that the directional differential of the fluxes at 1 AU in the IBEX energy ranges will provide important constraints on the properties of pickup protons in the inner heliosheath as well as on the nature of the heliospheric termination shock and the level of turbulence near the termination shock and in the inner heliosheath. It was also shown that imaging of the inner heliosheath in the ENA fluxes at 1 AU will allow us to establish important constraints on the global structure of the interface and parameters of the CHISM, such as the direction and magnitude of the interstellar magnetic field, the degree of ionization of hydrogen in the CHISM, and the number densities of protons and H atoms in the CHISM.

It is important to note that the results presented here are not ready for the direct comparison with future IBEX measurements of ENA fluxes. To perform the direct comparison with future IBEX data we plan to compute the directional differential ENA fluxes in IBEX energy ranges at 80 AU and at 1 AU. For the latter case we will employ a recent model of the neutral particle transport in the supersonic solar wind. This advanced model developed in our group (Katushkina and Izmodenov, Astron. Lett., submitted) takes into account processes of charge exchange, photoionization, electron impact ionization, solar gravity and solar radiation pressure. As a boundary condition at 80 AU this code employs the velocity distribution function obtained in self-consistent models. The computed fluxes at 1 or 80 AU, as well as a code for the neutral particle transport in the supersonic solar wind, will be available for the IBEX Science Operations Center (ISOC).

Acknowledgements We thank Vladimir B. Baranov for many fruitful and inspiring discussions that simulate the most of research reported in this paper. This work was supported in parts by RFBR grant 07-02-01101, RFBR-CNRS grant 06-02-72557, Program RAS "Plasma processes in the solar system" (chair—Zelenyi L.M.), and by Basic Research Program OEMMPU RAS. V.V.I. and E.A.P are also supported by the Dynastia Foundation. M.S.R., Y.G.M. and V.V.I., acknowledge the support by the Royal Society of London. Part of the presented work was done in the frame of a team of International Space Science Institute (ISSI) in Bern (Switzerland).

References

D.B. Aleksashov, V.B. Baranov, E.V. Barsky, A.V. Myasnikov, Astron. Let. **26**, 743–749 (2000)
D. Alexashov, V. Izmodenov, Astron. Astrophys. **439**, 1171–1181 (2005)
D.B. Alexashov, S.V. Chalov, A.V. Myasnikov, V.V. Izmodenov, R. Kallenbach, Astron. Astrophys **420**, 729–736 (2004)
D.B. Alexashov, V.V. Izmodenov, M. Opher, 37th COSPAR Scientific Assembly D13-0015-08; P025-TueWed (poster) (2008)
V.B. Baranov, V.V. Izmodenov, Fluid Dyn. **5**, 689–707 (2006)
V.B. Baranov, Y.G. Malama, J. Geophys. Res. **98**, 15157–15163 (1993)
V.B. Baranov, Yu.G. Malama, Space Sci. Rev. **78**, 305–316 (1996)
V.B. Baranov, N.A. Zaitsev, Astron. Astrophys. **304**, 631–637 (1995)
V.B. Baranov, M.G. Lebedev, I.G. Malama, Astrophys. J. **375**, 347–351 (1991)
V.B. Baranov, Space Sci. Rev. **143**, 449–464 (2009)
P.C. Brandt, E.C. Roelof, P. Wurz, S. Barabash, D. Bazell, R. DeMajistre, T. Sotirelis, R. Decker, Energetic Neutral Atom (ENA) Imaging of the Heliosheath: Spectral Characteristics and Implications for Shock Acceleration from Observations by the Neutral Particle Detector (NPD) on board Venus Express (VEX), Geophysical Research Abstracts, vol. 11, EGU2009-9588, 2009
L.F. Burlaga, N.F. Ness, M.H. Acuña, Astrophys. J. **642**, 584–592 (2006)
S.V. Chalov, Adv. Space Res. **35**, 2106–2114 (2005)
S.V. Chalov, H.J. Fahr, Energetic particles from the outer heliosphere appearing as a secondary pick-up ion component. Astron. Astrophys. **401**, L1–L4 (2003)
S.V. Chalov, H.J. Fahr, V.V. Izmodenov, Evolution of pickup proton spectra in the inner heliosheath and their diagnostics by energetic neutral atom fluxes. J. Geophys. Res. **108**, 1266 (2003). doi:10.1029/2002JA009492
S.V. Chalov, D.B. Alexashov, H.J. Fahr, Astron. Astrophys **416**, L31–L34 (2004)
S.V. Chalov, H.J. Fahr, Y.G. Malama, Unexpected sites of efficient stochastic acceleration in the inner heliosheath. Ann. Geophys. **25**, 275–580 (2007)
M.A. Gruntman, Anisotropy of the energetic neutral atoms in the heliosphere. Planet. Space Sci. **40**, 439–445 (1992)
M. Gruntman, Energetic neutral atom imaging of space plasmas. Rev. Sci. Instrum. **68**, 3617–3656 (1997)
M. Gruntman, V.V. Izmodenov, J. Geophys. Res. **109**, A12108 (2004)
M. Gruntman, E.C. Roelof, D.G. Mitchell, H.J. Fahr, H.O. Funsten, D.J. McComas, J. Geophys. Res. **106**, 15767–15782 (2001)
J. Heerikhuisen, N.V. Pogorelov, V. Florinski, G.P. Zank, J.A. le Roux, Astrophys. J. **682**, 679–689 (2008)
M. Hilchenbach, K.C. Hsieh, D. Hovestadt et al., Astrophys. J. **503**, 916–922 (1998)
M. Hilchenbach, K.C. Hsieh, D. Hovestadt, R. Kallenbach, A. Czechowski, E. Möbius, P. Bochsler, in *The Outer Heliosphere: The Next Frontiers*, ed. by K. Scherer, H. Fichtner, H. Fahr, E. Marsch (Pergamon, Elmsford, 2000), pp. 273–276.
K.C. Hsieh, M.A. Gruntman, Adv. Space Res. **13**(6), 131–139 (1993)
V.V. Izmodenov, Astrophys. Space Sci. **274**, 55–69 (2000)
V. Izmodenov, Space Sci. Rev. **97**, 385–388 (2001)
V.V. Izmodenov, Space Sci. Rev. **143**, 139–150 (2009)
V. Izmodenov, D. Alexashov, Astron. Lett. **29**, 58–63 (2003)
V.V. Izmodenov, D.B. Alexashov, ESA Special Publication SP-592, 351–357 (2005)
V.V. Izmodenov, D.B. Alexashov, AIP Conf. Proc. **858**, 14–19 (2006)
V.V. Izmodenov, R. Kallenbach (eds.), *Physics of the Heliospheric Boundaries*, vol. 5. ESA Publication, Scientific Reports of International Space Science Institute, Bern (2006)
V.V. Izmodenov, R. Lallement, J. Geiss, Astron. Astrophys. **344**, 317–321 (1999)
V.V. Izmodenov, M. Gruntman, Y.G. Malama, J. Geophys. Res. **106**, 10681–10690 (2001a)
V. Izmodenov, M. Gruntman, V. Baranov, H. Fahr, Space Sci. Rev. **97**, 413–416 (2001b)
V.V. Izmodenov, Yu.G. Malama, G. Gloeckler, J. Geiss, Astrophys. J. **594**, L59–L62 (2003)
V.V. Izmodenov, D.B. Alexashov, A.V. Myasnikov, Astron. Astrophys. **437**, L35–L38 (2005a)
V.V. Izmodenov, Yu.G. Malama, M.S. Ruderman, Astron. Astrophys. **429**, 1069–1080 (2005b)
V.V. Izmodenov, Y.G. Malama, M.S. Ruderman, Adv. Space Res. **41**, 318–324 (2008)
S.M. Krimigis, D.G. Mitchell, E.C. Roelof, P.C. Brandt, Energetic neutral atoms (ENA) from the termination shock/heliosheath? The view from 10 AU, Presentation at "Voyagers in the Heliosheath Observations, models, and plasmas physics", January 9–14, 2009, Kauai, Hawaii, USA
R. Lallement, E. Quemerais, J.L. Bertaux, S. Ferron, D. Koutroumpa, R. Pellinen, Science **307**, 1447–1449 (2005)
B. Lembege, J. Giacalone, M. Scholer et al., Space Sci. Rev. **110**, 161–226 (2004)

B.G. Lindsay, R.F. Stebbings, J. Geophys. Res. **110**, A12213 (2005)
J. Linsky, B. Wood, Astrophys. J. **463**, 254 (1996)
L.J. Maher, B.A. Tinsley, J. Geophys. Res. **82**, 689–695 (1977)
Y.G. Malama, V.V. Izmodenov, S.V. Chalov, Astron. Astrophys. **445**, 693–701 (2006)
D. McComas, F. Allegrini, P. Bochsler, M. Bzowski, M. Collier, H. Fahr, H. Fichtner, P. Frisch, H. Funsten, S. Fuselier, G. Gloeckler, M. Gruntman, V. Izmodenov, P. Knappenberger, M. Lee, S. Livi, D. Mitchell, E. Möbius, T. Moore, D. Reisenfeld, E. Roelof, N. Schwadron, M. Wieser, M. Witte, P. Wurz, G. Zank, in *AIP Conf. Proc. 719: Physics of the Outer Heliosphere* (2004), pp. 162–181
D. McComas, F. Allegrini, L. Bartolone, P. Bochsler, M. Bzowski, M. Collier, H. Fahr, H. Fichtner, P. Frisch, H. Funsten, S. Fuselier, G. Gloeckler, M. Gruntman, V. Izmodenov, P. Knappenberger, M. Lee, S. Livi, D. Mitchell, E. Möbius, T. Moore, S. Pope, D. Reisenfeld, E. Roelof, H. Runge, J. Scherrer, N. Schwadron, R. Tyler, M. Wieser, M. Witte, P. Wurz, G. Zank, in *Solar Wind 11/SOHO 16, Connecting Sun and Heliosphere*, vol. 592 of ESA Special Publication (2005), pp. 689–692
D.J. McComas, F. Allegrini, L. Bartolone, P. Bochsler, M. Bzowski, M. Collier, H. Fahr, H. Fichtner, P. Frisch, H. Funsten, S. Fuselier, G. Gloeckler, M. Gruntman, V. Izmodenov, P. Knappenberger, M. Lee, S. Livi, D. Mitchell, E. Möbius, T. Moore, S. Pope, D. Reisenfeld, E. Roelof, H. Runge, J. Scherrer, N. Schwadron, R. Tyler, M. Wieser, M. Witte, P. Wurz, G. Zank, in *Physics of the Inner Heliosheath*, ed. by J. Heerikhuisen, V. Florinski, G.P. Zank, N.V. Pogorelov, vol. 858 of American Institute of Physics Conference Series (2006), pp. 241–250
D.J. McComas et al., Space Sci. Rev. (2009, this issue)
E. Möbius, H. Kucharek, L. Petersen, M. Bzowski, L. Saul, P. Wurz, S.A. Fuselier, V. Izmodenov, D.J. McComas, H.R. Mueller, D.B. Alexashov, Space Sci. Rev. (2009, this issue)
H.-R. Muller, V. Florinski, J. Heerikhuisen, V.V. Izmodenov, K. Scherer, D. Alexashov, H.-J. Fahr, Astron. Astrophys. **491**, 43–51 (2008)
A.V. Myasnikov, D.B. Alexashov, V.V. Izmodenov, S.V. Chalov, J. Geophys. Res. **105**, 5167–5178 (2000)
E. Quemerais, V.V. Izmodenov, Astron. Astrophys. **396**, 269–281 (2002)

Space Sci Rev (2009) 146: 353–369
DOI 10.1007/s11214-009-9519-4

IBEX Education and Public Outreach

L.M. Bartolone · K. Carney · S.B. Cohen · J. Erickson · J. Gutbezahl · P.H. Knappenberger Jr. · D.J. McComas

Received: 6 August 2008 / Accepted: 29 April 2009 / Published online: 4 June 2009

Abstract The IBEX mission includes a comprehensive Education and Public Outreach (E/PO) program that develops programs and products in exciting themes from astronomy and space physics. With the active involvement of the Principal Investigator and several science team partners, it is overseen and implemented by the Adler Planetarium & Astronomy Museum. The program includes an internationally distributed planetarium show with accompanying informal education materials that are also accessible to individuals with special needs; a national Space Science Core Curriculum for grades 6–8 in collaboration with other NASA missions; a professional development program for teachers; and workshops that engage Hispanic and Native American students. Materials are made available for download or for order via the IBEX website: http://www.ibex.swri.edu/. Our program is developed, implemented and monitored for effectiveness by organizations with proven capabilities and

L.M. Bartolone (✉) · K. Carney · P.H. Knappenberger Jr.
Adler Planetarium & Astronomy Museum, Chicago, IL, USA
e-mail: lbartolone@adlerplanetarium.org
url: http://www.adlerplanetarium.org

K. Carney
e-mail: kcarney@adlerplanetarium.org

P.H. Knappenberger Jr.
e-mail: pknappenberger@adlerplanetarium.org

S.B. Cohen · J. Gutbezahl
Program Evaluation Research Group, Lesley University, Cambridge, MA, USA
url: http://www.lesley.edu/academic_centers/perg/staff.html

J. Erickson
Great Explorations in Math and Science, Lawrence Hall of Science, University of California, Berkeley, CA, USA
e-mail: jerick@berkeley.edu
url: http://www.lawrencehallofscience.org/gems/

D.J. McComas
Southwest Research Institute, San Antonio, TX, USA
e-mail: dmccomas@swri.edu

experience in their respective areas of E/PO. This paper describes these program elements in detail and includes the rationale and design process for this E/PO program.

Keywords GEMS · Space Education · Teaching · Curriculum · Education and Public Outreach · EPO · STEM · Astronomy Education · Planetarium · LASSO · Teacher Workshop · LANLCSSE · Interactives · TimeSpace · Adler · Lawrence Hall of Science · Science Literacy · Space Science Sequences

1 Overview

The Interstellar Boundary Explorer (IBEX) is a Small Explorer (SMEX) mission that launched October 19, 2008. SMEX missions are cost-capped, relatively inexpensive missions with focused science goals. IBEX is a remarkable mission of exploration and discovery that provides the first global images of the interstellar boundaries and unveils the physics of the heliosphere's interstellar interaction. This investigation fills in the critical missing piece for understanding the connection of our Sun and solar system to the Galaxy, fulfills requirements from recent NASA and National Academy plans, makes fascinating connections to astrophysical phenomena, and addresses a serious challenge facing manned exploration by studying the region that shields out the vast majority of galactic cosmic ray radiation. This breakthrough mission is achieved with an innovative mission design using a simple, high heritage science payload and spacecraft described in the other chapters of this volume. The single, focused objective of IBEX is to discover the global interaction between the solar wind and the interstellar medium.

The Education and Public Outreach (E/PO) program is an integral part of the IBEX mission. It supports NASA's strategic coordination framework, is designed to inspire the next generation of explorers in science and technology, and will engage the public in the excitement of innovative research. As a mission of exploration and discovery, the IBEX mission and E/PO program will provide the public with never before seen views of the boundary of the solar system.

The IBEX E/PO Program was designed using the process described below which has resulted in the following program elements: an internationally distributed planetarium show with accompanying informal education materials that are also accessible to individuals with special needs; a national Space Science Core Curriculum for grades 6–8 in collaboration with other NASA missions; a professional development program for teachers; and robotics workshops that engage Hispanic and Native American students. Materials are made available for download or for order via the IBEX website (http://www.ibex.swri.edu/), which also contains games and information about the mission including monthly feature articles on team members.

2 Education and Public Outreach Design

Designing for Education and Public Outreach is similar in many ways to designing a scientific investigation. Just as scientists follow a process of inquiry into a scientific concept, E/PO professionals also follow a process to discover effective methods for teaching concepts. The steps may not always occur in the same order, but in general, research is done into the established knowledge base of the field, a general research question is formulated, hypotheses are created, and then the strategic design process begins. A scientific experiment and E/PO program both begin with the creation of goals and requirements. Then the designers determine what evidence they need to discern if they have achieved those goals. Next, they develop a way to collect that evidence by designing a tool or technique. Various designs are prototyped, tested, revised and tested again. Ultimately, the designed tools or techniques are used to gather evidence that supports or changes an existing model of either a scientific concept or teaching method.

2.1 History and Informative Resources

Ever since humans have looked to the skies, they have been discussing and teaching each other about what they have learned. Although astronomy education has been around for a very long time, "the astronomy education and public outreach *profession* is [finally] coming into existence" (Fraknoi 2005, p. 19). Astronomy E/PO has borrowed largely from the field of science education and even more specifically from physics education research, however journals and general communications between practitioners in this emerging field have only recently become established.

One of the greatest challenges of E/PO professionals, and science educators in general, is understanding the needs of their targeted learners. Human beings are constantly constructing ways of making sense of the world around them. Sometimes their ideas about the world fit in with the current scientific understanding, but many times they do not. Researching these misconceptions and approaches for constructing ideas better aligned with current scientific thinking is part of what E/PO professionals must do as they design programs. Misconception research has been published in a number of different journals including, for example, the online journal *The Astronomy Education Review* (AER 2001), which provides an excellent source of information for E/PO programs.

Education and Public Outreach professionals also draw upon guiding documents for science education such as the National Science Education Standards (National Research Council 1996), the American Association for the Advancement of Science (AAAS) Benchmarks for Science Literacy (AAAS 1993) and the Project 2061 Atlas of Science Literacy Vols. 1 and 2 (Project 2061, 2001 and 2007). These publications outline standards for education as well as detailing concepts that should be taught at various ages. The Atlas contains concept maps that break down science concepts into components, and links the components together. In addition, it highlights misconception research relevant to the concepts.

The design of E/PO programs is often done following an approach described by Wiggins and McTighe in Understanding by Design (2005). In this approach, the designers of an educational experience must first determine the most essential part of a concept that a participant should retain 5–10 years after an experience. In other words, what is the core idea that should stick with a participant, regardless of how they use the information in their life? The next step is to figure out how an educator will know if their participants have grasped this concept; how will participants provide evidence that they understand the concept? After that, an assessment is designed to provide the structure for collecting the evidence about participant understanding. Finally, the activity that helps the participant achieve that understanding is designed. When the activity is prototyped with a group of participants, evaluation is conducted to determine if the participants can produce evidence that they understand the concept. If not, the educational designer needs to go back to the activity to redesign or modify the approach.

A large number of E/PO professionals are involved with NASA's Science Mission Directorate E/PO Program, set in motion by NASA's commitment to education, which started with the 1958 Space Act. Since many E/PO professionals are designing programs that are smaller components of the larger educational portfolio, NASA has compiled a number of documents describing E/PO history, how to become involved, resources, evaluation factors and evaluation reports of Science Mission Directorate efforts in E/PO. These documents are available at NASA's "Education and Public Outreach—Science Mission Directorate" webpage, http://nasascience.nasa.gov/researchers/education-public-outreach (NASA 2008a).

2.2 Designing the IBEX E/PO

The Adler Planetarium & Astronomy Museum joined the IBEX mission as the E/PO Lead institution during our initial proposal phase and has been leading this component of the mission ever since. We assembled a team of scientists and educators to help design our E/PO program. Collectively, we used our knowledge and expertise in astronomy education to create a proposal for the E/PO activities that brought together a national team of experts in program evaluation, curriculum design, professional development, electronic communications and educational design for individuals with special needs. Preliminary research assesed the existing programs and needs of target audiences that relate to the science and engineering of the IBEX mission, as well as the goals of NASA's overarching E/PO program. The results were used to create a program that was positioned to be exciting, useful and distinctive.

The following sections describe the purpose and goals of IBEX's E/PO program and the program elements that have been designed to achieve them. For those efforts that are already underway, more description of the formative evaluation and details are available than for other activities that are scheduled to begin later in the mission.

2.2.1 Purpose and Goals of the E/PO Program

The purpose of the IBEX E/PO program is to inspire the next generation of explorers by highlighting the exploration and discovery aspects of the IBEX mission and offering broad educational and societal benefits beyond the scientific opportunities provided by imaging the global interaction between the solar wind and the interstellar medium.

The goals of the Interstellar Boundary Explorer E/PO are to:

- Communicate the scientific goals and results of the IBEX mission to the public in various locations and formats in ways that make the science, technology, engineering and math relevant to the learner.
- Address the needs of students/individuals with learning disabilities.
- Involve minority communities in our educational efforts.
- Conduct educational research about the public understanding of the boundaries of the Solar System, existence and behavior of charged particles, and mixing of fluids.
- Contribute to educational endeavors of NASA's Science Mission Directorate E/PO Initiatives.

A number of program elements have been chosen to help a national audience achieve the goals of the IBEX E/PO Program. Those program elements include:

- Create and widely distribute a planetarium show that communicates the scientific goals and results of IBEX to the public and shares with them the excitement of this Explorer mission.
- Create educational products and demonstrations to be distributed with the show, including interactive software for use on electronic kiosks.
- Partner with the Special Needs Working Group (SNWG) and the International Dyslexia Association (IDA) to create educational products accessible to individuals with special needs; specifically dyslexia, visual-impairment and hearing-impairment.

- Produce and distribute Spanish language versions of our educational products and interactive software for teachers and students.
- Collaborate with Los Alamos National Labs Center for Space Science Exploration (LANLCSSE) staff to sustain and improve the Los Alamos Space Science Outreach (LASSO) teacher workshop, which targets teachers of Hispanic and Native American students in New Mexico and is now being expanded to include teachers of Hispanic students nationally and involve students in Science, Technology, Engineering and Mathematics (STEM) improvement robotics workshops.
- Collaborate with Lawrence Hall of Science and Great Explorations in Math and Science (GEMS) to create and fund the Space Science Core Curriculum Sequence for grades 6–8 based on National Science Education Standards.
- Create and maintain an educational website, which allows access to electronic versions of the educational products that we produce and distribution of products that cannot be downloaded.
- Determine the needs of our audiences and assess and evaluate the success of our program to address those needs.
- Ensure close alignment with NASA's educational endeavors by working closely with NASA's Science Mission Directorate E/PO Initiatives.

All of the goals and program elements have different sets of audiences and partners. These relationships are detailed in Fig. 1. The purple section in the top left describes top level NASA goals and shows that NASA Space Science Advisory Committee E/PO Task Force Recommendations (SScAC E/PO T/F Recms, NASA 2003) influenced the choice of IBEX E/PO goals (shown in yellow). The alignment between program elements and IBEX E/PO goals are shown with green squares. The deeper green squares (core) show a goal that is achieved by participation in one program element while the lighter green squares (supporting) contribute to the achievement of a goal but do not achieve them on their own. The orange thematic strands show how topics in IBEX E/PO content are spread through the program elements. Different program elements concentrate on different science content aspects of the mission. Some of these choices are made due to the timing of their creation and availability of IBEX data, for example the planetarium show will not help audiences to understand IBEX data because it is completed before IBEX data is released.

IBEX E/PO received 2% of the overall mission budget (excluding launch vehicle). Ten percent of that is dedicated for evaluation, and the program elements' total cost in descending order are: GEMS Space Science Curriculum, IBEX planetarium show, LASSO professional development program, educational products, website, and other program costs. At the time of publication of this paper, some program elements have been completed (for example, the GEMS Space Science Core Curriculum and the IBEX: Search for the Edge of the Solar System planetarium show) while others are still in the very early stages of development (materials for people with special needs and LASSO professional development program). The section lengths in this paper reflect both the total costs and completeness of the program elements.

2.2.2 E/PO Program Alignment

As part of NASA's education efforts, the IBEX E/PO program is designed to reflect recommendations of SScAC Task Force and with the Office of Space Science (OSS) Education Strategy and Implementation Plan. In addition, the E/PO team is involved with annual conferences and retreats in an effort to navigate NASA's changing education infrastructure. The

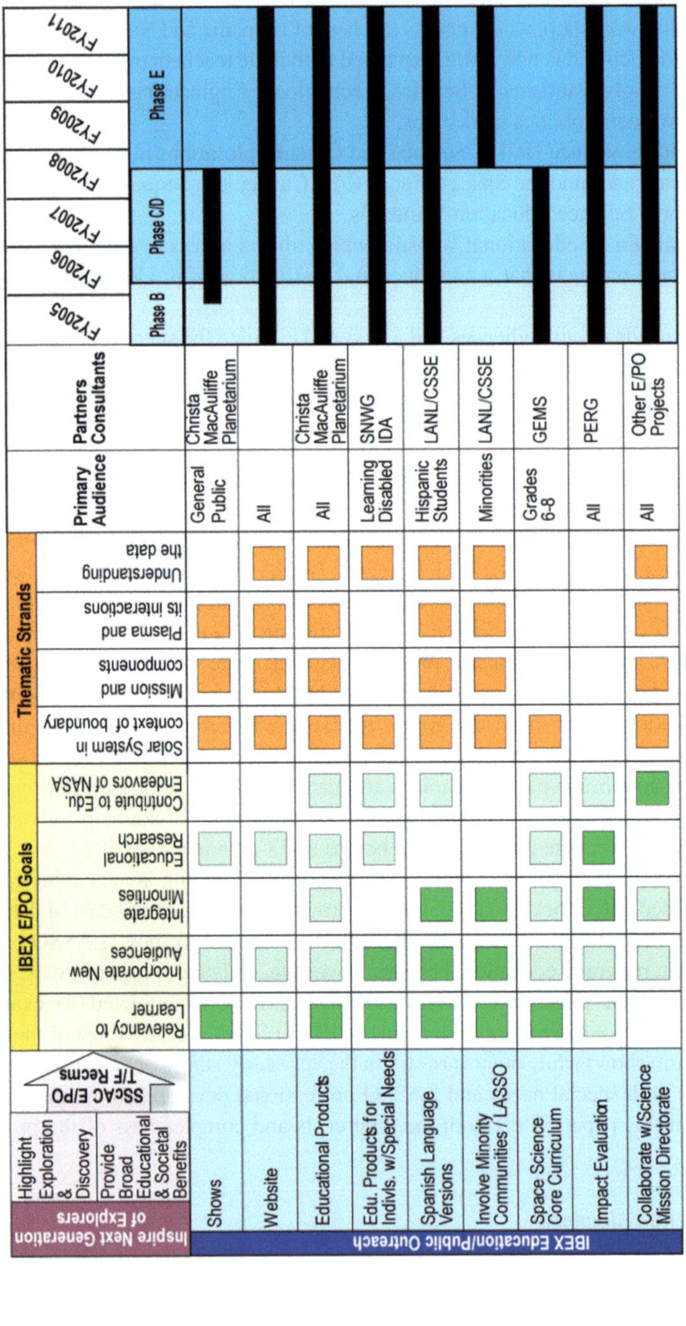

Fig. 1 IBEX E/PO overview table

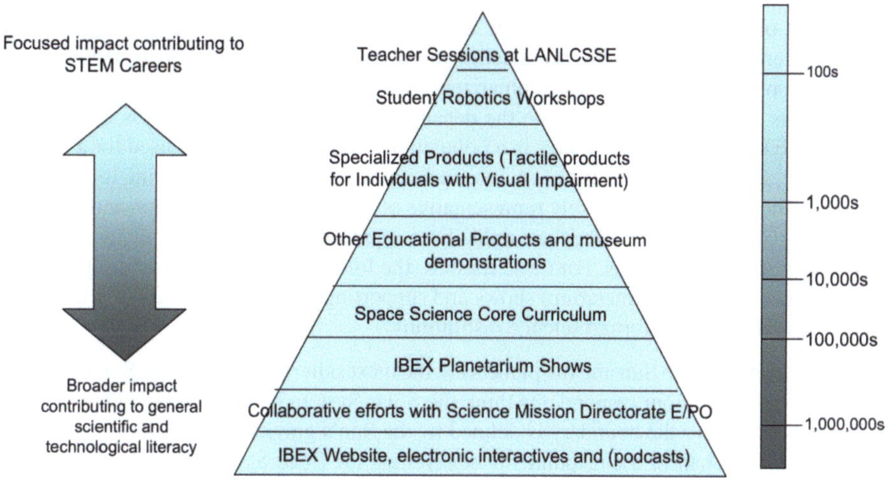

Fig. 2 Schematic of pyramid structure for IBEX E/PO program elements

current plan aligns with the document from Headquarters: *NASA Education Strategic Coordination Framework: A Portfolio Approach* (NASA 2007).

NASA's Education Office has the lead role in accomplishing the goal to "inspire and motivate students to pursue careers in science, technology, engineering, and mathematics" (STEM) and a strong supporting role in achieving the goal to "engage the public in shaping and sharing the experience of exploration and discovery." Special attention is paid to under-represented and under-served communities, in order to fill the "pipeline" with students throughout their education, who will then be well prepared to seek STEM careers, helping our nation to meet its workforce needs through the 21st century.

The IBEX E/PO program is aligned with these goals, and supports a suite of formal and informal educational and public outreach program elements that are specifically targeted at under-represented and under-served populations on all scales.

As shown in Fig. 2, IBEX E/PO program elements follow a "pyramid structure," with the program elements at the peak directly supporting STEM training, while those at the base are aimed at improving the science literacy of the general public. The peak elements focus on smaller groups of students through formal or informal settings, while the base elements reach a wide-spread audience that numbers in the millions.

3 Program Elements

3.1 Evaluation and Educational Research

Early in the E/PO program, the E/PO team met with the researcher from Program Evaluation and Research Group (PERG) at Lesley University to determine the scope of program evaluation that would take place over the life of the mission. Both parties agreed that the first step should be to conduct a needs assessment evaluation to determine what the audiences of IBEX's E/PO programs knew and were interested in about the science of the IBEX mission. Since a large portion of the activities were going to take place in an informal setting, PERG conducted a front-end evaluation, interviewing 90 visitors at 3 science museums about their knowledge of the Solar System, the Sun, and states of matter (including plasma), as well as their interest in unmanned space science missions and the people who work on them. Below are some interesting points summarized from the data. Due to the nature of the study, these results may be used to estimate the density of positive attitudes towards and knowledge of IBEX specific science topics in a population similar to the sample used for analysis. In this case, the sample consisted of science museum visitors who were willing to complete a survey. Their responses are likely representative of science museum visitors, and possibly the science-attentive public. They are unlikely to represent general public, many of whom do not visit science museums. For these reasons, the following study results were primarily used for developing the planetarium shows and supporting materials, the bulk of which will be distributed through informal science institutions.

- Data indicate that the Sun and the planets are the most salient features of the Solar System.
- About one-third of our respondents think the Solar System has a boundary; data suggest that people believe the boundary is defined by the Sun's gravitational pull.
- More than one-half of the respondents believe that the Solar System contains stars, constellations, or galaxies.
- Most respondents know basic information about the Sun: they are aware that it is a hot star, the largest object in the Solar System, with a gravitational pull that keeps the Earth in orbit, and that without the Sun, there would be no life on Earth.
- Data indicate that there is interest in the life cycle of the Sun; in particular, participants expressed curiosity about whether and when the Sun would die, and how that would affect life on Earth.
- Data indicate that there is a wide range of understanding about states of matter and atoms. In particular, our participants expressed confusion about what happens when an atom loses electrons; many indicated that they believe this to be fission and that it would cause an explosion.
- While most of our participants were familiar with the term "plasma," few had an understanding of it as an energized state of matter: many identified it with blood or television.
- In general, our participants are concerned about danger and safety. They believe that ionization is dangerous and are concerned with whether plasma is dangerous.
- Participants expressed interest in the engineering aspects of NASA missions, including spacecraft, robotics, communication with land-based teams, and repairs of unmanned ships.
- Participants also expressed interest in why NASA chooses certain missions, why people work in the field, and the types of scientific questions answered by NASA missions.
- Participants expressed less interest in the people working on missions than in the missions' details or the scientific knowledge gleaned from the missions.
- Participants indicated that they believe scientists' and engineers' intelligence and thought processes were different from their own.

Our E/PO team discussed the findings from the full report (Gutbezhal 2006) with the PERG team and used this information in preparing many of the activities described below.

The E/PO team also formed a plan for formative and summative evaluation with the PERG team. This plan includes focus groups and surveys on the TimeSpace planetarium show module, focus groups and surveys for the full-length planetarium show, focus groups (including groups of differently-abled users) for the website, rapid prototyping of educational products, observation and impact evaluation for teacher and student workshops at LANL and a summative evaluation of the entire program. Currently, unpublished memos and reports from the front-end evaluation, module focus groups and surveys, and full show focus groups and survey are available by contacting the IBEX E/PO Lead.

3.2 Planetarium Module and Full Show

The show production process began with the creation of an IBEX themed module that integrated into Adler's existing Digistar 3 planetarium show called TimeSpace. The TimeSpace show is an imaginative trip through space and time with several "stops" along the way. An extra "stop" 500 years in the future at the boundary of our Solar System was added to the already existing show in August of 2006. In the 5 minute module, the crew of a fictitious craft at the boundary of the Solar System celebrates the historic IBEX mission of 2008-9, which created the first maps of the boundary region allowing eventual travel to the boundary region and beyond. That module was used as a prototype for audience focus groups and surveys, which helped determine the content of our full-length production.

The full-length show, Search for the Edge of the Solar System, concentrates on the science of the Solar System boundary, how the IBEX mission explores the boundary, and the international team that works together to create and support the mission. The show begins as an electronic chat between two teens wondering where the Solar System ends. They do a search and discover information about the boundary region and the IBEX mission to study it. High quality video, interviews with team members, and full dome animation are incorporated into the show to help tell the story. Adler worked with the National Center for Supercomputing Applications at the University of Illinois, Urbana-Champaign to create visual representations of the Universe including the motion of stars through the Milky Way Galaxy.

As part of the full-length show production process, an 'animatic,' which consists of still pictures and simple narration, was created. This animatic was shown to 2 focus groups of high school students. Data from the focus groups was quite consistent. Participants found the show engaging and entertaining. Some participants were confused by the content, and would have liked more context for some of the ideas. They had several specific suggestions, which were taken into account when Adler staff revised the narration and visuals.

After the full-length show was almost complete in December 2008, it was shown in two formats, to small groups in a mini-dome and in a large theater. The small groups participated in focus groups, and the participants in the larger theater completed survey questions. The intent of this work was primarily to test the ability of the show to meet the following educational outcomes:

1. *After viewing the show, participants would KNOW that*:
   - There is only one star inside our Solar System.
   - The Solar System has boundaries.
   - One boundary is created by an interaction between charged particles coming from the Sun and the stuff that surrounds the Solar System.
   - The boundary changes as the Sun moves through the galaxy.
   - This boundary is invisible, but detectable by using specially designed tools.
   - The IBEX mission is investigating the boundary.
   - The Solar System boundary protects us from damaging rays.
   - NASA missions are complex, involving many diverse people, skills, and technologies from around the world.
2. *Participants would UNDERSTAND the following concepts*:
   - The Solar System has boundaries and components that are affected by forces inside and outside the boundaries.
   - IBEX is exploring the boundary.
   - "It takes a team" to explore.

3. *Participants are able to answer these ESSENTIAL QUESTIONS*:
   - Does the Solar System have a boundary? What is it like?
   - If the boundary changes, how will that affect the earth/me?
   - How can we investigate an invisible boundary?
   - What skills/traits/jobs are necessary to complete a mission?
4. *Participants would be able to demonstrate their ability to DO the following*:
   - State the Solar System has a boundary that contains only one Star.
   - Describe how the boundary is created.
   - Explain how IBEX will explore that boundary.
   - List at least one way the boundary or changing boundary affects them.
   - Express positive reactions about exploration of space.
   - Name at least one occupation/skill associated with unmanned space missions.
   - Express interest in learning more about the IBEX mission or science.

Results from the draft report include that after viewing the show, 91% of the participants answered correctly that the purpose of the IBEX mission is to investigate the boundary of the Solar System, 98% understood that the boundary is invisible to the naked eye and 97% responded that IBEX is collecting particles.

The front-end evaluation (Gutbezhal 2006) revealed another very interesting result: that "More than one half of the respondents believe that the solar system contains stars, constellations or galaxies." After viewing the show, participants were asked in a full-show survey, "How many stars does our Solar System contain?" and 75% answered correctly "one" on the open-ended question. We consider this to be a huge improvement on what can be a deeply held misconception about the organization of the Universe.

Adler has partnered with specialists to produce the show in other formats including digital portable dome format. The show will be translated into Spanish and will be closed captioned for the hearing impaired. Dome-masters for this show are available for the cost of shipping and handling, and a trailer can be viewed at http://ibex.swri.edu/planetaria/index.shtml. The IBEX show premiered March 2009 and is also now available to other planetaria.

3.3 Educational Products

Adler leads the design and content creation of several IBEX themed educational products, which include museum floor demonstration resources, educational posters/postcards for families, electronic materials for individuals with dyslexia, and tactile products for individuals with visual impairments. The educational content for the products is chosen using the results from the front end evaluation, conducted by PERG and the ability of these new resources to fill a hole in existing NASA educational material portfolios. The products are submitted to NASA's Science Mission Directorate education product review and communication review prior to production. Currently, eleven products have been prototyped and submitted for NASA review and more are in development. These approved products are available in print and as electronic copies at the IBEX website http://ibex.swri.edu/planetaria/index.shtml. Some of the current products include two posters, one lithograph, several floor demonstrations and activities, and two tactile diagrams with audio walkthrough. More products will be produced after launch and data release, describing the scientific results and discoveries of the IBEX mission.

3.4 Website

Adler has created and maintains an educational website where people can learn about the mission. It provides access to electronic versions of the educational products and distribution of products that cannot be downloaded. The address for the website is http://ibex.swri.edu (IBEX 2008). All sections of the website are section 508 compliant and accessible to screen-reading software applications. The IBEX website contains separate sections for students, educators, museums and researchers. The IBEX website will be mirrored in Spanish.

The IBEX web team produces interactive online activities that allow visitors to explore IBEX mission and science related content. So far, three animations/interatives have been created. An interactive timeline allows visitors to click through a timeline to view images relating to various mission milestones. The second interactive allows students to explore how adding energy to a substance changes the state of matter between solid, liquid, gas and plasma. The third interactive addresses a common educational misconception that the "Universe" is a jumble of objects with no organization or structure. This interactive takes advantage of the focus of the IBEX mission, the boundary region of the Solar System to help users "sort out" objects into realms of the Universe, for example objects that belong inside and outside of the Heliosphere.

Monthly highlights feature different individuals that work on the IBEX mission in order to describe the different types of jobs, preparations and backgrounds needed to make a mission successful. The highlights are emailed monthly to subscribers, and people wishing to subscribe can do so through the website. The website also contains information and links for researchers and IBEX team members. The website has had over 30,000 unique visitors over the past 3 years.

In addition to the main IBEX website, there are also IBEX pages on the NASA Portal, Facebook.com, and twitter.com.

3.5 GEMS Space Science Sequences

3.5.1 Introduction

The Space Science Sequences are two sets of curricula for grades 3–5 and 6–8 developed at Lawrence Hall of Science in partnership with NASA educational public outreach efforts. NASA partners included:

- IBEX Mission Education and Public Outreach.
- Kepler Mission Education and Public Outreach.
- Origins Education Forum/Hubble Space Telescope.
- Solar System Education Forum.
- Sun-Earth Connection Education Forum.

The Lawrence Hall of Science is a science and math education center at the University of California in Berkeley. It includes a hands-on museum and an interactive planetarium that serve schools and the general public, and is the home of many curriculum projects. One of these projects is Great Explorations in Math and Science (GEMS), which developed the Space Science Sequences, http://www.lhsgems.org/CurriculumSequences.htm. Since 1986 GEMS has developed supplemental math and science activities in units that require two to twelve sessions to teach in a school classroom. Many of these activities originated in programs that have been taught in the instructional labs and the planetarium at Lawrence Hall for many years. Great Explorations in Math and Science materials are known for their teacher-friendly instructions, use of simple materials, and extensive testing that insures that GEMS activities will be successful educational experiences in school classrooms.

Astronomical content has been included in GEMS materials from the very beginning, but formal collaboration with NASA did not begin until 1999 when GEMS developed *The Real Reasons for Seasons* with NASA's Sun-Earth Connection Educational Forum. Other successful collaborations followed, which resulted in the units *Living With a Star*, and *Invisible Universe*. With these and other GEMS units already developed it became apparent that GEMS units were not simply useful as an add-on to an astronomy curriculum, but that they could become a curriculum in their own right. Because GEMS units were originally designed to be supplemental activities, the process of creating a coherent sequence with comprehensive treatment of core concepts involved much more than simply concatenating the materials that were already developed.

3.5.2 Determining the Scope and Format of the Space Science Sequences

Teachers spend varying amounts of their classroom time on space science depending on their state standards and also on their own interests. The Space Science Sequences had to be devised to accommodate teachers who have no special interest in astronomy, but need to teach some basic required topics. A survey of teachers in the GEMS Network revealed that almost any school could realistically spend six weeks of its science time on space science during grades 3–5, and six more weeks during grades 6–8.

Having limited ourselves to six weeks of instruction for each sequence, we had to narrow the scope of topics so that each one we addressed could be taught effectively. This led to some hard choices, especially when it came to eliminating some well-loved topics such as constellations, astrobiology, and space travel. Choices were made based on the following:

- Common misconceptions.

Astronomy educators are very familiar with inaccurate ideas people hold about reasons for phases of the Moon, why the Earth has seasons, and how the Solar System and the Universe are structured. The Space Science Sequences address these and other misunderstood topics in depth.

- Space science research of the present and near future.

Space science is not simply a set of known facts and explanations. The partnership with NASA helped us to incorporate current questions and research into the Space Science Sequences. For example, students investigate how extra-solar planets could be discovered in the manner of the Kepler mission, and they read about the IBEX mission's exploration of the boundary of the heliosphere.

- State and National Standards.

There is wide variation in state standards, but we were able to apply the trends in standards to our curriculum design. For example, many state standards place understanding of phases of the Moon in the elementary grades. For this reason, we placed it in the sequence for grades 3–5 rather than in the middle school sequence, where it could also have been effectively addressed.

- Other requirements for basic space science literacy.

The structure of the Universe beyond the Solar System does not appear in most education standards until high school, if at all. NASA partners and developers at GEMS agreed that there is a place for this topic in a basic middle school astronomy curriculum. One reason is the high level of student interest and the flood of exciting new knowledge and ongoing research in this topic. Another reason is that instruction about the larger Universe supports learning about the more close-to-home topics, such as the structure of the Solar System, which *is* part of many middle school standards.

Each sequence has four units, with each unit focusing on an important topic over several class sessions.

Space Science Sequence for grades 3–5.
Unit 1 How Big and How Far.
Unit 2 Earth's Shape and Gravity.
Unit 3 How does the Earth Move?
Unit 4 Moon Phases and Eclipses.

Space Science Sequence for grades 6–8.
Unit 1 How Does the Sun Affect the Earth?
Unit 2 Why Are There Seasons?
Unit 3 The Solar System.
Unit 4 Beyond the Solar System.

Each unit within a sequence can be taught by itself, although ideas build through the units. This gives schools options for flexible implementation. All units in a sequence can be taught in the same grade, or the units can be spread out over the grade range for which the sequence was designed.

3.5.3 Development and Testing the Space Science Sequence for Grades 6–8

In October, 2003 NASA partners met in one group for the first time with GEMS staff at Lawrence Hall of Science to begin development of the Space Science Sequences. The overarching topics and the basic structure were decided upon fairly quickly. The fleshing out of a "concept storyline" that would be the foundation for the actual curriculum was a process that involved many more meetings over the next two years.

We began the process by using the principles of "backward design." This term, coined by Wiggins and McTighe in Understanding by Design (2005), describes a process wherein the goals of instruction are defined, then the assessments, and finally the methods of instruction. In their previous development of *The Real Reasons for Seasons*, GEMS had successfully tailored instruction to an assessment questionnaire, and we decided to pursue a similar strategy with the development of the Space Science Sequences. Having developed a list of key concepts for each unit, we developed assessment questionnaires for each unit. In practice we did not apply the principles of backward design rigidly. At various stages during actual development and testing we revised our key concepts and our assessments, making the process more iterative, rather than strictly backward design.

The pilot outlines of the 6–8 Space Science Sequence were tested in local schools by the GEMS staff. Activities were modified or discarded according to the success of the pilot tests. Trial versions of the four units were written for teachers across the country to test. Each unit was tested by 14 to 23 teachers. Feedback surveys from teachers and the results of the pre- and post-assessment questionnaires from students were analyzed. In general, the results showed that instruction was sound, but there were logistical concerns about timing and material management that make a great difference to teachers. Some activities were simplified, reordered, or rewritten to extend over more class sessions. In one case, a key activity was changed so much that we chose to have local teachers retest it in their classrooms before we wrote the final version.

3.5.4 Features of the Space Science Sequence For Grades 6–8

Assessment Each unit includes a brief pre- and post-questionnaire meant to measure student understanding of the overarching ideas of the units. They include a mixture of multiple choice and open ended questions. These questionnaires are not simply assessments for the teacher—they are learning tools for the students as well. As each unit progresses students are asked to reflect back on their answers to the pre-questionnaire and to consider whether and why their understanding has changed.

Figure 3 shows an example of an item from the questionnaire for Unit 4, Beyond the Solar System.

What is the order of these things from smallest to largest?
The Universe.
The Sun.
The Milky Way Galaxy.
The Solar System.
The Earth.

Within each unit there are multiple opportunities for assessment as students proceed through the activities. Rubrics are provided in the 6–8 Sequence for teachers to gauge students' learning. At critical junctures the Sequence provides extra activities for the students that a teacher can use if formative assessments indicate that students need more instruction before proceeding to other topics in the unit.

Focus on Scientific Models Students learn that scientists use models to understand, explain, and make predictions. Models presented in the 6–8 Sequence show scale, structure, and change over time in various astronomical systems. With each model that the students are involved with, they are invited to explore what is accurate and what is inaccurate, and how the models could be improved.

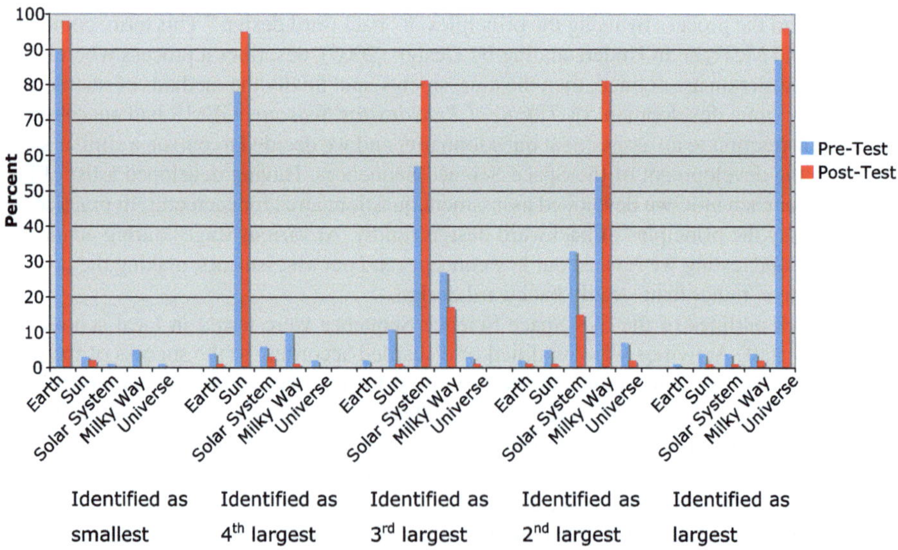

Fig. 3 GEMS field test data

Focus on Evidence and Discourse In each unit students have a chance to apply evidence from readings, research, and classroom activities and observations, to make a case for or against a statement or explanation. In Unit 2, *Why are There Seasons*?, they use scale models and data about temperatures and day-length around the world to discuss whether seasons are caused by the Earth's distance from the Sun. The evidence is against it, therefore in this case the discourse steers them toward an important explanation for seasons. In Unit 4, *Beyond the Solar System*, students discuss the probability of extra-terrestrial life, and whether and how there could be communication. In this case the discourse allows them to synthesize and apply what they have learned about the scale and structure of the Universe and the nature of stars and planetary systems.

Key Concepts and the Concept Wall Learning goals in the 6–8 Sequence are made explicit to students as the units progress. Key concepts are written out and posted for each session, building a wall of ideas about science content and the nature of science.

CD ROM The 6–8 Sequence comes with a CD ROM that gives the teacher alternative means of presentation as well as additional activities and resources. Many activities have group or class versions on the CD. For example, students explore models of the tilted Earth revolving around the Sun using spheres and light bulbs. The CD has an interactive computer model that students can explore individually or that the teacher can project for students to examine as a class to supplement their classroom experience. Field tests of the units in the 6–8 Sequence were conducted with and without the CD, and the results showed the CD made differences in student learning.

Educative Teacher Guide The teacher guide format allows teachers to easily access material that will help and inform their teaching. On each spread the left page has the instructions for presenting the activities, and the right page has "teacher considerations," which includes a variety of material. Rubrics for assessment, relevant science content, classroom management suggestions, alternative and supplementary activities, and general teaching strategies are all laid out on an as-you-need-it basis alongside the step-by-step instructions for the activities. Each unit also has a science background section with material that goes beyond the scope of the units to help teachers comfortably address student questions.

3.5.5 Dissemination of the Space Science Sequences

A GEMS Network has been established to support educators who use GEMS materials. Sites and centers in the GEMS Network are run by school districts, universities, and science centers to provide professional development and material support to teachers. The success of this network model has led us to start a similar network specifically for the Space Science Sequences. With the support of a NASA Shared Services Award to UC Berkeley's Space Sciences Lab and Lawrence Hall of Science, a project called Advancing Teacher Leaders in Space Science (ATLSS) was initiated in 2008. Four ATLSS sites for the 3–5 Sequence and four sites for the 6–8 Sequence have been selected for training and other professional support for educators.

3.6 LASSO Teacher and Student Workshops

Los Alamos National Laboratory Center for Space Science Exploration (LANLCSSE) will engage master teachers in sustained learning activities directly tied to IBEX thus supporting improved science, math, and technology content knowledge as well as life-long learning process skills. The Los Alamos Space Science Outreach (LASSO) science education effort adheres to an effective instructional model based on education research and cognitive theory. During the program, teachers will critically investigate projects through the examination of basic and advanced science concepts behind the project goals. The teachers will incorporate what they learn into standards-based lessons and activities that will be featured on the IBEX website. The student robotics workshops, also sponsored by LANLCSSE, will bring the excitement and education of engineering design to area schools.

3.7 Collaborative Programs

In order to ensure close alignment with NASA's educational endeavors, the IBEX E/PO team works closely with NASA's Science Mission Directorate (SMD) E/PO Initiatives. The E/PO Lead participates in monthly telecons and occasional meetings of collaborating SMD E/POs. IBEX contributes to the collective efforts of Heliophysics mission E/POs, such as Sun-Earth Day and International Year of Astronomy (IYA) 2009. Specifically, IBEX E/PO contributes to a number of NASA IYA presentations at national and regional conferences. IBEX is collaborating with several Heliophysics missions which received funding during their senior review process. The resulting program is called Heliophysics Educator Ambassadors and will train educator ambassadors to teach other teachers and their own students about Heliophysics. The first workshop is scheduled for summer of 2009 in Alaska. IBEX plans to make use of available NASA infrastructure to disseminate products, especially the Museum Alliance, and NASA's IYA website, http://astronomy2009.nasa.gov/ (NASA 2008b). IBEX is working together with the Heliophysics Forum to revise the "Sun-Earth Viewer" to include appropriate IBEX related resources.

4 Conclusion

A carefully designed Education and Public Outreach program can meet the needs of the critical stakeholders by leveraging diverse partnerships. Education and Public Outreach programs must be designed with evaluation as a key component in order to maximize and measure effectiveness. The IBEX E/PO program was designed with these principles in mind and will contribute significantly to the overall success of the NASA mission.

References

American Association for the Advancement of Science, *Benchmarks for Science Literacy* (Oxford University Press, London, 1993)

Astronomy Education Review (2001–2008). Retrieved May 23, 2008 from http://aer.noao.edu

A. Fraknoi, Steps and missteps toward an emerging profession. Mercury **34**(5), 19–25 (2005)

J. Gutbezhal, Interstellar Boundary Explorer (IBEX) Front End Evaluation Report. Unpublished report, Program Evaluation and Research Group, Lesley University (2006)

Interstellar Boundary Explorer E/PO Team, IBEX: Interstellar Boundary Explorer (2008). Retrieved May 23, 2008 from http://ibex.swri.edu

National Aeronautics and Space Administration, Implementing the Office of Space Science Education/Public Outreach Strategy: A Critical Evaluation at the Six-Year Mark (2003). Retrieved April 22, 2009 from http://nasascience.nasa.gov/educators/program-evaluation/OSS_EPO_Task_Force_Report.pdf

National Aeronautics and Space Administration, NASA Education Strategic Coordination Framework: A Portfolio Approach (2007). Retrieved May 23, 2008 from http://www.nasa.gov/audience/foreducators/topnav/materials/listbytype/Strategic_Coordination_Framework.html

National Aeronautics and Space Administration, Education and Public Outreach—Science Mission Directorate (2008a). Retrieved May 23, 2008, from http://nasascience.nasa.gov/researchers/education-public-outreach

National Aeronautics and Space Administration, International Year of Astronomy 2009 (2008b). Retrieved May 23, 2008 from http://astronomy2009.nasa.gov/

National Research Council, *National Science Education Standards* (National Academy Press, Washington, 1996)

Project 2061, *Atlas of Science Literacy* (American Association for the Advancement of Science and National Science Teachers Association, Washington, 2001)

Project 2061, *Atlas of Science Literacy*, vol. 2 (American Association for the Advancement of Science and National Science Teachers Association, Washington, 2007)

G. Wiggins, J. McTighe, *Understanding by Design Expanded*, 2nd edn. (Pearson Education, Upper Saddle River, 2005)

MIX
Papier aus verantwortungsvollen Quellen
Paper from responsible sources
FSC® C105338

If you have any concerns about our products,
you can contact us on
ProductSafety@springernature.com

In case Publisher is established outside the EU,
the EU authorized representative is:
Springer Nature Customer Service Center GmbH
Europaplatz 3, 69115 Heidelberg, Germany

Printed by Libri Plureos GmbH
in Hamburg, Germany